T0233721

Lecture Notes in Computer Science　　9418

Commenced Publication in 1973
Founding and Former Series Editors:
Gerhard Goos, Juris Hartmanis, and Jan van Leeuwen

Editorial Board

David Hutchison
  Lancaster University, Lancaster, UK
Takeo Kanade
  Carnegie Mellon University, Pittsburgh, PA, USA
Josef Kittler
  University of Surrey, Guildford, UK
Jon M. Kleinberg
  Cornell University, Ithaca, NY, USA
Friedemann Mattern
  ETH Zurich, Zürich, Switzerland
John C. Mitchell
  Stanford University, Stanford, CA, USA
Moni Naor
  Weizmann Institute of Science, Rehovot, Israel
C. Pandu Rangan
  Indian Institute of Technology, Madras, India
Bernhard Steffen
  TU Dortmund University, Dortmund, Germany
Demetri Terzopoulos
  University of California, Los Angeles, CA, USA
Doug Tygar
  University of California, Berkeley, CA, USA
Gerhard Weikum
  Max Planck Institute for Informatics, Saarbrücken, Germany

More information about this series at http://www.springer.com/series/7409

Jianyong Wang · Wojciech Cellary
Dingding Wang · Hua Wang
Shu-Ching Chen · Tao Li
Yanchun Zhang (Eds.)

Web Information Systems Engineering – WISE 2015

16th International Conference
Miami, FL, USA, November 1–3, 2015
Proceedings, Part I

 Springer

Editors
Jianyong Wang
Tsinghua University
Beijing
China

Wojciech Cellary
Poznan University of Economics
Poznan
Poland

Dingding Wang
Florida Atlantic University
Boca Raton, FL
USA

Hua Wang
Victoria University
Melbourne, VIC
Australia

Shu-Ching Chen
Florida International University
Miami, FL
USA

Tao Li
Florida International University
Miami, FL
USA

Yanchun Zhang
Victoria University
Melbourne, VIC
Australia

ISSN 0302-9743 ISSN 1611-3349 (electronic)
Lecture Notes in Computer Science
ISBN 978-3-319-26189-8 ISBN 978-3-319-26190-4 (eBook)
DOI 10.1007/978-3-319-26190-4

Library of Congress Control Number: 2015953784

LNCS Sublibrary: SL3 – Information Systems and Applications, incl. Internet/Web, and HCI

Springer Cham Heidelberg New York Dordrecht London
© Springer International Publishing Switzerland 2015
This work is subject to copyright. All rights are reserved by the Publisher, whether the whole or part of the material is concerned, specifically the rights of translation, reprinting, reuse of illustrations, recitation, broadcasting, reproduction on microfilms or in any other physical way, and transmission or information storage and retrieval, electronic adaptation, computer software, or by similar or dissimilar methodology now known or hereafter developed.
The use of general descriptive names, registered names, trademarks, service marks, etc. in this publication does not imply, even in the absence of a specific statement, that such names are exempt from the relevant protective laws and regulations and therefore free for general use.
The publisher, the authors and the editors are safe to assume that the advice and information in this book are believed to be true and accurate at the date of publication. Neither the publisher nor the authors or the editors give a warranty, express or implied, with respect to the material contained herein or for any errors or omissions that may have been made.

Printed on acid-free paper

Springer International Publishing AG Switzerland is part of Springer Science+Business Media
(www.springer.com)

Preface

Welcome to the proceedings of the 16th International Conference on Web Information Systems Engineering (WISE 2015), held in Miami, Florida, USA, in November 2015. The series of WISE conferences aims to provide an international forum for researchers, professionals, and industrial practitioners to share their knowledge in the rapidly growing area of Web technologies, methodologies, and applications. The first WISE event took place in Hong Kong, China (2000). Then the trip continued to Kyoto, Japan (2001); Singapore (2002); Rome, Italy (2003); Brisbane, Australia (2004); New York, USA (2005); Wuhan, China (2006); Nancy, France (2007); Auckland, New Zealand (2008); Poznan, Poland (2009); Hong Kong, China (2010); Sydney, Australia (2011); Paphos, Cyprus (2012); Nanjing, China (2013); and Thessaloniki, Greece (2014). This year, for a second time, WISE was held in North America, in Miami, supported by Florida International University (FIU).

WISE 2015 hosted several well-known keynote and invited speakers. Moreover, two tutorials were presented on the topics of building secure Web systems and accessing the Princeton Wordnet.

A total of 171 research papers were submitted to the conference for consideration, and each paper was reviewed by at least two reviewers. Finally, 53 submissions were selected as full papers (with an acceptance rate of 31 % approximately), plus 17 as short papers. The research papers cover the areas of big data techniques and applications, deep/hidden Web, integration of Web and Internet, linked open data, the Semantic Web, social network computing, social Web and applications, social Web models, analysis and mining, Web-based applications, Web-based business processes and Web services, Web data Integration and mashups, Web data models, Web information retrieval, Web privacy and security, Web-based recommendations, and Web search.

In addition to regular and short papers, the WISE 2015 program also featured three special sessions, including a special session on Data Quality and Trust in Big Data (QUAT 2015), a special session on Decentralized Social Networks (DeSN 2015), and an invited session.

QUAT is a qualified forum for presenting and discussing novel ideas and solutions related to the problems of exploring, assessing, monitoring, improving, and maintaining the quality of data and trust for "big data." It provides a forum for researchers in the areas of Web technology, e-services, social networking, big data, data processing, trust, and information systems and GIS to discuss and exchange their recent research findings and achievements. This year, the QUAT 2015 program featured six accepted papers on data cleansing, data quality analytics, reliability assessment, and quality of service for domain applications. QUAT 2015 was organized by Prof. Deren Chen, Prof. William Song, Dr. Xiaolin Zheng, and Dr. Johan Håkansson.

The goal of DeSN 2015 was to serve as a forum for researchers or professionals from both academia and industry to exchange new ideas, discuss new solutions, and share their experience in the design, implementation, analysis, experimentation, or measurement related to decentralized social networks. The DeSN 2015 program included two invited speakers, Dr. Sarunas Girdzijauskas and Dr. Bogdan Carbunar, and three accepted papers. The DESN 2015 co-chairs included Dr. Antoine Boutet, Dr. Sarunas Girdzijauskas, and Dr. Frederique Laforest.

The invited session included five research papers from leading research groups. Each invited paper featured a specific domain, with five papers covering recommender systems, demand trend prediction in cloud computing, deep learning, database security, and social network privacy.

We wish to take this opportunity to thank the honorary co-chairs, Prof. S.S. Iyengar and Prof. Marek Rusinkiewicz; the tutorial and panel co-chairs, Prof. Guandong Xu and Prof. Mitsunori Ogihara; the WISE challenge program co-chairs, Prof. Weining Qian and Qiulin Yu; the workshop co-chairs, Prof. Hill Zhu and Prof. Yicheng Tu; the publication chair, Prof. Hua Wang; the Local Organizing Committee co-chairs, Mr. Carlos Cabrera and Ms. Catherine Hernandez; the publicity co-chairs, Prof. Mark Finlayson, Prof. Giovanni Pilato, and Prof. Yanfang Ye; the registration chair, Mr. Steve Luis; the financial co-chairs, Ms. Lian Zhang and Ms. Donaley Dorsett; and the WISE society representative, Prof. Xiaofang Zhou. The editors and chairs are grateful to the website and social media masters, Mr. Steve Luis and Mr. Bin Xia, for their continuous active support and commitment, and Dr. Rui Zhou and Ms. Sudha Subramani for their effort in preparing the proceedings.

In addition, special thanks are due to the members of the International Program Committee and the external reviewers for a rigorous and robust reviewing process. We are also grateful to the School of Computing and Information Sciences of Florida International University and the International WISE Society for supporting this Conference. The WISE Organizing Committee is also grateful to the special session organizers for their great efforts to help promote Web information system research to broader domains.

We expect that the ideas that have emerged in WISE 2015 will result in the development of further innovations for the benefit of scientific, industrial, and societal communities.

November 2015

Jianyong Wang
Wojciech Cellary
Dingding Wang
Hua Wang
Shu-Ching Chen
Tao Li
Yanchun Zhang

Organization

Honorary Chairs

S.S. Iyengar — Florida International University, USA
Marek Rusinkiewicz — New Jersey Institute of Technology, USA

General Chairs

Shu-Ching Chen — Florida International University, USA
Tao Li — Florida International University, USA
Yanchun Zhang — Victoria University, Australia

Program Co-chairs

Wojciech Cellary — Poznan University of Economics, Poland
Dingding Wang — Florida Atlantic University, USA
Jianyong Wang — Tsinghua University, China

Publication Chair

Hua Wang — Victoria University, Australia

Local Arrangements Chairs

Ivana Rodriguez — Florida International University, USA
Catherine Hernandez — Florida International University, USA

Financial Co-chairs

Carlos Cabrera — Florida International University, USA
Lian Zhang — Florida International University, USA

Web and Social Media Masters

Steve Luis — Florida International University, USA
Bin Xia — Nanjing University of Science and Technology, China

WISE Society Representative

Xiaofang Zhou — University of Queensland, Australia

Tutorial and Panel Co-chairs

Guandong Xu   University of Technology Sydney, Australia
Mitsunori Ogihara   University of Miami, USA

WISE Challenge Program Chairs

Weining Qian   East China Normal University, China
Qiulin Yu   Ping An Technology (Shenzhen) Co., Ltd., China

Workshop Co-chairs

Hill Zhu   Florida Atlantic University, USA
Yicheng Tu   University of South Florida, USA

Publicity Co-chairs

Mark Finlayson   Florida International University, USA
Giovanni Pilato   Italian National Research Council (CNR), Italy
Yanfang Ye   West Virginia University, USA

Program Committee

Karl Aberer   EPFL, Switzerland
Markus Aleksy   ABB, Switzerland
Boualem Benatallah   University of New South Wales, Australia
Azer Bestavros   Boston University, USA
Antonis Bikakis   University College London, UK
David Camacho   Universidad Autonoma de Madrid, Spain
Barbara Catania   University of Genoa, Italy
Tiziana Catarci   Sapienza University of Rome, Italy
Richard Chbeir   LIUPPA Laboratory, France
Fei Chen   HP, USA
Enhong Chen   University of Science and Technology of China
Yueguo Chen   Renmin University of China
Gao Cong   Nanyang Technological University, Singapore
Alexandra Cristea   University of Warwick, UK
Schahram Dustar   Vienna University of Technology, Austria
Damiani Ernesto   University of Milan, Italy
Marie-Christine Fauvet   Joseph Fourier University of Grenoble, France
Bry Francois   Ludwig Maximilian University, Germany
Jun Gao   Peking University, China
Hong Gao   Harbin Institute of Technology, China
Yunjun Gao   Zhejiang University, China
Claude Godart   University Henri Poincare, France

Daniela Grigori	Laboratoire LAMSADE, University Paris-Dauphine, France
Hakim Hacid	Bell Labs, France
Armin Haller	CSIRO ICT Centre, Australia
Xiaofeng He	East China Normal University, China
Yuh-Jong Hu	National Chengchi University, Taiwan
Luke Huan	University of Kansas, USA
Jianbin Huang	Xidian University, China
Marta Indulska	University of Queensland, Australia
Pokorny Jaroslav	Charles University, Czech Republic
Yan Jia	National University of Defense Technology
Lili Jiang	Max Planck Institute for Informatics, Germany
Peiquan Jin	University of Science and Technology of China
Ruoming Jin	Kent State University, USA
Yiping Ke	Nanyang Technological University, Singapore
Fang Li	Shanghai Jiao Tong University, China
Xue Li	School of ITEE, University of Queensland, Australia
Chengkai Li	University of Texas at Arlington, USA
Lei Li	Samsung, USA
Mengchi Liu	Carleton University, Canada
Shuai Ma	Beihang University, China
Xiaofeng Meng	Renmin University of China
Mikolaj Morzy	Poznań University of Technology, Poland
Wilfred Ng	HKUST, China
Kjetil Nørvåg	Norwegian University of Science and Technology, Norway
Mitsunori Ogihara	University of Miami, USA
George Pallis	University of Cyprus, Cyprus
Dasgupta Partha	Arizona State University, USA
Zbigniew Paszkiewicz	PricewaterhouseCoopers, Cyprus
Olivier Pivert	ENSSAT, France
Weining Qian	East China Normal University, China
Tieyun Qian	Wuhan University, China
Jarogniew Rykowski	Poznań University of Economics, Poland
Rizos Sakellariou	University of Manchester, UK
Wei Shen	Nankai University, China
John Shepherd	UNSW, Australia
Dezhao Song	Thomson Reuters, USA
Dandan Song	Beijing Institute of Technology, China
Reima Suomi	University of Turku, Finland
Stefan Tai	KIT, Germany
Dimitri Theodoratos	New Jersey Institute of Technology, USA
Farouk Toumani	Limos, Blaise Pascal University, Clermont-Ferrand, France
Xiaojun Wan	Peking University, China
Guoren Wang	Northeast University, China

De Wang	Google, USA
Hua Wang	Victoria University, Australia
Wei Wang	Fudan University, China
Ingmar Weber	Qatar Computing Research Institute, Qatar
Xintao Wu	University of North Carolina at Charlotte, USA
Junjie Wu	Beihang University, China
Josiane Xavier Parreira	DERI - National University of Ireland, Galway, Ireland
Hayato Yamana	Waseda University, Japan
Yanfang Ye	West Virginia University, USA
Hongzhi Yin	University of Queensland, Australia
Tetsuya Yoshida	Nara Women's University, Japan
Jeffrey Xu Yu	Chinese University of Hong Kong, SAR China
Ge Yu	Northeast University, China
Xiaohui Yu	Shandong University, China
Qi Zhang	Fudan University, China
Wenjie Zhang	The University of New South Wales, Australia
Yanchun Zhang	Victoria University, Australia
Xiaofang Zhou	University of Queensland, Australia
Xingquan Zhu	Florida Atlantic University, USA
Lei Zou	Peking University, China

Contents – Part I

Contents – Part II

Grouping Product Aspects from Short Texts Using Multiple Classifiers

Daniel Kailer[1]([✉]), Peter Mandl[1], and Alexander Schill[2]

[1] Munich University of Applied Sciences, Lothstr. 64, 80335 Munich, Germany
dkailer@hm.edu, mandl@cs.hm.edu
[2] TU Dresden, Nöthnitzer Str. 46, 01187 Dresden, Germany
alexander.schill@tu-dresden.de

Abstract. In this paper we present and evaluate a classification model to group product aspects from short user comments, found as pros and cons in consumer review websites. Because of the distinct vocabulary used by consumers to describe the same aspects of a product, it is necessary to group pros and cons to support consumers' decision making. For this purpose we propose a supervised classification model, consisting of an ensemble classifier that combines a main text classifier (e.g. Naive Bayes) and several string-based classifiers. Furthermore we make use of WordNet as a domain independent ontology to detect semantically related words. Experimental results using pros and cons from five heterogeneous product groups show, that the proposed method outperforms existing approaches to group pros and cons from short texts. We also found that the reusable short comments from our sample follow a power law distribution, that is usually present in social tagging systems.

Keywords: Aspect-oriented opinion mining · Short text · Combining classifiers · Supervised learning

1 Introduction

The amount of user-generated opinions is steadily increasing on the Web. These opinions are often an important source of information to support the decision making of other consumers. A lot of opinions are expressed through customer reviews, where users describe what they like or dislike about a product in free text. Many research papers in the area of opinion mining have studied this type of opinion expression.

A less studied review format is the explicit statement of pros and cons of product aspects in the form of short comments. The main characteristic of these comments is that they are very short, sometimes only one word. Such pros and cons are found in consumer review websites like epinions.com or buzzillions.com. An advantage of this format is that consumers can quickly obtain the opinion about certain product aspects without reading a long textual review. A problem of this type of opinion expression is that consumers are still forced to browse

© Springer International Publishing Switzerland 2015
J. Wang et al. (Eds.): WISE 2015, Part I, LNCS 9418, pp. 1–15, 2015.
DOI: 10.1007/978-3-319-26190-4_1

through all pros and cons to obtain the overall opinion of product aspects. Review websites like buzzillions.com are trying to mitigate this problem, by allowing consumers to reuse pros and cons specified by other consumers and by displaying them in an aggregated manner (see Fig. 1).

The reuse and aggregation of pros and cons is however only partially helpful, because consumers often use a different vocabulary to describe the same aspect. This is illustrated in Fig. 1, where different expressions are used to describe the same aspect, for example the comments 'Sound' and 'Poor audio'. It is therefore necessary to group all similar aspect expressions to reduce the cognitive effort required by consumers to manually combine these expressions.

To address this problem, we present a supervised ensemble classifier to group product aspects from short comments containing pros and cons. The ensemble classifier consists of a main text classifier such as Naive Bayes or Support Vector Machine (SVM) and additional string-based classifiers, that try to compensate the weaknesses of the main text classifier. We further use WordNet [13] as a knowledge-based measure to find semantically related words.

The remainder of this article is organized as follows. In Sect. 2, we first discuss the related work relevant to our approach. We then present our ensemble classification model and the proposed string-based classifiers in Sect. 3. The results of our empirical evaluation on buzzillions.com are then presented and discussed in Sect. 4. Finally, we conclude with a summary of our contributions in Sect. 5.

Great picture quality (56)	Sound (17)
Picture (33)	Speakers (5)
Everything (13)	Poor audio (4)
3d (9)	Sound quality (4)
Price (9)	The sound (3)
Features (6)	Price (3)

Fig. 1. Overview of user-generated pros and cons for product aspects (source: http://www.buzzillions.com, last access: 03/29/2015)

2   Related Work

The work presented in this article is related to the research areas of aspect-oriented opinion mining and text classification. The task of aspect-based opinion mining is to identify and extract product aspects from consumer reviews and to determine their sentiment. According to Liu et al. the three most common review formats for specifying product aspects are [10]:

- Format F_{PC} - Reviewers can specify pros and cons for product aspects.
- Format F_R - Reviewers can write a free text review without explicitly specifying pros and cons.
- Format F_{RPC} - A combination of F_{PC} and F_R where reviewers write a review and additionally specify pros and cons.

The work in this paper addresses the grouping of product aspects specified in the format F_{PC}. The expression of product aspects can be either explicit or implicit [10, 14]. An example of an explicit aspect is *'great picture quality'*, where the aspect *'picture quality'* is part of the text. Implicit aspects on the other hand don't explicitly contain the aspect, for example the text *'very expensive'*, which implicitly describes the aspect *'price'*. The recognition of implicit aspects is especially important for the format F_{PC}, because there is little context available.

The previous work of Carenini et al. combines a supervised and an unsupervised method to group product aspects from reviews (format F_R). They reuse the unsupervised approach from Hu and Liu [6] to identify product aspects, where primarily nouns are considered as product aspects. Implicit aspects are not taken in consideration. Carenini et al. additionally use product taxonomies to add knowledge about the product groups to the algorithm and make use of WordNet to calculate the similarity between two terms. They evaluate their method for two product groups. A weakness of their approach is that there might not be an existing product taxonomy for every product group or that such a taxonomy must be modified to fit the actual use case, which is time consuming.

The work of Zhai et al. uses a semi-supervised classification model to group product aspects from reviews (format F_R). They use a Naive Bayes classifier and an expectation maximization algorithm with additional soft-constraints. The approach of Zhai et al. does not take implicit product aspects into account. Their grouping algorithm assumes that valid aspect expressions are already extracted by an existing system (e.g. [14]). The distributional context for every aspect expression is then extracted, i.e. words surrounding the aspect expression. The proposed window size is three, i.e. three words before and after the aspect expression (excluding stopwords) are used as context. This method of context extraction is useful for whole sentences, but less relevant for review format F_{PC}.

Regarding the review format F_{RPC}, the work of Guo et al. proposes an unsupervised classification approach to group product aspects [4]. They use the topic model Latent Dirichlet Allocation (LDA) to identify latent topics. Because this is an unsupervised approach, only the number of clusters (topics) and not the product groups themselves can be defined. The approach doesn't consider implicit aspects for grouping. According to the findings of Chen et al., the use of topic models is rather inappropriate for shorter texts [2], therefore we did not consider topic models for the review format F_{PC}.

An approach to group aspects from short pros and cons (review format F_{PC}) is proposed by Lu et al. [11]. They use clustering to group short user comments from the online auction website eBay.com. The considered aspects are however not product aspects, but aspects describing a seller or buyer. Lu et al. limit the number of clusters to 8, which is not appropriate for grouping product aspects, because there are often more than 8 aspects relevant for a product category.

The work that is most related to our work comes from Liu et al. [10] and addresses the extraction and grouping of product aspects from pros and cons (review format F_{PC}). Their approach is based on natural language processing (NLP) and uses a supervised rule mining approach to extract product aspects from pros and cons. Their method for grouping synonyms uses WordNet and is relatively conservative, yielding an F_1-score of 68.42 % (100 % precision and 52 % recall). We compare the results of Liu et al. with our results in Sect. 4.3.

3   Classification Model

3.1   Overview

Some works described in the previous section use an unsupervised classification approach to group product aspects. For our classification model however we use a supervised approach, because we argue that a user participation is essential to create meaningful product categories. This view is in line with Carenini et al. [1] and Zhai et al. [16], who also state that user participation and the inclusion of existing knowledge is needed for a meaningful aspect grouping.

The classification problem in our work can be seen as an ontology matching problem. The user-generated pros and cons can be considered as an uncontrolled vocabulary, whereas the predefined product aspects can be viewed as a controlled vocabulary or a lightweight ontology. This is illustrated in Fig. 2. When the reuse of existing pros and cons is possible (as shown in Fig. 1), the resulting short texts are comparable to *social tags* in social tagging systems. The uncontrolled vocabulary can then be seen as a user-generated taxonomy or folksonomy [5]. In a previous work we have already proposed a concept called Rated Tags, which allows the specification of product aspects as rateable tags, and evaluated it in a study in regard to the decision support of customers [8].

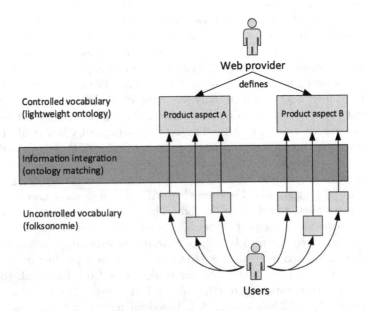

Fig. 2. Classification problem – transforming user-generated comments into predefined product aspects

Because we are using a supervised classification, we consider a set of classes C that represent the product aspects. A document set D contains all documents, whereas every document d_i is assigned a class c_j. For the classification process, the document set D is split into a training set L and a test set U, whereas the class information is withheld from the classifier in U. Each document d_i consists of a set of terms T_i and a sentiment s_i. Combining all terms of the training set L results in the vocabulary V (see Eq. 1).

$$V = \bigcup_{i=1}^{|L|} T_i \tag{1}$$

This results in a word vector Z_i of length $|V|$ for every document d_i. We argue that for short texts it is sufficient to denote only the occurrence or non-occurrence of a word. Therefore every vector Z_i can be considered as a binary vector, where $z_{ik} = 1$ if $v_k \in T_i$ and $z_{ik} = 0$ otherwise.

For our final vector we also take the sentiment of a document into consideration. As an example we take two classes c_1 and c_2, where documents of the former are rated mostly positively and the latter mostly negatively. If a document with a positive sentiment is to be classified, we consider it more likely that the document belongs to class c_1 than to class c_2. To express this, we add a binary attribute s_i to our word vector, that contains the value 0 for negative sentiments and the value 1 for positive sentiments. This results in a binary feature vector F with length $|V| + 1$. To illustrate this, we consider two documents l_1 and l_2, where l_1 describes a con with the text 'blurry picture quality' and l_2 describes a pro with the text 'great sound quality'. The resulting vocabulary V and the feature vectors F_1 and F_2 are displayed in Table 1.

Table 1. Exemplary feature vectors

V	blurry	picture	quality	great	sound	
F_i	Z_i					s_i
F_1	1	1	1	0	0	0
F_2	0	0	1	1	1	1

3.2 Ensemble Classifier

The main part of the classification model is an ensemble classifier, that is composed of a set of several classifiers K. Previous works have shown that the combination of different classifiers can improve the classification result [9,15]. Our model consists of a main text classifier and several string-based classifiers, which try to compensate the weaknesses of the main classifier. For the combination of the individual classifier results it is important how the classifiers output their results. In regard to this, Xu et al. state the following three output levels [15]:

- Abstract level: outputs only a final class.
- Rank level: outputs a ranking of classes, beginning with the most likely.
- Measurement level: outputs the probabilities for each class.

Our model uses the measurement level, which provides the most details for the classification results [15]. Therefore the requirement for each classifier k_n is to return a vector of probabilities $P_{k_n u_m}$ of length $|C|$ for any unclassified document u_m. We then take the average of all classifier probabilities to find the average probability of a class c_i, resulting in a vector $\Omega(u_m)$ containing all the probabilities for each class (see Eqs. 2 and 3).

$$\eta(c_i, u_m) = \frac{1}{|K|} \cdot \sum_{n=1}^{|K|} P_{k_n u_m}(c_i) \tag{2}$$

$$\Omega(u_m) = (\eta(c_1, u_m), \eta(c_2, u_m), ..., \eta(c_{|C|}, u_m)) \tag{3}$$

Finally, we determine the classes with the maximal average probability (see Eq. 4). If two or more classes have equal probabilities, the ensemble classification function $\gamma_E(u_m)$ outputs the first class of the candidate set X_{u_m}.

$$X_{u_m} = \{c_i \mid c_i \in C \wedge \eta(c_i, u_m) = arg\,max\,\Omega(u_m)\}$$
$$\gamma_E(u_m) = x_{u_m 1} \tag{4}$$

3.3 Augment Main Classifiers with WordNet

As a main classifier for text classification we propose Naive Bayes or Support Vector Machine (SVM). While Naive Bayes is relatively efficient and easy to implement, SVM can be seen as the current state of the art classifier for text classification. A drawback of SVM is its implementation complexity. When using a Naive Bayes classifier we propose the use of a multivariate Bernoulli Naive Bayes, instead of the usually for text classification used Multinomial Naive Bayes approach, because it performs better on short texts [12].

To support the main classifier, which relies on the feature vector, we propose the use of the lexical database WordNet. The shortness of the user-generated pros and cons can limit the performance of the main classifiers, for example when an unclassified document contains only one word, which is also not present in the vocabulary V. To support such scenarios we identify all words of an unclassified document u_m that are not in the vocabulary, i.e. $\exists t \in T_m : t \notin V$, and replace them with semantically related words, that are in the vocabulary.

The first step is to use a part-of-speech (POS)-tagger to identify the POS for the words of a document that are not in the vocabulary. This is required in order to use WordNet. For every term, we retrieve all senses of the word from WordNet, i.e. we do not perform a word-sense disambiguation. For every word-sense, we collect all its semantic relations (called *synsets* in WordNet), e.g. synonyms, antonyms etc. If a word from these semantic relations is contained in the vocabulary, it is used instead of the original word. If no related word is found, the original word remains unchanged (see Algorithm 1). This approach is especially helpful to detect synonymous implicit aspects, which are often appear as adjectives.

Algorithm 1. Replace unknown words in the vocabulary based on WordNet

```
 1: procedure AUGMENTDOCUMENT(d, V)
 2:     taggeddocument = tagDocumentWithPOS(d)
 3:     wordtokenlist = tokenizeToWords(taggeddocument)
 4:     for each word in wordtokenlist do
 5:         if word ∉ V then
 6:             candidate = findWordNetCandidate(word, V)
 7:             if candidate ≠ NULL then
 8:                 word = candidate   // replace word
 9:     Return wordtokenlist
10:
11: procedure FINDWORDNETCANDIDATE(wordWithPOS, V)
12:     synsets = findSynSets(wordWithPOS)
13:     for each synset in synsets do
14:         for each synword in synset do
15:             if synword ∈ V then
16:                 Return synword
17:     Return NULL
```

3.4 String-Based Classifiers

Additionally, we propose the use of different string-based classifiers to compensate for the weaknesses of the main text classifier. Based on the vector model, the main classifier treats word pairs like *'big'* and *'bigger'* as well as *'picture'* and the misspelled word *'pitcure'* as two different features. String-based classifiers on the other hand operate on strings or string parts and not on the vector model and are thus able to detect word similarities on character level. We propose three complementary string-based classifiers to determine the similarity of two texts, which are expressed as sets of terms T_x and T_y, where every term t_{x_i} is compared to every term t_{y_j}.

The first classifier checks if a part of t_{x_i} is contained in t_{y_j} and vice versa. This is shown in Eq. 5, where $\lambda_{t_{x_i}, t_{y_j}}$ denotes the similarity function.

$$\lambda_{t_{x_i}, t_{y_j}} = \begin{cases} \dfrac{|t_{x_i} \cap t_{y_j}|}{|t_{y_j}|} & \text{if } |t_{x_i}| \leq |t_{y_j}|, t_{x_i} \in t_{y_j} \\[2ex] \dfrac{|t_{x_i} \cap t_{y_j}|}{|t_{x_i}|} & \text{if } |t_{x_i}| > |t_{y_j}|, t_{y_j} \in t_{x_i} \\[2ex] 0 & \text{otherwise} \end{cases} \tag{5}$$

While the above classifier only checks for strings that are fully contained in another string, the second proposed classifier checks for partly overlapping strings. To avoid too many false positives, a constraint is that the compared strings must begin with the same characters, for example the strings 'easy' and 'ease'. The similarity function is shown in Algorithm 2.

Algorithm 2. Similarity for strings starting with the same characters

1: **procedure** COMMONSTARTSIMILARITY$(s1, s2)$
2:     $commonchars = 0$
3:     **for** $i = 1$ to $Min(|s1|, |s2|)$ **do**
4:         **if** $s1_i == s2_i$ **then** // i is the i-th character
5:             $commonchars = commonchars + 1$
6:         **else**
7:             Exit For
8:     Return $commonchars / Max(|s1|, |s2|)$

The final string-based classifier is based on the Levenshtein-distance, a well known edit distance. The use cases for this classifier are especially misspelled words like 'pitcure' instead of 'picture'. The edit distance of two strings t_{x_i} and t_{y_j} is expressed by the distance function $\delta(t_{x_i}, t_{y_j})$. Again, to avoid too many false positives we use a very conservative setting for this classifier. We only treat two strings as similar, if their edit distance is at most one. An edit distance is calculated as the number of edit operations, whereas an edit operation is either the insertion, replacement, deletion or transposition of characters. To calculate the similarity of two strings, we normalize the result of the distance function with respect to the shortest string and subtract it from 1 (see Eq. 6).

$$\lambda_{t_{x_i}, t_{y_j}} = \begin{cases} 1 - \dfrac{\delta(t_{x_i}, t_{y_j})}{Min(|t_{x_i}|, |t_{y_j}|)} & \text{if } Min(|t_{x_i}|, |t_{y_j}|) > 0, \delta(t_{x_i}, t_{y_j}) \leq 1 \\ 0 & \text{otherwise} \end{cases} \quad (6)$$

For the integration into the ensemble classifier, each string-based classifier returns a vector of probabilities of length $|C|$. For that, an unclassified document u_m is compared to each training document l_i. The similarity between u_m and l_i is treated as the probability that u_m belongs to the class c_i of l_i.

To optimize the quality of the above string-based classifiers, we calculate the optimal similarity threshold τ in the learning phase of the classifiers. For that, we compare all training documents L with each other and store the similarity λ for each pair. We further store the variable ε, which denotes whether the classes of the compared documents are equal or not, where $\varepsilon = 1$ for equal classes and $\varepsilon = 0$ otherwise. This results in a set O of length n, where n equals the number of comparisons, i.e. $n = \frac{|L|^2 - |L|}{2}$, $O = \{(\lambda_{1,2}, \varepsilon_{1,2}), (\lambda_{1,3}, \varepsilon_{1,3}), ..., (\lambda_{n-1,n}, \varepsilon_{n-1,n})\}$.

To determine the threshold τ where the classifier performs best, we calculate the F_1-score for each similarity and return the similarity with the maximal F_1-score. This is shown in Algorithm 3 (the abbreviations TP, FP and FN stand for True Positive, False Positive and False Negative respectively). The threshold τ is then later used in the classification phase. Only similarity results exceeding τ, i.e. $\lambda \geq \tau$, will be taken into consideration.

Algorithm 3. Determine similarity threshold τ where F_1-score is maximal

---

1: **procedure** GETBESTSIMILARITYTHRESHOLD(O)
2:     $\tau = 0.0; maxF1 = 0.0$
3:     **for** each *similarity* in O_λ **do**
4:         $TP = 0; FP = 0; FN = 0$
5:         **for** each (λ, ε) in O **do**
6:             **if** $\varepsilon == 1$ **then**
7:                 **if** $\lambda \geq$ *similarity* **then**
8:                     $TP = TP + 1$
9:                 **else**
10:                     $FN = FN + 1$
11:             **else if** $\lambda \geq$ *similarity* **then**
12:                 $FP = FP + 1$
13:         $F1 = calculateF1()$ // based on TP, FP and FN
14:         **if** $F1 > maxF1$ **then**
15:             $maxF1 = F1$
16:             $\tau = similarity$
17:     Return τ

3.5 Preprocessing

To reduce and normalize the feature vector, we propose the following preprocessing activities. First, special characters were removed, hyphens were replaced by white space and all characters were lowercased. Then the text was tokenized to process individual words. Stopwords (e.g.'the', 'of' etc.) were removed from the text. For normalization purposes, we determined the lemma for each word with the help of WordNet and used the Porter-stemmer[1] as stemming algorithm.

The above steps can be considered as basic preprocessing activities for text classification. We additionally added problem-specific preprocessing steps. Because pros and cons often contain adjectives describing the quality of an aspect, we removed negation prefixes to normalize these adjectives. For that, we used the regular expression `^(un|il|im|ir|in|dis|mis|non)(\w+)$` to detect possible candidate words. For every candidate, we used WordNet to check whether a word without the prefix exists, for example the word 'reliable' for the candidate 'unreliable'. If such a word exists, we removed the prefix from the candidate word. We further removed neutral adjectives (e.g.'good', 'bad' etc.), because we argue that they are too generic and can be misleading for text classification. For example 'good sound' and 'good picture' are not related, but both contain the neutral adjective 'good'.

4 Empirical Evaluation

4.1 Methodology

As the source for our sample data, we chose the consumer review website buzzillions.com, which provides more than 17 millions reviews[2] from different product

---

[1] http://tartarus.org/martin/PorterStemmer, last access: 03/15/2015.
[2] http://www.buzzillions.com/about, last access: 03/16/2015.

categories. Buzzillions.com allows consumers to create reviews in free text and to additionally specify and reuse pros and cons. In our study we are however only interested in the user-generated pros and cons. As mentioned in Sect. 3.1, these reusable pros and cons are comparable to social tags. Therefore we subsequently use the term *tags* as a generic term for pros and cons.

For the extraction of the sample tags, we chose the following five heterogeneous product categories:

- Digital camera
- GPS device
- TV
- Running shoes
- Baby seat (car)

For every category, we ordered the products from the website buzzillions.com descending by the number of customer reviews. From the most reviewed products we then randomly chose six products for each category. We developed a web crawler that automatically iterates through all reviews and extracts all pros and cons (if present). After the extraction of the tags it had shown, that the categories 'shoes' and 'baby seat' had significantly fewer distinct tags. Therefore we selected another six products for each of these categories based on the above mentioned method and added them to the previously extracted tags.

The final numbers of the extracted tags for each category are displayed in Table 2. For the evaluation only the distinct tags are relevant. However, the total number of tags is also interesting, because it shows that consumers tend to reuse existing tags and that the degree of reuse depends on the respective category.

Table 2. Overview for the extracted tags and assigned classes

Category	Distinct tags	Total tags	Number of classes
Camera	939	5203	23
GPS	1419	15016	21
TV	1480	7861	23
Shoes	947	20473	15
Baby seat	441	15778	13
Total	*5226*	*64331*	*95*

To later determine the performance of our classifier, it was necessary to manually classify every tag, i.e. a single class was assigned to every tag. This process is also called the creation of a *gold standard* or *ground truth* [12]. In order to do that, we created several classes for each product category (see Table 2). It must be noted that the creation of such classes is always of subjective nature and that different web providers might choose finer or coarser classes. The maximal number of classes for a category was 23. This number might seem high, but it is not uncommon as the study of Zhai et al. shows, where the category 'vacuum' contains even 28 classes [16].

For the evaluation of the classifier, we used the typical metrics, namely precision, recall and F_1-score. These are based on the number of true positives (TP), false positives (FP) and false negatives (FN). The formulas for precision (P), recall (R) and F_1-score are shown in Eqs. 7, 8 and 9 respectively.

$$P = \frac{TP}{TP + FP} \tag{7}$$

$$R = \frac{TP}{TP + FN} \tag{8}$$

$$F_1 = 2 \cdot \frac{P \cdot R}{P + R} \tag{9}$$

We report the macro-average precision, recall and F_1-score weighted by class size, i.e. these metrics were calculated per class and then averaged with respect to the number of instances per class. For the determination of the these metrics, we used a 10-fold stratified cross validation. In a k-fold cross validation, the document set is randomly split into k subsets (here $k = 10$), where $k - 1$ subsets are used as a training set and the remaining subset is used as test set. The evaluation process is executed k times where each of the k subsets is used once as test set. The cross validation was executed based on the data mining framework WEKA[3] in the version 3.6.12. The used baseline classifiers Bernoulli Naive Bayes, Multinomial Naive Bayes and Support Vector Machine (SMO algorithm) were also taken from WEKA[4].

4.2 Results

The results of the 10-fold cross validation are shown in Tables 3 and 4. Table 3 shows the baseline results, i.e. the results of the main text classifiers with basic preprocessing. Table 4 on the other hand shows the results for the proposed ensemble classifier, including the problem-specific preprocessing (see Sect. 3.5). The SVM-classifier achieved the highest average F_1-score for the baseline as well as in combination with the ensemble classifier. The application of the ensemble classifier increased the F_1-score of SVM from 74.67 % to 79.82 %, which is a relative increase of 6.9 %.

The average performance of the Bernoulli Naive Bayes is little worse than SVM, but outperforms the Multinomial Naive Bayes. However, through the combination of the Multinomial Naive Bayes with the string-based classifiers, the results are very close to the Bernoulli Naive Bayes. The relative increase of the F_1-score for Bernoulli Naive Bayes was 11.15 %, while the relative increase for the Multinomial Naive Bayes was 39.62 %, compared to the baseline results. A graphical comparison of the baseline and the ensemble classifier in regard to their average F_1-score is shown in Fig. 3.

---

[3] http://www.cs.waikato.ac.nz/ml/weka/ (last access: 03/16/2015).

[4] http://weka.sourceforge.net/doc.dev/weka/classifiers/Classifier.html (last access: 03/16/2015).

Table 3. Baseline results (MNB = Multinomial Naive Bayes, BNB = Bernoulli Naive Bayes, SVM = Support Vector Machine)

Category	Precision			Recall			F_1-score		
	MNB	BNB	SVM	MNB	BNB	SVM	MNB	BNB	SVM
TV	62.32 %	72.46 %	**77.78** %	58.78 %	70.88 %	**76.09** %	52.66 %	70.19 %	**76.38** %
Camera	64.42 %	72.83 %	**77.53** %	58.42 %	71.10 %	**75.78** %	55.35 %	70.47 %	**75.73** %
GPS	73.07 %	72.53 %	**75.47** %	62.44 %	72.30 %	**73.05** %	60.50 %	71.93 %	**73.24** %
Baby seat	65.19 %	70.58 %	**76.92** %	56.92 %	67.89 %	**72.24** %	54.48 %	68.07 %	**73.05** %
Shoes	68.03 %	75.09 %	**77.25** %	55.88 %	73.69 %	**75.18** %	51.25 %	73.13 %	**74.97** %
Average	66.61 %	72.70 %	**76.99** %	58.49 %	71.17 %	**74.47** %	54.85 %	70.76 %	**74.67** %

Table 4. Results of the ensemble classifier

Category	Precision			Recall			F_1-score		
	MNB	BNB	SVM	MNB	BNB	SVM	MNB	BNB	SVM
TV	83.55 %	84.70 %	**85.76** %	83.76 %	85.20 %	**85.84** %	82.88 %	84.59 %	**85.63** %
Camera	78.34 %	79.90 %	**80.90** %	77.74 %	79.91 %	**80.49** %	76.45 %	78.86 %	**80.02** %
GPS	75.03 %	76.46 %	**77.77** %	74.60 %	76.70 %	**77.02** %	74.07 %	76.25 %	**76.83** %
Baby seat	77.34 %	75.94 %	**80.22** %	75.83 %	75.51 %	**79.32** %	75.59 %	75.19 %	**79.23** %
Shoes	76.74 %	**78.94** %	78.24 %	74.72 %	**78.97** %	77.82 %	73.90 %	**78.34** %	77.39 %
Average	78.20 %	79.19 %	**80.58** %	77.33 %	79.26 %	**80.10** %	76.58 %	78.65 %	**79.82** %

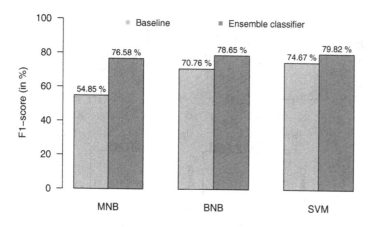

Fig. 3. Comparison of the average F_1-scores of the main text classifiers (baseline) and the ensemble classifier (MNB = Multinomial Naive Bayes, BNB = Bernoulli Naive Bayes, SVM = Support Vector Machine)

Additional to the classification results, we also analyzed the reuse of tags (pros and cons). We discovered that the distribution of the tags follows a power law distribution as shown in Fig. 4. We limited the number of distinct tags in Fig. 4 to 60 to ensure readability. Figure 4 shows that most categories have a stable (dominating) set of 10 to 20 tags. The remaining tags are mostly used once and form the long tail. To the best of our knowledge there is no existing work that shows that user-generated opinions (tags) in the form of pros and cons follow a power law distribution. Most existing research in this area only studied traditional tagging systems, for example the study of the social bookmarking service Delicious.com from Dellschaft and Staab [3].

4.3 Discussion

The results of the proposed ensemble classifier are promising. The classifier achieved higher F_1-scores for all heterogeneous categories than the baseline classifier. For the category 'GPS', the classifier performed worst. After analyzing the data we assume that the reason for this is that a very heterogeneous vocabulary was used in this category to describe the same aspect. Some exemplary comments from our sample are 'Crazy routes', 'Sometimes just wrong' or 'Says street after passing'. Such a descriptive vocabulary makes classification very hard.

As mentioned in Sect. 2, our work is comparable to the work of Liu et al. [10]. Their conservative grouping approach achieves a precision of 100 %, a recall of 52 % and therefore a F_1-score of 68.42 %. The proposed ensemble classifier achieved an F_1-score of 79.82 % and therefore outperforms the approach of Liu et al. However, both studies are not directly comparable, because they do not use the same sample. Additionally, Liu et al. only state that they "manually tagged a large collection of reviews" [10], but do not specify a concrete number.

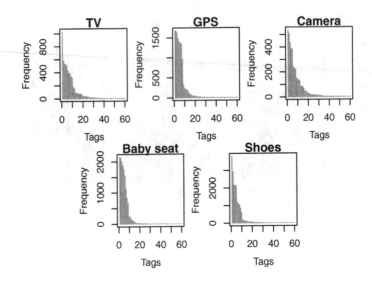

Fig. 4. Presence of a long tail, based on the usage of opinionated tags

Furthermore, compared to the heterogeneous sample in our study, their approach was only evaluated for electronic goods.

To evaluate the performance of our string-based classifiers, we compared them with the approach of Islam and Inkpen [7], who proposed three string-based classifiers for short text classification. We implemented these algorithms and used them instead of our proposed string-based classifiers as part of the ensemble classifier. This resulted in an average F_1-score of 78.15 %, which is a little less than the F_1-score of 79.82 %, which was achieved by using our string-based classifiers. Because short text classification is a difficult problem, we argue that even such a small improvement is a notable contribution.

5   Conclusion

This paper proposed the combination of several classifiers to group product aspects from very short texts in the form of pros and cons. We evaluated the approach with a sample from five heterogeneous product groups from the review website buzzillions.com. The F_1-scores of our classification model were superior to the baseline and outperformed existing approaches. The evaluation showed, that the Naive Bayes classifiers achieved an F_1-score that is comparable to the SVM-classifier, when combined with the proposed string-based classifiers. Furthermore, our sample showed that the reuse of pros and cons resulted in a power law distribution, that is typically present in social tagging systems.

References

1. Carenini, G., Ng, R.T., Zwart, E.: Extracting knowledge from evaluative text. In: Proceedings of the 3rd International Conference on Knowledge Capture. pp. 11–18. ACM (2005)
2. Chen, X., Li, L., Xu, G., Yang, Z., Kitsuregawa, M.: Recommending related microblogs: a comparison between topic and WordNet based approaches. In: Proceedings of the 26th AAAI Conference on Artificial Intelligence, pp. 2417–2418 (2012)
3. Dellschaft, K., Staab, S.: An epistemic dynamic model for tagging systems. In: Proceedings of the 19th Conference on Hypertext and Hypermedia, pp. 71–80. ACM (2008)
4. Guo, H., Zhu, H., Guo, Z., Zhang, X., Su, Z.: Product feature categorization with multilevel latent semantic association. In: Proceedings of the 18th ACM Conference on Information and Knowledge Management, pp. 1087–1096. ACM (2009)
5. Gupta, M., Li, R., Yin, Z., Han, J.: Survey on social tagging techniques. ACM SIGKDD Explor. Newslett. **12**(1), 58–72 (2010)
6. Hu, M., Liu, B.: Mining and summarizing customer reviews. In: Proceedings of the 10th ACM SIGKDD international conference on Knowledge discovery and data mining, pp. 168–177 (2004)
7. Islam, A., Inkpen, D.: Semantic text similarity using corpus-based word similarity and string similarity. ACM Trans. Knowl. Disc. Data (TKDD) **2**(2), 10:1–10:25 (2008)
8. Kailer, D., Mandl, P., Schill, A.: Supporting customers' decision making with rated tags. In: Proceedings of the 16th International Conference on Electronic Commerce, pp. 33–40. ACM (2014)
9. Kittler, J., Hatef, M., Duin, R.P., Matas, J.: On combining classifiers. IEEE Trans. Pattern Anal. Mach. Intell. **20**(3), 226–239 (1998)
10. Liu, B., Hu, M., Cheng, J.: Opinion observer: analyzing and comparing opinions on the web. In: Proceedings of the 14th International Conference on World Wide Web, pp. 342–351. ACM (2005)
11. Lu, Y., Zhai, C., Sundaresan, N.: Rated aspect summarization of short comments. In: Proceedings of the 18th International Conference on World Wide Web, pp. 131–140. ACM (2009)
12. Manning, C.D., Raghavan, P., Schütze, H.: Introduction to information retrieval. Cambridge University Press (2008)
13. Miller, G.A.: WordNet: a lexical database for english. Commun. ACM **38**(11), 39–41 (1995)
14. Popescu, A.M., Etzioni, O.: Extracting product features and opinions from reviews. In: Kao, A., Poteet, S.R. (eds.) Natural Language Processing and Text Mining, pp. 9–28. Springer, London (2007)
15. Xu, L., Krzyzak, A., Suen, C.Y.: Methods of combining multiple classifiers and their applications to handwriting recognition. IEEE Trans. Syst. Man Cybern. **22**(3), 418–435 (1992)
16. Zhai, Z., Liu, B., Xu, H., Jia, P.: Grouping product features using semi-supervised learning with soft-constraints. In: Proceedings of the 23rd International Conference on Computational Linguistics, pp. 1272–1280. ACL, August 2010

Time-Dependent Popular Routes Based Trajectory Outlier Detection

Jie Zhu[1], Wei Jiang[1], An Liu[1,2], Guanfeng Liu[1,2], and Lei Zhao[1,2(✉)]

[1] School of Computer Science and Technology, Soochow University, Suzhou, China
zjcomeon@gmail.com, jwpker@outlook.com
[2] Collaborative Innovation Center of Novel Software Technology
and Industrialization, Nanjing, Jiangsu, China
{anliu,gfliu,zhaol}@suda.edu.cn

Abstract. With the rapid proliferation of the GPS-equipped devices, a myriad of trajectory data representing the mobility of the various moving objects in two-dimensional space have been generated. In this paper, we aim to detect the anomalous trajectories from the trajectory dataset and propose a novel time-dependent popular routes based algorithm. In our algorithm, spatial and temporal abnormalities are taken into consideration simultaneously to improve the accuracy of the detection. For each group of trajectories with the same source and destination, we firstly design a time-dependent transfer graph and in different time period, we can obtain the top-k most popular routes as reference routes. For a pending inspecting trajectory in this time period, we will label it as an outlier if has a great difference with the selected routes in both spatial and temporal dimension. To quantitatively measure the "difference" between a trajectory and a route, we propose a novel time-dependent distance measure which is based on Edit distance in both spatial and temporal domain. The comparative experimental results with two famous trajectory outlier detection methods TRAOD and IBAT on real dataset demonstrate the good accuracy and efficiency of the proposed algorithm.

Keywords: Outlier detection · Time-dependent popular route · Trajectory pattern mining

1 Introduction

In recent years, the booming development of GPS-equipped portable devices has helped us gathering a huge amount of trajectory data. According to a report of a

This work was supported by the National Natural Science Foundation of China under Grant Nos. 61073061, 61003044, 61232006, and 61303019, the Natural Science Foundation of Jiangsu Province of China under Grant No. SBK2015021685, Jiangsu Provincial Department of Education of China under Grant No. 12KJB520017, the Doctoral Fund of Ministry of Education of China under Grant No. 20133201120012, and Collaborative Innovation Center of Novel Software Technology and Industrialization, Jiangsu, China.

ⓒ Springer International Publishing Switzerland 2015
J. Wang et al. (Eds.): WISE 2015, Part I, LNCS 9418, pp. 16–30, 2015.
DOI: 10.1007/978-3-319-26190-4_2

data research organization in China, there are about 66, 000 taxis in Beijing and about 1, 900, 000 passengers each day. Each carry generates one trajectory and there are about 69 million trajectories in one single year. Such a big dataset can help us understanding the cabbies' driving behavior, the city's traffic condition and so on. On this background, extensive researchers are encouraged in trajectory pattern mining, such as life pattern mining [1, 2], popular routes discovering [3, 4], transportation mode mining [5].

Trajectory outlier detection (TOD) is also a popular research topic in trajectory pattern mining. According to J. Han *et al.* [6], an outlier means a data object that is grossly different from or inconsistent with the remaining set of data. The trajectory outlier means a trajectory that has a great difference with most other trajectories in terms of some similarity metric.

Some TOD algorithms have been proposed. Each algorithm addresses certain aspects of abnormality. Among these TOD algorithms, the first impressive method is TRAOD (TRAjectoy Outlier Detection) [7], which firstly splits a trajectory into many trajectory partitions and then compares each trajectory partition with its neighbors to determine whether it is an outlying portion or not. The main advantage of TRAOD lies in the ability to detect outlying sub-trajectories. But because of its sub-trajectory detection strategy, TRAOD has a high time complexity of $O(n^2)$. Moreover, the detected result of TRAOD may be influenced by irrelated trajectories because it detects outliers in the whole dataset, as shown in Fig. 1. Recent years, another impressive method is IBAT (Isolation Based Anomalous Trajectory detection) [8]. IBAT focuses on the test trajectory and tries to separate it from the reset trajectories by randomly selecting points solely from the test trajectory. IBAT is more efficient than TRAOD because IBAT does not need to partition the trajectory and the time complexity is $O(n)$. But IBAT has a same insufficiency with TRAOD: both of them do not have enough attention on the travel time (departure time, arrival time and ongoing time). TRAOD does not take the time constraint into account and IBAT just assumes the travel time of the trajectories to be detected are in the same time range but there is no in-depth analysis in IBAT.

Taking the travel time into account can ensure more accurate detection result. Figure 2 shows an example of two groups of trajectories between two areas in different time. τ_o and τ_n are two trajectories that walk the same path. But τ_o is an outlier while τ_n is not because traffic condition changes over time. In other words, outliers' pattern is not static and usually changes with the time. To detect the time-dependent outliers, this paper proposes a novel TOD algorithm called *time-dependent popular routes based trajectory outlier detection* (TPRO).

TPRO detects outliers with the help of the popular routes. The popular routes represent the most trajectories' pattern, so it is a reasonable solution to detect outliers based on the popular routes. As we mentioned above, TPRO focuses on detecting the time-dependent outliers. So time-dependent popular routes are involved to achieve this goal.

TPRO does not partition the trajectories because when facing with a large dataset, efficiency is the first priority while sub-trajectory detection is time-consuming. Given a trajectory dataset, in order to eliminate the influence of

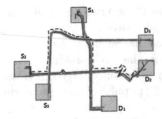

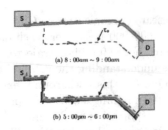

(a) 8 : 00am ~ 9 : 00am

(b) 5 : 00pm ~ 6 : 00pm

Fig. 1. A set of trajectories where S_i is the source area and D_i is the destination area. The $S_2 \to D_2$ dashed curve is actually a trajectory outlier. But each subpart of it has enough closed neighbors because of being deceived by $S_1 \to D_1$ and $S_3 \to D_3$ trajectories. TRAOD cannot identify this kind of outlier.

Fig. 2. Two groups of trajectories which start from S and end at D in different time. τ_o is an outlier in 8 : 00am ~ 9 : 00am because it has a great difference with other trajectories during this time. But the traffic condition changes when 5 : 00pm ~ 6 : 00pm. τ_n, walking the same path with τ_o, is a normal trajectory.

irrelevant trajectories, TPRO divides trajectories with the same source and destination (we call them relevant trajectories) into the same group. Then the dataset can be divided into many groups and detection is token group by group. During the detection, if a trajectory has a great difference with the popular routes during its travel time, this trajectory is classified as an outlier.

Despite that the meaning of the outlier is easy to understand, it is nontrivial to detect outlier based on the time-dependent popular routes. There are mainly two challenges in TPRO:

1. Each trajectory will be compared with its corresponding popular routes to judge whether it is an outlier or not. So given a trajectory (assume its departure time is t_s and arrival time is t_d), TPRO should efficiently retrieve the corresponding popular routes during the time of $t_s \sim t_d$.
2. When calculating the difference between a trajectory and its corresponding popular routes, not only the spatial info but also the temporal info (departure time, arrival time and ongoing time) should be taken into account.

In response to the first challenge, a time-dependent transfer graph is constructed in TPRO. This graph records how many trajectories have passed through each road in different time. With the help of the time-dependent transfer graph, TPRO can efficiently retrieve the top-k most popular routes in a user specified time range. And TPRO also puts forward the time-dependent edit distance to address the second challenge. The time-dependent edit distance not only takes the spatial distance into account but also considers the temporal distance.

The main contributions of this paper are as follows:

1. This paper presents a time-dependent popular routes based trajectory outlier detection algorithm, which takes both spatial and temporal abnormality into consideration and gives us a new solution in trajectory outlier detection.
2. We put forward an efficient popular routes query method in TPRO, which can efficiently retrieve the popular routes during a user specified time range.
3. We provide a real trajectory dataset in which the outliers have been labelled by user study.

The rest of this paper is organized as follows. A formal definition of our problem is given in Sect. 2. Section 3 gives a detailed statement of our solution and Sect. 4 shows our experiment's result. Section 5 gives a brief introduction of the related work. At last, a conclusion is given in Sect. 6.

2   Problem Definition

This part presents some prior definitions and gives a formal definition of the problem this paper focuses on.

Definition 1 *(Raw Trajectory). A raw trajectory $\tilde{\tau}$ is a time-ordered sequence of sampled points: $\tilde{\tau} = (\tilde{p}_1, \tilde{p}_2, \tilde{p}_3, ..., \tilde{p}_x)$. Each sampled point $\tilde{p}_i$ is represented by $\langle \tilde{l}_i, \tilde{t}_i \rangle$ where $\tilde{l}_i$ is a geographic coordinate and $\tilde{t}_i$ is the sampling time.*

It is hard to find a common path from a group of raw trajectories because of the discrete sampled points. So this paper preprocesses the dataset and map each raw trajectory into the road network to get a mapped continuous trajectory.

Definition 2 *(Road Network). A road network is a directed graph $G = (V, E)$ where V is a set of vertices representing road intersections and E is a set of edges representing road segments.*

We use v_i to represent a certain vertex in G. If v_i and v_j are two endpoints of a certain edge, then we have $\varphi(v_i, v_j) = 0$. If the edge's direction is $v_i \rightarrow v_j$, then it can be denoted as e^i_j. Otherwise, the edge can be denoted as e^j_i.

Definition 3 *(Mapped Trajectory). A mapped trajectory τ is a sequence of time-ordered road network locations. It can be denoted as $\tau = (p_1, p_2, p_3, ..., p_m)$. Each road network location p_i is represented as $\langle v_i, t_i \rangle$ where v_i a certain vertex in the road network and for all $i \in \{1, 2, 3, ..., m-1\}$ that $\varphi(v_i, v_{i+1}) = 0$. And t_i is the time τ passes v_i.*

Henceforth, we will only deal with the mapped trajectories. So for simplicity, we will drop the *mapped* qualifier. Thus trajectory in the rest of the article is short for mapped trajectory. After giving a definition of the trajectory, the time-dependent route is defined as follows.

Definition 4 *(Time-Dependent Route). A $v_1 \rightarrow v_m$ time-dependent route is denoted as $\gamma = (pf_1, pf_2, pf_3, ..., pf_m)$ and each $pf_i \in \gamma$ is represented as $\langle v_i, \bar{t}_i, freq_i \rangle$ where v_i represents a certain vertex in the road network and for all $i \in \{1, 2, 3, ..., m-1\}$ that $\varphi(v_i, v_{i+1}) = 0$. Meanwhile, $freq_i$ means how many trajectories have pass through v_i and $\bar{t}_i$ is the average pass time.*

For simplicity, the *time-dependent* qualifier will be dropped and route is short for time-dependent route in the rest of this paper. After giving a definition of the trajectory and the route, trajectory route distance function is put forward to indicate the difference degree between a trajectory and a route.

Definition 5 *(Trajectory Route Distance Function). A trajectory route distance function $\delta(\tau, \gamma)$ is a formula that can give a difference score between τ and γ.*

Based on above definitions, we give a formal definition of the outlier and the problem this paper focuses on next.

Definition 6 *(Outlier). Given a trajectory τ, a route set $R = \{\gamma_1, \gamma_2, ..., \gamma_k\}$, a trajectory route distance function δ and an anomalous score threshold θ, we can calculate the trajectory's anomalous score*

$$s_\tau = \sum_{i=1}^{k} w_{\gamma_i} \cdot \delta(\tau, \gamma_i) \tag{1}$$

where w_{γ_i} is the popularity weight of γ_i among the route set R. If $s_\tau > \theta$, then we say that τ is a θ-outlier on R and δ.

Problem: Given a trajectory dataset T, a route distance function δ and an anomalous score threshold θ, we need to get a trajectory set $T' = \{\tau_1, \tau_2, ..., \tau_n\}$ that satisfies: for all $\tau_i \in T'$, τ_i is a θ-outlier on its corresponding popular routes and δ.

3   TPRO Algorithm

This section introduces how TPRO solves the problem proposed above. Given a trajectory dataset, to eliminate influence of irrelevant trajectories, TPRO first divides the trajectories into the many groups according to their source and destination. Then after trajectory grouping, the detection is taken for each group respectively. In each group, we firstly construct a time-dependent transfer graph from the trajectories. Then with the help of this graph, the time-dependent popular routes querying can be more efficient. At last, we use a time-dependent edit distance based trajectory route distance function to judge whether a trajectory is an outlier or not.

3.1   Dataset Grouping

The source vertex and destination vertex of a trajectory τ is represented as $\tau.s$ and $\tau.d$. If we adopt the strategy that only trajectories starting at same vertex and ending at same vertex can be gathered into one group, we will find that each group has few trajectories. So we put forward the grid-equal-to relation to enlarge the particle size of source area and destination area.

Definition 7 *(Grid-Equal-To Relation). Given two number m, n, we can split the road network G into m × n size-equal grids. For two vertices v_i, v_j, if v_i and v_j fall into the same grid, then we say that v_i is m-n-grid-equal-to v_j. It can be denoted as $o(G, m, n, v_i, v_j) = 1$.*

For two certain trajectories τ_i and τ_j, after given the grid number m and n, if $o(G, m, n, \tau_i.s, \tau_j.s) = 1$ and $o(G, m, n, \tau_i.d, \tau_j.d) = 1$, they will be divided into the same group.

3.2   Construction of Time-Dependent Transfer Graph

After a certain group of trajectories with the same source and destination are mapped into the road network, we can get a subgraph of the road network (Fig. 3 shows an example of this subgraph). And for each vertex in this subgraph, we use a vertex frequency table (i.e. the table beside each vertex in Fig. 3) to record how many trajectories have pass through each vertex in different time range. This subgraph is called the time-dependent transfer graph (TTG).

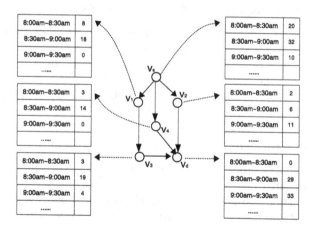

Fig. 3. An example of TTG. v_s is the source and v_d is the destination. Each table beside the vertex v_i is called vertex frequency table of v_i.

Form Fig. 3, we can see that the vertex frequency tables in TTG are in the same time interval (30 min in this example). This time interval is called TTG time interval and is denoted as Δt.

With the help of these vertex frequency tables, we can easily estimate how many trajectories have pass through a certain vertex during a user specified time range. For example, we can infer that there are 26 trajectories (8 trajectories during $8:00am \sim 8:30am$ and 18 trajectories during $8:30am \sim 9:00am$) have passed through v_1 during $8:00am \sim 9:00am$. In some cases, the use specified time range does not fully cover the vertex frequency table time ranges. Such as, what if we wan to know that how many trajectories have passed through v_1 during $8:10am \sim 9:00am$. From the TTG, we can know that there are 18 trajectories have passed through v_1 during $8:30am \sim 9:00am$, but we cannot infer how many trajectories during $8:10am \sim 8:30am$ directly. In such situation, we multiply the trajectories number by the proportion of the covered time range. Thus, the trajectories during $8:10am \sim 8:30am$ is

$$8 \times \frac{8:30am - 8:10am}{8:30am - 8:00am} = 8 \times \frac{20\,\text{min}}{30\,\text{min}} \approx 5$$

Obviously, more smaller the TTG time interval is, more accurate the inferred number is. But the space cost and time cost will increase.

And from the TTG in Fig. 3, we can also infer the average pass time of a certain vertex during a user specified time range. For example, there are 5 and 18 trajectories have passed through v_1 during $8:10am \sim 8:30am$ and $8:30am \sim 9:00am$ respectively. So the average pass time during $8:10am \sim 9:00am$ is

$$\frac{5 \times \frac{8:10am+8:30am}{2} + 18 \times \frac{8:30am+9:00am}{2}}{5+18} \approx 8:40am$$

3.3 Retrieving Time-Depended Popular Route

For a trajectory τ to be tested, TPRO compares it with the popular routes during $t_s \sim t_d$ (t_s represents the departure time and t_d represents the arrival time) to judge if it is an outlier. So this paragraph explains how to query the time-dependent popular routes with the help of TTG.

Assume that $t_s = 8:00am$ and $t_d = 9:00am$, we should find the top-k most popular routes during this time. First of all, we can traverse the TTG and calculate each vertex's trajectories number and the average pass time during $t_s \sim t_d$. Then all possible routes during $t_s \sim t_d$ are got as follows:

- $\gamma_1 = (\langle v_s, \ 8:33am, \ 52 \rangle, \langle v_1, \ 8:36am, \ 26 \rangle, \langle v_3, \ 8:41am, \ 22 \rangle, \langle v_d, 8:45am, 29 \rangle)$
- $\gamma_2 = (\langle v_s, \ 8:33am, \ 52 \rangle, \langle v_4, \ 8:40am, \ 17 \rangle, \langle v_d, \ 8:45am, \ 29 \rangle)$
- $\gamma_3 = (\langle v_s, \ 8:33am, \ 52 \rangle, \langle v_2, \ 8:38am, \ 8 \rangle, \langle v_d, \ 8:45am, \ 29 \rangle)$

Now that all routes have been got, we should judge which route is more popular. Inspired by Luo et al. [9], the route popularity and more-popular-than relation are proposed as follows.

Definition 8 (Route Popularity). *The popularity of a certain route $\gamma = (pf_1, pf_2, pf_3, ..., pf_m)$ can be represented as an ordered frequency sequence: $\rho_\gamma = (freq_{j_1}, freq_{j_2}, freq_{j_3}, ..., freq_{j_m})$, where:*

1. $\{freq_{j_1}, freq_{j_2}, freq_{j_3}, ..., freq_{j_m}\} \Leftrightarrow \{pf_1.freq, \; pf_2.freq, \; pf_3.freq, \; ..., \; pf_m.freq\}$
2. $freq_{j_1} \geq freq_{j_2} \geq ... \geq freq_{j_m}$

Definition 9 *(More-Popular-Than Relation). For two routes γ and γ', assume their popularity sequences are $\rho_\gamma = (freq_{j_1}, freq_{j_2}, freq_{j_3}, ..., freq_{j_m})$ and $\rho_{\gamma'} = (freq'_{j_1}, freq'_{j_2}, freq'_{j_3}, ..., freq'_{j_n})$. If one of the following statements holds:*

- *ρ_γ is prefix of $\rho_{\gamma'}$,*
- *or there exists a number $q \in \{1, 2, 3, ..., min(m, n)\}$ such that:*
  1. *$freq_{j_x} = freq'_{j_x}$ for all $x \in \{1, 2, 3, ..., q - 1\}$, if $q > 2$.*
  2. *$freq_{j_q} > freq'_{j_q}$*

then we say γ is more-popular-than γ', denoted as $\gamma \succeq \gamma'$.

According to Definitions 8 and 9, we have that $\rho_{\gamma_1} = (52, 29, 26, 22)$, $\rho_{\gamma_2} = (52, 29, 17)$ and $\rho_{\gamma_3} = (52, 19, 8)$. Obviously, $\rho_{\gamma_1} \succeq \rho_{\gamma_2} \succeq \rho_{\gamma_3}$. It means that γ_1 is more popular than γ_2 and γ_2 is more popular than γ_3. Assume that $k = 2$, then the top-k most popular routes during $8 : 00am \sim 9 : 00am$ are γ_1 and γ_2.

Luo *et al.* has proved that the selected popular routes by this method satisfy three key properties: suffix-optimal (i.e., any suffix of the popular route is also popular), length-insensitive (i.e., popular does not mean the shorter/longer the better), and bottleneck-free (i.e., popular routes should not contain infrequent vertices or edges) in [9].

3.4 Outlier Detection

After the top-k popular routes got, we compare the trajectory with each popular route. As we known, edit distance can represent two sequences' difference degree. But trajectory (or route) is not just vertex sequence, it also carries the temporal information. So we propose a time-dependent edit distance based trajectory route distance function to handle this problem.

Assume $\tau_{-1} = (p_1, p_2, p_3, ..., p_{m-1})$ is a sub-trajectory of $\tau = (p_1, p_2, p_3, ..., p_{m-1}, p_m)$ after remove the last point p_m. And $\gamma_{-1} = (pf_1, pf_2, pf_3, ..., pf_{n-1})$ is prefix of $\gamma = (pf_1, pf_2, pf_3, ..., pf_{n-1}, pf_n)$ after remove the last tuple pf_n. The trajectory route distance function in TPRO is defined as a recursive equation:

$$\delta(\tau, \gamma) = Min \begin{cases} \delta(\tau_{-1}, \gamma) + delete\_cost(p_m) \\ \delta(\tau, \gamma_{-1}) + delete\_cost(pf_n) \\ \delta(\tau_{-1}, \gamma_{-1}) + replace\_cost(p_m, pf_n) \end{cases} \tag{2}$$

where

$$delete\_cost(p_m) = \begin{cases} 0.5, & m < 2 \text{ or } p_m.v \neq p_{m-1}.v \\ 0, & otherwise \end{cases}$$

$$+ \begin{cases} 0.5, & m < 2 \text{ or } |p_m.t - p_{m-1}.t| > \Delta t \\ 0, & otherwise \end{cases} \tag{3}$$

$$delete\_cost(pf_n) = \begin{cases} 0.5, & n < 2 \text{ or } pf_n.v \neq pf_{n-1}.v \\ 0, & otherwise \end{cases}$$

$$+ \begin{cases} 0.5, & n < 2 \text{ or } |pf_n.\bar{t} - p_{n-1}.\bar{t}| > \Delta t \\ 0, & otherwise \end{cases} \tag{4}$$

$$replace\_cost(p_m, pf_n) = \begin{cases} 1, & p_m.v \neq pf_n.v \\ 0, & otherwise \end{cases} + \begin{cases} 1, & |p_m.t - pf_n.\bar{t}| > \Delta t \\ 0, & otherwise \end{cases} \tag{5}$$

If there is only one vertex in τ (or γ), which means that $m = 1$ (or $n = 1$), then we have that $\tau_{-1} = \phi$ (or $\gamma_{-1} = \phi$). Assume that we use $\tau.len$ and $\gamma.len$ to represent the number of vertices of τ and γ, then these two initial conditions in this recursive equation are

$$\delta(\tau, \phi) = \tau.len; \tag{6}$$

$$\delta(\phi, \gamma) = \gamma.len; \tag{7}$$

From Eqs. 3, 4 and 5, we can see that the $delete\_cost$ or the $replace\_cost$ can be broken down into the spatial cost and the temporal cost. If two vertices are different, the spatial cost is 0.5 (delete) or 1 (replace). Otherwise, the spatial cost is 0. When calculating the temporal cost, we have the aid of the TTG time interval Δt in Subsect. 3.2. If the time lag is larger than Δt, the temporal cost is 0.5 (delete) or 1 (replace). Otherwise, the temporal cost is 0.

The pseudo code of the time-dependent edit distance based trajectory route distance function is shown in Algorithm 1.

Algorithm 1. Trajectory Route Distance Function

Input: a trajectory τ, a route γ
Output: Distance between τ and γ
1: **DECLARE** int $DP[0..\tau.len][0..\gamma.len]$
2: **for** $i := 0$ to $\tau.len$ **do**
3:     $DP[i][0] = i$;
4: **end for**
5: **for** $j := 0$ to $\gamma.len$ **do**
6:     $DP[0][j] = j$;
7: **end for**
8: **for** $i = 1$ to $\tau.len$ **do**
9:     **for** $j = 1$ to $\gamma.len$ **do**
10:        $DP[i][j] = minimum($
11:            $DP[i-1][j] + delete\_cost(\tau.p_i)$,
12:            $DP[i][j-1] + delete\_cost(\gamma.pf_j)$,
13:            $DP[i-1][j-1] + replace\_cost(\tau.p_i, \gamma.pf_j)$
14:        $)$;
15:     **end for**
16: **end for**
17: **return** $DP[\tau.len][\gamma.len]$;

The trajectory route distance function can give a difference score between a trajectory and a popular route. But in most cases, there are more than one popular route between two areas. But it does not mean that each popular route has the same popularity degree. So we propose the popularity weight to represent how popular a route is among a set of routes.

Definition 10 *(Popularity Weight). Assume there is a route set $R = \{\gamma_1, \gamma_2, \gamma_3, ..., \gamma_k\}$ and for each $\gamma_i = (pf_1, pf_2, pf_3, ..., pf_m) \in R$, we have*

$$\gamma_i.sum = \sum_{a=1}^{m} \gamma_i.pf_a.freq \tag{8}$$

then the popularity weight of γ_i can be represented as

$$w_{\gamma_i} = \frac{\gamma_i.sum}{\sum_{b=1}^{k} \gamma_b.sum} \tag{9}$$

3.5 Time Complexity

The overall pseudo code of TPRO has been shown in Algorithm 2. From the pseudo code, we can see that the time complexity of TPRO is $O(k \cdot n)$ where k is the number of popular routes used to detect outliers and n is the number of trajectories in the dataset. In most cases, k is a small number (less than 10), so we can use approximation $O(n)$ for the time complexity.

4   Experiment Result

This section gives an exhibition of our experiment and the result. The first subsection gives an introduction to the experiment dataset and environment setting. The second subsection elaborates which criteria we use to evaluate our algorithm. Finally, we give an analysis on the experiment result.

4.1   Experiment Setting

The experiment is taken under a real-world dataset which contains $412,032$ trajectories. This dataset is collected from around $10,700$ taxis in BeiJing in 2012. We pick up about $1,300$ trajectories from the dataset and asked volunteers to manually label whether each trajectory is abnormal or not. This labeled dataset is used to evaluate the accuracy of TPRO.

The road network in our experiment contains about $165,000$ vertices and $226,000$ edges. And the road network is split into 120×130 grids in the grouping step[1]. Each grid's size is about 1.5 km × 1.5 km.

Our algorithm is implemented in cpp. The machine we use to accomplish the experiment has a quadcore Inter Core i5 CPU (3.2 GHz) and 8G memory. The operating system is Linux 3.13.0 x 86_64 and the compiler is g++ 4.8.2.

---

[1] That's to say m is set to 120 and n is set to 130 in the grouping step.

Algorithm 2. TPRO

Input: road network G, dataset T, grid number m and n, popular routes number k, score threshold θ, TTG time interval $\varDelta t$

Output: outlier set T'

1: //DATASET GROUPING
2: $grid = CreateGrid(G, m, n);$ // create $m \times n$ grids on the road network
3: $groups = \phi;$
4: **for** ecah $\tau \in T$ **do**
5:     $srcGrid = grid.getLocatedGrid(\tau.s);$
6:     $destGrid = grid.getLocatedGrid(\tau.d);$
7:     $groups[srcGrid, destGrid].add(\tau);$ // add trajectory to corresponding group
8: **end for**
9: //DETECTING IN EACH GROUP
10: **for** each $T^* \in groups$ **do**
11:     $ttg = CreateTTG(T^*, \varDelta t);$ // construct time-dependent transfer graph
12:     **for** each $\tau \in T^*$ **do**
13:         $s_\tau = 0;$
14:         $routes = GetTopKRoutes(ttg, k, \tau.t_s, \tau.t_d);$ // query popular routes
15:         **for** each $\gamma \in routes$ **do**
16:             $s_\tau += w_\gamma \cdot \delta(\tau, \gamma);$ // compare trajectory with each route
17:         **end for**
18:         **if** $s_\tau > \theta$ **then**
19:             $T'.add(\tau);$ // add to outlier set
20:         **end if**
21:     **end for**
22: **end for**
23: **return** $T';$

4.2 Evaluation Criteria

In practice, detection rate (the fraction of anomalous trajectories that are successfully detected) and false alarm rate (the fraction of normal ones that are predicted to be anomalous) are two important measures to evaluate the performance of an anomaly detection method. Obviously, a good outlier detection method should have a high detection rate and a low false alarm rate. After we plot the detection rate on y-axis and the false alarm rate on x-axis, we can get a curve called Receiver Operating Characteristic (ROC) [10] curve. The AUC [11] value is defined as the area under the ROC curve. For a randomly chosen normal trajectory τ_n and a randomly chosen anomalous trajectory τ_a, the AUC value is equal to the probability that $s_{\tau_a} > s_{\tau_n}$. Obviously, if the AUC value is close to 1, the outlier detection method is of high quality.

4.3 Results

In this section, we first analysis how the parameters affect the experiment result. Then we give a comparison between TPRO, TRAOD and IBAT from the accuracy and efficiency.

Varying Parameters. There are mainly three parameters in the detecting step of TPRO: score threshold θ, popular routes number k and TTG time interval Δt. So this paragraph elaborates how these three parameters affect the detection rate, false alarm rate and the process time[2].

Figure 4(a) shows how θ affects the detection rate, false alarm rate and process time when $k = 5$ and $\Delta t = 600$ s. As θ increasing, which means that the detection criterion becoming more conservative, the detection rate and the false alarm rate will fall. But the process time is stable. When $\theta \approx 1.0$, we have a high detection rate and a low false alarm rate.

k represents how many popular routes will be used to judge if a trajectory is an outlier. Of course, more popular routes are used, more accurate the result will be. But the process time will grow linearly because we must compare each trajectory with each popular route. Figure 4(b) shows how k affects the detection rate, false alarm rate and the process time when $\theta = 1.0$ and $\Delta t = 600$ s. It shows that as k increasing, the false alarm rate will fall and the process time will go up. But it has a small effect on the detection rate. When $k = 5$, we can have a high detection rate and a low false alarm rate. Meanwhile, the process time is acceptable.

Figure 4(c) shows how Δt affects the detection rate, false alarm rate and process time when $k = 5$ and $\theta = 1.0$. If Δt is too small, TPRO will overstate the temporal cost when calculating the distance between a trajectory and a route. So the false alarm rate is very high and will fall as Δt increasing. But more smaller Δt is, more accurate the selected popular routes are. But the popular routes query time will be more longer. So as Δt increasing, detection rate and the process time will both fall, too. When $\Delta t = 600$ s, we can have a low false alarm rate and a less process time. Meanwhile, the detection rate is acceptable.

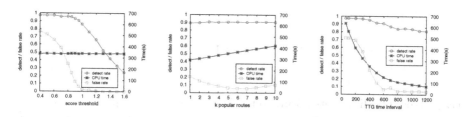

Fig. 4. Detection rate, false alarm rate and time cost under varying θ (Left), k (Middle) and Δt (Right)

TPRO vs. TRAOD and IBAT. This paragraph give a comparison between TPRO, TRAOD and IBAT. All of the three algorithms are tested in their best parameters, which are listed in Table 1.

Figure 5(a) shows the ROC curves of TPRO, TRAOD and IBAT. For better illustration, the ranges of false alarm rate and detection rate are set to $[0 \sim 0.5]$ and $[0.4 \sim 1]$. We can see that TPRO has a larger area under the ROC curve

[2] These three evaluating indicators are counted under the labeled dataset.

Table 1. Parameter setting of TPRO, TRAOD and IBAT

Algorithm	TPRO	TRAOD	IBAT
Parameters	$k = 5$, $\Delta t = 600\,\text{s}$	$D = 80$, $p = 0.95$	$m = 100$, $\psi = 256$

than TRAOD and IBAT. It means that TPRO has a better performance than TRAOD and IBAT in accuracy.

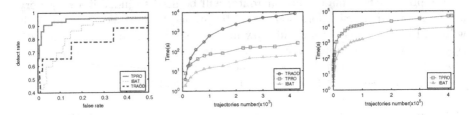

Fig. 5. ROC curves of TPRO, TRAOD and IBAT (Left). Efficiency under the small dataset (Middle) and large dataset (Right).

Figure 5(b) and (c) show the process time of TPRO, TRAOD and IBAT under different scale dataset. Because the time complexity of TRAOD is $O(n^2)$, which is very time consuming in larger dataset detection. So we only test it on the small dataset. From these two figures, we can see that the time cost of TPRO is between IBAT's and TRAOD's.

5   Related Work

Some related works are introduced in this part, which can be categorized into two groups. The first one focuses on trajectory outlier detection and the second one focuses on popular route mining.

Trajectory Outlier Detection: Some algorithms have been proposed to detect trajectory outlier, but each addresses certain aspects of abnormality. Lee *et al.* [7] put forward a group-and-detect framework and develop an algorithm called TRAOD. TRAOD splits a trajectory into various subparts (at equal intervals), then a hybrid of the distance-based and density-based approach is used to classify each subpart is abnormal or not. Chen *et al.* [8] propose an isolation based method, called IBAT. For a group of trajectories and a trajectory will be tested in this group, they randomly pick a point from the test trajectory and remove other trajectories which do not contain this point. This process is repeated until no trajectory is left or all the trajectories left contain all the points the test trajectory has. If the test trajectory is an outlier, this process will end very soon. And Li *et al.* [12] emphasis on historical similarity trends between data points. At each time step, each road segment checks its similarity with the other road segments, and the historical similarity values are recorded in a temporal neighborhood vector at each road segment. Outliers are calculated from drastic changes in these vectors [13]. Guan *et al.* [14] use a feature vector, such as $\langle direction, speed, angle, location \rangle$, to represent a trajectory segment and detect the outliers according to these features. Mohamad *et al.* [15] take the speed and turn directions into consideration. If a trajectory has an sudden speed change or some unexpected turns, it is an abnormal trajectory. Recently, some studies have used learning methods to identify anomalous trajectories [16,17]. But these methods usually need training data, which is inconvenient to label. There are also some works [18–20] have been done for the stream data detection or on-line detection.

Popular Route Mining: Finding the most desirable path has been a hot research topic for decades. Many works [21–23] have been done in finding the shortest/fastest path. But the popular route does not mean the shortest or fastest path. In most case, we prefer the most frequent path as the popular route. Lots of algorithms have been proposed for popular route searching. Zaiben *et al.* [3] introduce a transfer probability network to discover popular route from historical trajectories. They derive the probability of transferring from every significant location to the destination based on the historical trajectories, and the transfer probability is used as an indicator of popularity. The popularity of a route is defined as the product of transfer probabilities of all significant locations on the route. Luo *et al.* [9] also construct a network graph (called footmark graph) to mine frequent path. But they describe the edge frequency as the total number of trajectories passing through the edge. Then they define a descending edge frequency sequence to judge which path is more frequent. Another work, such as [4], aims at deriving routes from uncertain trajectory data.

6   Conclusions

In this paper, we propose a time-dependent outlier detection algorithm which is called TPRO. Given a trajectory dataset, we first divide the relevant trajectories into same group. Then for each group, we propose a time-dependent transfer graph to speed up querying time-dependent popular routes. We use a time-dependent edit distance to represent the difference score between a trajectory and a route. If a trajectory has a great difference with all of the selected popular routes, it's an outlier. We evaluate our method on a real-world dataset. The experiment result shows that our method has a better performance than TRAOD and IBAT.

In the future, we plan to enhance our algorithm in two directions. Firstly, although TPRO is more efficient than TRAOD, but TPRO is slower than IBAT in about ten times. So we want to improve TPRO on efficiency. Secondly, we will improve TPRO for on-line outlier detection.

References

1. Ye, Y., Zheng, Y., Chen, Y., Feng, J., Xie, X.: Mining individual life pattern based on location history. In: IEEE MDM, pp. 1–10 (2009)
2. Zheng, Y., Xie, X., Ma, W.-Y.: Geolife: A collaborative social networking service among user, location and trajectory. IEEE Data Eng. Bull. **33**, 32–39 (2010)
3. Chen, Z., Shen, H.T., Zhou, X.: Discovering popular routes from trajectories. In: IEEE ICDE, pp. 900–911 (2011)
4. Wei, L.-Y., Zheng, Y., Peng, W.-C.: Constructing popular routes from uncertain trajectories. In: ACM SIGKDD, pp. 195–203 (2012)
5. Zheng, Y., Liu, L., Wang, L., Xie, X.: Learning transportation mode from raw GPS data for geographic applications on the web. In: WWW, pp. 247–256 (2008)
6. Han, J., Kamber, M., Pei, J.: Data Mining: Concepts and Techniques. Morgan Kaufmann Publishers, San Francisco (2006)
7. Lee, J.-G., Han, J., Li, X.: Trajectory outlier detection: A partition-and-detect framework. In: IEEE ICDE, pp. 140–149 (2008)
8. Zhang, D., Li, N., Zhou, Z.-H., Chen, C., Sun, L., Li, S.: iBAT: detecting anomalous taxi trajectories from GPS traces. In: ACM UbiComp, pp. 99–108 (2011)
9. Luo, W., Tan, H., Chen, L., Ni, L.M.: Finding time period-based most frequent path in big trajectory data. In: ACM SIGMOD, pp. 713–724 (2013)
10. Fawcett, T.: An introduction to roc analysis. Pattern Recogn. Lett. **27**(8), 861–874 (2006)
11. Bradley, A.P.: The use of the area under the roc curve in the evaluation of machine learning algorithms. Pattern Recogn. **30**(7), 1145–1159 (1997)
12. Li, X., Li, Z., Han, J., Lee, J.-G.: Temporal outlier detection in vehicle traffic data. In IEEE ICDE, pp. 1319–1322 (2009)
13. Gupta, M., Gao, J., Aggarwal, C., Han, J.: Outlier detection for temporal data. Synth. Lect. Data Min. Knowl. Discov. **5**, 1–129 (2014)
14. Yuan, G., Xia, S., Zhang, L., Zhou, Y., Ji, C.: Trajectory outlier detection algorithm based on structural features. J. Comput. Inf. Syst. **7**(11), 4137–4144 (2011)
15. Mohamad, I., Ali, M., Ismail, M.: Abnormal driving detection using real time global positioning system data. In: Space Science and Communication (IconSpace), pp. 1–6. IEEE (2011)
16. Sillito, R.R., Fisher, R.B.: Semi-supervised learning for anomalous trajectory detection. In: BMVC, pp. 1–10 (2008)
17. Li, X., Han, J., Kim, S., Gonzalez, H.: Roam: rule-and motif-based anomaly detection in massive moving object data sets. In: SIAM SDM, pp. 273–284 (2007)
18. Yu, Y., Cao, L., Rundensteiner, E.A., Wang, Q.: Detecting moving object outliers in massive-scale trajectory streams. In: ACM KDD, pp. 422–431 (2014)
19. Bu, Y., Chen, L., Fu, A. W.-C., Liu, D.: Efficient anomaly monitoring over moving object trajectory streams. In: ACM SIGKDD, pp. 159–168 (2009)
20. Chen, C., Zhang, D., Castro, P.S., Li, N., Sun, L., Li, S., Wang, Z.: iBOAT: Isolation-based online anomalous trajectory detection. In: IEEE TITS(2013)
21. Gonzalez, H., Han, J., Li, X., Myslinska, M., Sondag, J.P.: Adaptive fastest path computation on a road network: a traffic mining approach. In: VLDB (2007)
22. Sacharidis, D., Patroumpas, K., Terrovitis, M., Kantere, V., Potamias, M., Mouratidis, K., Sellis, T.: On-line discovery of hot motion paths. In: ACM EDBT (2008)
23. Kanoulas, E., Du, Y., Xia, T., Zhang, D.: Finding fastest paths on a road network with speed patterns. In: IEEE ICDE, pp. 10–10 (2006)

Temporally Enhanced Ontologies in OWL: A Shared Conceptual Model and Reference Implementation

Sven de Ridder and Flavius Frasincar[✉]

Erasmus University Rotterdam,
PO Box 1738, 3000 DR Rotterdam, The Netherlands
svenderidder@gmail.com, frasincar@ese.eur.nl

Abstract. The temporal dimension has been recognized as an integral feature of many Semantic Web applications, but there are significant differences in how ontology authors choose to represent changes in time. We present a temporal conceptual model for OWL DL ontologies that allows the expression of *fluent properties*, i.e., properties that change in time, that is both representation-agnostic and serializable for the various available representation schemes. We also provide *Kala*, a reference implementation developed in Java, that can be used to generate temporal ontologies, convert between temporal ontologies in different representation schemes, and develop new applications such as temporal querying or visualization tools.

Keywords: Semantic Web · OWL · Time · Fluents · Kala

1 Introduction

The ability to identify trends and make predictions is of critical importance for successful trading in financial markets. The increase in prominence of sophisticated, low-latency algorithmic trading systems has spurred development of technologies such as news analytics for the timely extraction of information that is relevant to the identification of market opportunities [24]. The Semantic Web, and OWL in particular, provide the technology to represent, manage, share, and reason over self-describing data, but these representations tend to be *synchronic*, i.e., they lack the crucial time dimension.

One reason for the lack of temporality in existing ontologies is that, while much effort has gone into providing support for temporal features at the representational level, there seems to be a lack of a shared, representation-agnostic model at the conceptual level, which is where user requirements commonly need to be met. Here, we consider the *conceptual level* to be the level at which humans may express and interpret information that closely relates to the perceived real world, the level that captures the essential semantics of temporal ontologies. In contrast, we consider the *representational level* to describe the organization of the information for representation and storage as computer data, typically

© Springer International Publishing Switzerland 2015
J. Wang et al. (Eds.): WISE 2015, Part I, LNCS 9418, pp. 31–45, 2015.
DOI: 10.1007/978-3-319-26190-4_3

as entities and relationships between entities, supported by a representation-specific vocabulary to express the individual data items. We shall refer to the conceptualization of temporal ontologies as the *temporal conceptual model*, and to the specification on the representational level as the *temporal representation scheme*.

Existing representation schemes are generally not directly compatible, and one result of the focus on representation schemes is that it greatly reduces the interoperability of various temporal implementations. Given the high conversion barriers, the ability to share and reuse data — a core objective of the Semantic Web — suffers. Secondly, the authors and users of temporal ontologies are directly exposed to the details of the particular representation scheme, which makes the development and use of temporal ontologies a cumbersome, complex, and error-prone process. Lastly, the focus on specific representation schemes results in applications that operate on temporal ontologies becoming tightly bound to a particular logical structure and implementation. This discourages the development of such applications, because their potential audience will be limited to the users of a particular representation scheme. Examples of these applications are temporally-enhanced reasoning, querying, and visualization.

Our focus will be on the introduction of concepts that form the building blocks of the semantics for temporally enhanced ontologies. The temporal conceptual model is designed to be mappable to selected representation schemes in OWL DL; that is, these representation schemes can be expressed in the $\mathcal{SHOIN}(\mathcal{D})$ description logic, and are fully compatible with the $\mathcal{SROIQ}(\mathcal{D})$ description logic employed by OWL 2. The model is, itself, composed of two orthogonal partitionings: one that describes the *time domain*, while the other describes *fluent properties*. Fluent properties, first described in the earliest literature on computer learning and artificial intelligence by McCarthy and Hayes [21], represent properties and relationships that may change with time.

Fluent properties form a suitable focal point for the exploration of a temporal conceptual model for a number of reasons. Firstly, the concept is immediately familiar: one does not have to stretch the imagination to think of examples of properties and relationships that change over time; a person's address, employer, and even favorite soccer team are all subject to such change — in fact, it may be more difficult to conceive examples of properties and relationships that categorically do *not* change over time! Secondly, fluent properties are conceptually simple: in effect, they represent ternary relations that simply extend the familiar binary relations with a fixed role for the third operand, the time interval. Thirdly, they are useful; as we have argued above, fluent properties allow for the evolution of an ontology to be expressed, examined, queried, and visualized. Lastly, fluent properties are already supported, in some form, through existing representation schemes.

This paper is structured as follows. Section 2 presents background on temporal models in general and the current state of temporal ontologies in particular. After this, in Sect. 3, we present the formal description of the proposed temporal conceptual model. An example implementation is provided in Sect. 4, followed by an evaluation of the implementation in Sect. 5. Lastly, we give our concluding remarks and identify possible future work in Sect. 6.

2 Representations of Temporality

The topic of data temporality has enjoyed great prolificacy: the scientific literature is rich in discussions of temporality, from philosophical treatises of time to discussions of temporal infrastructures and reasoning. Historically, this interest stems from the importance of time in many real-world applications, from logging and scheduling systems to biomedical databases and algorithmic trading.

Much of the early research on data temporality has focused on the field of temporal databases, a topic with similarities to temporal ontologies, and one that is considerably more mature. In fact, a striking resemblance to the current state of temporal ontologies may be gleaned from past reports on the field of temporal databases. Pissinou et al., in their report [27] on a 1992 ARPA/NSF workshop that was aimed specifically at identifying problems within the field of temporal database technology, conclude that the many different custom extensions to the relational model, each intended to serve very specific user needs regarding temporal support, and the resulting lack of a common terminology, infrastructure, and conceptual model for temporal databases, are primary reasons for reduced adoption of temporal database technology; similarly, we find that the field of temporal ontologies faces the same issues. The researchers and participants also identify the ad-hoc nature of many applications extended to include temporal information and the understandable resistance to replace existing applications with full-fledged temporal database technology as obstacles in the development and adoption of a standard for temporal databases, and conclude that there is a need for *open architectures* that allow for easy conversions between different representations.

In response to such findings, a consensus temporal query language specification, *TSQL2* [28], was developed, but the specification, despite strong initial ISO interest, failed to catch on: by the time that SQL:1999 was formally published, the specification had failed to meet the committee's requirements and could not be included in the language standard, and SQL vendor interest waned. Eventually, however, a number of key ideas from TSQL2 found their way into the SQL:2011 specifications [22]. Of these ideas, the concepts of *valid-time* ("application-time tables") and *transaction-time* ("transaction-time tables") have proved particularly useful: valid-time marking the time that a fact is held to be *true*, and transaction-time marking the time that a fact is *known* in the database. The approach to temporality in databases, then, typically resolved to marking tuples with valid-time and transaction-time timestamps, and this formed the basis for the general temporal database model (see, e.g., the conceptual model by Jensen et al. [16], the survey of temporal databases by Özsoyoğlu and Snodgrass [26], and the survey of temporal entity-relationship models by Gregersen and Jensen [11]).

With the development of the Semantic Web and its primary languages, the Resource Definition Language (RDF) and the Web Ontology Language (OWL), came efforts to represent temporal information in these languages. The Temporal RDF [12] language extension and its related query language T-SPARQL [10] form the main solution approach to introduce time to RDF.

Unfortunately, Temporal RDF is not compatible with OWL DL because of its use of *RDF reification*. Compounding this problem is the snapshot-based entailment mechanism of Temporal RDF, which expands any temporal statement defined over a time interval into a series of temporal statements defined over each time instant contained by the interval, and the lack of available serializations, for example to RDF/XML.

Another approach is *ontology versioning* [17], in which "snapshots" of the ontology are created for each state of the ontology during its development. Unsurprisingly, this comes at the cost of significant data redundancy. Moreover, its support for particular classes of queries (e.g., "when is fact S true in the database?") is limited. However, the approach may also be used to model *transaction-time*, and may then be considered to be completely orthogonal to other (generally *valid-time*) approaches discussed here. The application of ontology versioning, therefore, may be appropriate in some cases where transaction-time needs to be modeled in addition to valid-time, but, perhaps, at low enough resolution so as to reduce the impact of data redundancy.

There have also been proposals to extend description logics with valid-time; see, e.g., the surveys by Artale and Franconi [2] and Lutz et al. [20]. Such temporal description logics are generally based on the $\mathcal{ALC}$ description logic [8]. These extensions are generally not compatible with OWL: the decidability of temporal description logics is compromised when the language is extended to the full description logics of $\mathcal{SHOIN(D)}$ for OWL DL or $\mathcal{SROIQ(D)}$ for OWL 2 [3]. Opting for temporal description logics would also mean giving up on the rich toolset developed for the OWL language, such as editors and reasoners.

Representation schemes for modeling temporality in OWL DL ontologies generally follow either the *reification*[1] approach or the *4D-fluents* approach. In the reification approach, a property instance is *reified*, that is, converted into a proper instance, and the original property's subject and object instances, or subject instance and datatype value, are then related to the newly reified relation through conventional property assertions to retain the information expressed by the original ⟨subject, property, object⟩ or ⟨subject, property, value⟩ triples. However, since we are now able to specify the reified property as the subject or object of additional triples, we effectively gain the ability to express properties that are ternary, quaternary, or generally *n-ary* in nature. The general approach of reification is, therefore, appropriately named *n-ary relations* [25].

At first glimpse, the reification approach seems appropriate for adding a ternary component, e.g., valid time, to any property assertion, and develop temporal ontologies based on temporally qualified properties. The reification approach is not without problems, however. One problem is the proliferation of objects, namely one for each reified property assertion; related to this is the problem of providing meaningful names to the reified properties, or, alternatively, dealing with objects that may not have meaningful names. Another problem is the reduction of OWL reasoning capabilities over ontologies with reified

---

[1] Note that the reification representation scheme is not the same as the RDF reification, the latter being not available in OWL DL.

properties; property semantics such as inverses or cardinalities are difficult or impossible to describe for reified properties in a general reasoning context.

In contrast to reification, the 4D-fluents approach [29] does not associate property assertions with valid-time intervals directly, but instead opts to have temporal properties hold between *timeslices* of entities, a timeslice being defined as the temporal facet of some entity as it "occupies" some interval in time. In order to be consistent, both subject and object timeslices must be *compatible*, that is, occupy the same interval in time. An important advantage of the 4D-fluents approach over the reification approach is that properties retain their semantics in reasoning contexts: for example, we may trivially define the inverse of a fluent property, as well as symmetry and transitivity; something that is not straight-forward in the reification approach. The 4D-fluents approach, however, suffers from worse object proliferation than the reification approach in the general case.

The 4D-fluents approach has inspired several implementations. *tOWL* [23] employs the 4D-fluents approach and combines it with concrete domains and Allen's interval algebra [1] to allow the expression of temporal restrictions. *SOWL* [5], a spatio-temporal representation, uses the 4D-fluents approach as its temporal component. A reinterpretation of the 4D-fluents approach is implemented by the *MUSING* project [19], which focuses on adoption of the approach in the context of a reasoning architecture. Baratis et al. propose the 4D-fluents approach, combined with Allen's interval algebra, as the basis for *TOQL* [4].

Both approaches employ strategies that force conceptual concessions that conflict with intuitive understanding: the reification representation scheme models properties as classes, property assertions as instances, and prevents the user from specifying qualifiers for property semantics; the 4D-fluents representation scheme retains the property semantics, but requires the user to view instances as "spacetime-worms" and accept the conceptual implications that such a view necessitates. Converting a synchronic ontology with only static properties to a temporal ontology with dynamic properties is thus a cumbersome, error-prone process, as is the conversion between representation schemes. The lack of work on conceptual models for temporal ontologies in the literature indicates the need for improvements in this area.

3   The Temporal Conceptual Model

In this section we describe the proposed temporal conceptual model. Section 3.1 describes the time model. Section 3.2 builds on the time model to present the fluents model.

3.1   The Time Model

The time model extends the OWL model by introducing time instants and time intervals (the so-called *temporal primitives*), as well as assertions that relate these primitives to one another or assign to them discrete timestamp values. These temporal primitives and assertions form the building blocks for time models of varying expressive power and complexity.

The time model allows primitives to be declared explicitly through *primitive declarations*. Such declarations may or may not translate to OWL class membership declarations when serializing to a representation scheme R, depending on whether R represents time instants or time intervals as individuals or, instead, represents them directly as datatype values. The following axioms declare the anonymous individual _:t1 to be a time instant, and the named individual period2013Q1 to be an interval. We use a syntax that closely resembles the OWL *Abstract Syntax* in order to concisely express concepts in a familiar way.

```
TimeDeclaration(TimeInstant(_:t1))
TimeDeclaration(TimeInterval(period2013Q1))
```

The *"before"* relation between time instants t_1 and t_2 can be explicitly expressed in the temporal conceptual model, as shown below:

```
TimeInstantRelationAssertion(_:t1 _:t2 <)
```

Interval endpoint assertions relate time intervals to time instants: they specify that some time interval i_1 starts at a time instant t_s or ends at a time instant t_e. To preserve consistency, the assertion of interval endpoints t_s and t_e as, respectively, the start and the end time instants of time interval i implies that $t_s < t_e$ holds. The following axioms declare the time interval period2013Q1 to start at time instants _:t1 and cal2013:jan1, and end at time instant _:t2. Furthermore, the time interval year2013 has the same start instant as period2013Q1. Note that _:t1 and cal2013:jan1 can be inferred to refer to the same time instant.

```
IntervalStartAssertion(period2013Q1 _:t1)
IntervalStartAssertion(period2013Q1 cal2013:jan1)
IntervalEndAssertion(period2013Q1 _:t2)
IntervalStartAssertion(year2013 _:t1)
```

Relations between time intervals are necessarily of a more complex nature than those between instants, because any two time intervals are not necessarily disjoint. Allen's work on interval relations [1] provides an algebra with useful qualities, and we have adopted the algebra to provide a mapping of relations between time intervals that is both jointly exhaustive and mutually exclusive. The time model allows interval relations to be expressed through the use of TimeIntervalRelationAssertions. As an example, consider the facts that _:i1 *meets* (m) _:i2, and that _:i1 *contains* (di) _:i3. The following statements assert these facts in the model:

```
TimeIntervalRelationAssertion(_:i1 _:i2 m)
TimeIntervalRelationAssertion(_:i1 _:i3 di)
```

In order to support timestamp values, we introduce *instant time assertions*. We have chosen to use the xsd:dateTime type to represent timestamp values, as it is already commonly used in OWL ontologies; however, the rather more complex time types introduced by the OWL-Time ontology [14] may also be considered in later versions. The InstantTimeAssertion represents the association of a time instant with a particular timestamp. For example, to express that _:t1 is associated with June 5th, 2013, 6:42:23 PM in the Central European Summer Time (CEST: UTC+02:00) time zone, we declare the following assertion:

```
InstantTimeAssertion(_:t1
   "2013-06-05T18:42:23.000+02:00"^^xsd:dateTime)
```

3.2   The Fluents Model

The fluents model extends the OWL model by allowing the expression of *fluent properties*. These fluent properties resemble the standard OWL object and datatype properties, but assertions of these properties are additionally qualified with intervals from the time model previously introduced, indicating the (valid-time) interval that the fact is held to be true.

Like regular properties, fluent properties are *named resources*; that is, they can be referenced through identifiers that, in turn, may be extended to full URIs. They should not be considered OWL *entities*, however, for the simple reason that fluent properties are not part of the OWL specifications. It is important to note that the sets of regular properties and fluent properties are disjoint: no regular property may be used as a fluent property in the temporal conceptual model, and vice versa. One compelling reason for this separation is semantics: when a particular property is recognized to be a fluent property, its ability to change its value over time can be seen to be an intrinsic quality; declaring the property in a non-temporal context removes this ability and creates a contradiction. Another, more technical, reason is that neither the reification nor the 4D-fluents representation schemes allow for regular properties to be used as fluent properties, and vice versa: in the reification scheme, this would result in an OWL DL property and an OWL DL class sharing the same identifier, which is strictly prohibited by the standard; and in the 4D-fluents representation scheme, this would violate the domain and range restrictions on fluent properties: fluent datatype properties are restricted to domains of timeslice individuals, and fluent object properties are restricted to both domains and ranges of timeslice individuals.

The temporal conceptual model provides *fluent property declarations* to allow the ontology author to declare fluent properties explicitly. The manner in which these declarations are expressed in the eventual representation schemes is dependent on the details of the particular serialization. As an example, consider the following axioms, which declare a fluent object property ceoOf and a fluent datatype property hasTitle:

```
FluentsDeclaration(FluentObjectProperty(ceoOf))
FluentsDeclaration(FluentDataProperty(hasTitle))
```

The ontology author may wish to specify domain and range restrictions for fluent properties to restrict their use in ways that enforce correctness. These work similar to domain and range restrictions on regular properties: i.e., the domain of fluent object and datatype properties may be restricted to any *class expression*, as may the range of fluent object properties, and the range of fluent datatype properties may be restricted to any *data range*. The hasTitle fluent datatype property, for example, may be restricted as follows:

```
FluentDataPropertyDomain(hasTitle Person)
FluentDataPropertyRange(hasTitle DataOneOf(
 "Mr."^^xsd:string"Mrs."^^xsd:string"Ms."^^xsd:string))
```

As with regular properties, sequences of multiple domain or range restrictions on fluent properties are interpreted to represent the *intersection* of those domain or range restrictions.

A fluent object property assertion expresses the fact that a relation holds between two individuals during a particular time interval. Similarly, a fluent datatype property assertion expresses the fact that an attribute value holds for a particular individual during a particular time interval.

As fluent property assertions are at the heart of the temporal conceptual model, we shall provide a formal definition. Let us first define the concept of a *snapshot reduction* of a temporal ontology:

Definition 1. *Let $\mathcal{T} = \langle \mathcal{T}_p, < \rangle$, where $\mathcal{T}_p$ is a set of time instants and $<$ is a binary relation on the set $\mathcal{T}_p$ that is at least a strict partial order, be a linear, ordered time domain. Let $\mathcal{O}^{\mathcal{T}}$ be a temporally enhanced ontology over $\mathcal{T}$. Then, a snapshot reduction $\mathcal{O}_t^{\mathcal{T}}$ of $\mathcal{O}^{\mathcal{T}}$ at time $t \in \mathcal{T}_p$ is a non-temporal ontology that represents $\mathcal{O}^{\mathcal{T}}$ at time t.*

We can now formally define fluent object property assertions as follows:

Definition 2. *Let $\mathcal{C}$ be the set of all class expressions. Let $\mathcal{T} = \langle \mathcal{T}_p, < \rangle$, where $\mathcal{T}_p$ is a set of time instants and $<$ is a binary relation on the set $\mathcal{T}_p$ that is at least a strict partial order, be a linear, ordered time domain. Let $\mathcal{O}^{\mathcal{T}}$ be a temporally enhanced ontology over $\mathcal{T}$. Let $f^{OP} : D \longrightarrow R$, $D \subseteq \mathcal{C}$, $R \subseteq \mathcal{C}$ be a fluent object property, and let $OP : D \longrightarrow R$ be the non-temporal interpretation of f^{OP}. Then, a fluent object property assertion, $fa^{OP} = \langle s, f^{OP}, o, i \rangle$, $s \in D$, $o \in R$, $i = \langle t_s, t_e \rangle$, $t_s \in \mathcal{T}_p$, $t_e \in \mathcal{T}_p$, $t_s < t_e$ is said to hold between t_s and t_e if there exists an object property assertion $a^{OP} = \langle s, OP, o \rangle$ in every snapshot reduction $\mathcal{O}_t^{\mathcal{T}}$ of $\mathcal{O}^{\mathcal{T}}$, $t_s \leq t < t_e$.*

The formal definition of fluent datatype property assertions follows similarly, but is omitted for reasons of conciseness.

Using fluent property assertions, the ontology author may now express facts about individuals that only hold during a particular time interval. For example, in order to express that sam was the CEO of ibm during interval _:i1, and that mary was to be addressed as "Ms." during interval _:i2, we add the following axioms:

```
FluentObjectPropertyAssertion(sam ceoOf ibm _:i1)
FluentDatatypePropertyAssertion(mary hasTitle
 "Ms."^^xsd:string _:i2)
```

4   Reference Implementation

To illustrate the use of our temporal conceptual model, we have developed *Kala* – Kālá being the Sanskrit word for time – as a proof-of-concept implementation.

Kala is a Java Application Programming Interface (API) to aid the development of applications that need to create, manipulate, or query temporal ontologies at the conceptual level. It is based on the OWL API [15] and aims to have a design and interface that is familiar to users of that library. Like the OWL API, it provides an axiom-centric view of the (temporal) ontology, as opposed to the RDF triple-centric view of APIs such as Jena [6]. This axiom-centric view allows developers to utilize the library without concern for representation issues, in particular those related to parsing and serialization. Additionally, extending the OWL API will aid the later development of plug-ins for Protégé [18], a mature and widely-used ontology editor.

In order to guide the extension of the OWL API with temporal constructs, we pose two strong requirements:

1. The newly introduced temporal constructs must be completely separated from the constructs already supported by the OWL API in order to ensure the orthogonality of the non-temporal ontology and the temporal model; and
2. Kala must be compatible with custom implementations of the OWL API.

The orthogonality of the non-temporal components of the ontology and the temporal model is crucial for the correct functioning of the temporal model as an abstraction. We cannot, for example, decide that time instants and time intervals are subclasses of OWL individuals: time instants and time intervals are not necessarily represented as individuals in the representation scheme, and we need to restrict the operations that are permitted on these special entities in order to ensure that the temporal model will always have a correct mapping to the representation schemes. A statement such as, for example, "Bob is a friend of time instant t_1" does not make much sense in this context. Similar arguments hold for the fluent properties.

The compatibility with custom implementations of the OWL API is similarly crucial in the context of such developments as OWL database backends (for an examplar OWL database backend, see OWLDB [13]). In order to support custom implementations, we implement the extended functionality through the use of the *Decorator* design pattern [9]. We implement the additional Kala functionality by providing Decorators for the following three OWL API interfaces:

The Ontology. The OWL API views an OWLOntology, essentially, as a collection of OWLAxioms and OWLAnnotations. It provides methods to query these axioms and annotations, directly or through convenience methods, and collaborates with other objects to change the contents of this internal collection. Introducing new categories of axioms, then, necessitates extending the OWLOntology in such a way that it can also store these new axioms, but without altering its existing behavior.

The Data Factory. The OWL API OWLDataFactory presents the interface for producing the entities, class expressions, and axioms that form the building blocks of the OWL ontology. The OWLDataFactory follows the *Factory* design pattern [9]. We extend the existing OWLDataFactory to support the construction of the entities and axioms of the temporal conceptual model.

The Ontology Manager. The `OWLOntologyManager`, lastly, is responsible for the creation, loading, saving, and manipulation of ontologies. Since we need to create a new type of ontology, the temporal ontology, we will need to extend the behavior of the `OWLOntologyManager` so that it can properly manage these temporal ontologies.

Parsing and serialization are performed through the `Parser` and `Serializer` interfaces, respectively. Each has a representation-scheme-specific specialization (e.g., `FluentsSerializer`), and all are initialized with data on the capabilities and vocabulary of the chosen representation through the `RepresentationScheme` object. The `Parser` follows the *Builder* design pattern [9]: it iterates through a provided `OWLOntology` object that represents the temporal ontology using a particular representation scheme, and builds and modifies structures that together represent the eventual `TemporalOntology` as it goes. The `Serializer`, on the other hand, is implemented as a *Visitor* that visits every `TemporalAxiom` in the ontology and serializes it as one or more OWL axioms. The insight here is that all `TemporalAxioms` can be serialized independently of one another (something that is not true for *parsing*, where generally multiple OWL axioms must be parsed in combination in order to extract a single `TemporalAxiom`). The `Serializer` internally stores the set of generated axioms, and produces an `OWLOntology` from this set when the `createOntology()` method is called.

In order to raise awareness of the project and, hopefully, spur future developments, we have made the source code for Kala available on GitHub [7]. Due to space limitations, the details of the parsing and serialization algorithms can be found online at http://tinyurl.com/qhw273a.

5   Evaluation

We evaluate the temporal conceptual model — and Kala, our reference implementation — by representing a complex business process. The choice of a fitting scenario for our evaluation needs to satisfy three requirements. Firstly, the scenario must, naturally, convey temporal semantics that necessitate the utilization of a temporal model. Secondly, the complexity of the scenario must be sufficient to allow a range of expression types to be illustrated. Lastly, the scenario must present a relevance for the economic domain to prove its utility. One scenario fitting these requirements is found in the literature on tOWL. Illustrating the expressive power of tOWL, Milea et al. [23] present a historical account of the Leveraged Buy-Out (LBO) process for *Alliance Boots GmbH*, a multinational pharmaceuticals and retailing group that was formed through the 2006 merger of the *Boots group* and *Alliance UniChem*. In what was to be the largest LBO in European business history, two hedge funds, *Kohlberg Kravis Roberts & Co* (KKR) and *Terra Firma*, vied to acquire the company during the months of March and April of 2007, with the former emerging with the winning bid and, thus, the acquisition of Alliance Boots. Section 5.1 shows the modeling of this LBO process in the proposed conceptual model and Sect. 5.1 shows its serialization in two representation schemes.

5.1   Modeling the LBO Process

For reasons of conciseness, we refer the reader to the original paper [23] for the detailed representation of the LBO example, and limit ourselves to the definitions of fluent properties at the TBox level and the assertions of fluent properties and temporal entities at the ABox level. The authors identify four main stages of the LBO process: 1. Early Stage, 2. Due Diligence, 3. Bidding, and 4. Acquisition. These stages are pairwise disjoint in time and jointly exhaustive over the lifetime of the LBO process. Furthermore, the LBO process does not necessarily progress through the stages in a linear manner, and it may be aborted at any point in time.

TBox Level. At the TBox level we find the conceptual information representing the Company type and its subtypes, HedgeFund and Target; the Stage type and its subtypes, EarlyStage, DueDiligence, and so on; and the various restrictions that enforce the validity of the representation. The TBox level is also where the fluent properties are defined. We present two such fluent properties: earlyStage, which temporally relates any LBOProcess to its EarlyStage; and inStage, which temporally relates any Company to any Stage of an LBOProcess in which it is involved. Finally, we present their respective restrictions. Using the syntax introduced in Sect. 3.2:

```
FluentsDeclaration(FluentObjectProperty(earlyStage))
FluentObjectPropertyDomain(earlyStage LBOProcess)
FluentObjectPropertyRange(earlyStage EarlyStage)

FluentsDeclaration(FluentObjectProperty(inStage))
FluentObjectPropertyDomain(inStage Company)
FluentObjectPropertyRange(inStage Stage)
```

ABox Level. The ABox contains the assertional information related to any particular LBO process. Here, we find representations of participating companies, alliance_boots, kkr, and terrafirma; the instantiation of the LBO process we wish to represent, lbo1; and its various stages, es1 and so on. Additionally, we represent the temporal relations between these entities, define the temporal primitives involved, and assign discrete time values. To represent, for example, the following news snippet of information:

Buyout firm Terra Firma mulls Boots bid
(Sun Mar 25, 2007 8:42 EDT)

we use the syntax introduced in Sects. 3.1 and 3.2:

```
TimeDeclaration(TimeInterval(i1))
TimeDeclaration(TimeInstant(t1))
IntervalStartAssertion(i1 t1)
InstantTimeAssertion(t1
  "2007-03-25T08:42:00-04:00"^^xsd:dateTime)
```

```
FluentObjectPropertyAssertion(lbo1 earlyStage es1 i1)
FluentObjectPropertyAssertion(alliance_boots inStage es1 i1)
FluentObjectPropertyAssertion(terrafirma inStage es1 i1)
```

5.2   Serialization to OWL DL Representation Schemes

We present the serializations of the temporal ontology discussed above to OWL
DL representations of both the reification approach as well as the 4D-fluents
approach below.

The Reification Representation. In the reification representation scheme,
the declaration of a fluent property is serialized as a class declaration and property declarations for its subject and object relations. The serializer also specifies property restrictions for the subject and object relations and sets domain
and range restrictions for the subject and object relations. An example for the
earlyStage fluent object property is given below.

```
Class(earlyStage partial restriction(holds(someValuesFrom Interval)))
SubClassOf(earlyStage
  restriction(earlyStageToProcess(someValuesFrom LBOProcess)))
SubClassOf(earlyStage
  restriction(processToEarlyStage(someValuesFrom EarlyStage)))

ObjectProperty(earlyStageToProcess domain(earlyStage)
  range(LBOProcess))
Func(earlyStageToProcess)

ObjectProperty(processToEarlyStage domain(earlyStage)
  range(EarlyStage))
Func(processToEarlyStage)
```

Fluent property assertions are *reified* as anonymous individuals. Linking the
subject, object, and interval to the fluent property is then done through a series
of regular property assertions. The example below shows the serialization of
linking the lbo1 to its es1 through a fluent object property assertion:

```
Individual(_:a1 type(earlyStage) value(earlyStageToProcess lbo1)
  value(processToEarlyStage es1) value(holds i1))
```

The 4D-fluents Representation. The 4D-fluents representation differs from
the original tOWL example [23] in a small number of ways. Firstly, the explicit
class declarations for timeslice classes do not appear in the output. In the original
serialization, for example, the class of all timeslices of the Early Stage is defined
as in the following snippet (presented in OWL Abstract Syntax), which also
shows the use of the explicit class description in the specification of an object
property:

```
Class(EarlyStage_TS complete
  restriction(timeSliceOf(someValuesFrom EarlyStage)))

ObjectProperty(earlyStage
  domain(LBOProcess_TS)
  range(EarlyStage_TS))
```

This explicit class description is then used in axioms such as object property range restrictions. We do not generate such explicit class declarations for the timeslice classes and use the property restrictions directly in the expressions of such axioms, instead:

```
ObjectProperty(earlyStage
  domain(restriction(timeSliceOf(someValuesFrom LBOProcess))
  range(restriction(timeSliceOf(someValuesFrom EarlyStage)))
```

This alternate specification of the earlyStage property, however, conveys the same information as the original, but in a more compact manner.

At the ABox level, we introduce timeslices for every individual that participates in a fluent property. These timeslices are currently represented by *anonymous individuals* in the serialization, but otherwise follow the same serialization pattern as the original tOWL example. Timeslices are introduced for every *unique* combination of the individual participating in a fluent property and the time interval over which that fluent property holds. If an individual participates in multiple fluent properties over the same time interval, no additional timeslices are created, but instead, the same timeslice is reused for each fluent property. As an example, consider the timeslice relating to lbo1:

```
Individual(_:a1 type(TimeSlice) value(timeSliceOf lbo1)
  value(earlyStage _:a2) value(dueDiligence _:a3)
  value(bidding _:a4) value(abort _:a5))
```

6   Conclusion

In this paper, we have shown how the problem of defining the temporal conceptual model for OWL DL ontologies can be broken down into the definition of a *time model* that is *orthogonal* to the static ontology, and a *fluents model* that combines aspects of the static ontology and the time model to define the fluent properties. The concept of *orthogonality* was of key importance in allowing the concepts and expressivity of the time model and the fluents model to be established independently, which further allows the temporal conceptual model to be both simple and expressive. We have developed a reference implementation, Kala, with the goal of extending the OWL API with the additional constructs in a manner that closely resembles the design philosophy behind that library.

As future work, Kala could be extended in several ways. Adding the ability to produce a *snapshot reduction*, i.e., a regular OWL DL ontology that is representative of a temporal ontology at a particular time, would allow the temporal ontology to interact with existing tools such as OWL DL reasoners and query languages. Similarly, the access methods and internal data structures could be improved to provide greater convenience for users, for example by allowing the inspection of the evolution of a particular property's values over time. Such extensions could, finally, be used to develop applications on top of Kala. Tools such as temporal reasoners, temporal query languages, and temporal visualizations can benefit from Kala, as the functionality provided by these would represent a powerful argument to justify the use of the more complex temporal ontologies. For the ontology author, plugins for Integrated Development Environments such as Protégé, based on Kala, would allow the convenient development of temporal ontologies without considering the details of a particular representation scheme.

References

1. Allen, J.F.: Maintaining knowledge about temporal intervals. Commun. ACM **26**(11), 832–843 (1983)
2. Artale, A., Franconi, E.: A survey of temporal extensions of description logics. Ann. Math. Artif. Intell. **30**(1–4), 171–210 (2000)
3. Artale, A., Lutz, C.: A correspondence between temporal description logics. J. Appl. Non-Class. Logics **14**(1–2), 209–233 (2004)
4. Baratis, E., Petrakis, E.G.M., Batsakis, S., Maris, N., Papadakis, N.: TOQL: temporal ontology querying language. In: Mamoulis, N., Seidl, T., Pedersen, T.B., Torp, K., Assent, I. (eds.) SSTD 2009. LNCS, vol. 5644, pp. 338–354. Springer, Heidelberg (2009)
5. Batsakis, S., Petrakis, E.G.M.: SOWL: spatio-temporal representation, reasoning and querying over the Semantic Web. In: 6th International Conference on Semantic Systems (I-Semantics 2010), p. 15. ACM Press, New York (2010)
6. Carroll, J.J., Dickinson, I., Dollin, C., Reynolds, D., Seaborne, A., Wilkinson, K.: Jena: implementing the Semantic Web recommendations. In: 13th International World Wide Web Conference on Alternate Track Papers & Posters, pp. 74–83. ACM Press, New York (2004)
7. De Ridder, S.: owl-kala. http://www.github.com/owl-kala/owl-kala
8. Donini, F.M., Lenzerini, M., Nardi, D., Schaerf, A.: Reasoning in description logics. Princ. Knowl. Represent. **1**, 191–236 (1996)
9. Gamma, E., Helm, R., Johnson, R., Vlissides, J.: Design Patterns: abstraction and reuse of object-oriented design. Springer, Berlin (2001)
10. Grandi, F.: T-SPARQL: A TSQL2-like temporal query language for RDF. In: 1st International Workshop on Querying Graph Structured Data (GraphQ 2010) Organized in Conjunction with the 14th East-European Conference on Advances in Databases and Information Systems (ADBIS 2010), pp. 21–30 (2010)
11. Gregersen, H., Jensen, C.S.: Temporal entity-relationship models: a survey. IEEE Trans. Knowl. Data Eng. **11**(3), 464–497 (1999)
12. Gutierrez, C., Hurtado, C.A., Vaisman, A.: Introducing time into RDF. IEEE Trans. Knowl. Data Eng. **19**(2), 207–218 (2007)
13. Henß, J., Kleb, J., Grimm, S., Bock, J.: A database backend for OWL. In: 5th International Workshop on OWL: Experiences and Directions (OWLED 2009) (2009)
14. Hobbs, J.R., Pan, F.: An ontology of time for the Semantic Web. Trans. Asian Lang. Inf. Process. **3**(1), 66–85 (2004)

15. Horridge, M., Bechhofer, S.: The OWL API: a Java API for working with OWL 2 ontologies. In: 5th International Workshop on OWL: Experiences and Directions (OWLED 2009) (2009). http://ceur-ws.org/Vol-529/owled2009_submission_29.pdf
16. Jensen, C.S., Soo, M.D., Snodgrass, R.T.: Unifying temporal data models via a conceptual model. Inf. Syst. **19**(7), 513–547 (1993)
17. Klein, M., Fensel, D.: Ontology versioning on the Semantic Web. In: International Semantic Web Working Symposium (SWWS 2001), pp. 75–91. Springer, Berlin (2001)
18. Knublauch, H., Fergerson, R.W., Noy, N.F., Musen, M.A.: The Protégé OWL plugin: an open development environment for semantic web applications. In: McIlraith, S.A., Plexousakis, D., van Harmelen, F. (eds.) ISWC 2004. LNCS, vol. 3298, pp. 229–243. Springer, Heidelberg (2004)
19. Krieger, H.U., Kiefer, B., Declerck, T.: A framework for temporal representation and reasoning in business intelligence applications. In: Hinkelmann, K. (ed.) AI Meets Business Rules and Process Management. Papers from AAAI 2008 Spring Symposium, pp. 59–70. AAAI Press, Stanford (2008)
20. Lutz, C., Wolter, F., Zakharyashev, M.: Temporal description logics: a survey. In: 15th International Symposium on Temporal Representation and Reasoning (TIME 2008), pp. 3–14. IEEE Computer Society (2008)
21. McCarthy, J., Hayes, P.: Some Philosophical Problems from the Standpoint of Artificial Intelligence. Stanford University Press, Stanford (1968)
22. Melton, J.: Information technology - database languages - SQL. part 2: Foundation (SQL/Foundation). Standard (ISO/IEC 9075-2:2011) (2011)
23. Milea, V., Frasincar, F., Kaymak, U.: tOWL: a temporal web ontology language. IEEE Trans. Syst., Man Cybern. Part B, Cybern. (IEEE T-SMC-Part B) **42**(1), 268–281 (2012)
24. Mitra, G., Mitra, L.: The Handbook of News Analytics in Finance. Wiley, Chichester (2011)
25. Noy, N., Rector, A., Hayes, P., Welty, C.: Defining n-ary relations on the Semantic Web. W3C Working Group Note, April 2006. http://www.w3.org/TR/swbp-n-aryRelations/
26. Özsoyoğlu, G., Snodgrass, R.T.: Temporal and real-time databases: a survey. IEEE Trans. Knowl. Data Eng. **7**(4), 513–532 (1995)
27. Pissinou, N., Snodgrass, R.T., Elmasri, R., Mumick, I.S., Özsu, T., et al.: Towards an infrastructure for temporal databases: report of an invitational ARPA/NSF workshop. ACM SIGMOD Rec. **23**(1), 35–51 (1994)
28. Snodgrass, R.T.: The TSQL2 Temporal Query Language. Kluwer, Boston (1995)
29. Welty, C., Fikes, R., Makarios, S.: A reusable ontology for fluents in OWL. In: Bennett, B., Fellbaum, C. (eds.) 4th International Conference on Formal Ontology in Information Systems (FOIS 2006), pp. 226–236. IOS Press, Amsterdam (2006)

HV: A Feature Based Method for Trajectory Dataset Profiling

Wei Jiang[1], Jie Zhu[1], Jiajie Xu[1,2], Zhixu Li[1,2], Pengpeng Zhao[1,2],
and Lei Zhao[1,2]([✉])

[1] School of Computer Science and Technology, Soochow University, Suzhou, China
jwpker@outlook.com, zjcomeon@gmail.com
[2] Collaborative Innovation Center of Novel Software Technology
and Industrialization, Nanjing, Jiangsu, China
{xujj,zhixuli,anliu,zhaol}@suda.edu.cn

Abstract. The pervasiveness of location-acquisition and mobile computing techniques have generated massive spatial trajectory data, which has brought great challenges to the management and analysis of such a big data. In this paper, we focus on the trajectory dataset profiling problem, and aim to extract the representative trajectories from the raw trajectory as a subset, called profile, which can best describe the whole dataset. This problem is very challenging subject to finding the most representative trajectories set by trading off the profile size and quality. To tackle this problem, we model the features of the whole dataset from the aspects of density, speed and the directional tendency. Meanwhile we present our two kinds of methods to select the representative trajectories by the global heuristic voting (HV) function based on the feature model. We evaluate our methods based on extensive experiments by using a real-world trajectory dataset generated by over 12,000 taxicabs in Beijing. The results demonstrate the efficiency and effectiveness of our methods in different applications.

Keywords: Spatial databases · Trajectoy · Data profiling

1 Introduction

The advance in location-acquisition technology has generated a myriad of spatial trajectories recording the mobility of various moving objects, such as people, vehicles, and animals. Such trajectories offer us unprecedented information to understand moving objects and locations, fostering a broad range of applications in traffic flow monitoring [1], path planning [2,3], predictive queries [4,5]

This work was supported by the National Natural Science Foundation of China under Grant Nos. 61073061, 61472263, 61402312, and 61402313, the Natural Science Foundation of Jiangsu Province of China under Grant No. SBK2015021685, Jiangsu Provincial Department of Education of China under Grant No. 12KJB520017, and Collaborative Innovation Center of Novel Software Technology and Industrialization, Jiangsu, China.

© Springer International Publishing Switzerland 2015
J. Wang et al. (Eds.): WISE 2015, Part I, LNCS 9418, pp. 46–60, 2015.
DOI: 10.1007/978-3-319-26190-4_4

and pattern mining [6]. However, the storage and management of such massive data is always very expensive due to the large data scale. Furthermore, most existing analyses and mining algorithms based on the historical trajectories are memory-resident, which is difficult to load the whole dataset into memory. Thus, an efficient trajectory dataset profiling method is essential to simplify the raw dataset to a small but representative profile as the synthetic dataset, which encapsulates the mobility patterns hidden in the whole dataset and speeds up above data analysis tasks.

Many previous works focus on the compression of the GPS points in one trajectory, which approximate a raw trajectory to a polyline with the subset of the important points [7–9]. In other words, the key of these methods is to discard some points with negligible errors from an original trajectory. All of these methods, however, pay all attention to transforming a trajectory into an approximate representation, which may lose some useful information such as speed and turns, and lead to the sparsity of raw trajectories that will create uncertainty in sampling-rate sensitive applications such as map-matching [10].

Different from previous works, we choose a complete trajectory as the smallest unit instead of a point, and extract the representative trajectories as the profile to the following mining task. [11] has proposed an unsupervised sampling method for trajectories, which can find a representative sub-trajectory subset of the whole dataset. Nevertheless, they take only consideration of the spatial-temporal similarity than other advanced features, which may be not enough for some real applications. For instance, the route planning [12,13] would need the speed feature of the trajectories to describe the congestion of the road network.

In this paper, we propose a feature based method of trajectory dataset profiling to meet the demand of different applications. To measure the quality of the profiling, we model the trajectory dataset features from three different angles. (1) The area density of a dataset is the number of the trajectories passing by each area, which reflects the trajectory aggregation in space. Meanwhile, it occupies an important place in some density sensitive applications like trajectory clustering [14] and frequent pattern mining [15]. (2) The area speed is the average speed of the trajectories crossing each area, which is widely applied to the trajectory analysis for driving, such as intelligent transportation [13] and traffic analysis [16]. (3) The directional tendency is the migration direction of most trajectories staying in this area, which can reflect the tendency of the traffic migration in each area. Thus, it is useful in some recommendation and prediction [12,13]. We take into consideration the above three features to evaluate the data quality of the profiling.

Based on above feature model, we transform the profiling of trajectory dataset into a multi-objective combination optimization problem. In other words, we want to pick up the optimal trajectory combination to minimize the error of the above three features. Obviously, an exhaustive algorithm would search for all possible solutions in order to get the optimal trajectory subset, leading to an intolerable cost. We propose a heuristic voting (HV) algorithm, which assesses each trajectory by a global heuristic voting function, and then greedily extracts the most representative trajectories into the profile by a non-dominated selection. This method, however, may fail into a local optimal combination because the addition of new trajectories may worsen the previously selected trajectories. To tackle this problem in a global view, we propose an optimized algorithm HV-II that is the extension of the simulated annealing (SA) algorithm [17]. In detail, it initializes some profiles by the unbiased random sampling and improves these sets by a series of heuristic evolution based on the voting function.

An extensive experimental study has been conducted on a real trajectory dataset generated by over 12,000 taxicabs in Beijing. We evaluate our profiling algorithms by using the datasets into different applications, and the results demonstrate the efficiency and effectiveness of our methods.

The contributions of our works are summarized below:

1. We introduce a novel problem of trajectory dataset profiling, aiming at extracting the representative trajectories as the profile to speed up several analyses and mining tasks in the field.
2. To measure the quality of the profile, we model the feature of the trajectory dataset from three different angles. According to this model, we transform our problem into a multi-objective combination optimization problem.
3. We propose two algorithms to tackle this problem. The HV algorithm selects the most representative trajectory into the profile set by a heuristic voting function. And the HV-II algorithm improves the HV algorithm by extending the SA algorithm.
4. We operate a variety of experiments using a real trajectory dataset in order to evaluate our approaches thoroughly.

The rest of the paper is organized as follows. Section 2 presents the problem statement of trajectory dataset profiling. Sections 3 and 4 describe the details of the HV and HV-II algorithm respectively. Then a comprehensive experiment evaluation is present in Sect. 5. Finally, we discuss the related works in Sect. 6 and concludes the whole paper in Sect. 7.

2   Problem Statement

Let us assume a dataset $D = (\tau_1, \tau_2, \cdots, \tau_n)$ of n trajectories, where τ_i denotes the $i - th$ trajectory of dataset, $i \in \{1, 2, \cdots, n\}$. Each trajectory is represented by a sequence of position points $p_i = (x_i, y_i, t_i)$, denoted as $\tau_i = (p_1, p_2, \ldots, p_m)$, where (x_i, y_i) is the position in the 2D Euclidean space at time stamp t_i. The dataset S is said to be a profile of D if S is the form of $(\tau_{s_1}, \tau_{s_2}, \cdots, \tau_{s_k})$ where $k \leq n$ and $\tau_{s_i} \in D$. That means the profile S is a subset of the raw trajectories.

In this paper, we adopt an approximate representation of trajectories, motivated by [16], by transforming trajectories into a tensor $A_{\tau_i} \in \mathbb{R}^{X*Y*F}$, with two dimensions standing for spatial and feature respectively. As shown in Fig. 1, we assume a regular grid of equal rectangular cells with user-defined size (e.g. $100 * 100 \ m^2$). An entry $A_{\tau_i}(x, y, k) = c$ denotes the $k - th$ feature value of the trajectory τ_i in grid cell $c_{x,y}$. In this way, we can speed up computations, at the same time preserving the mobility pattern of each trajectory by the reasonable definition of trajectory feature model. In our work, we transform a trajectory into a feature trajectory with three parts: density, speed and direction. The density is the points number of this trajectory in each cell $c_{x,y}$. The speed is the average speed of the trajectory in each area. And the direction is the main movement direction of this trajectory in each area. Formally:

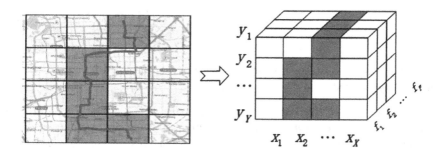

Fig. 1. Trajectory approximate representation

Definition 1. *Feature Trajectory. Give a regular grid $G_{X,Y}$ consisting of the cells $c_{x,y}$ ($1 \leqslant x \leqslant X, 1 \leqslant y \leqslant Y$) with the cell size θ and a trajectory $\tau_i = (p_1, p_2, \ldots, p_m)$ in D, the feature trajectory $A_{\tau_i} \in \mathbb{R}^{X * Y * 3}$ of τ_i is a tensor, where the entry $A_{\tau_i}(x, y, 1)$ is the density defined as the number sampling points of τ_i in $c_{x,y}$, the $A_{\tau_i}(x, y, 2)$ is the average traveling speed in $c_{x,y}$ and the $A_{\tau_i}(x, y, 3)$ is the main direction defined as the average angle of each segment $\overline{p_i p_{i+1}}$. Each angle, denoted by $\theta(\overline{p_i p_{i+1}})$, is defined to be the angle of an anti-clockwise rotation from the positive $x - axis$ to a vector from p_i to p_{i+1}. Thus the direction $A_{\tau_i}(x, y, 3)$ falls into $[0, 2\pi)$.*

This technique allows us to view all trajectories in D as a fixed tensor A_{τ_i}, where each value of the entry corresponds to a feature attribute of the trajectory crossing by each cells. In addition, by the choice of the reasonable grid size in the different application, a whole trajectory dataset D can be also transformed into a tensor $A_D \in \mathbb{R}^{X * Y * F}$ in Definition 2, which is composited by each trajectory tensor A_{τ_i}. The A_D reflects the advanced traffic condition hidden in the raw data, such as the aggregation degree, road congestion and migration tendency, which is crucial in many trajectory mining and pattern recognition tasks.

Definition 2. *Feature Trajectory Dataset. Give the grid $G_{X,Y}$ and the trajectory dataset D, the feature trajectory dataset $A_D \in \mathbb{R}^{X * Y * 3}$ of D is a tensor, where the entry $A_D(x, y, 1)$ is the density defined as a ratio of the number sampling points of all trajectories in $c_{x,y}$ to the size of D, the $A_D(x, y, 2)$ is the average traveling speed of all trajectories in $c_{x,y}$ and also the $A_D(x, y, 3)$ is the average main direction of all trajectories in D.*

In order to estimate the feature error between D and S, we propose the measurement taking account of the above three features respectively by the root mean square error (RMSE) metric. More formally:

Definition 3. *Feature Error Measurement. Give a regular grid $G_{X,Y}$ and the dataset D and S, the feature error measurement $\mathcal{E} = (\mathcal{E}_1, \mathcal{E}_2, \mathcal{E}_3)$, where $\mathcal{E}_i$ is doted by Eq. 1.*

$$\mathcal{E}_i(D, S) = \sqrt{\frac{\sum_{x=1}^{X} \sum_{y=1}^{Y} (1 - A_S(x, y, i)/A_D(x, y, i))^2}{\sum_{x=1}^{X} \sum_{y=1}^{Y} 1}} \qquad (1)$$

Now the problem of feature based trajectory dataset profiling (FTDP) is formalized as follows.

Definition 4. *FTDP Problem. Give a trajectory set D, the size of grid cells θ and the size k of the profile S, find the profile S from the raw dataset D by minimizing three feature error measurement $\mathcal{E} = (\mathcal{E}_1, \mathcal{E}_2, \mathcal{E}_3)$ respectively.*

3   Heuristic Voting Algorithm

In this section, we describe the *Heuristic Voting*(HV) algorithm to tackle the FDTP problem defined in Definition 4. Recall the goal of the profiling is to pick up k representative trajectories as the profile S with minimizing each feature error $\mathcal{E}_i$. The complexity of an exhaustive algorithm would search for all possible solutions, which will result in a prohibitive cost $\Theta\binom{n}{k}$.

In order to reduce the complexity of calculation, we present approximative algorithm HV. The key idea of HV is to select the top-k representative trajectories evaluated by a voting function greedily. More detail, in each iteration step, we vote each trajectory in different feature importance depending on the current feature demand of S. Then the trajectories whose the voting value is non-domination with others are selected and HV adds these trajectories into S to update the feature demand.

The heuristic voting function evaluates each trajectory from three different feature angles. On the density side, a trajectory may first be selected if it passes the hight density regions of the trajectory dataset D. However this strategy will lead to the high density regions oversampled. We propose a bidirectional heuristic strategy by maximizing a formula (see Eq. 2), which takes into account the density $A_D(x, y, 1)$ as well as the density $A_S(x, y, 1)$ in profiling S. In this way, the set S should contain the trajectories crossing the high density regions and, at the same time, cover the whole space as possible.

$$H_d(\tau_i) = \sum^{c_{x,y} \in \tau_i} A_D(x, y, 1) \cdot (1 - A_S(x, y, 1)) \qquad (2)$$

On the speed side, the trajectory should be first selected that can increase the similarity of average speed in two dataset intuitively. Our method is shown in Eq. 3, which has two components. $\Delta_\nu(x, y)$ means the distance of average speed between D and S in grid cell $c_{k,l}$ in current state, and $\overline{\Delta_\nu(x, y)}$ means the distance after we pick this trajectory τ_i into S. In this work, we have selected to use the continuous function of *Gaussian* kernel to smoothen the distance, which is widely used in a variety of applications in the field of the machine learning. The control parameter $0 \leq \sigma \leq 1/2$ tunes how fast the function decreases with distance. The larger value of $H_\nu(\tau_i)$ indicate the trajectory τ_i has a larger probability to be selected, which can reduce the speed distance efficiently.

$$\Delta_\nu(x, y) = \exp\left[-\frac{(A_D(x, y, 2) - A_S(x, y, 2))^2}{2\sigma^2}\right]$$

$$\overline{\Delta_\nu(x, y)} = \exp\left[-\frac{(A_D(x, y, 2) - \overline{A_S(x, y, 2)})^2}{2\sigma^2}\right] \quad (3)$$

$$H_\nu(\tau_i) = \sum^{c_{k,l} \in \tau_i} \overline{\Delta_\nu(x, y)} \cdot (1 - \Delta_\nu(x, y))$$

As the same with the speed, the larger value of the voting function $H_\theta(\tau_i)$ in Eq. 4 indicate the trajectory τ_i has a larger probability to be selected, which can reduce the direction error rapidly, where $\overline{A_S(x, y, 3)}$ also means the direction angle after we add this trajectory into S. We use the cosine of angle distance to assess the direction error due to it's easily calculated by the vector.

$$H_\theta(\tau_i) = \sum^{c_{k,l} \in \tau_i} \cos(A_D(x, y, 3) - \overline{A_S(x, y, 3)}) \cdot (1 - \cos(A_D(x, y, 3) - A_S(x, y, 3))) \quad (4)$$

Now we have a strategy to vote the representative trajectories in three different features, and the most important trajectory should be pick into S to update the feature status of S. However, a trajectory may not have the highest voting value in all of feature side. To tackle this problem, we propose a sample selection method to selete the trajectories whose voting value is non-dominated with others depending on the trajectory domination, which defined as follows.

Definition 5. *Trajectory Domination. Give a dataset D and the profile S, a trajectory τ_i dominates another trajectory τ_j if to every feature heuristic voting function has $H_*(\tau_i) \geq H_*(\tau_j)$ and has at least one feature satisfy $H_*(\tau_i) > H_*(\tau_j)$, denote as $\tau_i \succ \tau_j$.*

Algorithm 1 shows the detail of the selection method. After the initialization (line 1), we compute the dominance relation between each trajectory and others, where S_{τ_i} is the set of trajectories dominated by τ_i, and N_{τ_i} means the number of trajectory dominate τ_i (line 3–10). Then we add the trajectory τ_i into set F if τ_i is non-dominated by other trajectories and delete all trajectories dominated by τ_i from T (line 12–13). In this way, the HV algorithm, shown in Algorithm 2, votes each trajectory by the heuristic voting strategy in each step (line 3–5), where the voting value can expresses the gain of similarity of two set in different features when we add this trajectory. Then the selection method select the feature non-dominate trajectory as the most feature representative trajectories into

---

Algorithm 1. Selection

Require: trajectory dataset T
Ensure: the non-dominated trajectories set F
1: $F \leftarrow \emptyset$
2: **for** each trajectory $\tau_i \in$ T **do**
3:     $S_{\tau_i} \leftarrow \emptyset, N_{\tau_i} \leftarrow 0$
4:     **for** each trajectory $\tau_j \in$ T **do**
5:         **if** $\tau_i \succ \tau_j$ **then**
6:             $S_{\tau_i} \leftarrow S_{\tau_i} \cup \{\tau_j\}$
7:         **else**

8:             $N_{\tau_i} \leftarrow N_{\tau_i} + 1$
9:         **end if**
10:     **end for**
11:     **if** $N_{\tau_i} = 0$ **then**
12:         $F \leftarrow F \cup \{\tau_j\}$
13:         $T \leftarrow T - S_{\tau_i}$
14:     **end if**
15: **end for**
16: **return** F;

Algorithm 2. HV Algorithm

Require: trajectory dataset D and the size k of profile.
Ensure: the profile set S
1: $S \leftarrow \emptyset, C \leftarrow D$
2: **while** $|S| < k$ **do**
3:     **for** each trajectory $\tau_i \in C$ **do**
4:         calculate the three value of the heuristic voting function $H_*(\tau_i)$
5:     **end for**
6:     $S \leftarrow S \cup (F \leftarrow Selection(C))$
7:     $C \leftarrow C - F$
8:     update the feature tensor A_S
9: **end while**
10: **return** S;

---

the profile S (line 6–8). Finally, we update the feature tensor A_S after adding trajectories.

We now present, in Lemma 1, the time complexity of the HV algorithm.

Lemma 1. *The time complexity of HV in Algorithm 2 is $\Theta(\frac{k}{d} \cdot (nl + n^2))$, where n is the number of trajectory of D, k is the size of profile S, d is the average size of non-dominated trajectories in one step and l is the average number of crossing grid cell per trajectory.*

Proof 1. *We need to select k trajectories to add into S. The HV algorithm selects the most representative trajectories $\frac{k}{d}$ time iterations. The cost of each iteration step has two part: one is the calculation the heuristic voting value $H_*(\tau_i)$ that is $\Theta(nl)$, and other is the selection method which need the $\Theta(n^2)$ cost. Therefore, the cost of the whole algorithm is $\Theta(\frac{k}{d} \cdot (nl + n^2))$.*

Compared with the random sampling technique that constructed profile randomly, HV provides a deterministic solution to fit the most representative trajectories greedily, which can ensure reduce the feature error in every iteration. However, this algorithm may fail to find the optimal profile and trapped in local optimal trajectory combinations due to the previous selected trajectories may get worse with the addition of new trajectories. To overcome this problem, we present a optimized method HV-II by extending the SA algorithm in Sect. 4.

4   Extended Heuristic Voting Algorithms

As mentioned in Sect. 3, HV algorithm will trap into the local optimal profiling due to the heuristic voting strategy depend on the current features tensor A_S, and the previous selected trajectory can't remove from S which may get worse after the other trajectories selected. To tackle this problem, we present our HV-II algorithm, extended by simulated annealing(SA) [17], to get a approximate global profiling.

The key idea of HV-II is to improve a good approximative profile to the global optimum by a series of local heuristic evolution. In detail, we first initialize a archive set $List_A$ by unbiased random sampling, which comprises of a series profiles S_i no-dominated in each feature error $\mathcal{E}_i$, defined in Definition 6, and selected one randomly as the current set $S_{current}$. In next, we use a series of iteration to evolve $S_{current}$. In each iteration, S_{new} is the evolution of $S_{current}$ by a local heuristic evolution depended on the voting function. Then S_{new} is accepted by checking the dominance relationship with all profile in $List_A$. The structure of the HV-II is shown in Algorithm 3. The parameters that need to be set a priori are mentioned below.

- *iter*: The number of iterations of the whole processing procedure.
- B_{min}: The minimum size of the archive set $List_A$, which is equal to the minimum number of non-dominate profiles.
- B_{max}: The maximum size to which the archive set $List_A$ may be filled before clustering is used to reduce its size to B_{min}.
- T_{max}: The maximum (initial) temperature of SA.
- α: The cooling rate of SA.

Definition 6. *Profile Dominance. Give two profiles S_k and S_m, S_k dominate S_m if to every feature error has $\mathcal{E}_i(S_k) \leq \mathcal{E}_i(S_m)$ and has at least one feature satisfy $\mathcal{E}_i(S_k) < \mathcal{E}_i(S_m)$, denote as $S_k \succ S_m$.*

The different step of the HV-II algorithm are now explained in detail.

4.1   Initialization

The algorithm begins with the initialization of a number of profile solutions as the archive set $List_A$. Each of these solutions is produced by the random sampling, accepting a new profile in $List_A$ only if it is non-dominated with all solution in $List_A$. This is continued for a number of iterations. Thereafter, the non-dominated solutions that are obtained are stored in $List_A$, up to the size of $List_A$ equals B_{min}. That means, initially, there are a B_{min} number of solutions non-dominated each other in archive set $List_A$. Then we select one from $List_A$ as the current profile $S_{current}$ randomly.

Algorithm 3. HV-II Algorithm

Require: trajectory dataset D, the profile size k and all above parameters
Ensure: the profile set S
  1: Initialization of $List_A$, $S_{current} \leftarrow random(List_A)$
  2: **for** $i \leftarrow 1$ to $iter$ **do**
  3:     $S_{new} \leftarrow$ the evaluation of $S_{current}$ by local heuristic evolution
  4:     Check the domination relation of S_{new} and all profile in $List_A$
  5:     **if** S_{new} dominates $j(j \geq 1)$ sets in $List_A$ **then** /* Case 1 */
  6:         $List_A \leftarrow List_A \cup \{S_{new}\}$, $S_{current} \leftarrow S_{new}$
  7:         Remove all the j sets dominated by S_{new}
  8:     **end if**
  9:     **if** S_{new} is no-dominated with all the sets in $List_A$ **then** /* Case 2 */
 10:         $List_A \leftarrow List_A \cup \{S_{new}\}$, $S_{current} \leftarrow S_{new}$
 11:         **if** $|List_A| > B_{max}$ **then**
 12:             Cluster $List_A$ to B_{min} of clusters
 13:         **end if**
 14:     **end if**
 15:     **if** S_{new} is dominated by $j(j \geq 1)$ sets in $List_A$ **then** /* Case 3 */
 16:         Accept S_{new} with the probability $prob = \frac{1}{1+exp(\Delta\mathcal{E}_{avg}/T)}$
 17:     **end if**
 18: **end for**
 19: **return** $S \leftarrow S_{current}$

4.2   Local Heuristic Evolution

Give a initialized profile set $S_{current}$, the HV-II will generate a smaller feature error solution by evolve it. In other word, HV-II need allow us to preserve the most representative trajectories and exchange the redundant trajectory of $S_{current}$ with other more important one as possible. More detail, we generate a solution S_{rand} by random sampling and construct a candidates trajectories set S_{cand} comprised of the trajectories in $S_{current}$ and S_{rand}. Then we get a new solution S_{new} by executing HV algorithm in Sect. 3 and the important trajectories of S_{rand} will add into S_{new}.

4.3   The Main Process

The evolution profile S_{new} is acceptable by checking the domination relation between S_{new} and all profiles in $List_A$, three different cases may arise. These are enumerated below.

– *Case 1:* S_{new} dominates $j(j \geq 1)$ solutions of the $List_A$. In this case, the S_{new} is selected as the $S_{current}$, and added to the $List_A$. All the dominated profiles dominated by S_{new} are removed from the $List_A$.
– *Case 2:* S_{new} is non-dominated with every profiles in the $List_A$. In this case, the S_{new} is also selected as the $S_{current}$ and added to the $List_A$. In case the $List_A$ becomes overfull ($|List_A| > B_{max}$), clustering $List_A$ is performed to reduce the size to B_{min} by k-means algorithm.

– *Case* 3: S_{new} is dominated by $j(j \geq 1)$ profiles of the $List_A$. The S_{new} is acceptable based on the principle of SA [17]. In our work, a given temperature T, a new profile is selected with a probability

$$prob = \frac{1}{1 + exp(\Delta \mathcal{E}_{avg}/T)} \tag{5}$$

where $\Delta \mathcal{E}_{avg} = \sum_{i=1}^{j}(\Delta \mathcal{E}(S_i, S_{new}))/j$ and $\Delta \mathcal{E}(S_a, S_b)$ is the feature error distance of profile S_a and S_b, defined in Eq. 6, where N is number of features and R_i is the range of the $i - th$ feature estimated the initialization. We will set the S_{new} to the $S_{current}$ and added to the $List_A$ if we accept S_{new} with the probability $prob$.

$$\Delta \mathcal{E}(S_a, S_b) = \Pi_{i=1, \mathcal{E}_i(S_a) \neq \mathcal{E}_i(S_b)}^{N}(|\mathcal{E}_i(S_a) - \mathcal{E}_i(S_b)|/R_i) \tag{6}$$

4.4 Complexity

The complexity of basic operations in HV-II are as follows:

1. $List_A$ initialization: $\Theta(B_{min})$.
2. The evolution of $S_{current}$:$\Theta(\frac{k}{d} \cdot (2kl + (2k)^2))$, where $k >> d, l$, the complexity can note as $\Theta(k^3)$.
3. Procedure to check the domination status between S_{new} and the $List_A$ members:$\Theta(N \cdot B_{min})$, where N is the number of features.
4. Clustering procedure:$\Theta(B_{max})$.

Therefore, overall complexity of HV-II becomes $\Theta(B_{min} + iter \cdot (k^3 + N \cdot B_{min} + B_{max})) \approx \Theta(iter \cdot k^3)$ when $k >>$ other parameters. Note that, the complexity of HV-II is not relevant to the size n of D which is efficient in profiling massive trajectory dataset.

5   Experimental Study

In this section, we first present the experimental settings, including the dataset and experiment environment, the parameters and evaluation approaches in our experiment. Then we report the major results with some discussions.

5.1   Experimental Settings

Dataset. We conduct our experiments on a real dataset that are collected from around 12,000 taxis. This dataset contains over 1.6 million trajectory records in Beijing, China, collected during an one-week period in November, 2014. The sampling rate of these trajectories ranges form 40 s to 2 min. In our experiment, We picked up about 21.3 thousand trajectories from the spacial region ranges between 116.077°–116.694° longitude and 39.689°–40.148° latitude in one hour.

Experiment Environment. We implement all algorithms in Java. The machine we use to accomplish the experiment has a quad-core Inter Core i5 CPU (3.2 GHz) and 8 G memory. The operating system is Windows 7 and Java VM version is 1.8.0_20 64-Bit.

Parameters. The parameters introduced in our approach are: (a) The size k of profile set S; (b) The grid cell size θ; (c) The initial temperature T_{max} sets to $T_{max} = 1$ and the cooling rate α sets to $\alpha = 0.995$ in HV-II in our experiments, which affects the accept probability; (d)The number of iterations $iter$ in HV-II sets to 200 in our experiments, which affects the effectiveness and the execution time; (d) the limit of the archive set $List_A$ is set to $B_{min} = 20$ and $B_{max} = 50$ in our experiment.

Evaluation Criteria. Our profiling algorithm is evaluated both in terms of running time and data quality. The running time is measured using the actual program execution time. The data quality is measured using two metrics the *Average of Feature Errors* (AFE) and the *Variance of Feature Errors* (VFE).

5.2   Experimental Result

In our first experiment, we fix the grid cell size $\theta = 100(m)$, and compare our profiling algorithm with the random sampling technology by scaling the profiling rating, defined as $c = k/n$. As Fig. 2a shown, the AFE will decrease with the increasement of the profile size, and HV-II is the most effective one, which has only below 20 % feature error after discarding 70 % trajectories from D. Figure 2b shows the change of the VFE. Different with the unstable random method, the HV-II and HV will decrease with the profile size, which can indicate these two methods keep relative balance in each features. The execution efficiency is shown in Fig. 2b. When the size $k \ll n$, HV-II is more efficient than HV due to the cost of HV-II is irrelevant with the raw dataset size n, and HV-II will slow with the profiling rate increased due to a large number of evolution iterations.

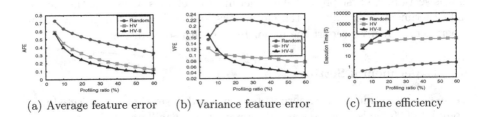

(a) Average feature error   (b) Variance feature error   (c) Time efficiency

Fig. 2. Performance w.r.t profiling ratio

Then we fix the profiling rate $c = 20\%$ and demonstrate the performance under different the grid size θ in Fig. 3. As θ increasing, the AFE will decrease in

Fig. 3a due to the reduction of the granularity demand in each features. Meanwhile, the VFE of each features will increase in Fig. 3b due to the sensitiveness of each feature is different in grid constraint. As observed from Fig. 3c, the grid size only affect efficiency of the calculation of feature error and it have no effect with large data volume.

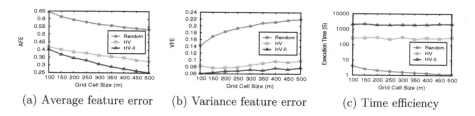

(a) Average feature error   (b) Variance feature error   (c) Time efficiency

Fig. 3. Performance w.r.t grid size

The *Trajectory Clustering* is an important application to discover the common movement patterns. We assess the ability of our method to preserves patterns extracted the sub-trajectory clustering algorithm in [14], which is a typical clustering techniques having as output different mobility (centroid of a trajectory cluster). We generate a profile set with sample rate $c = 30\%$ and grid cell size $\theta = 100(m)$, and the visualization of clusters in Fig. 4a and b clearly shows our approach succeeds to preserve the sketch with various density center and each patterns, before and after the profiling. We evaluate the clustering quality using two metrics: the number of the clusters *Clusters* and the sum of squared error *QMeasure* of each cluster which defined in [14]. The Fig. 4c shows the clustering performance of out methods, where HV-II has same performance with raw dataset both the *Clusters* and *QMeasure*. We repeated the experiment several time tuning the parameters of this algorithms and with various profile sizes and the conclusion remained the same.

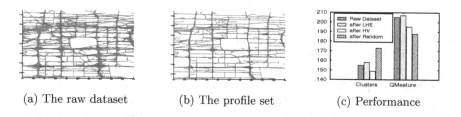

(a) The raw dataset   (b) The profile set   (c) Performance

Fig. 4. The result of the clustering

The *Route Recommend* is also a common application to find a fastest (quickest) route to the user depend on the historical trajectories. The vast trajectories are mapped to the road network by *Map-Matching* algorithm [10] to obtain the information of the traffic condition. We assess the ability of our profiling method in T-Drive system in [13], which use a time-dependent landmark graph generated by the trajectories to estimate the travel time of different route. First, Fig. 5a shows the number of covered road after map-matching algorithm in each dataset before and after profiling. Then T-Drive need to select top-k most frequent path to build the landmark graph, and we use the *Jaccard* distance between raw dataset and each profiling methods to measure the data quality shown in Fig. 5b. Finally, we execute the query of time-dependent fastest rout in each dataset and the error rate of estimated travel time, defined as $\mathcal{E}_t = (T_D - T_S)/T_D$, is shown in Fig. 5c, where HV-II has the lowest error than others.

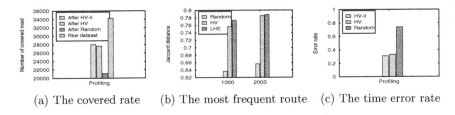

   (a) The covered rate   (b) The most frequent route   (c) The time error rate

Fig. 5. The result of route recommend

6 Related Work

Trajectory profiling is very important to the mining of the massive trajectory dataset. It constitutes a quite challenging issue, so few related works have been carried out. To the best of our knowledge, it is only one explorative method, sub-trajectory sampling techniques, that has been proposed in [11].

In [11] the authors propose a method for trajectory segmentation and sampling method based on the representativeness of the (sub-)trajectories. They segment a trajectory depending on the density and trajectory similarity information, and use a greedy optimization approach by a global voting scheme to find the most representative sub-trajectories. However, this approach would not cover our problem due to some important area traffic information will be discarded, such as popular area, frequent pattern and road speed.

Other exciting works focus on the trajectory compression by approximating a polyline with a subset of the points from the raw trajectory. The Douglas-Peucker algorithm [7] approximates a trajectory by a line segment and recursively selects the point contributing the biggest error as a split point. Recently, a direction-preserving simplification algorithm [9] is proposed that preserves direction information based on angular difference and keeps position information simultaneously. All of above works focus on using location points with as little as possible to replace the entire single trajectory.

The representative trajectory [14] was initially presented in sub-trajectories clustering, which designed a partition-and-group framework, which partitions trajectories into line segments and then build groups for those close segments.

A naive solution for the trajectory profiling would first cluster the dataset and select the centroids(representative trajectory) of each cluster as the profiling sets. The aim of clustering, however, is to partition the data into groups and not to downsize the trajectory dataset. In addition, such a solution would provide a single representation associated with each cluster, that means, we can't selected a real representative trajectory set from one cluster.

Summarizing, although very interesting as a problem and with great potentials in the trajectory management and mining domain, to the best of our knowledge, there is no related work on the trajectory dataset profiling taking the complex trajectory features into consideration.

7   Conclusions

This paper makes a profound research on a novel problem of feature based trajectory dataset profiling, which contributes to the more effective management of trajectory database and speeds up the existing trajectory mining tasks. Beside, we put forward two algorithms: HV and HV-II. HV is to select the most feature representative trajectories according to the current status by the heuristic voting function greedily. HV-II optimizes HV by the extension of the simulated annealing algorithm, and each iteration will improve the profile by the heuristic evolution. The extensive experiments using a real-world trajectory dataset demonstrate the efficiency and effectiveness of our methods. In the future, we are intended to apply more effective trajectory features to our profiling. Thus, our method will be applicable to more a wide range of trajectory tasks.

References

1. Li, X., Han, J., Lee, J.-G., Gonzalez, H.: Traffic density-based discovery of hot routes in road networks. In: Papadias, D., Zhang, D., Kollios, G. (eds.) SSTD 2007. LNCS, vol. 4605, pp. 441–459. Springer, Heidelberg (2007)
2. Zheng, Y., Zhang, L., Xie, X., Ma, W.-Y.: Mining interesting locations and travel sequences from GPS trajectories. In: Proceedings of the 18th International Conference on World Wide Web, pp. 791–800. ACM (2009)
3. Chen, Z., Shen, H.T., Zhou, X.: Discovering popular routes from trajectories. In: 2011 IEEE 27th International Conference on Data Engineering (ICDE), pp. 900–911. IEEE (2011)
4. Monreale, A., Pinelli, F., Trasarti, R., Giannotti, F.: Wherenext: a location predictor on trajectory pattern mining. In: Proceedings of the 15th ACM SIGKDD International Conference on Knowledge Discovery and Data Mining, pp. 637–646. ACM (2009)
5. Xue, A.Y., Zhang, R., Zheng, Y., Xie, X., Huang, J., Xu, Z.: Destination prediction by sub-trajectory synthesis and privacy protection against such prediction. In: 2013 IEEE 29th International Conference on Data Engineering (ICDE), pp. 254–265. IEEE (2013)
6. Giannotti, F., Nanni, M., Pinelli, F., Pedreschi, D.: Trajectory pattern mining. In: Proceedings of the 13th ACM SIGKDD International Conference on Knowledge Discovery and Data Mining, pp. 330–339. ACM (2007)

7. Douglas, D.H., Peucker, T.K.: Algorithms for the reduction of the number of points required to represent a digitized line or its caricature. Cartographica: Int. J. Geogr. Inf. Geovis. **10**(2), 112–122 (1973)
8. Trajcevski, G., Cao, H., Scheuermanny, P., Wolfsonz, O., Vaccaro, D.: On-line data reduction and the quality of history in moving objects databases. In: Proceedings of the 5th ACM International Workshop on Data Engineering for Wireless and Mobile Access, pp. 19–26. ACM (2006)
9. Long, C., Wong, R.C.-W., Jagadish, H.V.: Direction-preserving trajectory simplification. Proc. VLDB Endow. **6**(10), 949–960 (2013). http://dx.doi.org/10.14778/2536206.2536221
10. Lou, Y., Zhang, C., Zheng, Y., Xie, X., Wang, W., Huang, Y.: Map-matching for low-sampling-rate GPS trajectories. In Proceedings of the 17th ACM SIGSPATIAL International Conference on Advances in Geographic Information Systems, pp. 352–361. ACM (2009)
11. Panagiotakis, C., Pelekis, N., Kopanakis, I., Ramasso, E., Theodoridis, Y.: Segmentation and sampling of moving object trajectories based on representativeness. IEEE Trans. Knowl. Data Eng. **24**(7), 1328–1343 (2012)
12. Yuan, J., Zheng, Y., Zhang, C., Xie, W., Xie, X., Sun, G., Huang, Y.: T-drive: driving directions based on taxi trajectories. In: Proceedings of the 18th SIGSPATIAL International Conference on Advances in Geographic Information Systems, pp. 99–108. ACM (2010)
13. Yuan, J., Zheng, Y., Xie, X., Sun, G.: T-drive: enhancing driving directions with taxi drivers' intelligence. IEEE Trans. Knowl. Data Eng. **25**(1), 220–232 (2013)
14. Lee, J.-G., Han, J., Whang, K.-Y.: Trajectory clustering: a partition-and-group framework. In: Proceedings of the 2007 ACM SIGMOD International Conference on Management of Data, pp. 593–604. ACM (2007)
15. Wei, L.-Y., Zheng, Y., Peng, W.-C.: Constructing popular routes from uncertain trajectories. In: Proceedings of the 18th ACM SIGKDD International Conference on Knowledge Discovery and Data Mining, pp. 195–203. ACM (2012)
16. Wang, Y., Zheng, Y., Xue, Y.: Travel time estimation of a path using sparse trajectories. In: Proceedings of the 20th ACM SIGKDD International Conference on Knowledge Discovery and Data Mining, pp. 25–34. ACM (2014)
17. Kirkpatrick, S., Vecchi, M., et al.: Optimization by simulated annealing. Science **220**(4598), 671–680 (1983)

A Dynamically Extensible Open Cross-Document Link Service

Ahmed A.O. Tayeh[(✉)] and Beat Signer

Web and Information Systems Engineering Lab, Vrije Universiteit Brussel,
Pleinlaan 2, 1050 Brussels, Belgium
{atayeh,bsigner}@vub.ac.be

Abstract. Since the introduction of the term hypertext in the early 1960s, the goal has been to link, annotate as well as transclude parts of documents. However, most existing document linking approaches show some shortcomings in terms of the offered link granularity and cannot easily be extended to support new document formats. More recently, we see new document formats such as the Office Open XML (OOXML) standard which facilitate the linking to parts of certain document formats. We present a dynamically extensible open cross-document link service enabling the linking and integration of arbitrary documents and multimedia content. In our link browser, emerging document formats are supported via visual plug-ins or by integrating third-party applications via gateways. The presented concepts and architecture for dynamic extensibility improve the document life cycle in so-called cross-media information spaces and enable future-proof cross-document linking.

Keywords: Cross-document linking · Dynamic link service extensibility

1 Introduction

Most existing document formats only provide a simple embedded unidirectional link model for defining associations between documents [16,19]. While many document formats support the linking to third-party documents, it is normally not possible to address and link to parts of these documents. For example, in a PDF document we can create hyperlinks to an entire Word document via a URI but we cannot easily link to specific parts within this document. There is no doubt that the advent of the Extensible Markup Language (XML) in combination with the link model defined by XLink has been a major step towards advanced linking on the Web. However, XLink only deals with XML-based documents and does not support other document formats or models.

We recently witnessed enhancements in various document formats for the linking of documents. For example, in the current PDF specification [1] links can be created between pieces of information stored in a PDF document via so-called GoToE actions. Furthermore, OOXML [2] hyperlinks enable the addressing of parts of other OOXML documents. Nevertheless, as argued by Tayeh

© Springer International Publishing Switzerland 2015
J. Wang et al. (Eds.): WISE 2015, Part I, LNCS 9418, pp. 61–76, 2015.
DOI: 10.1007/978-3-319-26190-4_5

and Signer [19], the support of cross-document linking for arbitrary document formats is a challenging task which requires concrete information about each document format to be linked. Furthermore, it asks for a revision of a given document format specification in order to support a new document type. We aim for a link service that is flexible enough to support existing as well as emerging document formats.

The extensibility of a link service is an essential feature and should form an integral part of any document linking service. However, nowadays we can see this extensibility feature only implemented for link resolution in web browsers. When an HTML link points to a third-party document (e.g. a PDF document), based on the target's MIME type the web browser calls a specific plug-in to visualise the document. Nevertheless, the link service extensibility should not be limited to resolving links to entire documents but also allow users to create links between snippets of information in arbitrary document formats. Furthermore, a link service should take into account that users rely on proprietary applications (e.g. Microsoft Word) to author and visualise specific types of documents and address the challenge of seamlessly integrating these third-party applications. Last but not least, a linking service has to be dynamically extensible. Imagine, that each time we navigate to a new document type on the Web, we would have to install a new version of the Web browser. This would definitely be a big burden for any user. We can outline various reasons why a link service should be dynamically extensible. First of all, it is not feasible to extend or redeploy the existing service and user interface every time a new document format has to be supported. Further, each user might only make use of a small subset from the multitude of existing document formats. Rather than having a monolithic link service that supports all document formats, users should be able to dynamically extend the link service on demand in order to support their preferred document formats. Finally, offering cross-document linking features to proprietary third-party applications should not ask for changes to the core of these systems since this might not be accepted by their creators.

In this paper, we present recent extensions and adaptations of our open cross-document link service in order to deal with dynamic extensibility. We begin in Sect. 2 by providing a brief overview of our link service and highlight some of its earlier shortcomings with respect to dynamic extensibility. The essential requirements for an extensible link service are outlined in Sect. 3. The concepts and architecture to support dynamic extensibility are discussed in Sect. 4. In Sect. 5 we compare our link service solution with the existing body of work. After a critical discussion of the presented solution and an outline of future research directions, we provide some concluding remarks.

2   Cross-Document Link Service

Most document formats offer an embedded unidirectional link model, implying that only the owner of a document can add new links to the document. Furthermore, a target document is not aware of any links that have been defined pointing to it from one or multiple source documents. Thereby, in most document formats the offered link models are far away from the "non-sequential writing" definition of hypertext [14] where at any given time pointers can be added to documents to direct a reader to a different section, paragraph or another (part of a) document. In order to overcome the shortcomings of existing document link models, in some of our earlier work we have presented an open cross-document link service [19]. The link service offers a plug-in architecture for integrating different document formats. The link model of our cross-document link service is based on the Resource Selector Link (RSL) hypermedia metamodel [18]. RSL is based on the principle idea of linking arbitrary entities, whereby an entity can either be a resource, a selector or a link. A resource represents the base unit for a given media type, such as an image, a video or a complete document. A selector is always related to a specific resource and is used to address parts of the resource. Finally, a link is a bidirectional association between one or multiple source entities and one or multiple target entities. In our cross-document link service, new document formats are supported by implementing *data plug-ins* that extend the RSL resource and selector concepts and contain information on how to address resources (documents) as well as selectors (anchors) attached to documents in a given document format. For example, in a data plug-in for the HTML document format, an HTML resource (document) can be represented as a URI pointing to a web resource and its selector might be represented via an XPointer-like expression.

The visualisation component of our link service consists of a link browser to visualise the supported document formats. The user interface further offers the necessary GUI actions for performing the basic CRUD (create, read, update and delete) operations on a link. For each document format to be visualised in the link browser, a *visual plug-in* has to be implemented. A visual plug-in for a given document format needs to visualise documents as well as their anchors and has to provide the necessary functionality to create, update and delete anchors. Our cross-document link service currently supports the XML, text and PDF document formats as well as general multimedia content such as images via the corresponding data and visual plug-ins. It is worth mentioning that our link service can not only deal with various document formats but also general multimedia content such as images or movies. A selector of a movie resource can, for example, be defined as a timespan. The underlying RSL metamodel is general enough to support resources and selectors of different media types and has been used before to support links across movies and web pages [17]. The visual plug-ins for various document formats visualise the documents in a Java Swing `JPanel` which can also be used to visualise arbitrary multimedia content.

We further provide a proof of concept implementation integrating a third-party application in the form of the Google Chrome browser which acts as a client for our link service. The Google Chrome plug-in has been implemented by using the Google Chrome API[1]. The communication between the link service and the Google Chrome application plug-in has been realised via a special HTML

---

[1] https://developer.chrome.com/extensions.

gateway component. The HTML gateway implements the WebSocket communication protocol and translates any messages between the cross-document link service and the Google Chrome plug-in. Thereby, users are able to create bidirectional hyperlinks between supported document formats. Figure 1 shows a bidirectional hyperlink between a PDF document and a JPEG image which are visualised on the left and right hand side of the link browser component as well as bidirectional link between a PDF document and an HTML document visualised in the third-party Google Chrome web browser.

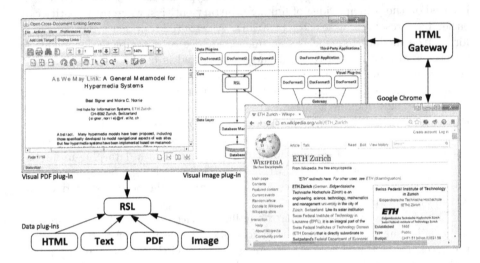

Fig. 1. Cross-document link browser with third-party application

The existing cross-document link service showed a number of shortcomings in terms of extensibility. In order to illustrate these shortcomings, we use the scenario of a user who installed the link service with support for the text, PDF, XML and HTML formats. Our user is a researcher often reading Word and PDF documents and would therefore like to add support for cross-document links between PDF and Word documents. Unfortunately, there is no online repository offering plug-ins for different document formats and our user cannot find and download the necessary plug-ins for Word. However, the researcher is lucky because they found some information on how to implement data and visual plug-ins. Moreover, they have access to the code of the HTML gateway. Our user decided to ask a developer who is familiar with Microsoft's Office Developer Tools to implement the necessary plug-ins for supporting the Word document format. When checking the code of the gateway, the developer realised that they cannot implement a new plug-in since the coupling between the communication protocol as well as the message handling in the gateway asks for a good understanding of the communication among the link service components. Furthermore, some classes and packages used in the gateway are not available online and are not accessible via public modifiers to be used outside of the HTML plug-in. Therefore, they decided that supporting a visualisation of Word documents and their anchors within the link browser side by side with PDF documents is enough. Unfortunately, even if the developer can provide an implementation of the necessary data and visual plug-ins for Word, we still need a way that the link service can discover these new plug-ins. This scenario highlights that even if a link service offers a decent plug-in architecture, we still have to address the issue of dynamic extensibility. Users should be able to extend the link service without the need for a new release of the link service supporting the new document format. In the remaining part of this paper we describe the new essential concepts necessary for realising a dynamically extensible link service.

3   Requirements

From our previous experience with the cross-document link service and through an analysis of the shortcomings of existing link services, we derived the following fundamental requirements for an extensible cross-document link service.

Flexible and Extensible Link Service Architecture. Most annotation and link services are not based on explicit link models. They often contain a mixture of conceptual and technical hard-coded link models. The fact that document formats have different document models means that anchors and selections can be defined and addressed in different ways. For example, an anchor in a tree-document model can be defined via an XPointer-like expression while an anchor in a linear text-only document model might be defined by its start and end indices. A link model should be flexible enough to deal with different document formats whereas the extensibility deals with emerging document formats to be supported at a later stage. Most annotation and link services have to be redeployed whenever a new media type should be supported. We strongly believe that the link service user interface should be extensible and support new document formats without redeploying the entire link service.

Support Multiple Document Formats. This requirement can be divided into two sub-requirements: *the link service should be able to support existing document formats* and *should be able to deal with emerging document formats*. A link service should not be restricted to a fixed set of predefined document formats. However, a link service that adheres to the first requirement of a flexible and extensible link service architecture does not necessarily have to satisfy this requirement. For example, the previous version of our link service addressed the first requirement but the support for a new document format still required some intervention of the link service provider. An extensible link service should offer a simple mechanism to allow third-party developers or users to integrate additional document formats without a redeployment of the core link service.

Easy Integration of Third-Party Applications. Supporting third-party applications should be taken into account in any successful link service. Otherwise, the link service would have to provide the authoring/editing of third-party documents and be appealing enough to convince users to abandon their preferred third-party applications. Indeed, this is not practical as users might want to continue using the applications they are familiar with. Nowadays, most proprietary third-party document processing applications come along with their own Software Development Kit (SDK) in order to be extended on demand with some extra functionality. An extensible link service should benefit from a third-party application's extensibility rather than forcing third-party application vendors to rewrite their applications. Plug-ins or add-ins can be implemented for these third-party applications in order to provide visual handles for creating and editing anchors in the supported document format.

Flexible Communication Channels. The support for third-party applications asks for communication across different protocols. The APIs and SDKs of some third-party applications limit their plug-ins or add-ins to a specific set of communication protocols. For example, Google's Chrome Extension API and extensions for other web browsers only support WebSocket communication with third-party applications whereas TCP sockets are the default communication protocol for third-party desktop applications. An extensible link service should support the multitude of existing communication protocols since otherwise it might not be possible to integrate certain third-party applications.

Customisable Link Service. The previous requirements imply that the different link service components are extensible via data plug-ins, visual plug-ins or third-party application plug-ins. It is not practical to push users to install all plug-ins at once given that they might not use most of the supported document formats or third-party applications. Therefore, end users should be able to customise their link service by installing only the plug-ins for the document formats that are really needed. Customisability of the link service means that the link service is extensible on demand. The LaTeX environment which has been used to write this article is a good example for on-demand extensibility via various packages to support extra functionality. This on-demand extensibility not only saves storage space but also increases the overall performance. In order to successfully support on-demand customisation, the availability of the plug-ins should be ensured via a central online plug-in repository.

Plug-in Versioning. Different document format specifications are often updated to support new features. Moreover, third-party applications are normally updated with new features to either support the new document format specification or to enable new features in the application itself. Therefore, new versions of plug-ins for some document formats are expected to be published and the link service should offer some plug-in versioning mechanism.

4   Dynamically Extensible Link Service

The overall architecture of our link service is outlined in Fig. 2. Components depicted with solid lines have been presented in earlier work [19], while some of the dashed *gateways, plug-in tracking, online repository* and *communication*

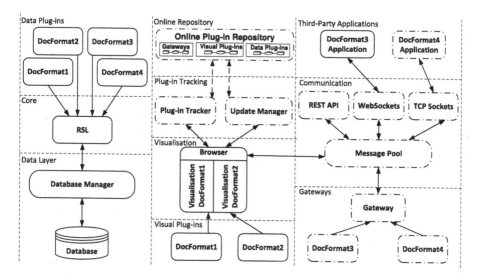

Fig. 2. Cross-document link service architecture

components are playing a central role for the dynamic link service extensibility described in this paper.

In our link service, the gateway and communication components play an important role for integrating third-party applications. For each document format to be integrated via a third-party application, a gateway handling the messages with the third-party application plug-in has to be provided. The third-party application plug-in can use any communication protocol supported by the link service. The plug-in tracking component consists of a plug-in tracker and an update manager. These two components are responsible for keeping track of the installed plug-ins as well as for installing new plug-ins on demand by communicating with the online plug-in repository.

The OSGi framework [10] plays an important role in achieving our link service's dynamic extensibility. The OSGi specification defines a dynamic modular system for the Java programming language but the deployment of an OSGi-based application does not differ from regular Java applications. More details about the motivation for using the OSGi framework in our link service can be found in [19]. Conceptually, the OSGi framework consists of three layers, including the *module layer*, the *life cycle layer* and a *service layer*. The module layer is responsible for packaging and sharing the application code. Each module of an application is called a *bundle* and corresponds to a Java JAR file with some extra metadata in the form of a manifest file. The life cycle layer controls specific modules at execution time. The interaction and communication between installed modules is managed by the service layer. Each component depicted by a rectangle in the link service is a bundle. All data, visual and gateway plug-ins are bundles with different metadata. The life cycle layer is extensively used by the plug-in tracker for dynamic extensibility.

In order to get a better understanding of the roles played by the different components, we come back to the scenario presented earlier, but this time we assume that the user installed the extensible version of the link service. When the user wants to add support for the Word document format, they can either open the online repository web page to search for Word document plug-ins or search via the link service's update manager interface. The user learns that there are two options for the Word document format. The first extension visualises documents and their anchors in the link browser. This means that the extension consists of a data plug-in and a visual plug-in. A second extension supports the visualisation directly within Microsoft Word. This extension consists of a data plug-in, a gateway as well as the necessary Microsoft Word application add-in. In both cases, the user must install the data plug-in which has some metadata stored in its manifest file in order to be correctly identified and used by the link service components. The update manager reads the metadata, the plug-in is downloaded via a secure shell protocol and installed by the plug-in tracker. The plug-in tracker does the necessary work to inject the plug-in into the running link service. If the user wishes to visualise their documents within the generic link browser, the visual plug-in for the Word document type has to be installed.

In the case that the user wants to visualise their documents in Microsoft Word, they have to install the gateway plug-in in the link service and follow the instructions provided for the Microsoft Word add-in in order to extend Microsoft Word. The gateway plug-in is installed based on the same mechanism used for data and visual plug-ins. After installing the plug-ins, the user should be able to create hyperlinks in Word documents and link them to any of the already supported document formats. Let us assume that after three months the developer of the third-party Word plug-in has fixed some bugs and implemented some new nice visualisation as well as other features. They upload a new version of the plug-in to the online repository. The update manager will inform the user that a new version of a Word plug-in is available and wait for a confirmation to install the new version. In the following subsections we elaborate on the different components necessary for the dynamic extensibility described in the scenario.

4.1 Gateways

The gateway component is introduced in the link service in order to overcome the limitation of third-party application integration introduced earlier by the link service. The gateway component is the most essential component to integrate existing third-party applications and it is flexible enough to integrate any new third-party application. A third-party application extension (plug-in) acts as a client to the link service and can be implemented in any programming language supported by the third-party application SDK. The component contains an interface that provides the abstract methods needed to translate any message exchanged with a third-party application plug-in. Message translation simply involves the marshalling and unmarshalling of Java objects. JSON messages sent by an external third-party plug-in are unmarshalled into Java objects by the corresponding gateway and Java objects are marshalled to JSON objects by the gateway to be sent to its corresponding external third-party application plug-in. Two things are worth mentioning here. First, the gateway component can easily be extended for a specific document format, since third-party developers are not required to understand the different communication channels and processes among the link service components. Second, the link service does not provide a general JSON representation for messages that should be exchanged with external third-party application plug-ins, but it rather asks developers to form these objects. For multiple reasons, we provide developers the freedom to marshal and unmarshal the objects and send as much information as they want from the link service to third-party applications. First of all, the objects to be marshalled or unmarshalled represent information about documents and anchors in a specific document format. This information is introduced by the data plug-in of a given document format and we cannot anticipate what information the object contains. The link service treats all objects as entities which is the general representation of RSL resources and selectors. Nevertheless, third-party developers are aware that entity objects received by the gateway from other link service components must be instances of the specific document format and contain some additional information. It further enables developers to send arbitrary information to external third-party application plug-ins for rich link visualisation.

Listing 1.1. Gateway interface

```
abstract long getResourceId(JSONObject msg);
abstract JSONObject openDocument(Resource res,
    HashSet <Anchor> anchors, Anchor entityHighlight);
abstract long getTargetEntityID(JSONObject msg);
```

In a gateway plug-in, the developer has to provide a class implementing the gateway interface. Some of the methods of the gateway interface are shown in Listing 1.1. The `getResourceId()` method is called by the link service in order to return the ID of the resource (document) included in the JSON message. The second method is used to serialise a Java object to a JSON message which can be sent to the external third-party application plug-in to open a document. The method receives the resource (document), the set of anchors contained within the document and optionally a specific anchor to be highlighted in the case that the document has to be visualised as a result of a link that has been followed. It is worth mentioning that the anchor object in the link service contains information about a link source which is either a selector or a complete document. It further contains information about the targets which can either be other documents or selectors of documents. The anchor object contains enough information about each target such as its MIME type, its ID as well as the contained document in the case of a selector, enabling the development of rich visual plug-ins. For example, multi-target links can be represented via a pop-up menu in order to give the user the flexibility to navigate to any target document.

4.2   Plug-in Metadata and Repository

The different types of plug-ins (e.g. data, visual and gateway plug-ins) require a mechanism to differentiate between them in order to correctly use them in the link service and to correctly inject them when extending the link service. We have exploited the OSGi manifest file to correctly identify each document format plug-in. Aside from the specific OSGi metadata required by any OSGi bundle, different document format plug-ins must contain specific metadata to be a valid extension for the link service and to be correctly identified by the link service. Based on the type of the plug-in (i.e. data, visual or gateway), different metadata keys and values should be included in the plug-in. The `Extension-Name`, `Extension-Mime` and `Extension-Type` metadata is required for all types of plug-ins. This metadata provides information about the MIME type (e.g. `text/html` or `application/pdf`) supported by the plug-in, its name and its type. A plug-in developer should maintain the consistency of the MIME type provided in a document format plug-in. In other words, a data plug-in and a visual plug-in for a given document format must have the same MIME type. The same holds for data and gateway plug-ins in the case of a third-party application integration.

The link service's user interface contains an abstract class defining the necessary functions to visualise a document as well as to perform the CRUD operations for links in a given document format. A visual plug-in must extend the abstract class of the user interface component. Furthermore, each gateway has to implement the gateway interface. The link service can communicate with a visual plug-in to visualise documents or for CRUD operations on a link by instantiating the class that extends the abstract user interface class. Furthermore, the link service can communicate with a gateway plug-in in order to marshal or unmarshal objects by instantiating the class that implements the gateway interface. In order to instantiate these classes, the classpath is stored in the `Extension-Class` metadata included in the corresponding visual or gateway plug-in. Finally, the online repository provides simple interfaces to search for plug-ins and for uploading new plug-ins.

4.3   Plug-In Tracking

The plug-in tracking consists of the plug-in tracker and the update manager. The update manager is responsible for keeping track of the available document format plug-ins in the online repository by reading the metadata for every plug-in. Users can interact with the update manager by using its GUI in order to search for plug-ins of different document formats. Moreover, the update manager notifies the plug-in tracker about the availability of any document format. Last but not least, the update manager is in charge of downloading the different plug-ins.

The plug-in tracker is responsible for installing, keeping track of and managing the different plug-ins in the link service. The plug-in tracker's extender pattern listens for any OSGi bundles (plug-ins) being started or stopped in the link service. When a plug-in is installed, the tracker checks whether it is an extension based on the predefined extension metadata and performs the necessary operations to integrate it in the link service. In the case of a data plug-in, the tracker adds the MIME type to the list of supported document formats. If a visual plug-in is installed, the plug-in tracker checks whether the data plug-in for the same document format (MIME) has already been installed. In the case that the data plug-in is missing or the user does not confirm the installation of the data plug-in, the visual plug-in will not be installed. If the data plug-in is already installed or installed after a user's confirmation, the plug-in tracker notifies the user interface component that a new visual plug-in exists which in turn injects the new plug-in into the link browser user interface. As a result, the user can see that the new document format has successfully been integrated in the link service and can start using it. For gateway plug-ins, the plug-in tracker will maintain the availability of its data plug-in with the same mechanism used when installing a visual plug-in and add it to the list of supported gateways.

4.4   Communication Protocols

In order to support as many third-party applications as possible, our link service communication component supports three different communication protocols. TCP sockets and WebSockets are used for full duplex communication channels and a REST API can be used as a fall-back solution for third-party application SDKs not offering full duplex communication. JSON messages coming from third-party applications through different communication protocols are centrally managed via the message pool component. The message pool also keeps track of all active third-party application plug-ins and their sessions in order to forward the JSON messages produced by different gateways.

Note that each JSON message exchanged with a third-party application plug-in contains a `command` key with one value for the predefined request values. For example, the `create` value is used in some messages sent by the external plug-ins and informs the message pool that the user wants to create a link in a document visualised in its third-party application. The `showTarget` value is contained in some messages sent by the external plug-ins and tells the message pool that the user clicked on a link in a document visualised in the third-party application. In this case, the JSON message should contain information about the specific link target. The message pool then asks the corresponding gateway to return the ID of the intended target by passing the JSON message to the `getTargetEntityId()` gateway method shown earlier in Listing 1.1. When the link service requires an external third-party application plug-in to open a document with its selectors, it sends the request to the message pool. The message pool then asks the corresponding gateway to marshal the object to a JSON message containing the `command` key with the `openDocument` value and the third-party application plug-in will know how to process the message.

Messages coming from third-party application plug-ins are forwarded to the user interface component via the message pool after they have been unmarshalled to Java objects by the corresponding gateway. The message pool can identify the correct gateway by using the MIME type defined by the communication session in the handshake process. Moreover, when a message has to be sent from the link service to the external third-party application plug-in, the message pool requests the gateway to marshal the message to a JSON message before forwarding it to the correct communication protocol with the active session.

5 Related Work

Based on the six requirements introduced earlier in Sect. 3, we present a comparison of some existing link services and annotation tools in Table 1. Each requirement is mapped to one dimension (column) in the comparison table, expect for the second requirement which is mapped to the two 'Cross-Document Linking' and 'Emerging Document Formats' dimensions. The former evaluates whether a link service supports cross-document linking between multiple existing document formats, while the latter evaluates whether a link service is extensible and might support emerging document formats. Further, the last row in the table presents our dynamic link service. We use the ✓ symbol to illustrate that a feature is supported whereas the (✓) symbol means that there is only limited support for a given feature or the feature is supported but with some major drawbacks.

Table 1. Comparison of existing link services and annotation tools

Link Service	Extensible Architecture	Cross-Document Linking	Emerging Document Formats	External Applications	Flexible Channels	Customisability	Plug-in Versioning
Intermedia	(✔)	✔	✗	✗	✗	✗	✗
Sun's Link Service	(✔)	✔	(✔)	(✔)	✗	✗	✗
Microcosm	✗	✔	✗	(✔)	✗	✗	✗
Annotea Solutions	✗	✗	✗	✗	✗	✗	✗
MADCOW	✗	✗	✗	✗	✗	✗	✗
FAST	(✔)	✗	✗	(✔)	✗	✗	✗
Dynamic Link Service	✔	✔	✔	✔	✔	✔	✔

Open hypermedia systems such as Intermedia [9], Sun's link service [15] and Microcosm [11] addressed the limitations of embedded links by managing links separately from the linked documents in so-called linkbases. Intermedia supports the linking across five different document formats but shows a number of shortcomings. Even though Intermedia is based on a layered architecture, it is not evident how it can be extended to support additional document formats. Moreover, Intermedia was intended to be used as a complete authoring tool and not purely as a link service. This implies that any document format that would like to profit from Intermedia's linking features has to be visualised and authored with Intermedia's viewers. Furthermore, features such as the integration of third-party applications, customisation and versioning have not been considered in Intermedia.

Sun's link service—a pure link service providing a protocol to communicate with external applications—shows two major shortcomings. First, the link service forms a monolithic component with a core link model that is not extensible. In other words, more advanced forms of links cannot be supported without a redeployment of the link service. Second, the protocol comes in the form of a program library that has to be included in any external application in order to communicate with the link service. This implies that third-party applications have to be rewritten to benefit from the features offered by the link service.

Microcosm supports linking for Microsoft applications and further offers some nice features including generic links or the dynamic linking of documents. Nevertheless, it is not evident how Microcosm could be extended in order to support some of the other important features. It is worth mentioning that a number of open hypermedia systems have been used to enrich the Web with external links by considering the Web as a client for these open hypermedia solutions [4,6].

The XLink standard supports so-called extended links which can be stored in linkbases and be used to realise bi- and multi-directional links. A link service making use of the XLink model and XPointer expressions can only support the linking across four types (MIME) of XML documents [8]. Nevertheless, most web link services and applications that make use of XLink, such as XLinkProxy [7], solely support HTML documents. On the other hand, RSL is flexible and provides a number of features that XLink lacks such as user rights management, context resolvers and overlapping links. Note that recent Semantic Web technologies and XML promote the concept of linked data [12] on the Web.

A number of systems such as the W3C's Amaya[2] web browser implement the Annotea standard [13]. MADCOW [5] is another web annotation plug-in for Microsoft Internet Explorer that uses a client-server architecture allowing users to store their annotations on dedicated remote servers. MADCOW goes beyond the functionality provided by Annotea-based tools and offers the possibility to annotate richer media types such as images and videos. Nevertheless, most of these tools and standards adopt the simple annotation concepts (e.g. notes or comments) and do not support the creation of hyperlinks between existing content. Even if the linking of existing content is supported and the extensibility is addressed, these solutions are limited to the features offered by XLink.

More recently, various digital libraries management systems (DLMSs) have incorporated interactive annotation features that facilitate discussions among researchers. Whereas in most of the DLMSs the annotation features are offered by built-in components, the Flexible Annotation Service Tool (FAST) [3] has been developed to be a stand-alone annotation tool in order to offer its services to multiple DLMSs. Even though FAST offers simple annotation features rather than cross-document linking and also lacks some of the other features, it is based on an interesting extensible architecture. FAST consists of two main components,

---

[2] http://www.w3.org/Amaya/.

the core annotation service and a number of gateways (interfaces). Each gateway is connected to a different DLMS and ensures that the DLMS gets access to the core annotation service offered by FAST. Hence, any DLMS can benefit from FAST's features by developing a new FAST gateway. The flexible integration of different DLMSs with FAST can be considered as third-party integration.

6   Discussion and Future Work

To the best of our knowledge, the presented link service is the first dynamically extensible and customisable link service. The extensible open cross-document link service is a future-proof linking solution for arbitrary document formats and multimedia content types. Its dynamic extensibility further allows third-party developers and end users to support any document format and multimedia content type without the intervention of the link service provider. To address the preferences of the end users, the integration of different document formats and multimedia content types can either happen within the link browser or in their preferred third-party applications. The presented dynamically extensible link service further deals with updates for specific document formats or third-party applications by providing a mechanism for maintaining different versions of the same document format plug-in. Furthermore, the link service represents an ideal platform for investigating innovative forms of cross-media linking.

The manageability and maintainability of links has always been an issue in hypermedia systems. This includes broken links, the consistency of links when the linked documents evolve or the management of link metadata in collaborative environments such as Google Docs. The RSL links used by our link service are bidirectional and therefore the link service is able to solve the issue of broken links by removing any link target from a document when the link source has been deleted from the source document. The management of links in evolving documents is still a hot topic. Even though we are planning to apply more than one mechanism for addressing this problem, we have currently adopted a simple document archiving solution of linked documents which also helps to address the broken link problem in the case of missing documents.

We are currently working on the integration of third-party applications such as Microsoft Word, PowerPoint or Acrobat Reader. Moreover, we are investigating a model for cross-media document formats where content can easily be transcluded from various document formats and multimedia types. In the near future, we plan to evaluate the usability of the presented link service in a user study. Last but not least, we are working on an enhanced desktop environment that exploits the links defined via the link service in combination with some data mining algorithms to enhance the search and retrieval of desktop documents.

7   Conclusion

We have presented a dynamically extensible link service enabling the linking across arbitrary document formats and media types. In contrast to existing link services and link models, our solution supports the integration of new document formats without having to apply any changes to the core of the link service or graphical link browsers and without a redeployment of the link service. Based on the concepts of data plug-ins, visual plug-ins, gateways as well as third-party application add-ins, emerging document formats can either be supported within our link browser or via any third-party application that has been extended to communicate with a document format-specific gateway. The presented dynamically extensible cross-document link service acts as a research platform for investigating document and link management as well as maintainability in so-called cross-media information spaces. Our solution might further inspire other link service providers to reconsider the dynamic extensibility of their approaches.

References

1. Adobe Portable Document Format Reference, 6th edn. Version 1.7, Adobe Systems Incorporated, November 2006
2. Standard ECMA-376: Office Open XML File Formats, 3rd edn. ECMA International, June 2011
3. Agosti, M., Ferro, N.: A system architecture as a support to a flexible annotation service. In: Türker, C., Agosti, M., Schek, H.-J. (eds.) Peer-to-Peer, Grid, and Service-Orientation in Digital Library Architectures. LNCS, vol. 3664, pp. 147–166. Springer, Heidelberg (2005)
4. Anderson, K.M., Taylor, R.N., Whitehead Jr., E.J.: Chimera: hypermedia for heterogeneous software development environments. ACM Trans. Inf. Syst. **18**, 211–245 (2000)
5. Bottoni, P., Civica, R., Levialdi, S., Orso, L., Panizzi, E., Trinchese, R.: MADCOW: a multimedia digital annotation system. In: Proceedings of AVI 2004, Gallipoli, Italy, May 2004
6. Bouvin, N.O.: Unifying strategies for web augmentation. In: Proceedings of Hypertext 1999, Darmstadt, Germany, February 1999
7. Ciancarini, P., Folli, F., Rossi, D., Vitali, F.: XLinkProxy: external linkbases with xlink. In: Proceedings of DocEng 2002, McLean, USA, November 2002
8. DeRose, S., Maler, E., Daniel, Jr., R.: XML Pointer Language (XPointer) Version 1.0, January 2001
9. Haan, B.J., Kahn, P., Riley, V.A., Coombs, J.H., Meyrowitz, N.K.: IRIS hypermedia services. Commun. ACM **35**(1), 36–51 (1992)
10. Hall, R., Pauls, K., McCulloch, S., Savage, D.: OSGi in Action: Creating Modular Applications in Java. Manning Publications, Greenwich (2011)
11. Hall, W., Davis, H., Hutchings, G.: Rethinking Hypermedia: The Microcosm Approach. Kluwer Academic Publishers, Boston (1996)
12. Heath, T., Bizer, C.: Linked Data: Evolving the Web into a Global Data Space. Morgan and Claypool Publishers, San Rafael (2011)
13. Koivunen, M.-R.: Semantic authoring by tagging with annotea social bookmarks and topics. In: Proceedings of SAAW 2006, Athens, Greece, November 2006
14. Nelson, T.H.: Literary Machines. Mindful Press, Sausalito (1982)
15. Pearl, A.: Sun's link service: a protocol for open linking. In: Proceedings of Hypertext 1989, Pittsburgh, USA, November 1989

16. Signer, B.: What is wrong with digital documents? a conceptual model for structural cross content composition and resuse. In: Proceedings of ER 2010, Vancouver, Canada, November 2010
17. Signer, B., Norrie, M. C.: A framework for cross-media information mangement. In: Proceedings of EuroIMSA 2005, Grindelwald, Switzerland, February 2005
18. Signer, B., Norrie, M.: As we may link: a general metamodel for hypermedia systems. In: Parent, C., Schewe, K.-D., Storey, V.C., Thalheim, B. (eds.) ER 2007. LNCS, vol. 4801, pp. 359–374. Springer, Heidelberg (2007)
19. Tayeh, A.A.O., Signer, B.: Open cross-document linking and browsing based on a visual plug-in architecture. In: Benatallah, B., Bestavros, A., Manolopoulos, Y., Vakali, A., Zhang, Y. (eds.) WISE 2014, Part II. LNCS, vol. 8787, pp. 231–245. Springer, Heidelberg (2014)

Twitter Sarcasm Detection Exploiting a Context-Based Model

Zelin Wang[1,2]([✉]), Zhijian Wu[1,2], Ruimin Wang[3], and Yafeng Ren[2]

[1] State Key Laboratory of Software Engineering, Wuhan University, Wuhan, China
[2] School of Computer, Wuhan University, Wuhan, China
whwzl@whu.edu.cn
[3] International School of Software, Wuhan University,
Bayi Road, Wuhan 430072, China

Abstract. Automatically detecting sarcasm in twitter is a challenging task because sarcasm transforms the polarity of an apparently positive or negative utterance into its opposite. Previous work focus on feature modeling of the single tweet, which limit the performance of the task. These methods did not leverage contextual information regarding the author or the tweet to improve the performance of sarcasm detection. However, tweets are filtered through streams of posts, so that a wider context, e.g. a conversation or topic, is always available. In this paper, we compared sarcastic utterances in twitter to utterances that express positive or negative attitudes without sarcasm. The sarcasm detection problem is modeled as a sequential classification task over a tweet and his contextual information. A Markovian formulation of the Support Vector Machine discriminative model as embodied by the SVM^{hmm} algorithm has been employed to assign the category label to entire sequence. Experimental results show that sequential classification effectively embodied evidence about the context information and is able to reach a relative increment in detection performance.

Keywords: Sarcasm detection · Sentiment classification · Support vector machine · Sequential classification

1 Introduction

Sentiment analysis in twitter has been one of the most popular research topics in NLP (Natural Language Processing) in the past decade, as shown in several recent surveys [1,2]; The goal of sentiment analysis to automatically detect the polarity of a twitter message, while sarcastic or ironic statement transforms the polarity of an apparently positive or negative utterance into it opposite. So it is very important to differentiate sarcastic utterance from the utterances that express positive or negative attitudes without sarcasm [3]. Sarcasm detection is considered to be an important aspect of language which deserves special attention given its relevance in fields such as sentiment analysis and opinion mining [4].

© Springer International Publishing Switzerland 2015
J. Wang et al. (Eds.): WISE 2015, Part I, LNCS 9418, pp. 77–91, 2015.
DOI: 10.1007/978-3-319-26190-4_6

The issue of automatic sarcasm detection in twitter has been addressed in the past few years, the sarcasm detection is usually considered as a classification problem, previous approaches mostly relied on features modeling to the single tweet. These methods did not leverage contextual information regarding the author or the tweet to improve the performance of the task. However, tweets are filtered through streams of posts, so that a wider context, e.g. a conversation or topic, is always available.

Considering the following tweet from our dataset: *"@syydsand @gretchlol this seems like a lie. do you no longer associate with yourself?"* Did the author intend this tweet sarcastic? Without additional context it is difficult to know. But if we peruse the author's conversational context which is shown in Fig. 1, we can reasonably inter that this tweet was intended sarcastically.

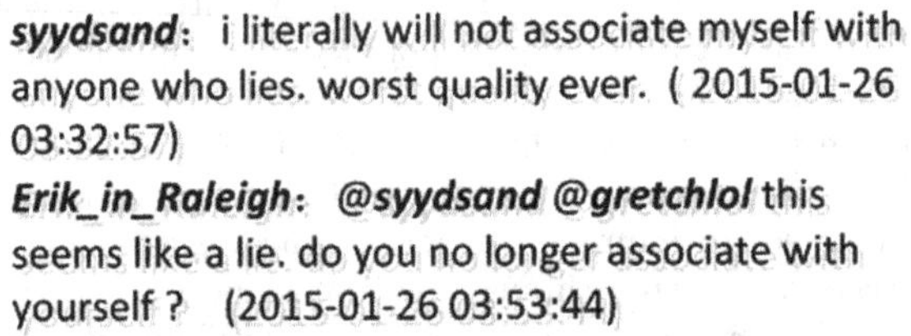

> **syydsand**: i literally will not associate myself with anyone who lies. worst quality ever. (2015-01-26 03:32:57)
>
> **Erik_in_Raleigh**: *@syydsand @gretchlol* this seems like a lie. do you no longer associate with yourself ? (2015-01-26 03:53:44)

Fig. 1. A tweet and its conversation-based context, the content in parentheses represents the posting time of this tweet

According to the example in Fig. 1, we know the contextual information is benefit to improve the detection performance. This motivated us to detect sarcasm in twitter by using different contextual information.

In this paper, we focus on message-level sarcasm detection on English Twitter using a context-based model along three lines: first, we introduce three different types of contextual information, that are *conversation*, as chains of tweets that are reply to the previous one, *posting history*, also chains of tweets that are come the same author, and *topic*, built based on the same hashtag. Then we focus on feature modeling of tweets, they will also account for contextual information. Finaly, we introduce a more complex classification model that works over an entire tweet sequence and not on one tweet at a time. From a computational perspective, a target tweet and its context are arbitrarily long sequence of messages, ordered according to time with the target tweet being the last. The SVM^{hmm} learning algorithm [5,6] has been employed, as it allow to classify a tweet within an entire sequence. While SVM based classifiers allow to recognize the category label from one specific tweet at a time, the SVM^{hmm} learning algorithm collectively labels all tweet in a sequence.

The contributions of this paper are as follows:

– To the best of our knowledge, the context-based model is proposed to identify sarcasm in twitter for the first time.

- Results show the context-based model can improve the performance for twitter sarcasm detection.
- Results show the history-based context can improve the performance of sentiment analysis in twitter.

2 Related Work

There has recently been a flurry of interesting work on sarcasm detection [7–11]. In these work, verbal irony detection has mostly been treated as a standard text classification task, some innovative approaches specific to detect irony have been proposed.

Carvalho et al. (2009) [12] created an automatic system for detecting sarcasm relying on emoticons and special punctuation, they focused on detection of ironic style in newspaper articles. Veale and Hao (2010) [13] proposed an algorithm for separating ironic from non-ironic similes, detecting common terms used in this ironic comparison. Reyes et al. (2012) [14] have recently proposed a model to detect sarcasm in Twitter, they defined four groups of features: signatures, unexpectedness, style, and emotional scenarios. Moreover, Barbieri and Saggion (2014) proposed a novel linguistically motivated set of features to detect irony in twitter, the features take into account frequency, written/spoken difference, sentiments, ambiguity, intensity, synonymy and structure, experimental results show their model achieves state-of-the-art performance.

There are also a few computational models that detect sarcasm on Twitter and Amazon [3,7,15]. Davidov et al. (2010) proposed a semi-supervised identification, they used 5-fold cross validation on their kNN-like classifiers and obtained 55 % in F-measure on the Twitter dataset. Gonzalez-Ibanez et al. (2011) experimented with Twitter data divided into three categories, they used two classifiers-support vector machine (SVM) with sequential minimal optimization (SMO) and logistic regression, they used various combinations of unigrams, dictionary-based features and pragmatic factors to achieve the better performance. The work of Riloff et al. (2013) detected one type of sarcasm: contrast between a positive sentiment and negative situation. They used a bootstrapping algorithm to acquire lists of positive sentiment phrases and negative situation phrases from sarcastic tweets.

To our knowledge, however, no previous work on sarcasm detection has designed the model which leverages contextual information regarding the author or tweet. But this is very necessary in some cases, some sarcastic utterances can not be recognized by the lack of contextual information. In this paper, we modeled the sarcasm detection problem as a sequential classification task over tweet and his contextual information (one or more tweets, representing conversation, related topic, or posting history). A Markovian formulation of the Support Vector Machine discriminative model as embodied by the SVM^{hmm} algorithm has been employed to assign the category label to entire sequences. Experimental results prove that sequential classification effectively embodied evidence about the contextual information and is able to reach a increment in F1 measure.

3   Dataset

The aim of this paper is to estimate the contribution of the context-based model, existing state-of-the-art approaches neglect the contextual information, so that the datasets with labeled contexts are not available. In this section, we will introduce the dataset used in our method.

3.1   Basic Dataset Construction

In Twitter, people post message of up to 140 characters. Apart from plain text, a tweet may contain references to other users (@user), URLs, and hashtags (#hashtag) which are tags assigned by the user to identify topic or sentiment. Previous work [7,16] also showed that human judge other than the tweets' authors, achieve low levels of accuracy when trying to classify sarcastic tweets. So we argue that using hashtags labeled by their authors of the tweets produces a better quality dataset. In other words, the best judge of whether a tweet is intended to be sarcastic is the author of the tweet. To build the dataset including negative (N), sarcastic (S), and positive (P) tweets, we used a Twitter Streaming API[1] to collect tweets that express sarcasm (#sarcasm, #sarcastic, #irony, #ironic), positive sentiment (e.g. #happy, #joy), and negative sentiment (e.g. #sadness, #angry, #frustrated), respectively. To reduce some noisy tweets, we remove the following tweets:

- we applied automatic filtering to remove retweets, duplicates, quotes, spam, tweets written in language other than English;
- we filtered all tweets where the hashtags were not located at the very end of the message[2].

Finally, we get the 1500 tweets in each of the three categories, sarcastic, positive and negative, which each category includes 500 tweets, respectively. Meanwhile, we remove the hastage which can represent the sarcastic, negative or positive, all 1500 tweets are called basic dataset.

It is worth noting that we can build a classifier to detect sarcasm in twitter based on simple or complex feature modeling. But this paper aim to apply context-based model to improve detection performance. So the classifier (not employing contextual information) built in basic dataset is used to a baseline classifier. Next, we introduce how to get the contextual information and to determine the category label of the context.

3.2   Context Generation

For a tweet, its contextual information is usually embodied by the stream of this tweet, we get the following three contextual information for each tweet in basic dataset using Twitter API:

---

[1] http://dev.twitter.com/docs/streaming-apis.

[2] To address the concern of Davidov et al. (2010) that tweets with #hashtags are noisy.

- **History-based Context**: An entire tweet sequence can be derived including the multiple tweets preceding the target tweet that are from the same author.
- **Conversation-based Context**: An entire tweet sequence can be derived including the multiple tweets preceding the target tweet that represent the interactive information with other users;
- **Topic-based Context**: An entire tweet sequence can be derived including the multiple tweets preceding the target tweet that contain the same hashtag.

After the extraction of contextual information, we obtain three types of contextual information for each tweet in basic dataset, It is worth noting that not all tweets in basic dataset have contextual information. Finally, we get the contextual information for each tweet in basic dateset, statistical information is showed in Table 1.

Table 1. Basic dataset and contextual information

Category	Basic	History	Conversation	Topic
Negative	500	2224	73	972
Sarcastic	500	2321	267	614
Positive	500	2229	113	1032
Total	1500	6774	453	2618

In Table 1, Basic represents the basic dataset, History represents the history-based context, Conversation represents the conversation-based context and Topic represents the topic-based context. The numbers of tweets are shown in columns 2, 3, 4, and 5, respectively, column 2 represents the basic dataset and column 3–5 represents different contextual information, column 2 (basic dataset) includes 1500 tweets, while column 3–5 represents the subsets of target tweets for which the history-based, conversation-based and topic-based context, respectively, was available. History-based contexts are 6774 tweets (column 3), and topic-based contexts contain 2618 tweets (column 5), while conversation-based contexts only include 453 tweets (column 4).

3.3   Dataset Annotation

To get the dataset with labeled contexts, we need to determine the category label (negative, sarcastic, positive) for each tweet from all contexts. Manual annotation is time-consuming and laborious, so we use a multi-class classifier which is trained based on basic dataset to predict the category label of the tweet in contexts. The disadvantage of this method is that it introduces noise which some contexts will be mislabeled, but it is a realistic solution to determine the category label of contextual information. Experimental results also show the sequential classification get the better performance, though there are some mislabeled tweets in the sequences.

After determining the category label of all tweets from contexts, we can devise the context-based model to detect sarcasm. In the following section, we will prove that sequential classification approach embedding contextual information can get the better performance than multi-class approach (not employing any context). It is worth noting that the dataset (basic dataset and tweets from contexts) used in the paper are automatically constructed without relying on any manually coded resource.

4   The Proposed Approach

This paper proposes a context-based model that exploits the contextual information to detect sarcasm in twitter. Firstly, we formalize three different contextual information which may improve the performance of the task. Secondly, we take consideration of feature modeling of all tweets, we not only use the simple and classical method (Bag of Word, BoW), and use the feature modeling to the closely related nature of social media text. Finally, we introduce the multiple classification approach ($SVM^{multiclass}$) and the sequential classification approach (SVM^{hmm}) to detect sarcasm in twitter.

4.1   Generating Different Contexts

Based on the nature of the twitter, we use the following three types of contextual information.

Conversation-Based Context. In twitter, a target tweet may be the part of the conversation. If we can get the conversational information preceding the target tweet. We are more likely to judge the category label of target tweet with the help of conversational information. Specially, for each tweet $t_i \in \mathcal{T}$, let $r(t_i) : \mathcal{T} \to \mathcal{T}$ be a function that returns either the tweet to which t_i is a reply to, or $null$ if t_i is not a reply. Then, the conversation-based context $\Upsilon_i^{C,l}$ of tweet t_i is the sequence of tweet iteratively built by applying function $r(t_i)$, until l tweets have been selected or $r(t_i) = null$, where l is the number of limiting the size of the context.

History-Based Context. The previous tweets (we called history tweets) about the author of a tweet can reflect author's attitude towards some events or people. History tweets should be useful to improve the detect performance. Specially, for target tweet t_i, an entire tweet sequence can be derived including the l tweets preceding the target tweet t_i that contain the same author. Let $t_i \in \mathcal{T}$ to be a tweet, the history-based context $\Omega_i^{H,l}$ is the sequences of tweets, l is the number of context from the posting history of the author for a target tweet t_i.

Topic-Based Context. In twitter, hashtag represents a topic which can be discussed by other users. We select the tweets with the same hastag in the time window as the third context. Specially, for a target tweet t_i, an entire tweet sequence can be derived including the l tweets preceding the target tweet t_i that contain the same hashtag set. Let $t_i \in \mathcal{T}$ be a tweet and $t(t_i) : \mathcal{T} \rightarrow \mathcal{P}(\mathcal{H})$ be a function that returns the entire hashtags set $H_i \subseteq \mathcal{H}$ observed into t_i. Then, the topical context $\Gamma_i^{T,l}$ for a tweet t_i is a sequence of the most recent l tweets t_j such that $H_i \cap H_j \neq \emptyset$, i.e. t_j and t_i share at least one hashtag, and t_j has been posted before t_i.

For different contexts, a specific context size l can be imposed by focusing only on the last l tweets of the sequences. According to the above method, we get the three different types of contexts.

4.2 Feature Engineering

For a tweet, different approaches of feature modeling have been used in many work [6,17]. This paper aims at applying the context-based model to detect sarcasm in twitter, we use the following two types of feature modeling methods to represent a tweet, respectively.

Bag of Word. The bag of word (BoW) is the simple method which describes the lexical overlap tweets, thus represented as vectors, whose dimensions corresponding to the different words. Components denote the presence or not of the corresponding word in twitter. Even if it is simple, the BoW model is one of the most informative representations in sentiment analysis and text classification [18].

Word Cluster. The disadvantage of Bow is the sparsity of the word space. Meanwhile, twitter message belongs to social media so that there are many nonstandard word in twitter message, e.g. *be4* (before), *2gether* (together) and *loveee* (love). These nonstandard words make the space more sparse. So we presented another word representations based on word clusters to explore shallow semantic meanings and reduced the sparsity of the word space.

Owoputi et al. (2013) obtained hierarchical word clusters via Brown clustering [19,20] on a large set of unlabeled tweets[3]. The algorithm partitions the words into a base set of 1,000 clusters, and induces a hierarchy among those 1,000 clusters with a series of greedy agglomerative merges that heuristically optimize the likelihood of a hidden Markov model with a one-class-per-lexical type constraint. In their word cluster, many variants of standard words are considered as the same class or closed-class, including pronouns (u = "you") and prepositions (be4 = "before"). if we use this word cluster to represent a tweet, we can get only 1000 dimensions vectors. This word clusters provided by CMU pos-tagging tool[4] were used to represent a tweet. For each tweet we recorded the number of words from each cluster, resulting in 1000 features.

---

[3] This method is found from Liang (2005), https://github.com/percyliang/brown-cluster.

[4] http://www.ark.cs.cmu.edu/TweetNLP/.

4.3 Modeling Sarcasm Detection as a Sequential Classification Problem

For a tweet and its contexts, once different feature representations are available, a sequential classification approach, based on the SVM^{hmm} [5] will be introduced, as an explicit account of different contexts. To prove the effectiveness of sequential classification approach, we first discuss a multi-classification schema (named $SVM^{multiclass}$) proposed in [21], as the baseline in this paper.

The Multi-class Approach. The $SVM^{multiclass}$ schema [21] is applied to implicitly compare all category labels and select the most likely one, using the multi-class formulation described in [22]. The algorithm acquires a specific function $f_y(x)$ for each category label $y \in \mathcal{Y}$, where $\mathcal{Y} = \{negative, sarcastic, positive\}$. Given a feature vector $x \in \mathcal{X}$ representing a tweet t_i, $SVM^{multiclass}$ allows to predict a specific category label $y^* \in \mathcal{Y}$ by applying the discriminant function $y^* = argmax_{y \in \mathcal{Y}} f_y(x)$, where $f_y(x) = w_y * x$ is a linear classifier associated to each category label y. Given a training set $(x_1, y_1) \ldots (x_n, y_n)$, the learning algorithm determines each classifier parameters w_y by solving the following optimization problem:

$$min \quad \frac{1}{2} \sum_{i=1\ldots k} ||w_i||^2 + \frac{C}{n} \sum_{i=1\ldots n} \eta_i$$
$$s.t. \forall i, \forall y \in \mathcal{Y} : x_i \cdot w_{y_i} \geq x_i \cdot w_y$$
$$+ 100 \Delta(y_i, y) - \eta_i$$

(1)

where C is a regularization parameter that trades off margin size and training error, while $\Delta(y_i, y)$ is the loss function that returns 0 if $y_i = y$, and 1 otherwise.

The Sequential Classification. The category label prediction of a target tweet can be seen as a sequential classification task over this tweet and its context, and the SVM^{hmm} algorithm can be thus applied. Given an input sequence $\mathbf{x} = (x_1 \ldots x_n) \subseteq \mathcal{X}$, where $\mathbf{x}$ is a tweet and its context, e.g. the conversation-based, topic-based or history-based context, x_i is a feature vector representing a tweet, the model predicts a label sequence $y = (y_1 \ldots y_l) \in \mathcal{Y}^+$ after learning a linear discriminant function $F : \mathcal{P}(\mathcal{X}) \times \mathcal{Y}^+ \to \mathcal{R}$ over input/output pairs. The labeling $f(\mathbf{x})$ is thus defined as: $f(\mathbf{x}) = argmax_{\mathbf{y} \in \mathcal{Y}^+} F(\mathbf{x}, \mathbf{y}, \mathbf{w})$. It is obtained by maximizing F over the response variable $\mathbf{y}$, for a specific given input $\mathbf{x}$. i.e. $F(\mathbf{x}, \mathbf{y}, \mathbf{w}) = < \mathbf{w}.\phi(\mathbf{x}, \mathbf{y}) >$. As ϕ extracts meaningful properties from an observation/label sequence pair $(\mathbf{x}, \mathbf{y})$, in SVM^{hmm}, it is modeled through two types of features: interactions between attributes of the observation vectors x_i and a specific label y_i (i.e. **emissions** of x_i by y_i) as well as interactions between neighboring labels y_i along the chain (**transitions**). In other words, ϕ is defined so that the complete labeling $\mathbf{y} = f(\mathbf{x})$ can be computed efficiently from F, using a Viterbi-like algorithm, according to the linear discriminant function

$$\mathbf{y}^* = argmax_{\mathbf{y} \in \mathcal{Y}+} \{ \sum_{i=1...l} [\sum_{j=1...k}$$

$$(x_i \cdot w_{y_{i-j}}) + \phi_{tr}(y_{i-j}, ..., y_i) \cdot w_{tr}] \} \tag{2}$$

In the training phase, giving training examples $(\mathbf{x}^1, \mathbf{y}^1)...(\mathbf{x}^n, \mathbf{y}^n)$ of sequence of feature vectors $\mathbf{x}^j$ with their correct tag sequences $\mathbf{y}^j$, SVM^{hmm} solves the following optimization problem

$$min \quad \frac{1}{2} \sum_{i=1...k} ||w_i||^2 + \frac{C}{n} \sum_{i=1...n} \eta_i$$

$$s.t. \quad \forall y, n : \{ \sum_{i=1...l} (x_i^n \cdot w_{y_i^n}) + \phi_t r(y_{i-1}^n, y_i^n)$$

$$\cdot w_{tr} \} \geq \{ \sum_{i=1...l} (x_i^n \cdot w_{y_i^n}) + \phi_t r(y_{i-1}^n, y_i^n) \tag{3}$$

$$\cdot w_{tr} \} + \triangle(y^n, y)$$

where $\triangle(y^n, y)$ is the loss function, computed as the number of misclassified labels in the sequence, $(x_i \cdot w_{y_i})$ represents the emissions and $\phi_{tr}(y_{i-1}, y_i)$ represents the transitions, Indeed, through SVM^{hmm} learning the category label for the target tweet is made dependent on its context. The markovian setting thus acquires pattern across tweet sequences to recognize the category label even for truly ambiguous tweets.

5   Experiments

The aim of this paper is to estimate the contribution of the proposed model in performance based on different scenarios, whereas different contexts (e.g. conversation) are possibly made available or just singleton tweet, with no context, are targeted.

5.1   Experimental Setup

A first experiment has been run to validate the effectiveness of contextual information over tweets. Based on basic dateset and contextual information, the different settings are adopted corresponding to different classification approaches:

- *multi-class*: Based on basic dataset, *multi-class* approach ($SVM^{multiclass}$) is applied, which does not require any context and can be considered as a baseline.
- *conversation*: Based on basic dataset and conversation-based context, *conversation* refers to the sequential tagging classifier (SVM^{hmm}) observing the conversation-based context. The training and test of the classifier is here run with different context sizes (1, 3 or 5), by parameterizing l in $\varUpsilon_i^{C,l}$;

– *history*: Based on basic dataset and history-based context, *history* refers to the sequential tagging classifier (SVM^{hmm}) observing the history-based context. Different context sizes (1, 3 or 5) have been considered, by parameterizing l in $\Omega_i^{H,l}$;
– *topic*: Based on basic dataset and topic-based context, *topic* refers to the sequential tagging classifier (SVM^{hmm}) observing the topic-based context. Different context sizes (1, 3 or 5) have been considered, by parameterizing l in $\Gamma_i^{T,l}$.

In our experiment, the performance evaluation is always carried out against one target tweet. We use the 10-fold cross validation to evaluate performance. Performance scores are reported in terms of precision, recall and F-measure.

5.2   Experimental Results

Based on BoW, experimental results of sarcasm detection are showed in Table 2, we can know that *multi-class* (not employing any context) can get 52.67 % in F-measure. We can get the better performance by using sequential classification approaches. When we use the history-based context, the performance of SVM^{hmm} will improve with the increment of the number of sizes, we can get 58.32 % in F-measure when l is set to 5. For the conversation-based context, the performance will be improved 2 % when l is set to 1, 3, or 5. This tells us that the conversation-based context is very effective and stable. If we use the topic-based context, the proposed approach will experience a performance drop in F-measure, but the precision have a big improvement, we will discuss it later.

Table 2. Evaluation results using BoW

Methods	Precision(%)	Recall(%)	F-Value(%)
multi-class	50.16	55.38	52.67
history-1	46.29	64.08	53.75
history-3	45.01	69.64	54.68
history-5	51.46	67.27	**58.32**
conversation-1	42.07	80.14	55.15
conversation-3	41.14	84.97	55.34
conversation-5	40.93	84.86	55.22
topic-1	59.67	39.62	47.62
topic-3	65.79	17.86	28.10
topic-5	62.16	38.74	47.73

Based on word cluster, experimental results of sarcasm detection are showed in Table 3, we can know that *multi-class* (not employing any context) can get 54.54 % in F-measure, there is 2 % improvement than *multi-class* based on Bow.

This tells us that word cluster is very effective. Meanwhile, We can get the better performance by using sequential classification approaches. When we use the history-based context, the performance of SVM^{hmm} will improve with the increment of the number of sizes, we can get 60.32 % in F-measure when l is set to 5. For the conversation-based context, the performance will be improved 2 % when l is set to 1, 3, or 5. This tells us that the conversation-based context is very effective and stable. Like the Bow, the topic-based context can not get the better performance in F-measure, but get the high precision.

Compared with the current best system **CURRENT** [23] in which use a complex set of linguistically motivated, easy-to-computer features from the single tweet, context-based model outperforms the current system. The main reason is that our model uses the contextual information which is very useful to detect sarcasm in twitter.

Table 3. Evaluation results using word cluster

Methods	Precision(%)	Recall(%)	F-Value(%)
multi-class	51.85	58.20	54.54
history-1	47.94	65.00	55.18
history-3	46.15	72.80	55.96
history-5	53.68	70.40	**60.32**
conversation-1	43.12	83.00	56.72
conversation-3	42.08	87.20	56.74
conversation-5	41.60	87.80	56.42
topic-1	64.05	38.00	47.56
topic-3	67.90	15.60	25.20
topic-5	62.60	39.80	48.50
CURRENT	52.37	58.63	55.31

5.3   Experimental Analysis

To analyze the impact of different context, we compute the precision of the sequences in which the length is greater than 2. In other words, we only care to the target tweet which have context. This paper aims at detecting sarcasm in twitter, so we analyze the target tweet (500 tweets) which its category label is sarcastic in basic dataset. l is set to 5, their related information are shown in Table 4.

In Table 4, NUMBER represents the number of the context, SEQUENCE represents the number of the sequences which are accurately predicted, TAR-GET represents the number of sequences in which the target tweet is accurately predicted, P1 represents the proportion of S in N, and P2 is the proportion of T in N.

Table 4. Evaluation results of the sarcastic tweets including the contexts

Context	NUMBER	SEQUENCE	TARGET	P1(%)	P2(%)
history	498	66	355	13 %	71 %
conversation	172	92	172	53 %	100 %
topic	131	2	35	2 %	27 %

For a sarcastic tweet t_i with contextual information. Based on Table 4, we can know that the tweet t_i can be predicted to sarcastic with 71 % probability if this tweet has history-based context. The tweet t_i will be predicted as sarcastic with 100 % probability if this tweet has conversation-based context. The tweet t_i will be predicted as sarcastic with 27 % probability if this tweet has topic-based context. Based on these analysis, this can explain the reason that topic-based context can not improve the detection performance. Meanwhile, we can know that the conversation-based context is the best effective to detect sarcasm in twitter. In our previous experiment, the reason that the performance from history-based context is better than the conversation-based context dues to because the number of sarcastic tweets having history-based context is 498 in all 500 tweets, but the number of sarcastic tweets having conversation-based context is only 172 in all 500 tweets.

5.4 Parameter Sensibility

In Tables 2 and 3, the history-based context can improve the detection performance. With the increment of l, the performance will be improved, so we need find the best l for this type of context, based on this type of context, we experiment the performance of the proposed approach about l from 1 to 20. The result shows in Fig. 2, we can know that our model can get the best performance when l is set to about 5.

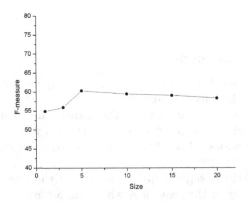

Fig. 2. The performance of the model on different context sizes

5.5 Experimental Results About Sentiment Analysis in Twitter

In our context-based model, the history-based context can get the best performance for sarcasm detection in twitter. For sentiment analysis in twitter, previous work [6] has not been developed the model to exploit the history-based context. In this section, we discuss the impact of the proposed model to the performance of sentiment analysis in twitter.

Table 5. Evaluation results of sentiment analysis in twitter

Methods	Negative			Positive			Macro-F(%)
	P(%)	R(%)	F(%)	P(%)	R(%)	F(%)	
multi-class	67.34	69.20	68.28	68.68	66.40	67.33	67.70
history-1	68.68	72.60	70.34	70.57	64.40	67.00	68.67
history-3	68.77	73.20	70.28	70.84	64.60	67.45	**69.13**
history-5	69.99	69.60	69.56	68.88	67.40	67.90	68.73

We delete all tweets in basic dataset which its category label is sarcastic, meanwhile, we delete all contextual tweets for sarcastic tweets. All other settings are same to sarcasm detection. There is a two classification problem (negative and positive) of sentiment analysis in twitter. Based on feature modeling of word cluster, experimental results are showed in Table 5. The *multi-class* approach (not employing any context) can get 67.70 % in Macro-F. The SVM^{hmm} can get the better performance (69.13 %) than *multi-class* approach. Results show that the history-based context can improve the performance of sentiment analysis in twitter.

6 Conclusion

In this paper, the role of contextual information in sarcasm detection over Twitter is investigated. We modeled the sarcasm detection problem as a sequential classification task over target tweet and its context. A Markovian formulation of the Support Vector Machine discriminative model as embodied by the SVM^{hmm} algorithm has been employed to assign the category label to entire sequence. Results show that sequential classification effectively embodied evidence about the contextual information and is able to reach a relative increment in detection performance. It is worth noting that our proposed approach does not require manually coded resources.

References

1. Liu, B., Zhang, L.: A survey of opinion mining and sentiment analysis. In: Aggarwal, C.C., Zhai, C. (eds.) Mining Text Data, pp. 415–463. Springer, New York (2012)

2. Tsytsarau, M., Palpanas, T.: Survey on mining subjective data on the Web. Data Min. Knowl. Discov. **24**(3), 478–514 (2012)
3. Davidov, D., Tsur, O., Rappoport, A.: Semi-supervised recognition of sarcastic sentences in Twitter and Amazon. In: Processdings of the Fourteenth Conference on Computational Natural Language Learning (CoNLL 2010), Uppsala, Sweden, pp. 107–116 (2010)
4. Pang, B., Lee, L.: Opinion mining and sentiment analysis. Found. Trends Inf. Retireval **2**(1–2), 1–135 (2008)
5. Altun, Y., Tsochantaridis, I., Hofmann, T.: Hidden Markov support vector machines. In: Processdings of the International Conference on Machine Learning (ICML 2003), Washington, pp. 3–10 (2003)
6. Vanzo, A., Crose, D., Basili, R.: A context-based model for sentiment analysis in Twitter. In: Processdings of the 25th International Conference on Computational Linguistics: Technical Papers (COLING 2014), Dublin, Ireland, pp. 2345–2354 (2014)
7. Gonzalez-Ibanez, R., Muresan, S., Wacholder, N.: Identifying sarcasm in Twitter: a closer look. In: Proceedings of the 49th Annual Meeting of the Association for Computational Linguistics: Human Language Technologies (ACL-HLT 2011), Portland, Oregon, pp. 581–586 (2011)
8. Filatova, E.: Irony and Sarcasm: corpus generation and analysis using crowdsourcing. In: Language Resources and Evaluation, pp. 392–398 (2012)
9. Burfoot, C., Baldwin, T.: Automatic satire detection: are you having a laugh? In: Joint Conference of the 47th Annual Meeting of the Association for Computational Linguistics and the 4th International Joint Conference on Natural Language Processing of the Asian Federation of Natural Language Processing (ACL-IJCNLP 2009), Singapore, pp. 161–164 (2009)
10. Tepperman, J., Traum, D., Narayanan, S.: "Yeah Right": Sarcasm recognitio for spoken dialogue Systems. In: Proceedings of the 9th International Conference on Spoken Language Processing, Antwerp, Belgium (2006)
11. Tsur, O., Davidov, D., Rappoport, A.: Semi-supervised recognition of sarcastic sentences in online product reviews. In: AAAI Conference on Weblogs and Social Media, Atlanta, Georgia, pp. 107–116 (2010)
12. Carvalho, P., Sarmento, L., Silva, M.J., de Oliveira, E.: Clues for detecting irony in user-generated contents: oh...!! it's so easy; -). In: Proceedings of the 1st CIKM Workshop on Topic-Sentiment Analysis for Mass Opinion, New York, NY, pp. 53–56 (2009)
13. Veale, T., Hao, Y.: Detecting ironic intent in creative comparisons. In: Proceedings of the 19th Europen Conference on Artificial Intelligence (ECAI 2010), The Netherlands, Amsterdam, pp. 765–770 (2010)
14. Reyes, A., Rosso, P., Veale, T.: A multidimensional approach for detecting irony in Twitter. Lang. Resour. Eval. **47**(1), 239–368 (2012)
15. Riloff, E., Qadir, A., Surve, P., Silva, L.D., Gilbert, N., Huang, R.: Sarcasm as contrast between a positive sentiment and negative situation. In: Proceedings of the 2013 Conference on Empirical Methods in Natural Language Processing (EMNLP 2013), Seattle, USA, pp. 704–714 (2013)
16. Wallace, B.C., Choe, D.K., Kertz, L., Charniak, E.: Humans require context to infer ironic intent (so computers probably do, too). In: Processdings of the 52nd Annual Meeting of the Association for Computational Linguistics (ACL 2014), Baltimore, Maryland, USA, pp. 512–516 (2014)

17. Nakov, P., Rosenthal, S., Kozareva, Z., Stoyanov, V., Ritter, A., Wilson, T.: Semeval-2013 task 2: sentiment analysis in Twitter. In: Second Joint Conference on Lexical and Computational Semantics (*SEM), Proceedings of the Seventh International Workshop on Semantic Evaluation(SemEval 2013), Atlanta, Georgia, USA, vol. 2, pp. 312–330 (2013)
18. Pang, B., Lee, L., Vaithyanathan, S.: Thumbs up? Sentiment classification using machine learning techniques. In: Preceedings of the Conference on Empirical Methods in Natural Language Processing (EMNLP 2002), Philadelphia, pp. 79–86 (2002)
19. Owoputi, O., O'Conor, B., Dyer, C., Gimpel, K., Schneider, N., Smith, N.A.: Improving part-of-speech tagging for online conversational text with word clusters. In: The 2013 Conference of the North American Chapter of the Association for Computational Linguistics: Human Language Technologies (NAACL 2013), Atlanta, pp. 380–390 (2013)
20. Brown, P.F., deSouza, P.V., Mercer, R.L., Della Pietra, V.J., Lai, J.C.: Class-based n-gram models of natural language. Comput. Linguist. $18(4)$, 467–479 (1992)
21. Joachims, T., Finley, T., Chun-Nam, Y.: Cutting-plane training of structural SVMs. Mach. Learn. $77(1)$, 27–59 (2009)
22. Crammer, K., Singer, Y.: On the algorithmic implementation of multiclass kernel-vector machines. J. Mach. Learn. Res. 2, 265–292 (2001)
23. Barbieri, F., Saggion, H.: Modelling irony in Twitter. In: Proceedings of the Student Research Workshop at the 14th Conference of the Europen Chapter of the Association for Computational Linguistics, Gothenburg, Sweden, pp. 56–64 (2014)

Ranked Reverse Boolean Spatial Keyword Nearest Neighbors Search

Hailin Fang[1], Pengpeng Zhao[1]([✉]), Victor S. Sheng[2], Zhixu Li[1], Jiajie Xu[1], Jian Wu[1], and Zhiming Cui[1]

[1] School of Computer Science and Technology, Soochow University, Suzhou 215006, People's Republic of China
{hlfang,ppzhao,zhixuli,xujj,jianwu,szzmcui}@suda.edu.cn
[2] Computer Science Department, University of Central Arkansas, Conway, USA
ssheng@uca.edu

Abstract. Recently, Reverse k Nearest Neighbors (RkNN) queries, returning every answer for which the query is one of its k nearest neighbors, have been extensively studied on the database research community. But the RkNN query cannot retrieve spatio-textual objects which are described by their spatial location and a set of keywords. Therefore, researchers proposed a RSTkNN query to find these objects, taking both spatial and textual similarity into consideration. However, the RSTkNN query cannot control the size of answer set and to be sorted according to the degree of influence on the query. In this paper, we propose a new problem Ranked Reverse Boolean Spatial Keyword Nearest Neighbors query called Ranked-RBSKNN query, which considers both spatial similarity and textual relevance, and returns t answers with most degree of influence. We propose a separate index and a hybrid index to process such queries efficiently. Experimental results on different real-world and synthetic datasets show that our approaches achieve better performance.

Keywords: Reverse k Nearest Neighbor · Spatial keyword search · Ranking

1 Introduction

In recent years, a Reverse k Nearest Neighbors (RkNN) [4,23] query has attracted great attention in the database research community. Then it has been extensively applied in business, such as marketing, decision support, data mining and resource allocation. RkNN returns a set of answer data points for which a query data point is their k nearest neighbors. However, the shortcoming of the RkNN query is that it is unable to control the size of answers returned. Then answer set may be none or a larger number of answer objects, since the returned objects are not sorted. Lee et al. [11] proposed Ranked Reverse Nearest Neighbor Search called RRNN query to solve the problem. RRNN query can retrieve t most influence data points according to the influence degree of each point to the query.

© Springer International Publishing Switzerland 2015
J. Wang et al. (Eds.): WISE 2015, Part I, LNCS 9418, pp. 92–107, 2015.
DOI: 10.1007/978-3-319-26190-4_7

But with the rapid development of mobile internet, a great amount of spatio-textual objects are generated everyday by mobile devices. All of these objects contain both location and text descriptions. However, RRNN methods only consider the spatial similarity, they cannot retrieve the objects having both spatial and textual similarity. Lu et al. [14] proposed a RSTkNN query, which returns an object containing spatial location and text description. However, RSTkNN query could not control the size of the answer set and could not determine the answers influence degree of each answer to the query. In this paper, we first define a new problem called Ranked Reverse Boolean Spatial Keyword Nearest Neighbors (Ranked-RBSKNN) query, which considers both spatial similarity and textual relevance, and returns the top t most affected objects. Since in a real-world environment, spatial similarity is not sufficient to measure the influence between two objects, textual similarity also plays an important role. For instance, a manager intends to promote a product, she/he wants to find potential customers nearby according to user's preferences. In this case, it is very important that we need to return a fix number of potential customers who are most affected on a product and take into account textual and spatial similarity. For example, Fig. 1 illustrates a simple example of Ranked-RBSKNN. As shown in Fig. 1(a), both squares and circles represent spatio-textual objects in Euclidean space. The data points that contain the keywords of a query present in circles, otherwise in squares. Each object has location and text description, we aim to find two nearest potential customers who are most affected by products. In Fig. 1(a), the star shows the query location of a mall, and its text description $\{seafood, discount\}$ is shown as the last row in Fig. 1(c). The rest rows in Fig. 1(c) present the text descriptions of the data points containing the keywords of the query (i.e., the circles in Fig. 1(a)). Note that we don't enumerate square data points in Fig. 1(c). We first apply the RSTkNN to find objects, using both the location of the mall and its text description in a query. As shown in Fig. 1(b), when we set $k = 1$ no answers returns. When we set $k = 2$, only one data point p_7 is returned. This shows that the RSTkNN cannot control the size of the answer set. However, we sometimes need to return a fix number of answers. Thus, we need to solve the Ranked-RBSKNN query problem.

In this paper, we propose a new method for this problem. The basic straight forward method could use RSTkNN iteratively to return top t answers by increasing the query parameter k, which starts from 1. It is obvious that its efficiency is very poor. In order to improve the query processing efficiency, we first propose a separate index structure called SIS. SIS first uses an inverted index to obtain candidates, and then uses an R-tree index to compute the influence degree of candidates. The inverted index is used to filter out invalid objects that can not satisfy the text constraint of a query and add valid objects to a candidate set. The R-tree index is used to compute influence degree of each candidate with its corresponding query, and then rank these candidates according to their degree of influence. Then we design another new method, a hybrid index structure. The hybrid index structure utilizes a signature file and the inverted index in R-tree. Specifically, it creates a signature for each node in R-tree and an additional

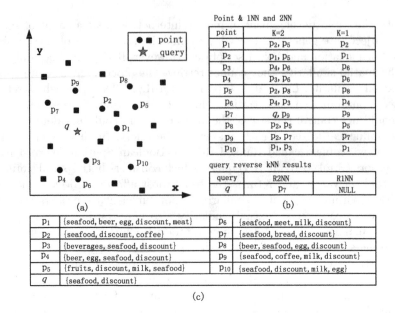

Fig. 1. An example of Ranked-RBSKNN

integrated inverted index for leaf nodes. To simplify our description since then, we denote this tree as InvSR-tree. The InvSR-tree query can use signatures and inverted indexs to filter a great amount of invalid objects and obtained a few candidates, so that it can improve the query efficiency tremendously. Our theoretical and experimental results on real-world and synthetic datasets show the efficiency of our proposed two methods. To summarize, we have made following contributions in this paper.

First, we define a novel query problem, a Ranked Reverse Boolean Spatial Keyword Nearest Neighbors Query called Ranked-RSTNN, which returns top t most affected answers in terms of both spatial similarity and text relevance.

Second, we propose two novel solutions (i.e., SIS and InvSR-tree) to solve the Ranked-RBSKNN query efficiently. In the two solutions, both the textual relevance and the spatial similarity are taken into consideration in a query processing, and answers are returned according to their influence degree on the query. In this paper, half-planes are used to determine which point has the greatest influence on a query. Besides, the degree of influences is used in our solutions, instead of the spatial distance similarity.

Third, extensive experiments are conducted on different real-world and synthetic datasets to evaluate the efficiency of our proposed methods.

The rest of this paper is organized as follows. Section 2 defines some basic concepts. Sections 3 and 4 elaborate our proposed approaches. Section 5 conducts the evaluation on different data sets and analyze of our experimental results. Section 6 reviews the related work. Finally we make a conclusion in Sect. 7.

2   Problem Statement

In this section, we first define some basic concepts used in this paper. The RkNN [4,23] query returns a set of data points for which the query q is one of its k-nearest neighbors among the data set. But RkNN cannot retrieve the spatio-textual objects. Therefore, Lu et al. [14] first proposed a RSTkNN query to retrieve the spatio-textual objects.

Definition 1. RSTkNN query [14]: Given a data set P, a query point q, the RSTkNN query returns a set of objects for which q is one of its k most similar objects among the data set P, i.e., $RSTkNN(q, k, P) = \{p | p \in P \land q \in STkNN(p, k, P)\}$

Definition 2. Spatio-Textual Object: A spatio-textual object is normally expressed by a point with its location information and text description denoted as $q = \{q.loc, q.term\}$, which is described in a two-dimension space.

From Definition 1, we can find that the RSTkNN query cannot determine the number of objects to be returned and the answers are not sorted. Thus, we define a novel query Ranked-RBSKNN query, i.e., the Ranked-RBSKNN query. Ranked-RBSKNN query can find objects that have the query q as one of the k most similar neighbors among all objects in database, where the similarity metric combines the spatial and textual similarity and the answers are sorted. The spatial degree of influence id defined in Definition 3 to quantify the influence of a query q on a spatio-textual object p.

Definition 3. Degree of Influence: Given spatio-textual object dataset P and a query q, the ϕ_p is denotes the degree of influence of the q on a object p, $p \in P$, $q.term \in p.term$. The degree of influence $\phi_p = |R|$, $R \subseteq P$, where $|R|$ is cardinality of R. $\forall p_i \in R$, we have $dist(p_i, q) \leq dist(p, q)$ and $q.term \in p_i.term$.

Definition 4. Ranked Reverse Boolean Spatial keyword NNs query (Ranked-RBSKNN query): Given a dataset P where all objects are spatio-textual objects, a query point q and a parameter t denote the number of answers. Ranked-RBSKNN query returns a set S, $|S| = t$, $S \subseteq P$, such that $\forall p \in S$, $\forall x \in (P - S)$, $\phi_p < \phi_x$.

3   The Separate Indexing Structure

To the best our knowledge, none of existing methods can directly support the Ranked-RBSKNN query. We can propose a straightforward method for processing the RSTkNN query, called k-iterative. The k-iterative is to iteratively invoke the RSTkNN method by increasing the query parameter k from 1 until t most influenced answers are obtained. Specifically, given a query q shown in Fig. 1(a) and its text description shown as the last row in Fig. 1(c), our aim is to find two objects containing query keywords and most affected on the query. First, we execute the RST1NN, check all objects containing query keywords and compute its degree of influence. If ϕ_p of the object p has the minimum score among all objects in the dataset and contains all query keywords, p will be returned as an answer. After running RST1NN, we only obtain one answer. However, our aim is to find two answers. Therefore, we increase the parameter k to 2 and execute RST2NN query. It should be noticed that the answers of the RST2NN query subsumes that the RST1NN query in the previous run. This process repeats with incrementing k at each iteration until the size of the returned answer set is equal to t. After that we sort the returned objects according the degree of influences on the query. As we know, RSTkNN computing cost is very expensive, thus the effective of k-iterative is very poor. It needs a new approach to solve this problem.

In this section, we put forward a new separate indexing structure called SIS, which uses the inverted index and R-tree separately to solve the Ranked-RBSKNN query taking both the spatial similarity and the text relevance into consideration. The SIS algorithm first uses the inverted index to find all the objects whose text descriptions contain the keywords of the query, and adds these objects into candidate set. Then, we use the R-tree index to compute the degree of influence of these candidates on spatial dimension. In this paper, inspired by paper [11,18] we use half-planes to determine whose ϕ_p need updating. For instance, as shown in Fig. 2 given a query q and objects $p_1, p_2, ..., p_n$, a perpendicular bisector $\perp_{(p,q)}$ between p and q divides the space into two half-planes. Let $H_q(p,q)$ denote the half-plane that contains q, and let $H_p(p,q)$ denote the half-plane that contains point p. Every point inside the half-plane $H_q(p,q)$ must be closer to q than to p and the ϕ_p of these points need increase by one. In other words, q prunes every point that lies in $H_p(p,q)$.

Figure 2 shows the process of using half-planes to prune the invalid objects and compute the degree of influence on the query. There are four data points and three index nodes in the plane and their text descriptions as shown in Fig. 1(c). The keywords set of the query is $q.term = \{beer, egg\}$ and the index nodes include $N_1, N_2, and\ N_3$. Assume that only N_2 meets the text constraint of the query. The keywords of all points is shown in Fig. 1(c), we can see that only two points p_1, p_4, and the index node N_2 meet the contain all keywords of query. The next step is to compute the influence degree ϕ_p of the candidates. First, we use the perpendicular bisector $\perp_{(p_1,q)}$ to divide the plane into two half-planes and we can find that the distance between N_2 and p_1 is shorter than between N_2 and q. So the impact of p_1 is larger than q and the ϕ_p of N_2 needs to increment 1. The point p_4 also satisfies the text constraint, but p_4 is in $H_q(p_1,q)$ half-plane, p_4 is much closer to the query location. In other words, the impact of p_4 on the query is larger than that of p_1. Therefore, the ϕ_p of p_4 doesn't need to increment 1. In summary, the ϕ_p of p_1, p_4, N_2 are updated to 1, 1, and 2 respectively. Second, we use the perpendicular bisector $\perp_{(p_4,q)}$ and its two half-planes $H_q(p_4,q), H_{p_4}(p_4,q)$ to update the ϕ_p of p_1, p_4, N_2 to 1, 1, and 2 respectively. The ϕ_p of p_4 is the same, so the distance between p_1 and q is less than that between p_4 and q (i.e., $dist(p_1,q) < dist(p_4,q)$), the point p_1 is

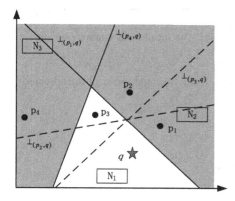
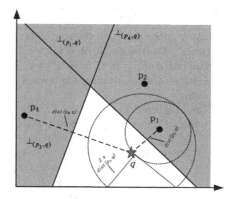

Fig. 2. Perpendicular bisector   **Fig. 3.** Example of check point

the most influence answer on the query and is returned. In the above example, the algorithm needs to examine all objects which may cause a large processing overhead. But we can find if the distance of each unexamined object p' to the query q (denoted as $dist(p', q)$) is greater than twice of the distance between p and q, then the ϕ_p of a data point p will be finalized. Because p' cannot be closer than q to p. Figure 3 reflects that the distance from the data point p_2 to the query or from p_4 to the query, either of which is greater than $2 \times dist(p_1, q)$. We can see that only p_1 is closer than other points, so the checked could be finalized. The above statement illustrates that neither of these two distances are closer than the distance from q to p_1.

The SIS query is shown in the Algorithm 1, to begin with we use the inverted index technique to prune the entire text dimension first and to find all the objects that contain all query keywords. After that, we use the obtained points in the candidate set and exploit the R-tree to compute the ϕ_p of the each candidate. Firstly, the priority queue, a candidate set, a finalized candidate set and a half-plane set denoted as P, C, F, H respectively are initialized at line 1. We will finalize the checking ϕ_p of each point in lines 6–8 in advance and get one answer point into the finalized set F. Each elements from the queue could be a node or a point. If it is a node, we will update ϕ_p of it and add it to the queue at lines 9–15. If it is a point, we will check whether it is in the $InvL$ set. If it is in this set then we will create half-planes and update other ϕ_p for others in the queue P and the candidate set C at lines 16–18. The algorithm will terminate when the priority queue is empty or t answers are obtained.

4   The Hybrid Indexing Structure

The SIS method can reduce the size of candidates by using the inverted index, which helps prune objects that do not contain the query keywords. After that,

Algorithm 1. $SIS\_Query(q, t)$

Input: a query point(q), the root index of P$(root)$, the number of answer objects(t).

Output: t Ranked result objects.

```
 1: Initialize :P := ∅, C := ∅, F := ∅, H := ∅;
 2: InvL ← getAllCandidate (q.term, InvFile); //use inverted index obtained candi-
    dates
 3: Enqueue(root, 1) to P
 4: while  P is not empty And t > 0 do
 5:     (ε, φ_p) ← dequeue(P);
 6:     for  each (p, φ_p) ∈ C do
 7:         if (dist(p, q) ≤ mindist(q, ε)/2) then
 8:             C ← C − {(p, φ_p)};   F ← F ∪ {(p, φ_p)};
 9:     if ε is an index node then
10:         for  each c ∈ ε'children do
11:             φ_c ← 1;
12:         for  each h ∈ H do
13:             if c inside h then
14:                 φ_c ← φ_c + 1
15:         Enqueue(c, φ_c)toP
16:     else if ε is a point then
17:         if ε is contain in InvL  then
18:             algorithm 2 23 − 33 lines;
        return
```

---

we only need to compute the ϕ_p of objects that satisfy the constraint of the text description. The SIS method not only enhances the pruning ability on text dimension but also reduce the number of candidates to computing degree of influences on spatial similarity. It means that SIS can improve the performance significantly and the efficiency is much higher than the k-iterative. However, the SIS algorithm has to examine all objects that contain all keywords. When there are a great deal of objects containing the keywords of the query and the query also has a large number of keywords, SIS won't perform very well. This is because SIS needs to retrieve a large number of objects, which results in a high computation cost on finding the intersection of the candidates for each query keyword. In this section, we come up with a new hybrid index called InvSR-tree to solve the Ranked-RBSKNN query. It can overcome the shortcoming of the SIS method. Briefly, we use a R-tree to establish the spatial index and integrate a signature into each node of the R-tree. Besides using the signature index, we also integrate the inverted index into R-tree leaf nodes. The details of InvSR-tree are as follows.

Figure 4 shows the distribution of spatio-textual objects and Fig. 5 shows the InvSR-tree index structure. We create a signature for each node of R-tree that we can quickly prune invalid objects on the text dimension. In order to improve the query efficiency, we establish an inverted index for each leaf node of R-tree. Thus, in the leaf nodes we can use the inverted index precisely to prune invalid points. This leads to retaining a small portion of the points, which meet the

Algorithm 2. $InvSRtree\_Query(q,t)$

Input: a query point(q), the root index of P$(root)$, the number of answer objects(t).
Output: t Ranked result objects.
1: **Initialize :**$P := \emptyset$, $C := \emptyset$, $F := \emptyset$, $H := \emptyset$;
2: $Enqueue(root, 1)$ to P
3: **while** P is not $empty$ And $t > 0$ **do**
4:     $(\varepsilon, \phi_p) \leftarrow dequeue(P)$;
5:     **for** each $(p, \phi_p) \in C$ **do**
6:         **if** $(dist(p, q) \leq mindist(q, \varepsilon)/2)$ **then**
7:             $C \leftarrow C - \{(p, \phi_p)\}$; $F \leftarrow F \cup \{(p, \phi_p)\}$;
8:     **if** ε is a non-leaf node **then**
9:         $boolean flag = matchText (\varepsilon.sig, q.sig)$;
10:       **if** $flag = true$ **then**
11:           **for** each $c \in \varepsilon' children$ **do**
12:               $\phi_c \leftarrow 1$;
13:               **for** each $h \in H$ **do**
14:                  **if** c $inside$ h **then** $\phi_c \leftarrow \phi_c + 1$;
15:               $Enqueue(c, \phi_c)$ to P
16:     **else if** ε is a leaf node **then**
17:         $cands = getCandsByLeafInvFile (q.term)$;
18:         **for** each $c \in cands$ **do**
19:           $\phi_c \leftarrow 1$;
20:           **for** each $h \in H$ **do**
21:              **if** c $inside$ h **then** $\phi_c \leftarrow \phi_c + 1$
22:         $Enqueue(c, \phi_c)$ to P
23:     **else if** ε is a data point **then**
24:         $H \leftarrow H \cup \{H p_\varepsilon(\varepsilon, q)\}$;
25:         **for** each (p, ϕ_p) in P **do**
26:           **if** p $inside$ $HP_\varepsilon(\varepsilon, q)$ **then** $\phi_p \leftarrow \phi_p + 1$
27:         **for** each (p, ϕ_p) in C **do**
28:           **if** p $inside$ $HP_\varepsilon(\varepsilon, q)$ **then** $\phi_p \leftarrow \phi_p + 1$;
29:         $C \leftarrow C \cup \{(e, \phi_p)\}$;
30:     **determine** $m = min(\phi_p)$ $with$ $p \in P \cup C$;
31:     **for** each $(p, \phi_p) \in F$ **do**
32:         **if** $\phi_p \leq m$ **then** $F \leftarrow F - (p, \phi_p)$;
33:         **Output** (p, ϕ_p); $t = t - 1$;
    return

---

text constraint on the query. After obtaining candidates via the inverted index filtering, we can only compute the ϕ_p of a fraction of candidates through R-tree. In our index, the query processing cost is greatly reduced.

An InvSR-tree Algorithm for Ranked-RBSKNN. The InvSR-tree algorithm for RSTkNN is shown in Algorithm 2. It considers both text relevance and spatial similarity, which simultaneously prunes on the spatial and text dimensions. It starts with P filled with the root of the R-tree index. For the objects

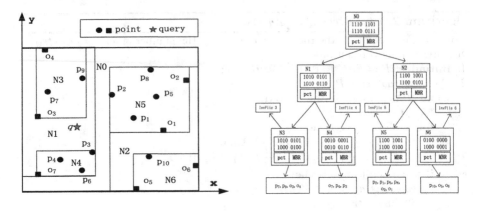

Fig. 4. R-tree of data point   **Fig. 5.** The InvSR-tree index structure

in the queue, we iteratively dequeue and handle each element in the queue separately with three possibilities, such as a non leaf node, a leaf node and a spatial object respectively in lines 8, 16, 23. Thereafter, it iteratively takes out the head element of the queue P, and examines it with the query point. If it is a non leaf node, the method $matchText\,(\varepsilon.sig, q.sig)$ will be used to check whether the node matches the query signature. If it matches, the ϕ_p of the node will be initialized and updated at lines 9–15. If the element is a leaf node, we will check whether the signature matches the query. If it matches, we will load the inverted index of the node from disk, find all points containing the query keywords and add them queue at lines 17–22. If the element is a data point, a half-plane HP will be created and preserved in H at line 24. Next, once all pending data points and index nodes in P and all data points in C fall inside this half-plane, their ϕ_p will be increased by one at lines 25–28. Finally, c will kept in C as a candidate at line 28. The algorithm will terminate when the priority queue is empty or the query parameter t is less than zero. At lines 6–7, we will finalize in advance to check whether the point has been added into the finalized set and removed from candidate set C. When the priority queue is empty or t answers are returned the query process will terminate.

5   Experimental Evaluation

In this section, we will investigate our proposed methods SIS and InvSR-tree on different datasets by comparing with baseline method k-iterative in terms of I/O cost and response time under various numbers of results and query keywords.

5.1   Setup

The experiments are conducted on three different datasets which are CD, GN, and SYN respectively. The characteristics of three dataset are summarized in

Table 1. Summary of the three Datasets

Data	Description	Total of Objects	Avg Terms per Object	Size (MB)
CD	Real data of California	102,004	10	12.7
GN	Real-life dataset of U.S	605,793	28	80.2
SYN	Synthetic dataset	372,325	69	120.8

Table 1. Let us briefly explain the three datasets. The dataset CD combines a real spatial object data at California and a real document collection from California in DBpekdia (wiki.dbpedia.org). The dataset GN is a real-life dataset obtained from the U.S. Board on geographic Names (geonames.usgs.gov), where each object is associated with its text description (a number of words) and with its geographic location coordinates. The synthetic dataset SYN consists of a number of objects. There objects are randomly chosen picture. Each picture has a location, and acts as its corresponding keywords. Our experiments are executed on a computer with 3.0 GHz CPU inter processor with 4 G RAM, running Linux. All algorithms were implemented in Java.

5.2 Experimental Results

As we said before, we evaluate our proposed approaches for the Ranked-RBSKNN query in terms of I/O cost and response time under various numbers of results and query keywords. The experimental are shown as follows.

I/O Cost: Figure 6 demonstrates the number of I/O accesses of all approaches on the three datasets CD, GN and SYN. From the Fig. 6, we can see that the I/O costs increase with the increment of query parameter t, which due to that the rise of t needs the expansion of search range in data space. As we know, I/O cost reported caused by accessing both nodes of R-tree and the disk data pages of text index files. Figure 6 also shows that our methods (i.e., SIS and InvSR-tree) perform much better than the baseline k-iterative. This is because k-iterative has to perform the RSTkNN query more than once for each spatio-textual object. Therefore it needs to repeatedly access the disk data pages. In SIS, the I/O cost consists of accessing the tree nodes and the inverted files of all the textual. Too many candidates are generated although SIS filters some object using inverted files on the textual. This results in a great I/O cost in accessing the tree nodes during computing the degree of influences on spatial dimension. The InvSR-tree also needs to access the nodes of R-tree and the inverted index of each leaf node. However, it is more efficient, since the number of nodes of R-tree accessed in InvSR-tree is much less than in SIS. Many invalid objects are pruned by signature. It only needs to read a few nodes and the inverted files of leaf nodes, which are constrained by the query keywords an location. From Fig. 6 we can see that the I/O cost of InvSR-tree is consistently than SIS.

Response Time Under Various Numbers of Results: In our experiments of evaluating the response time of the three methods with various number of results, the number of keywords in a query is fixed as three. The impact of

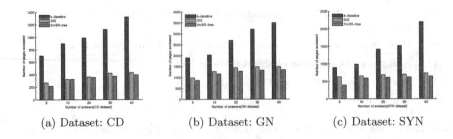

(a) Dataset: CD   (b) Dataset: GN   (c) Dataset: SYN

Fig. 6. The I/O cost on three different datasets

the number of keywords in a query will be evaluated in the next section. The response time of three methods on three datasets are shown in Fig. 7. From Fig. 7, we can see that both SIS and InvSR-tree methods have a better performance than that of the k-iterative method on three different datasets. This is because the k-iterative needs to spend more time on the query processing until returning t answers. Between the SIS and InvSR-tree, InvSR-Tree performs much better. This is because the SIS algorithm has to examine all objects that contain all keywords in a query. It needs to spend more time in the query processing and has a large number of candidates to compute their similarity scores. The reason why many candidates can not be returned as the answers is that each of them contains the keywords of query while the degree of influence on the query is very small. In contrast, the InvSR-tree uses the signatures to prune all subtrees, which don't contain all query keywords. And it only visits the relevant nodes of R-tree in ascending order of distance. In addition, we use the inverted files to prune the invalid objects of leaf nodes and reduce the number of candidates. Besides, the response time of k-iterative increases quickly with the increment of the number of results. However, the response time of both SIS and InvSR-tree only slowly increase with the number of results.

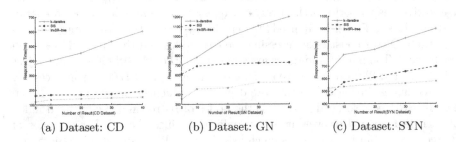

(a) Dataset: CD   (b) Dataset: GN   (c) Dataset: SYN

Fig. 7. Various numbers of results on three datasets

Response Time Under Various Numbers of Keywords: We further evaluate the response time of the three methods on three datasets, by increasing the number of keywords in a query. The experimental results are shown in Fig. 8.

From Fig. 8, we can see that InvSR-tree method performs the best, followed by SIS and k-iterative. This is because, the InvSR-tree prunes the invalid object by signatures. It uses the signature to determine whether the root node of InvSR-tree contains the query keywords. It also prunes all the subtree which dosen't match with the query signature. In addition, we can also see that with the increment of the number of query keywords, the response time of both SIS and k-iterative grows gradually. However, the response time of InvSR-tree reduces with the increment the number of query keywords. The reason is that the InvSR-tree prunes more unrelated objects through eliminating more nodes of R-tree. On the contrary, SIS and k-iterative need to check whether an object contains any keyword in the keyword list. The more keywords, the more checking time it needs. From the Fig. 8, we can see that the response time of all algorithms need more. As the number of keywords increases, they need to process more words.

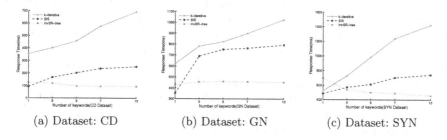

(a) Dataset: CD   (b) Dataset: GN   (c) Dataset: SYN

Fig. 8. Various numbers of keywords on three datasets.

In all, our experimental results show that our proposed two approaches (i.e., SIS and InvSR-tree) and query processing outperform the basic straight forward method k-iterative. In particular, the InvSR-tree index structure exploits the advantages of the R-tree pruning on the spatial dimensions and on the text dimensions uses the signature and inverted files pruning invalid objects. It can improves the query efficiency tremendously.

6 Related Works

The rank-aware query processing has drawn extensive attention of researchers on the database research community, such as top-k queries [1,3,8,15,24,30], reverse top-k queries [1,5,19–21,25], and reverse rank-aware queries [9–11,13, 29]. Top-k queries finding k objects with the highest scores that match queries the best has been studied extensively for past decades [1,8,15]. TA is most popular method of top-k queries and many various of them have been proposed to improve the efficiency, such as BPA and BPA2 [1]. Reverse top-k queries, retrieving all data points for which one treats a given query belongs to their top-k result set. Vlachou et al. proposed several solutions for monochromatic and bi-chromatic reverse top-k queries, namely RTA and GRTA [19]. Most recently,

Vlachou et al. further improve their solution for reverse top-k queries using branch-and-bound method without accessing each user's preferences to process reverse top-k query [20]. In [1], the author use the reverse top-k query find the most influential objects than preferred by more customers. For the first time, top-k queries from the perspective of the individual users. Reverse top-k queries have been studied mainly from the perspective of manufacturers, it is different.

Given a product, the reverse top-k query may return a number of customers who like the given product in the top-k result set. But, a few hot products may return some customers via reverse top-k query, a large proportion of products cannot find any matching customers. Inspired by this observation, researchers proposed a new kind query called Ranked-Reverse kNN (RRNN). For a given product, it retrieves t data points from dataset which has the high influence on a query, where t the number of answers number. Lee et al. [11] first proposed k-Counting and k-Browsing approaches to solve RRNN queries. Both solutions are able to deliver results progressively. Zhang et al. [29] proposed a R-kRanks query is very similar RRNN query, but the RRNN use Euclidean distance to describe the metric spatial similarity between two points, while R-kRanks uses the inner product to describe the matching degree between a query and a customer vector. However, the existing approaches on the rank-ware reverse kNN queries are not considered both spatial and textual information.

Recently, spatial keyword queries considering the spatial and text information have received significant attention (e.g., [2,6] for a comprehensive survey). For example, top-k spatial keyword queries [16,17,22,27,28,31], boolean spatial keyword queries [7,12,26], and so on. However, most of them don't consider the reverse kNN query with text information. Lu et al. [14] first defined the reverse spatial and textual k nearest neighbor query that consider both spatial similarity and textual relevance. They proposed a hybrid index tree called IUR-tree, which effectively combines spatial similarity and text relevance. The IUR-tree uses R-tree to establish the spatial index and integrate to textual vectors in each node. It computes the spatial score and the text score by R-tree and textual vectors. They present the detail of IUR-tree and index the hybrid information and proposed RSTkNN algorithms to quickly retrieves the spatio-textual objects from the dataset. But the RSTkNN algorithms cannot control the answer set size and determine the each answer degree of influence on the query. The number of answers returned is uncertain. Sometimes, it no answers or large number answers. In this paper, we first defined new problem Ranked-RBSKNN queries and propose two methods SIS and InvSR-tree to solve the problem. Our approaches consider both spatial similarity and textual relevance and return answers according to their degrees of influence on the query. Our experimental results on different datasets show the efficiency of our proposed approaches.

7 Conclusion

In this paper, we make the first defined a novel problem Ranked Reverse Boolean Spatial Keyword Nearest Neighbors query, namely the Ranked-RBSKNN query.

It ranks and retrieves the t most influence answers in term of both spatial similarity and textual relevance. We propose a separate index called SIS and a hybrid index called InvSR-tree to solve this kind of queries. In order to compare with our proposed approaches, we use the k-iterative method to retrieve answers of query. As is shown the experiments, we can see that both SIS and InvSR-tree methods have a better performance than that of the k-iterative method on three real-life and synthetic datasets. Extensive experiments on both synthetic and real-life datasets demonstrate the effectiveness of our algorithms.

As for the future works, there are two possible pieces of work. The first one is to consider top-k into the Ranked-RBSKNN query. Because the top-k query can comprehensive consideration the textual relevance and spatial similarity. The objects returned is k top-most relevant spatio-textual objects which are ranked based on a combination of both the spatial and textual similarity. For many objects which contain partial keywords of query but have high spatial similarity may be more useful to user. Compared to boolean query, more candidates will be examined in the query processing of top-k query. The second one is to take into account text relevance, spatial proximity and temporal of spatio-textual objects. Such two directions are helpful to handle large-scale data.

Acknowledgment. This work was partially supported by Chinese NSFC project (61170020, 61402311, 61440053, 61402312), and the US National Science Foundation (IIS-1115417).

References

1. Akbarinia, R., Pacitti, E., Valduriez, P.: Best position algorithms for top-k queries. In: Proceedings of the 33rd International Conference on Very Large Data Bases, pp. 495–506. VLDB Endowment (2007)
2. Cao, X., Chen, L., Cong, G., Jensen, C.S., Qu, Q., Skovsgaard, A., Wu, D., Yiu, M.L.: Spatial keyword querying. In: Atzeni, P., Cheung, D., Ram, S. (eds.) ER 2012. LNCS, vol. 7532, pp. 16–29. Springer, Heidelberg (2012)
3. Chaudhuri, S., Gravano, L.: Evaluating top-k selection queries. In: VLDB, vol. 99, pp. 397–410 (1999)
4. Cheema, M.A., Lin, X., Zhang, W., Zhang, Y.: Influence zone: efficiently processing reverse k nearest neighbors queries. In: 2011 IEEE 27th International Conference on Data Engineering (ICDE), pp. 577–588. IEEE (2011)
5. Cheema, M.A., Shen, Z., Lin, X., Zhang, W.: A unified framework for efficiently processing ranking related queries. In: EDBT, pp. 427–438 (2014)
6. Chen, L., Cong, G., Jensen, C.S., Wu, D.: Spatial keyword query processing: an experimental evaluation. Proc. VLDB Endowment **6**(3), 217–228 (2013)
7. De Felipe, I., Hristidis, V., Rishe, N.: Keyword search on spatial databases. In: IEEE 24th International Conference on Data Engineering, ICDE 2008, pp. 656–665. IEEE (2008)
8. Ilyas, I.F., Beskales, G., Soliman, M.A.: A survey of top-k query processing techniques in relational database systems. ACM Comput. Surv. (CSUR) **40**(4), 11 (2008)

9. Kriegel, H.P., Kroger, P., Renz, M., Zufle, A., Katzdobler, A.: Incremental reverse nearest neighbor ranking. In: IEEE 25th International Conference on Data Engineering, ICDE 2009, pp. 1560–1567. IEEE (2009)

10. Lee, K.C., Ye, M., Lee, W.C.: Reverse ranking query over imprecise spatial data. In: Proceedings of the 1st International Conference and Exhibition on Computing for Geospatial Research & Application, p. 17. ACM (2010)

11. Lee, K.C., Zheng, B., Lee, W.C.: Ranked reverse nearest neighbor search. IEEE Trans. Knowl. Data Eng. **20**(7), 894–910 (2008)

12. Li, G., Wang, Y., Wang, T., Feng, J.: Location-aware publish/subscribe. In: Proceedings of the 19th ACM SIGKDD International Conference on Knowledge Discovery and Data Mining, pp. 802–810. ACM (2013)

13. Lian, X., Chen, L.: Probabilistic inverse ranking queries in uncertain databases. VLDB J. Int. J. Very Large Data Bases **20**(1), 107–127 (2011)

14. Lu, J., Lu, Y., Cong, G.: Reverse spatial and textual k nearest neighbor search. In: Proceedings of the 2011 ACM SIGMOD International Conference on Management of Data, pp. 349–360. ACM (2011)

15. Marian, A., Bruno, N., Gravano, L.: Evaluating top-k queries over web-accessible databases. ACM Trans. Database Syst. (TODS) **29**(2), 319–362 (2004)

16. Rocha-Junior, J.B., Gkorgkas, O., Jonassen, S., Nørvåg, K.: Efficient processing of top-k spatial keyword queries. In: Pfoser, D., Tao, Y., Mouratidis, K., Nascimento, M.A., Mokbel, M., Shekhar, S., Huang, Y. (eds.) SSTD 2011. LNCS, vol. 6849, pp. 205–222. Springer, Heidelberg (2011)

17. Rocha-Junior, J.B., Nørvåg, K.: Top-k spatial keyword queries on road networks. In: Proceedings of the 15th International Conference on Extending Database Technology, pp. 168–179. ACM (2012)

18. Tao, Y., Papadias, D., Lian, X.: Reverse knn search in arbitrary dimensionality. In: Proceedings of the Thirtieth International Conference on Very Large Data Bases, vol. 30, pp. 744–755. VLDB Endowment (2004)

19. Vlachou, A., Doulkeridis, C., Kotidis, Y., Norvag, K.: Reverse top-k queries. In: 2010 IEEE 26th International Conference on Data Engineering (ICDE), pp. 365–376. IEEE (2010)

20. Vlachou, A., Doulkeridis, C., Nørvåg, K., Kotidis, Y.: Identifying the most influential data objects with reverse top-k queries. Proc. VLDB Endowment **3**(1–2), 364–372 (2010)

21. Vlachou, A., Doulkeridis, C., Nørvåg, K., Kotidis, Y.: Branch-and-bound algorithm for reverse top-k queries. In: Proceedings of the 2013 ACM SIGMOD International Conference on Management of Data, pp. 481–492. ACM (2013)

22. Wu, D., Yiu, M.L., Jensen, C.S., Cong, G.: Efficient continuously moving top-k spatial keyword query processing. In: 2011 IEEE 27th International Conference on Data Engineering (ICDE), pp. 541–552. IEEE (2011)

23. Yang, S., Cheema, M.A., Lin, X., Wang, W.: Reverse k nearest neighbors query processing: experiments and analysis. Proc. VLDB Endowment **8**(5), 605–616 (2015)

24. Yi, K., Yu, H., Yang, J., Xia, G., Chen, Y.: Efficient maintenance of materialized top-k views. In: Proceedings of the 19th International Conference on Data Engineering, pp. 189–200. IEEE (2003)

25. Yu, A., Agarwal, P.K., Yang, J.: Processing a large number of continuous preference top-k queries. In: Proceedings of the 2012 ACM SIGMOD International Conference on Management of Data, pp. 397–408. ACM (2012)

26. Zhang, C., Zhang, Y., Zhang, W., Lin, X.: Inverted linear quadtree: efficient top k spatial keyword search. In: 2013 IEEE 29th International Conference on Data Engineering (ICDE), pp. 901–912. IEEE (2013)
27. Zhang, D., Chan, C.Y., Tan, K.L.: Processing spatial keyword query as a top-k aggregation query. In: Proceedings of the 37th International ACM SIGIR Conference on Research & Development in Information Retrieval, pp. 355–364. ACM (2014)
28. Zhang, D., Tan, K.L., Tung, A.K.: Scalable top-k spatial keyword search. In: Proceedings of the 16th International Conference on Extending Database Technology, pp. 359–370. ACM (2013)
29. Zhang, Z., Jin, C., Kang, Q.: Reverse k-ranks query. Proc. VLDB Endowment **7**(10), 785–796 (2014)
30. Zheng, K., Fung, P.C., Zhou, X.: K-nearest neighbor search for fuzzy objects. In: Proceedings of the 2010 ACM SIGMOD International Conference on Management of Data, pp. 699–710. ACM (2010)
31. Zheng, K., Su, H., Zheng, B., Shang, S., Xu, J., Liu, J., Zhou, X.: Interactive top-k spatial keyword queries. In: 2015 IEEE 31st International Conference on Data Engineering (ICDE), pp. 423–434. IEEE (2015)

GraSS: An Efficient Method
for RDF Subgraph Matching

Xuedong Lyu[1], Xin Wang[1,4(✉)], Yuan-Fang Li[2],
Zhiyong Feng[1], and Junhu Wang[3]

[1] School of Computer Science and Technology, Tianjin University, Tianjin, China
{lyuxd,wangx,zhiyongfeng}@tju.edu.cn
[2] Faculty of Information Technology, Monash University, Melbourne, Australia
yuanfang.li@monash.edu
[3] School of Information and Communication Technology,
Griffith University, Nathan, Australia
j.wang@griffith.edu.au
[4] State Key Laboratory for Novel Software Technology,
Nanjing University, Nanjing, China

Abstract. Resource Description Framework (RDF) is a standard data model of the Semantic Web, and it has been widely adopted in various domains in recent years for data and knowledge representation. Unlike queries on relational databases, most of queries applied on RDF data are known as *graph queries*, expressed in the SPARQL language. Subgraph matching, a basic SPARQL operation, is known to be NP-complete. Coupled with the rapidly increasing volumes of RDF data, it makes efficient graph query processing a very challenging problem. This paper primarily focuses on providing an index scheme and corresponding algorithms that support the efficient solution of such queries. We present a subgraph matching query engine based on the FFD-index which is an indexing mechanism encoding a star subgraph into a bit string. A SPARQL query graph is decomposed into several star query subgraphs which can be efficiently processed benefiting from succinct FFD-index data structure. Extensive evaluation shows that our approach outperforms RDF-3X and gStore on solving subgraph matching.

Keywords: RDF · Subgraph isomorphism · Graph-based index

1 Introduction

RDF [9] is a W3C specification as a standard data model to describe machine-understandable information on the Semantic Web. An RDF dataset is a set of *triples*, each of which is of the form (s, p, o) where s is the subject, p the predicate, and o the object. An RDF dataset can be represented as a directed labeled graph. Large volumes of RDF data have been published in various fields with the development of Linked Data [1], containing over 52 billion triples as in 2012 and rapidly growing.[1]

---

[1] http://www.w3.org/wiki/SweoIG/TaskForces/CommunityProjects/
LinkingOpenData.

© Springer International Publishing Switzerland 2015
J. Wang et al. (Eds.): WISE 2015, Part I, LNCS 9418, pp. 108–122, 2015.
DOI: 10.1007/978-3-319-26190-4_8

Queries over RDF graphs can be expressed in the standard SPARQL [7] query language and its extensions, generally exhibit a much higher complexity. Subgraph matching, also known as subgraph isomorphism, is a widely known NP-Complete problem [10]. A SPARQL query often consists of multiple star structures as subqueries [8]. Hence, efficient evaluation of SPARQL queries, especially over large RDF graphs, critically depends on the efficient processing of star subqueries. We then propose **GraSS** (Converting **Gra**ph Querying into **S**tar **S**ubgraph Matching), an efficient approach to process subgraph matching over RDF data. GraSS stores and indexes sets of star subgraphs of data graphs. A query is also regarded as a combination of several star query subgraph, as shown in Fig. 1. Star query will be processed as a unit to avoid most of the join operations inside the query. The join operations are only involved on shared variables, e.g., $?var1$ in Fig. 1.

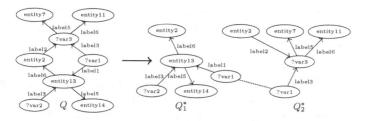

Fig. 1. Query graph Q decomposed into subgraphs Q_1^* and Q_2^* containing two shared vertex.

GraSS is based on the theoretical and experimental extensions of our previous work FFD-index [17]. FFD-index is designed to efficiently process star queries. It maintains all star subgraphs of a large RDF graph and denotes each subgraph with a bit string *fingerprint*. The main contributions of this paper can be summarized as follows:

- We decompose both data graphs and query graphs into sets of star subgraphs. Star subgraph is handled as a unit to avoid costly join operations.
- We encode a star subgraph into a fingerprint. Fingerprints preserve information about labels and directions of edges, which is used to effectively reduce the search space of subgraph matching.
- We perform extensive experiments to evaluate GraSS on both synthetic and real world datasets. Our comprehensive experimental results show that GraSS significantly outperforms the state-of-the-art RDF data management system RDF-3X [11] and gStore [21] in terms of query response time.

The rest of this paper is organized as follows. After related work is introduced in Sects. 2, 3 defines the preliminary concepts. Section 4 presents the technique of fingerprint and the FFD-index framework. Section 5 discusses the details of query processing of GraSS. Experimental results are shown in Sect. 6. Finally, Sect. 7 concludes the paper.

2 Related Work

Recently, the subgraph isomorphism problem has drawn a great deal of attention [2,4,6]. Some algorithms [5,15] have been proposed to address this problem, but they do not rely on any index scheme, thus inevitably, they cannot be applied on large-scale graphs due to the high time complexity. In order to speed up the subgraph matching over a large graph or a large amount of graphs, a number of related works turn to employ some indexing methods.

Triple-based Indexing. RDF-3X [11] takes the extensive use of B^+-trees as its basic index structures. It maintains all six possible permutations of the three items in a triple by building all 6-way indexes. Its performance depends on the query optimizer which takes advantages of statistics to generate bushy join trees. However, RDF-3X only employs the multi-way join operations to implement graph queries which may cause a high overhead of time. In other words, it focuses on the storage and indexing in terms of individual triples. Thus, large amount of join operations become an unavoidable cost. Triple-based index schemes are also employed by [12,16].

Feature-based Indexing. Discriminative subgraphs/substructures are employed to construct the feature-based indexes [3,18,19]. In gIndex [18], the authors propose "discriminative ratio" to measure the discriminative power of a small subgraph. gIndex has a better filtering performance than GraphGrep since subgraph preserves more structural information than paths. GADDI [19] indexes the NDS distance (Neighboring Discriminating Substructure distance) between pairs of neighboring vertices in the graph. Most of feature-based indexing methods depend on a costly preprocessing in which a mining technique is implemented to extract some discriminating substructures.

Structure-based Indexing. GRIN [14] uses graph partitioning and distance information to construct the index for graph queries. Its index is a balanced binary tree with each of its nodes containing a set of triples. However, GRIN keeps the index in memory which evidently puts a limit on the size of storage. A path-based technique, GraphGrep [13], enumerates all the paths shorter than a fixed max length as features of a graph. Due to splitting graph into paths, some structural information is lost, thus many false positive answers may probably be returned leading to a large candidate set.

In order to address problems aforementioned, we propose GraSS for fast subgraph matching on disk. We enhance the expressive power of *vertex signature* [20,21] and then propose the concept of *fingerprint* for star subgraph, which can describe a subgraph and be used as a filter to help to prune search space. Our method has something in common with gStore [21]. However, gStore tags each vertex with a signature and matches signatures of data vertices and query vertices one by one, while we tag graphs by calculating the fingerprint of each star subgraph, which allows matching a graph at once.

3 Preliminaries

In this paper, we restrict our discussion on *directed labeled graphs*. According to the W3C RDF standard [9], both subjects (denoted as S) and predicates (P) can be represented by URIs, while objects (O) can be represented by either URIs or literals. Let U and L denote the set of URIs and the set of literals respectively. Thus, we have $S \subseteq U$, $P \subseteq U$, and $O \subseteq U \cup L$. An RDF triple is of the form (s, p, o) where $s \in S$, $p \in P$, and $o \in O$.

Definition 1. (RDF GRAPH) *An RDF dataset $T = \{t \mid t \in S \times P \times O\}$ is a set of triples that can be modeled as a directed labeled graph $G = (V, E, \Sigma, l)$, where V is a finite set of vertices and $E \subseteq V \times V$ is a finite set of edges. $\Sigma = U \cup L$ is a set of labels. The labeling function $l : V \cup E \to \Sigma$ maps each vertex or edge to a label in Σ. $S = \{s \mid s = l(v), v \in V, \exists \langle v, u \rangle \in E\}$, $P = \{p \mid p = l(\langle v_i, v_j \rangle), \langle v_i, v_j \rangle \in E\}$, $O = \{o \mid o = l(v), v \in V, \exists \langle u, v \rangle \in E\}$.*

Definition 2. (QUERY GRAPH) *A basic graph pattern can be modeled as a directed labeled graph $Q = (V', E', \Sigma', l')$. V' is a finite set of vertices and $E' \subseteq V' \times V'$ is a finite set of edges, where $\Sigma' = U \cup L \cup VAR$ is a label set where VAR is a set of variables. The labeling function $l' : V' \cup E' \to \Sigma'$ maps each vertex or edge to a label in Σ'.*

A *Query Graph* is similar to an *RDF Graph* except that the nodes and edges in a query graph can be labeled with variables besides URIs and literals.

Definition 3. (STAR RDF SUBGRAPH) *Given a subgraph $G^* = (V, E, \Sigma, l)$ of an RDF Graph G, it is a Star RDF Subgraph if there is exactly one vertex $v \in V$ that satisfies: $\forall u \in V - v$, $\langle v, u \rangle \in E$ or $\langle u, v \rangle \in E$. v is called central vertex.*

In other words, a *star RDF graph* is a set of edges that share a central vertex regardless of the directions of these edges.

Definition 4. (SUBGRAPH ISOMORPHISM) *Given two graphs $G = (V, E, \Sigma, l)$ and $\overline{G} = (\overline{V}, \overline{E}, \overline{\Sigma}, \overline{l})$, a subgraph isomorphism from G to $\overline{G}$ can be denoted as an injective function $f : V \to \overline{V}$, such that $\forall u, v \in V$, if $\langle u, v \rangle \in E$ then $\langle f(u), f(v) \rangle \in \overline{E}$, $l(u) = l(f(u)), l(v) = l(f(v))$, and $l(\langle u, v \rangle) = l(\langle f(u), f(v) \rangle)$.*

We also refer to *subgraph isomorphism* as *subgraph matching*. Actually, a relaxed definition is used in the following of this paper. Since the labels of vertices and edges of a query graph may be variables, we assume that the condition $?var = \sigma$ is always true, where $?var \in VAR$ and $\sigma \in \Sigma$.

4 FFD-Index

In this section, we will introduce the encoding technique and the index scheme of FFD-index. Given an RDF (query) graph G, G can always be partitioned into a set of star RDF subgraphs. For each vertex v in G, there is at least one edge starting from or ending at v. Therefore, all adjacent edges of v form a v-centered star subgraph.

4.1   Subgraph Fingerprint

For a star subgraph $G^* = (V, E, \Sigma, l)$ with a vertex $v \in E$ as its central vertex, all the central vertex v's adjacent edges are encoded into a bit-string. $l(e)$ and $l(u)$ denote the labels of v's adjacent edge e and the corresponding vertex u respectively. The bit-string of $l(e)$ and $l(u)$ is represented by $Encode_{edge}(\langle e, u \rangle)$.

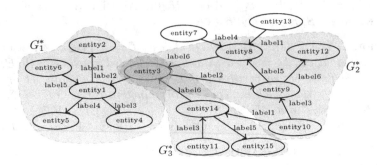

Fig. 2. RDF graph G. For simplicity, we use "label" and "entity" instead of real URIs to denote vertices and edges.

Suppose that the length of bit-string is $|Encode_{edge}(\langle e, u \rangle)| = m + n$. The first m bits are used to represent $l(e)$ and the rest n bits to represent $l(u)$. With a set of hash functions H, $\overline{m}$ out of the first m bits and $\overline{n}$ out of the other n bits are set to be "1" ($\overline{m} < m$, $\overline{n} < n$).

Taking the first m bits for example, $\overline{m}$ different hash functions h_i ($1 \leq i \leq \overline{m}$) are used to generate $\overline{m}$ integers $h_i(l(e))$. Then for each $h_i(l(e))$, the ($h_i(l(e))$ mod m)-th bit of first m bits is set to be "1". Obviously, there are at most $\overline{m}$ bits set to be "1" leaving the others "0". A similar procedure is applied for the last n bits to encode $l(u)$.

The fingerprint of G^* is composed of two parts, $fp_{out}(G^*)$ generated from E_{out}, the set of outgoing edges, and $fp_{in}(G^*)$ from E_{in}, the set of incoming edges. Formally, $fp_{out}(G^*) = Encode_{edge}(\langle e_1^{out}, u_1^{out} \rangle) \mid \dots \mid Encode_{edge}(\langle e_k^{out}, u_k^{out} \rangle)$, where $|E_{out}| = k$, $e_i^{out} \in E_{out}$ ($1 \leq i \leq k$), and $u_i^{out} \in V$. u_i^{out} is the corresponding vertex of e_i^{out} and "|" denotes the bitwise-or operator. $fp_{in}(G^*) = Encode_{edge}(\langle e_1^{in}, u_1^{in} \rangle) \mid \dots \mid Encode_{edge}(\langle e_l^{in}, u_l^{in} \rangle)$, where $|E_{in}| = l$, $e_j^{in} \in E_{in}$ ($1 \leq j \leq l$) and $u_j^{in} \in V$. These two bit-strings are concatenated together to form the fingerprint of G^*, $fp(G^*) = fp_{out}(G^*) \circ fp_{in}(G^*)$, where "$\circ$" is the concatenation operator. Figure 3 demonstrates how to generate a fingerprint of G_2^* in Fig. 2.

4.2   Indexing Scheme

We propose a compact index scheme optimized for star RDF subgraphs composed of five indexes and a *neighborhood table*. An entire RDF graph is partitioned into a set of star subgraphs stored in the neighborhood table, each subgraph is labeled with its fingerprint. Given a star RDF graph $G^* = \{V, E, \Sigma, l\}$

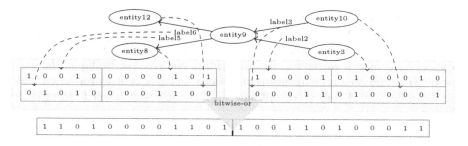

Fig. 3. The procedure of generating the fingerprint of G_2^* in Fig. 2. $m = 5$, $n = 7$, $\overline{m} = 2$, $\overline{n} = 2$. The dashed arrows denote $Encode_{edge}$ function.

cv	FgPrt: fingerprint	AdjSet: set of adjacent edges
ent1	101011110010 · 100110110010	$(0, lab1, ent2)$, $(1, lab2, ent3)$, $(0, lab3, ent4)$,$(0, lab4, ent5)$, $(1, lab5, ent6)$
ent3	100011001001 · 010100100101	$(0, lab2, ent1)$, $(0, lab2, ent9)$, $(1, lab6, ent8)$, $(1, lab6, ent14)$
ent8	010100100010 · 101110011011	$(1, lab1, ent13)$, $(1, lab5, ent9)$, $(0, lab6, ent3)$, $(1, lab4, ent7)$
ent9	110100001101 · 100110100011	$(1, lab2, ent3)$, $(0, lab5, ent8)$, $(0, lab6, ent12)$, $(1, lab3, ent10)$
ent14	110100110010 · 001111100001	$(1, lab1, ent10)$, $(0, lab5, ent15)$, $(1, lab3, ent11)$, $(0, lab6, ent3)$

(a) Neighborhood Table T For G

$\langle s, p \rangle$	DegMap
$\langle ent3, lab2 \rangle$	$\langle ent1, 5 \rangle$
	$\langle ent9, 4 \rangle$
$\langle ent6, lab5 \rangle$	$\langle ent1, 5 \rangle$
$\langle ent8, lab6 \rangle$	$\langle ent3, 4 \rangle$
$\langle ent14, lab6 \rangle$	$\langle ent3, 4 \rangle$
$\langle ent10, lab3 \rangle$	$\langle ent9, 4 \rangle$
$\langle ent7, lab4 \rangle$	$\langle ent8, 4 \rangle$
$\langle ent13, lab1 \rangle$	$\langle ent8, 4 \rangle$
$\langle ent9, lab5 \rangle$	$\langle ent8, 4 \rangle$
$\langle ent10, lab1 \rangle$	$\langle ent14, 4 \rangle$
$\langle ent11, lab3 \rangle$	$\langle ent14, 4 \rangle$

(b) SP_star

Fig. 4. Neighborhood table T and SP_star index. *lab* and *ent* are short for *label* and *entity*.

and a star query graph $Q^* = (V', E', \Sigma', l')$, finding a target subgraph is to find a matching between Q^* and a subgraph of G^*.

FFD-index is designed to leverage both structural information and labels to efficiently answer queries. Figure 4(a) shows the neighborhood table T for the RDF graph G in Fig. 2. The key column of T is the labels of central vertices. The set of adjacent edges is denoted as $AdjSet$, of which each element is depicted as a triple $(direction, l(e), l(u))$, where $direction$ denotes the direction of the edge.

If v, the central vertex of Q^*, is fixed, the candidate subgraph can be fetched by looking up the neighborhood table with the key $l(v)$. Otherwise, if v is variable, we have to find candidates according to Q^*'s labels and structural information. Suppose that $e \in E'$ is an incoming edge of Q^* from a neighbor vertex u to the central vertex v. Since u and e can be either fixed or variable, there are four possible combinations of their labels. Therefore, based on how many elements are fixed out of two (except the central vertex) triple elements, we devise five indexes: SP_star, OP_star, S_star, P_star, and O_star. Due to the limited space, only SP_star index for G is shown in Fig. 4(b). The key column, $\langle s, p \rangle$, of SP_index is incoming edges of central vertices. For example, the first row of SP_index in Fig. 4(b) means that both 5-degree central vertex $entity1$ and 4-degree central vertex $entity9$ have an incoming edge labeled $label2$ from $entity3$.

4.3 Index Construction/Update

The FFD-index can be constructed efficiently. First, as mentioned in Sect. 4.1, we encode each triple (s, p, o) of RDF dataset into a bit-string and refresh the fingerprints of neighborhoods and degrees of s and o. Second, the new fingerprints and degrees are inserted/updated into the neighborhood table. Finally, for each triple, five $(key, value)$ records are inserted into SP_star, OP_star, S_star, P_star, and O_star respectively. Algorithm 1 shows the index construction procedure with the time complexity of $\mathcal{O}(|R|)$.

A remarkable feature of FFD-index is that every star subgraph is encoded and stored individually, which makes updates easy to manage. Thus, updates can be done by enriching neighborhood sets, increasing degrees of central vertices, and recomputing some bits of fingerprints.

Algorithm 1. Constructing FFD-Index

Input: R: the RDF Dataset, T: the Neighborhood Table
Output: T, SP_star, OP_star, S_star, P_star, and O_star

1: **for** each RDF triple $r \leftarrow (s, p, o)$ in R **do**
2:     Represents r as an entire RDF graph, $G_r \leftarrow \{r\}$;
3:     Let s be the central vertex of G_r, $fp_s \leftarrow fp_{out}(G_r) \circ fp_{in}(\emptyset)$;
4:     Get $t_s \leftarrow \langle s, AdjSet_s, FgPrt_s \rangle$ from T;   // If t_s does not exists, create it in T.
5:     $AdjSet_s \leftarrow AdjSet_s \cup \{(0, p, o)\}$;   // Add $(0, p, o)$ into $AdjSet_s$.
6:     $FgPrt_s \leftarrow FgPrt_s \mid fp_s$;   // "|" denotes bitwise-or operation.
7:     Update t_s to T;
8:     Let o be the central vertex of G_r, $fp_o \leftarrow fp_{out}(\emptyset) \circ fp_{in}(G_r)$;
9:     Get $t_o \leftarrow \langle o, AdjSet_o, FgPrt_o \rangle$ from T;
10:    $AdjSet_o \leftarrow AdjSet_o \cup \{(1, p, s)\}$;   // Add $(1, p, s)$ into $AdjSet_o$.
11:    $FgPrt_o \leftarrow FgPrt_o \mid fp_o$;
12:    Update t_o to T;
13:    Get $item_{sp} \leftarrow \langle \langle s, p \rangle, DegMap \langle cv, degree \rangle \rangle$ from SP_star;
14:    $DegMap(o) \leftarrow DegMap(o) + 1$;   // Increase the degree of o by 1.
15:    Update $item_{sp}$ to SP_star;
16:    $\cdots$ // Similar procedures for OP_star, S_star, P_star, and O_star are omitted.
17: **end for**

---

5   Query Processing

In this section, we show how GraSS evaluates a graph query against the FFD-index. We start by showing how to extract star subgraphs from an entire query graph. As shown in Fig. 1, the outcome of extraction should satisfy the restriction that each vertex of original query graph appears in at least one extracted subgraph. The vertex which appears in more than one subgraph will involve a join operation, denoted by the dashed line in Fig. 1. To reduce the number of join operations, we prefer extracted star subgraph to contain as fewer variable vertices as possible, and whose central vertices are of higher degrees. A simplified version of this algorithm is shown in Algorithm 2, which only considers degrees of central vertices.

Algorithm 2. Decomposing Query Graph

Input: $Q = \{V', E', \Sigma', l'\}$
Output: L: The List of Star Subgraphs;
  1: Sort vertices on degree in descending order: $L_{sort} = \{v_1, \cdots, v_n\}$;
  2: $i \leftarrow 1$;
  3: **while** $E' \neq \emptyset$ **do**
  4:     $E' \leftarrow E' - nb(v_i)$;   // $nb(v_i)$ denotes the set of neighborhood edges of v_i.
  5:     $L.add(Q^*_{v_i})$;   // $Q^*_{v_i}$ denotes the star subquery whose central vertex is v_i.
  6:     $i \leftarrow i + 1$;
  7: **end while**

By using a method similar to generating the fingerprint of a star subgraph, a star query subgraph is also encoded into a bit-string, named *query fingerprint*. Furthermore, the query fingerprint will act as a filter to find the candidate set.

For convenience, star query subgraphs are classified into two categories in this paper, *CVVQ* (Central Vertex Variable Query) and *CVFQ* (Central Vertex Fixed Query), according to whether the label of central vertex v is variable or fixed.

Definition 5. *A star query is called a CVVQ if its central vertex is a variable, otherwise it is called a CVFQ.*

5.1   Processing CVVQs

For a CVVQ Q^*, we generate the query fingerprint $fp(Q^*)$ using the similar method designed for generating the fingerprint of star subgraph demonstrated in Fig. 3. The only difference is that when the label of an edge or a vertex is a variable, the bit string corresponding to this label will be set 0. If G^* is a candidate of Q^*, $fp(Q^*)$ and $fp(G^*)$ should satisfy: if i-th bit of $fp(Q^*)$ is 1, then the i-th bit of $fp(G^*)$ must be 1. We call this condition "$fp(G^*)$ *covers* $fp(Q^*)$", denoted as $fp(G^*) \succ fp(Q^*)$. Fingerprint covering is necessary but not sufficient for declaring that Q^* matches G^*.

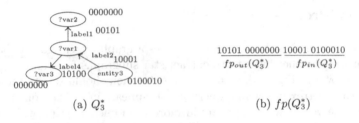

(a) Q_3^* (b) $fp(Q_3^*)$

Fig. 5. Query graph Q_3^* and $fp(Q_3^*)$

Two stages are performed to confirm that a subgraph is a candidate of a CVVQ. First, the degree of the central vertex of a candidate must be no less than that of the query graph. Second, the fingerprint cover condition must be satisfied.

For example, for the query Q_3^* in Fig. 5 and the graph G in Fig. 2. Consider the edge $\langle entity3, label2, ?var1 \rangle$ of Q_3^* in Fig. 5(a). It matches two $label2$ edges that start from $entity3$ in G. The corresponding record shown is in the first row of SP_star index in Fig. 4(b). Both G_1^* and G_2^* have the central vertex degree larger than that of Q_3^*. But only G_1^* satisfies the fingerprint cover condition. Here, we have $fp(G_1^*) = 101011110010 \cdot 100110110010$ and $fp(G_2^*) = 110100001101 \cdot 100110100011$. Since the third and fifth bits of $fp(Q_3^*)$ are 1, while these two positions in $fp(G_2^*)$ are occupied by 0, therefore, we get $fp(G_1^*) \succ fp(Q_3^*)$ but $fp(G_2^*) \nsucc fp(Q_3^*)$.

Algorithm 3. $GenerateCandidateSet_{SP}$

Input: SP_star, $Q^* = \{V', E', \Sigma', l'\}$: A CVVQ, $\langle l'(u), l'(e) \rangle$: Fixed $l'(u)$ and $l'(e)$
Output: C^*: The Candidate Set Induced by $\langle l'(u), l'(e) \rangle$
1: $C^* \leftarrow \emptyset$;
2: Get $\langle\langle l'(u), l'(e) \rangle, DegMap\langle centralvertex, degree \rangle\rangle$ from SP_index;
3: **for** each $\langle cv, deg \rangle \in DegMap\langle centralvertex, degree \rangle$ **do**
4:     **if** $deg \geqslant |E'|$ **then**
5:         $t_{cv} \leftarrow \langle cv, FgPrt_{cv}, AdjSet_{cv} \rangle$;   // Get t_{cv} from T according to cv;
6:         **if** $FgPrt_{cv} \succ fp(Q*)$ **then**
7:             $C^* \leftarrow C^* \cup \{t_{cv}\}$;
8:         **end if**
9:     **end if**
10: **end for**

The query algorithm is shown in Algorithm 3 with a time complexity of $\mathcal{O}(|DegMap|)$, where $|DegMap|$ is the number of star subgraphs with an incoming $l(e)$ edge staring from u. In Algorithm 3, we first fetch $DegMap\langle cv, degree \rangle$ for $\langle l'(u), l'(e) \rangle$ from SP_star index. Then, for each $\langle cv, deg \rangle \in DegMap\langle cv, degree \rangle$, line 4 requires that the degree of a candidate's central vertex must be larger than that of the star query's central vertex. Finally, the neighborhood table T is examined to validate the fingerprint covering condition, as shown by line 6.

Continuing with the example above, two more candidate sets can be generated based on the other two edges of G_1^* and P_star index. The final candidate set C_{final}^* must be a subset of the intersection of C_i^* $(1 \leqslant i \leqslant |E'|)$.

5.2   Processing CVFQs

One important thing to note is that each vertex in an RDF graph is represented by a unique URI. Hence, if v is the central vertex of a star subgraph G_i^*, there is no other star subgraph $G_j^*(i \neq j)$ that takes v as its central vertex. Then, if Q^* is a CVFQ, it is clear that there is at most one star subgraph in the neighborhood table matched by Q^*, which means the size of the candidate set for Q^* will not be larger than 1. So there is no need to calculate the fingerprint of Q^* and validate the fingerprint cover condition. Actually, we only check whether there exists a record $\langle l(v), AdjSet_{l(v)}, FgPrt_{l(v)} \rangle$ in the neighborhood table and the number of degrees of the central vertex. The detailed checking procedure is shown in Algorithm 4.

Algorithm 4. $GenerateCandidateSet_{CVFQ}$

Input: T: The Neighborhood Table, $Q^* = \{V', E', \Sigma', l'\}$: A CVFQ, $l'(v)$: The Label
    of Central Vertex
Output: C^*: The Candidate Set for Q^*
1: $C^* \leftarrow \emptyset$;
2: $t_{l'(v)} \leftarrow \langle l'(v), AdjSet_{l'(v)}, FgPrt_{l'(v)} \rangle$;   // Get $t_{l'(v)}$ from T according to $l'(v)$.
3: **if** $!isEmpty(t_{l'(v)}) \wedge |AdjSet_{l'(v)}| \geqslant |E'|$ **then**
4:      $C^* \leftarrow C^* \cup \{t_{l'(v)}\}$;   // Add $t_{l'(v)}$ into candidate set C^*.
5: **end if**

According to Definition 4, if a query graph Q can match an RDF subgraph G, each edge of Q should match to some edge of G. If the number of edges of Q^* is larger than that of G^*, then G^* must not be matched by Q^*. As lines 3–4 of Algorithm 4 show, the candidate should be dropped.

5.3   Verification

Each candidate, for either a $CVVQ$ or a $CVFQ$, will be verified to determine whether it can be matched by the query graph. Suppose that $G^* = \{V, E, \Sigma, l\}$ is an element in the candidate set for star query $Q^* = \{V', E', \Sigma', l'\}$, where $|E'| = k$. For each edge $e_i' \in E'$, a candidate edge set C_i' is associated with it. We select edges from E that can match e_i' and then put them into C_i'. P is the Cartesian product result of C_i' $(1 \leqslant i \leqslant k)$, $P = C_1' \times \cdots \times C_i' \times \cdots \times C_k'$.

As an undesirable situation, one edge $e \in E$ can be matched by more than one edge in E', e.g., e_i' and e_j' $(i \neq j)$ match e simultaneously, $C_i' \cap C_j' = \{e\}$. In Fig. 6(a), both triple patterns $\langle ?var2, label6, entity3 \rangle$ and $\langle entity14, ?var1,$

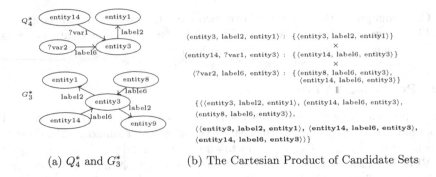

(a) Q_4^* and G_3^* (b) The Cartesian Product of Candidate Sets

Fig. 6. Query graph Q_4^* and the Cartesian Product of Candidate Sets

$entity3\rangle$ of Q_4^* can match $\langle entity14, label6, entity3 \rangle$ of G_3^*. According to Definition 4, one edge in an RDF graph can only be matched by at most one edge in the query graph. So we have to remove some elements from P to guarantee that no edge is matched by different triple patterns simultaneously. In Fig. 6(b), the element in bold is to be removed. We refer to this sort of elements as *overlapping elements*.

Algorithm 5. *Verification*

Input: $Q^* = \{V', E', \Sigma', l'\}$, $G^* = \{V, E, \Sigma, l\}$
Output: M: The Set of Matches From Q^* To G^*
1: $k \leftarrow |E'|$, $M \leftarrow \emptyset$;
2: **for** each $e_i' \in E'$ **do**
3:     Extract edge(s) from E that can match e_i': $C_i' \subseteq E$;
4: **end for**
5: $P \leftarrow \prod\limits_{i=1}^{k} C_i'$;   // P is the Cartesian product of C_i'
6: **for** each $t \in P$ **do**
7:     **if** t is not an overlapping element **then**
8:         $M \leftarrow M \cup \{t\}$;
9:     **end if**
10: **end for**

The verification algorithm is shown in Algorithm 5. For each edge in the query graph, we first find all edges that match it, as shown by line 2–3. Then, the Cartesian product is calculated and examined to guarantee there is no overlapping element.

6 Experiments

All our experiments were carried out on an Intel Quad-Core CPU@2.8 GHZ machine with 8 GB memory running Ubuntu 12.04 64-bit. Cassandra is used as the underlying data store of GraSS. We used both the synthetic dataset, LUBM[2], and the real dataset, YAGO (YAGO2_core[3]), in our experiments comparing GraSS with RDF-3X and gStore. The query size (the number of triples contained by a query) applied on YAGO and LUBM is designed ranging from 3 to 15. For each query size, we built a set of 20 queries by randomly selecting subgraphs from YAGO or LUBM and replacing some URIs with variables.

The length of fingerprints and the number of bits set to be 1 are adjusted based on the scale of data set, which is a trade-off between the expressiveness and the storage space. In the following experiments, we simplify the strategy by making the length a fixed value. We set $m = 30$, $n = 70$, and $\bar{m} = \bar{n} = 3$ for both LUBM and YAGO.

6.1 Performance on Synthetic Dataset

LUBM is a standard benchmark developed to evaluate the performance of Semantic Web repositories, which consists of a university domain ontology. As a feature of RDF data, large amount of URIs share the same prefixes, which allows a significant compression. We only store the common prefixes once to reduce the size of indexes. Figure 7(a) shows that after compressing GraSS is almost equivalent to RDF-3X in index size.

The different sizes of five indexes of GraSS, shown in Fig. 7(b), are determined by the nature of LUBM that the number of outgoing edges is larger than that of incoming edges for star subgraphs statistically. OP_star index and O_star index are distinctly larger than SP_star index and S_star index. Therefore, in our experiments, if a CVVQ contains a fixed incoming edge, the response time is generally very short. GraSS takes more time than RDF-3X and gStore in processing small star queries ($\leqslant 4$ edges). But when the scale of query graph increases, GraSS exhibits better performance. Figure 7(c) reports the performance comparison of the three methods. We use *pruning ability* to represent the ratio of average number of candidates after and before the validation of fingerprint covering condition. To evaluate the pruning ability, we generated another set of queries, elements of which are all star-shaped graphs ranging in size from 3 to 9. Figure 7(d) exhibits the average number of candidates before and after fingerprint covering validation. The ratio of false positive candidates filtered out by fingerprint is quite large, up to 80 % in some extreme situations.

6.2 Performance on Real Dataset

The RDF graph of the YAGO dataset contains 9.6 million vertices and 33 million edges. To have an overview of YAGO, an analysis about degree is given in Table 1, the average degree of vertices is 6.83. The time for constructing FFD-index for YAGO on Cassandra is about 23.2 min, which is a little longer than that of

---

[2] http://swat.cse.lehigh.edu/projects/lubm/.

[3] http://www.mpi-inf.mpg.de/departments/databases-and-information-systems/ research/yago-naga/yago/archive/.

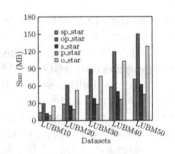

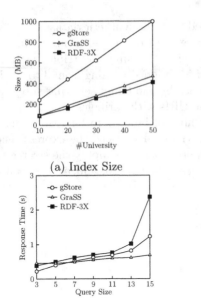

(a) Index Size

(b) GraSS Index Size

(c) Avg Response Time

(d) Pruning Ability

#triples	# BEF. VLD.	# AFT. VLD.
3	73.1	20.2
4	81.2	35.8
5	63.6	17.5
6	23.2	2.6
7	17.2	3.8
8	10.8	2.1
9	7.2	3.7

Fig. 7. Performance on LUBM

RDF-3X but much shorter than that of gStore. In terms of index size on disk, 2.5 GB taken by GraSS is less than half of that taken by gStore and basically the same as RDF-3X.

Table 1. Statistics about YAGO

#Vertices	#Edges	Avg degrees	Size in. nt format	GraSS		RDF-3X		gStore	
				CONST time	Size	CONST time	Size	CONST time	Size
9656346	32962704	6.83	4.6 GB	23.2 min	2.5 GB	15.1 min	2.7 GB	50.1 min	5.7 GB

Generally speaking, the number of costly join operations carried out by RDF-3X increases with the number of edges of a star graph query. In contrast, for GraSS, more edges mean more information, thus more bits of the query graph's fingerprint will be set "1" that makes the fingerprint more efficient in pruning search space. The average query response time on YAGO is shown in Fig. 8(a). Figure 8(b) shows the pruning ability of fingerprint on YAGO.

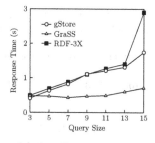

#triples	# BEF. VLD.	# AFT. VLD.
3	211.8	71.1
4	103.6	20.0
5	140.2	23.5
6	27.3	5.2
7	17.7	1.2
8	15.2	2.4
9	15.3	2.0

(a) Avg Response Time   (b) Pruning Ability

Fig. 8. Performance on YAGO

7   Conclusion

Subgraph matching is a widely-known NP-complete problem. Efficient evaluation of star subgraph matching is specifically critical for query answering over large-scale RDF data as it is a very common query pattern. Existing approaches usually handles star subgraph matching by translating it into a number of join operations, which is costly and inefficient.

In this paper, we present GraSS, a SPARQL query engine based on a novel subgraph encoding and index scheme, FFD-index, to reduce the search space to achieve improved query answering performance. In FFD-index, each star subgraph is compactly represented by a bit string fingerprint that preserves some essential structural information and label information.

Extensive experiments have been conducted on synthetic benchmark dataset LUBM and the real-world dataset YAGO. On both datasets, for queries consisting of a few big stars, GraSS is significantly faster than RDF-3X and gStore.

Acknowledgement. This work is supported by the National Natural Science Foundation of China (61100049), the National High-tech R&D Program of China (863 Program) (2013AA013204), and the Australia Research Council (ARC) Discovery grants DP130103051.

References

1. Bizer, C., Heath, T., Berners-Lee, T.: Linked data - the story so far. Int. J. Semant. Web Inf. Syst. **5**(3), 1–22 (2009)
2. Bunke, H.: Graph matching: theoretical foundations, algorithms, and applications. In: Proceedings of Vision Interface, pp. 82–88 (2000)
3. Cheng, J., Ke, Y., Ng, W., Lu, A.: Fg-index: towards verification-free query processing on graph databases. In: Proceedings of SIGMOD, pp. 857–872. ACM (2007)
4. Conte, D., Foggia, P., Sansone, C., Vento, M.: Thirty years of graph matching in pattern recognition. Int. J. Pattern Recogn. Artif. Intell. **18**(03), 265–298 (2004)

5. Cordella, L.P., Foggia, P., Sansone, C., Vento, M.: A (sub) graph isomorphism algorithm for matching large graphs. IEEE Trans. Pattern Anal. Mach. Intell. **26**(10), 1367–1372 (2004)
6. Fortin, S.: The Graph Isomorphism Problem. Technical Report, University of Alberta, Canada (1996)
7. Harris, S., Seaborne, A.: SPARQL 1.1 Query Language. W3C Recommendation (2013)
8. Kim, H., Ravindra, P., Anyanwu, K.: From SPARQL to mapreduce: the journey using a nested triplegroup algebra. Proc. VLDB Endow. **4**(12), 1426–1429 (2011)
9. Klyne, G., Carroll, J.J., McBride, B.: RDF 1.1 Concepts and Abstract Syntax. W3C Recommendation (2014)
10. Michael, R.G., David, S.J.: Computers and Intractability: A Guide to the Theory of NP-Completeness. WH Freeman Co., San Francisco (1979)
11. Neumann, T., Weikum, G.: RDF-3X: a RISC-style engine for RDF. Proc. VLDB Endow. **1**(1), 647–659 (2008)
12. Papailiou, N., Tsoumakos, D., Konstantinou, I., Karras, P., Koziris, N.: H2RDF+: An efficient data management system for big RDF graphs. In: Proceedings of SIGMOD, pp. 909–912. ACM (2014)
13. Shasha, D., Wang, J.T., Giugno, R.: Algorithmics and applications of tree and graph searching. In: Proceedings of PODS, pp. 39–52. ACM (2002)
14. Udrea, O., Pugliese, A., Subrahmanian, V.S.: GRIN: a graph based RDF index. In: Proceedings of the National Conference on Artificial Intelligence, vol. 22 (2007)
15. Ullmann, J.R.: An algorithm for subgraph isomorphism. J. ACM **23**(1), 31–42 (1976)
16. Weiss, C., Karras, P., Bernstein, A.: Hexastore: sextuple indexing for semantic web data management. Proc. VLDB Endow. **1**(1), 1008–1019 (2008)
17. Lyu, X., Wang, X., Li, Y.-F., Feng, Z.: FFD-Index: an efficient indexing scheme for star subgraph matching on large RDF graphs. In: Liu, A., Ishikawa, Y., Qian, T., Nutanong, S., Cheema, M.A. (eds.) DASFAA 2015 Workshops. LNCS, vol. 9052, pp. 240–245. Springer, Heidelberg (2015)
18. Yan, X., Yu, P.S., Han, J.: Graph indexing: a frequent structure-based approach. In: Proceedings of SIGMOD, pp. 335–346. ACM (2004)
19. Zhang, S., Li, S., Yang, J.: GADDI: Distance index based subgraph matching in biological networks. In: Proceedings of EDBT, pp. 192–203. ACM (2009)
20. Zou, L., Chen, L., Yu, J. X., Lu, Y.: A novel spectral coding in a large graph database. In: Proceedings of EDBT, pp. 181–192. ACM (2008)
21. Zou, L., Mo, J., Chen, L., Özsu, M.T., Zhao, D.: gStore: answering SPARQl queries via subgraph matching. Proc. VLDB Endow. **4**(8), 482–493 (2011)

FreeS: A Fast Algorithm to Discover Frequent Free Subtrees Using a Novel Canonical Form

Israt J. Chowdhury[(⊠)] and Richi Nayak

Science and Engineering Faculty,
School of Electrical Engineering and Computer Science,
Queensland University of Technology, Brisbane, Australia
{israt.chowdhury,r.nayak}@qut.edu.au

Abstract. Web data can often be represented in free tree form; however, free tree mining methods seldom exist. In this paper, a computationally fast algorithm *FreeS* is presented to discover all frequently occurring free subtrees in a database of labelled free trees. *FreeS* is designed using an optimal canonical form, BOCF that can uniquely represent free trees even during the presence of isomorphism. To avoid enumeration of false positive candidates, it utilises the enumeration approach based on a tree-structure guided scheme. This paper presents lemmas that introduce conditions to conform the generation of free tree candidates during enumeration. Empirical study using both real and synthetic datasets shows that *FreeS* is scalable and significantly outperforms (i.e. few orders of magnitude faster than) the state-of-the-art frequent free tree mining algorithms, *HybridTreeMiner* and *FreeTreeMiner*.

Keywords: Web data · Free tree · Canonical form · Enumeration approach

1 Introduction

In the Web domain, graphs and trees are commonly used data structures for modelling information with complex relations. Free trees - the connected, acyclic and undirected graphs - have become popular for presenting such data due to having unique properties [1–4]. For obtaining useful structural information, free tree mining provides a good compromise between the more expressive but computationally harder general graph mining and the less expressive but faster sequence mining. As a middle ground between these two extremes, free trees have been widely used for representing and mining data in diverse areas including web, bioinformatics, computer vision and networks. For example, in analysis of molecular evolution, an evolutionary free tree, called phylogeny, can describe the evolution history of certain species [5]. In bioinformatics various useful patterns can be treated as free trees during pattern mining [4]. In computer networking, multicast free trees have been mined and used for packet routing [6]. Web access logs represented as free trees give interesting insight about the user browsing behaviour without a specific point of entry [7].

The process of finding frequent subtrees incurs high cost due to the inclusion of expensive but unavoidable steps like frequency counting and candidate subtrees generation. Frequency counting step often requires subtree isomorphism checking which is

© Springer International Publishing Switzerland 2015
J. Wang et al. (Eds.): WISE 2015, Part I, LNCS 9418, pp. 123–137, 2015.
DOI: 10.1007/978-3-319-26190-4_9

computationally hard, even known as NP-complete problem in graph mining algorithms [4]. Exponential and redundant candidate generation is another problem. During candidate generation, determining a "good" growth strategy is critical as there can be many possible ways to extend a candidate subtree. These problems become worse in free trees, due to being less-constrained structurally, in comparison to other tree forms such as ordered and unordered. With these complexities involved, only a few free tree mining algorithms are available in the literature. Chi et al. developed an apriori-like algorithm *FreeTreeMiner* [8] as well as an enumeration tree based algorithm *HybridTreeMiner* [1] to discover frequent free subtrees in a database of free trees. Rückert et al. [4] and Zhao et al. [3] have proposed algorithms for mining frequent free trees from a graph database. These algorithms generate large number of false positives (i.e., invalid candidate subtrees) during enumeration that need to be pruned in the frequency counting step. This causes high processing time. Moreover, the necessity of performing isomorphism checking to avoid redundant candidate tree generation and false frequency counting causes additional computational complexity.

In this paper, we propose an algorithm, *FreeS* which is a fast and accurate method for mining frequent free induced subtrees in a database of labelled free trees. First, we propose a unique representation of free trees by introducing a new order-independent *balanced optimal canonical form* (BOCF) that can effectively handle the subtree isomorphism problem. We introduce conditions to conform free tree candidate generation in their BOCFs for which the necessary proofs are also provided. Second, we propose a *tree-structure guided scheme based enumeration* approach that only generates valid candidate subtrees. To the best of our knowledge, *FreeS* is the first algorithm that uses the underlying tree-structure information to avoid invalid subtree generation while mining frequent free subtrees. Because of using the optimal canonical form and tree-structure guided scheme based enumeration, *FreeS* does fast processing. Our experiments with both synthetic and real-life datasets confirm that *FreeS* is faster by few orders of magnitude than two leading free tree mining algorithms, *Hybrid-TreeMiner* and *FreeTreeMiner* (abbreviated as HBT and FTM respectively).

2 Preliminaries

Let a graph constitute a set of nodes $V = \{v_1, v_2, \ldots, v_n\}$ and a set of edges $E = \{(v_i, v_j) | v_i, v_j \in V\} = \{e_1, e_2, \ldots, e_{n-1}\}$. A labelled graph has a set of labels Σ, where a function $L : V \cup E \rightarrow \sum$ maps nodes with unique labels. A graph is connected but acyclic when it has at least one node that is connected to the rest of the graph by only one edge, which is leaf. For our purposes, the class of connected acyclic labelled graphs is of special interest, which is also called free tree, an unrooted unordered tree-like structure. In this paper, we denote a free tree with n nodes as n-free tree.

Let two free trees be t and T. t is a subtree of T if t can be obtained from T by repeatedly removing one degree nodes from its structure. Free trees t and T are *isomorphic* to each other if a bijective mapping exists between their set of nodes that preserves node labels, edge labels and also reflects the tree structures.

Let T_{db} be a database where each transaction is a labelled free tree. The problem of frequent free tree mining is to discover the complete set of frequent free subtrees. If tree

$T \in T_{db}$ has a subtree isomorphic to subtree t, that indicates T has an *occurrence* of t in its structure. Formally we define the support of subtree t in T_{db} using the concept of occurrence as follows,

$$Occurrence(t, T) = \begin{cases} 1 & \text{if } t \text{ exists in } T \\ 0 & \text{otherwise} \end{cases} \tag{1}$$

$$Support\,(t, T_{db}) = \sum\nolimits_{T \in T_{db}} Occurence(t, T) \tag{2}$$

The subtree t is called frequent if $Support\,(t, T_{db}) \geq minsup$ where $minsup$ is user-defined minimum support threshold.

In this paper, in a free tree, two adjacent nodes v_i and v_j with same label are defined as *equivalent nodes*, denoted by $v_i \cong v_j$. The *weight* of a node v_i is defined as the total number of its equivalent nodes and denoted by w_i (as shown in Fig. 1). Using weights, we represent free trees of a database in a concise manner for further processing. Figure 1 shows an example of two free trees and their corresponding weighted representations by combining equivalent nodes (highlighted using different color patterns).

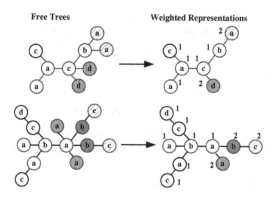

Fig. 1. Equivalent nodes and the condensed weighted representations of free trees. (Tree nodes are represented using labels. The edge labels are ignored in this paper.)

3 Canonical Form for Labelled Free Trees

A *Canonical Form* (CF) of a tree is a representative form that can consistently represent many equivalent variations of that tree into one standard form [8, 9]. Several CFs have been proposed for rooted tree representations using traversing algorithms such as *depth-first-search* (DFS) or *breadth-first-search* (BFS) [8]. However, defining CF for free trees is non-trivial as it requires handling the vast variants that a free tree can have, i.e., the isomorphism problem. Due to the inherent structural flexibility (e.g., undefined root node and no direction among sibling nodes), there are more ways to represent a free tree than that of a rooted tree. A canonical form is critical for appropriate representation and efficient processing of free trees, because it ensures finding a common pattern amongst free trees. Before we define CF of free trees, we explain the process for unordered rooted trees and extend it to free trees.

3.1 Why Canonical Form is Needed for Free Trees?

A *rooted tree* has a distinguished root node. A rooted tree that preserves order among the sibling nodes is called *rooted ordered*. This type of trees can easily be represented uniquely by using either the depth-first or the breadth-first string representations [8]. They do not face isomorphism. Two ordered trees will be similar iff all of its properties are identical; no variation is possible in similar rooted ordered trees [2]. Whereas, two similar unordered trees can have different orders among sibling nodes and these trees are called isomorphic trees. A free tree is also an unordered tree. The chance of having isomorphic trees in a database of free tree is very high due to the flexible property of being unrooted and unordered. Representing free trees using a systematic approach is non-trivial but critical to ensure its proper indexing for further processing and knowledge discovery.

Optimal Order: We will now briefly describe the concept of *optimal order* that is the basis of the proposed canonical form. An optimal order of a tree is an order obtained by the *balance optimal tree search (BOS)* algorithm [10–12] that traverses a rooted labelled tree uniquely, without the presence of sibling order information. Unlike existing traversal strategies [9], this algorithm works based on optimization instead of enforcing a left-to-right order among siblings. Three heuristics are applied recursively in this traversing algorithm to find out the optimum traversing path of a tree. Heuristic 1 identifies a potential node during the traversal process. Heuristics 2 and 3 select the best node if multiple nodes are identified as candidates for traversal.

Heuristic 1. *After the root node traversal, the children of the root node, i.e., $\{v_i, v_j, \ldots, v_k\}$ with weights $\{w_i, w_j, \ldots, w_k\}$ become eligible for traversing. The traversal order of these eligible nodes will be prioritized according to their ascending weights. The node with the highest weight is chosen first.*

Heuristic 2. *If two or more nodes $\{v_i, v_j, \ldots, v_k\}$ have the same maximum weight (i.e. maximum weight $= MAX\{w_i, w_j, \ldots, w_k\}$), the next node in the traversal order is selected based on the maximum number of their children (i.e., fan-out).*

Heuristic 3. *If two or more nodes hold the maximum weight with equal number of children, the traversal order will be prioritized using the minimum lexicographical order.*

The *optimal order* is unique even for trees that are isomorphic. This property is advantageous for mining frequent labelled free trees. For a free tree, several rooted ordered tree variations are possible only by changing the position of root node and the order among sibling nodes. An example can be seen in Fig. 2, where a free tree is treated as rooted unordered tree with root node "v_a" (Fig. 2a). Considering v_a as root node, several ordered variations of this free tree are shown in Fig. 2(b–e).

According to the BOS algorithm [10] the unique optimal traversal order of all these equivalent ordered trees will be "$v_a, v_b, v_c, v_d, v_c, v_f$". In contrast, the BFS or DFS

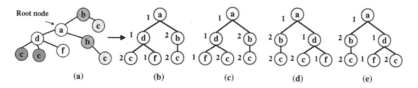

Fig. 2. Four rooted ordered trees obtained from the same rooted unordered tree.

traversal [8] will provide different traversing order for each equivalent ordered tree because of its structure dependent strategy. It is desirable to obtain a unique canonical form of an ordered tree representation; however, it is absolutely critical to obtain a single canonical form for all equivalent variations of a free tree to allow efficient indexing for further processing. The proposed optimal traversal strategy is based on optimization and is not sensitive to the structural changes. It gives the same optimal traversing order for all equivalent ordered trees that originate from a same free tree.

3.2   Balanced Optimal Canonical Form of Free Labelled Trees

If we can uniquely define root node of a free tree, then the optimal order can be used to define its canonical form. In this paper, we propose a two-step process for defining the canonical form of free trees. First, we normalize a free tree into the rooted unordered tree by fixing a root node and then we define the canonical form as well as canonical string.

Normalization: This step includes a systematic approach to define a root node in a free tree. Following the commonly used technique [1–3], all the leaf nodes along with their incident edges in the free tree are removed at each step until a single node or two adjacent nodes are left. The tree with a single remained node is called a *central tree* and, the tree with a pair of remaining nodes is called a *bicentral tree* [1]. With the remaining single node, this node becomes the root of the free tree. With the remaining two nodes, we apply *heuristic* 3 to obtain the root; therefore the node with minimum lexicographically ordered label becomes the root node.

The overall normalization takes $O(|T|)$ time, where $|T|$ is the number of nodes in the free tree. Figure 3 shows the process of obtaining the root node from the free trees.

Canonical Form and String: After the free tree is normalized to a rooted unordered tree, the balanced optimal canonical form can be defined as follows:

Definition 1 (*Balanced Optimal Canonical Form*). For a rooted labelled unordered tree, the balanced optimal canonical form is its optimal order of node labels along with corresponding weights.

A *canonical string representation* for labelled trees is equivalent to, but simpler than, canonical forms which facilitates frequency counting of trees in a database. For a balanced optimal canonical string encoding, we introduce four unique symbols $+1$, -1, $+2$ and -2 to specify directions on depth and breadth. More specifically, $+1$ and -1 are used to represent forward and backward travel towards depth between child and parent nodes; $+2$ and -2 are used to represent forward and backward travel towards breadth

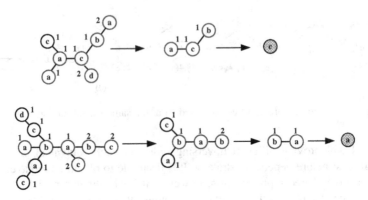

Fig. 3. Process of finding a root node in free trees

between sibling nodes respectively. We assume that none of these symbols are included in the alphabet of node labels. The canonical string representation of the rooted unordered tree is achieved by a guided record of sibling nodes,–*"under a parent node, a new node will always be recorded in a breadthwise direction from the existing rightmost sibling node."*

Example: For all the equivalent trees in Fig. 2 with the unique optimal order "$v_a, v_b, v_c, v_d, v_c, v_f$" the balanced optimal string representation of these trees will be "$1v_a, +1, 2v_b, +1, 2v_c, -1, +2, 1v_d, +1, 2v_c, -2, 1v_f$". Similarly, the optimal canonical string of the free tree in Fig. 4(a) will be "$1v_c, +1, 2v_d, +2, 1v_a, +2, 1v_b, +1, 2v_a, -1, -2, +1, 1v_a, +2, 1v_c$" and for the tree in Fig. 4(b) will be "$1v_a, +1, 2v_b, +2, 2v_a, -2, +1, 2v_c, -1, +2, +2, 1v_b, +1, 1v_a, +2, 1v_c, +2, 1v_a, -2, -2, +1, 1v_c, -1, +2, +1, 1v_d$".

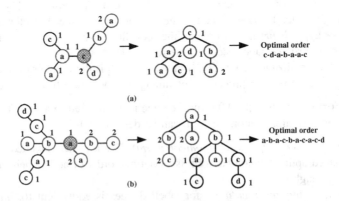

Fig. 4. Balanced optimal canonical form of free tree.

The isomorphic free trees can be successfully tracked because of having the same balanced optimal string representation. This ensures correct frequency counting for the processing of frequent subtrees. During the mining process, tree structural information such as level, weight, fan-out is stored that allows to differentiate the same alphabet appearing in different position. For sorting the optimal order it requires $O(|T| \log |T|)$ complexity, where $|T|$ is the number of nodes in a tree.

The balanced optimal canonical forms of free tree and rooted unordered tree embrace an interesting relationship which is described under Lemma 1. This relation is a fundamental step for growing the enumeration tree of free trees.

Lemma 1. *Balanced optimal canonical form of a free tree is always the balanced optimal canonical form of a rooted unordered tree; however, the reverse is not true.*

Proof. Consider a free tree T, with $v_1, v_2, \ldots, v_n$ nodes, with its balanced optimal canonical form t_{v1} that has a normalized root v_1. The n-number of different rooted unordered trees can be derived in their balanced optimal canonical forms $t_{v1}; t_{v2}; \ldots; t_{vn}$ by changing the position of root in T. Only one of the BOCFs of these rooted unordered trees will have the same BOCF as the free tree, e.g. t_{v1}.

Prior to detailing our *FreeS* algorithm, we add following two lemmas that introduce important conditions which are essential to hold true during candidate free subtree enumeration through the BOCF representation of fee trees. Fist we give the definitions of tree dimensions including depth, height and level as [13].

Definition 2 (*Depth, Height, Level of Node*). For node v_i of a tree T, depth is the length of the unique path from that node towards the root node, denoted by $d(T, v_i)$. The height $h(v_i)$ of node v_i is the longest path from that node to a leaf. The height H of a tree is the height of root node, $h(v_0)$. The level of a node v_i in a tree T is defined as $Lv(T, v_i) = H - d(T, v_i)$.

Lemma 2. *Balanced optimal canonical form of a rooted unordered tree T with two nodes is the balanced optimal canonical form of a free tree iff the root node has lexicographically minimum label.*

Proof. T is a rooted unordered tree with two nodes, where v_0 is root and v_1 is its child. The balanced optimal canonical form will be generated based on its optimal order, i.e., "v_0, v_1". Let us consider *case* 1, where root node v_0 has lexicographically minimum label. In this case treating T as free tree will end up having same canonical form as the rooted unordered tree, since a free tree considers the node with lexicographically minimum label as the center. Now consider *case* 2, where label of root node v_0 is higher than v_1. In this case the canonical form of free tree will be different than the rooted unordered tree, since v_1 will be the center instead of v_0.

Lemma 3. *Balanced optimal canonical form of a rooted unordered tree, T with 3 or more nodes and height H is the balanced optimal canonical form of a free tree iff the following conditions hold:*

1. *The root has at least 2 children;*
2. *The root node has lexicographically smaller label than the labels of its children; and*
3. *One branch or subtree induced by a child of the root has a leaf node, v_i positioned at level $Lv(T, v_i) = 0$ (bottom level of the tree) and at least another branch or one subtree induced by another child of the root has a leaf node, v_j positioned at level $Lv(T, v_i) \leq 1$ (at most one level up than the last level).*

Proof. For a rooted unordered tree T in its balanced optimal canonical form, we denote the root of T by v_0 and the children of v_0 by $v_1; \ldots; v_m$. Let us consider *case* 1. Tree T has 3 or more nodes and v_0 has only one child. It indicates that the rest of the nodes are appeared in that tree as child nodes of the immediate child of the root node. The node v_0 will be removed in the first step of finding center/bicenter. Consequently, v_0 cannot be the center or one of the bicenters. Therefore condition 1 will be held in this case. Let us consider *case* 2 when the root node v_0 has more than one child. This indicates that the leaf node of a subtree induced by one of $v_1; \ldots; v_k$ is at the bottom level of tree T. Assume this child to be v_j. If none of the subtrees induced by other child node of v_0 has a leaf node at the bottom level or second last level of tree T, then v_0 cannot be the center or one of the bicenters. This is because the center (or the bicenter) must be a node (or nodes) of the subtree induced by v_j. Without the loss of generality, we assume the subtree t_{v1} induced by v_1 has a leaf node at the bottom level of tree for which the path from root is H. The subtree t_{v2} induced by v_2 has a leaf node either at the last level or second last level. Therefore the path of that leaf node from root is either H or $H - 1$. Now $2H$ or $2H - 1$ will be the length of path considering from the bottom-level leaf of t_{v1} to the bottom-level leaf of t_{v2} which makes v_0 as the center or one of the bicenters of the free tree. Therefore, condition 3 holds. Besides in case 2, it is essential to hold the condition 2 true, when T turns out to a bicentral tree and v_0 will only become the center if it has lexicographically minimum label.

4   Frequent Free Subtree Mining Algorithm: *FreeS*

FreeS consists of two main steps: (1) candidate subtree generation using the enumeration tree; and (2) frequency counting to determine frequent subtrees.

4.1   Candidate Subtree Generation Using Enumeration Tree

Using the proposed balanced optimal canonical form of free trees and other tree structural information from a database, we define an enumeration tree that lists all subtrees in T_{db}, in their balanced optimal canonical forms. Since the underlying tree structure information is used for defining the enumeration tree, it is called tree-structure guided scheme based enumeration. To the best of our knowledge, *FreeS* is the first algorithm where this enumeration approach is used to generate candidate free trees.

Tree-Structure Guided Scheme Based Enumeration Tree: The task here is enumerating a complete and non-redundant list of candidate subtrees from a given database. A candidate enumeration technique can generate both valid and invalid candidates. A candidate subtree is called valid if it exists in the considered database [11]. It is desirable to enumerate only the valid subtrees in order to reduce the computational efforts, instead of generating all possible candidates and prune invalid subtrees later. The tree-structure guided scheme based enumeration allows invalid subtrees, which will never be significant in spite of being frequent, to be excluded from counting the number of candidate trees. It utilizes the tree structural information such as level, weight and fan-out of nodes, which are learned from a given database, in determining a valid subtree. This information is obtained after the free trees are normalized to rooted unordered trees. Instead of testing whether a tree actually exist in the database that is computationally expensive, a subtree is considered valid if it conforms to the tree structural information.

Extending the Enumeration Tree: The right-path extension and join operations have been used to grow the enumeration tree. Previous research has shown that the right-path extension produces a complete and non-redundant candidate generation [1, 8, 14]. However, the use of extension alone for growing enumeration tree can be inefficient because the number of potential growth may be very large, especially when the cardinality of alphabets for node labels is large [1, 8]. This shortcoming necessitates of using a join operation; however, it often generates invalid subtrees. *FreeS* controls it by using the tree-structure guided scheme based enumeration. The basis of growing the enumeration tree of free trees is as follows: *By removing the last leg (node along with edge), i.e., the rightmost leg at the bottom level, of a $(n + 1)$-free tree BOCF will result in the BOCF for another n-free tree.* The definitions of two operations for extending the enumeration tree are as follows.

Definition 3 (*FreeS-extension*). For node v_i (fan-out $\neq 0$) of a n-free tree in its balanced optimal canonical form t_v, an extension is possible by applying every frequent node label v_j that has a level equal to $Lv(t_v, v_i) - 1$. This extension operation will result in another balanced optimal canonical form t'_v of a new $(n + 1)$-free tree, with v_j child of v_i, in the enumeration tree iff conditions of Lemmas 2 and 3 are held. Further extension is possible from this new right-most node v_j iff conditions are fulfilled again.

Before giving the definition of *FreeS-join* operation, we define *equivalent group*.

Definition 4 (*Equivalent group*). If two BOCFs t_v and t'_v of two n-free trees have equal height H and common first n-1 nodes (along with labels and weights), they are considered as equivalent group, denoted by $t_v \cong t'_v$. Only the nth node of each of these trees, that appear last in their canonical forms, are different.

Definition 5 (*FreeS-join*). Join operation is a guided extension between two free trees in BOCFs t_v and t'_v, that are members of an equivalent group, $t_v \cong t'_v$. Assume, v_i and v_j are the corresponding right-most node of t_v and t'_v, where $w_i > w_j$ or, $w_i = w_j$ with v_i lexicographically sorts lower than v_j. By joining v_j in t_v at the position of $Lv(t_v, v_i) - 1$ will result in a new $(n + 1)$ node balanced optimal canonical form of free tree, denoted by $t_v \odot t'_v$, of the same height as tree t_v.

The join operation does not change the height or the level position of leaf nodes of a newly generated candidate tree, therefore Lemmas 2 and 3 are not considered. As in the tree-structure guided approach, the enumeration tree growth is guided by the prior learned tree structure information. Therefore only valid subtrees are expected to be generated as candidate trees.

Consider an example database in Fig. 5, where for minimum support 1, we compare the enumeration tree (Fig. 5b) used by *FreeS* with the enumeration tree (Fig. 5c) used by the *HybridTreeMiner* (HBT) method [1]. HBT also uses the right-path extension and join operations for growing the enumeration tree, but, these are defined using a different canonical form (Breadth First Canonical Form) [8], whereas we use BOCF

and the tree-structure guided scheme for growing the enumeration tree. The dotted
rectangles in (Fig. 5c) show the generation of invalid subtrees in HBT. We only show a
small part of the enumeration tree for HBT. If it is continued, it will grow in a much
bigger size and will result in much higher numbers of invalid subtrees. In contrast,
Fig. 5(b) is the complete enumeration tree of the considered database for *FreeS*.

It can be clearly seen that the *FreeS* enumeration tree generates much less candidates
in comparison to HBT enumeration tree because of producing only valid subtrees.
Generation of invalid subtrees causes extra memory space and then, pruning of these
subtrees causes additional computational cost for existing methods.

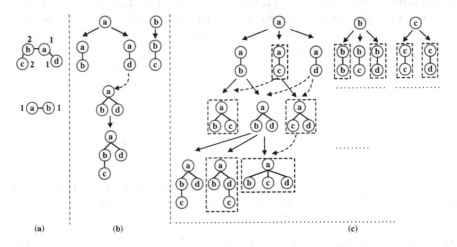

(a)   (b)   (c)

Fig. 5. Sample database of labelled free trees (a), enumeration tree for free trees using tree
structure guided scheme in *FreeS* (b) enumeration tree using the approach from HBT algorithm
(c) (the dotted line with arrow is showing the candidates that are generated using join operations
in HBT, and the dotted rectangle is showing the invalid candidate tree).

4.2   Frequency Counting

For counting frequency we modified the method described in [1, 8], which is basically
an apriori like frequency counting that gives the exact support measure of each candidate subtree by maintaining an occurrence list. We used a catching technique to make
the process of keeping occurrence list more efficient, which is *"stopped counting tree
when the ID counter reaches the min support"*, therefore the occurrence list becomes
smaller than usual.

Figures 6 and 7 list the overall enumeration approach and the *FreeS* algorithm. The
process of frequent subtree mining is initiated by scanning the database T_{db}, where free
trees are stored as BOCF strings along with weight, level and fan-out information of
each node. The set of frequent subtrees of size 1 is generated and the *Enumeration*
method (in Fig. 7) is called recursively for generating the candidates of larger sized
subtrees. The frequency of every resultant candidate tree is computed. The full pruning
is also performed to ensure downward-closure lemma [15]. But full pruning is

Input: Balanced optimal canonical form strings of labelled free trees present in a database T_{db}; level, weight and fan-out information of each node, minimum support (*minsup*) threshold.

Output: All frequent free subtrees.

1. $Result \leftarrow \emptyset$;
2. $Frq1 \leftarrow$ the set of all frequent subtrees of size 1;
3. $Frq2 \leftarrow \emptyset$;
4. **while** $Frq1 \neq \emptyset$ **do**
5.   **for all** $c \in Frq1$ **do**
6.   **if** *fan-out*(c) $!= 0$
7. $Candidate \leftarrow Enumeration\ (c, Frq1, level, weight, fan-out\ )$;
8.   **end if**
9.   **for all** $\mathcal{E}' \in Candidate$ **do**
10.   **if** support $(\mathcal{E}') \geq minsup$ **then**
11. $Frq2 \leftarrow Frq2 \cup \mathcal{E}'$;
12.   **end if**
13.   **end for**
14.   **end for**
15. $Frq1 \leftarrow Frq2$;
16. $Result \leftarrow Result \cup Frq1$;
17. $Frq2 \leftarrow \emptyset$;
18. **end while**
19. **return** $Result$

Fig. 6. High level pseudo code of *FreeS* algorithm.

Enumeration $(l_k, Frq1, level, weight, fan-out)$

1. $Output \leftarrow \emptyset$;
2. **for all** $\mathcal{E} \in Frq1$ **do**
3.   Enumerate candidate l_{k+1} by adding $\mathcal{E}$;   /* Using *FreeS-extension* */
4. $Output \leftarrow Output \cup l_{k+1}$;
5.   **end if**
6. **end for**
7. **for all** equivalent groups in $Output$ **do**
8. $l_{k+2} \leftarrow l_{k+1} \odot l'_{k+1}$;   /* Using *FreeS-join* and $l_{k+1} \cong l'_{k+1}$ */
9. $Output \leftarrow Output \cup l_{k+2}$;
10. **end for**
11. **return** $(Output)$

Fig. 7. High level pseudo code of candidate generation.

expensive; therefore to accelerate this process we cease the frequency checking for a subtree belong to $(K - 1)$ set as soon as the K subtree is found frequent.

5   Empirical Analysis

The efficacy of *FreeS* is shown by conducting systematic experiments using both real-life and synthetic datasets. *FreeS* is benchmarked with the most relevant and leading algorithms *FreeTreeMiner* (FTM) [8] and *HybridTreeMiner* (HBT) [1] which are designed to mine frequent free subtrees from a database of labelled free trees. All experiments have been done on a 2.8 GHz Intel Core i7 PC with 8 GB main memory and running the UNIX operating system.

CSLOGS: This real-life dataset has been widely used in evaluating various tree mining algorithms. CSLOGS [14] contains web access trees of the CS department of Rensselaer Polytechnic Institute during one month. There are a total of 59,691 transactions and 13,209 unique node labels (corresponding to the URLs of the web pages).

Figure 8(a) shows that *FreeS* can find the same amount of subtrees in significant lesser time than its counterparts. Results show that below a certain support threshold (0.25 %) the number of frequent trees explodes that causes huge memory consumption for HBT and consequently, the software automatically aborts the process. For calculating support of free trees, HBT uses occurrence list that makes the process faster, but, it is responsible for high memory usage too. *FreeS* performs this step within the memory size even for smaller minimum support threshold such as 0.15 % because of using modified occurrence list. FTM does not suffer from the memory exhaustion problem though; however the run time increases drastically for smaller supports due to the lack of efficient frequency counting and inclusion of the expensive apriori candidate generation.

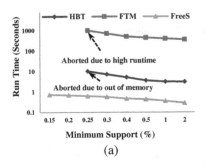
(a)

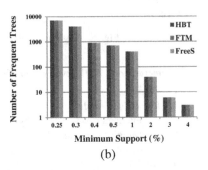
(b)

Fig. 8. Run time comparison (a) and completeness test (b) using CSLOGS data (a $\log_{10}$ scale is used in Y axis).

The runtime performance of *FreeS* is few orders of magnitude better than HBT and FTM due to several reasons. (1) *FreeS* uses tree-structure guided based enumeration tree that allows enumerating only valid subtrees. (2) BOCF is defined to enumerate only one free tree for either of central or bicentral free trees, hence the occurrence list only keeps record of one tree. (3) A catching technique assists in keeping the occurrence list shorter. On the other hand, HBT can't avoid generating invalid candidate subtrees during enumeration, which results in extra memory consumption. HBT may also enumerate two free trees from a bicentral tree because of the supplementary canonical form concept [1]. Consequently, it will keep record of both trees which increases the size of the occurrence list.

Results in Fig. 8(b) show that *FreeS* extracts the same amount of frequent patterns as the other state-of-the-art methods. The tree model guided enumeration employed in

FreeS does not generate any invalid trees but does not miss on any valid trees. All three algorithms satisfy the completeness property and do not miss any frequent patterns since they all used full pruning (downward closure lemma), not an opportunistic pruning. This shows the accuracy of *FreeS* in finding subtrees.

Synthetic Data Sets: We conducted few more experiments using synthetic datasets with varied properties to support all of the above findings. The synthetic data sets were generated by a tree generator as described in [14]. The dataset called D1 is created using following parameters: the number of labels $L = 10$, the number of vertices in the master tree $M = 100$, the maximum depth $D = 10$, the maximum fan-out $F = 5$ and the total number of subtrees $T = 5000$. Such characteristics reflect the properties of web-browsing but not of very large databases. Result in Fig. 9(a) shows that *FreeS* requires less runtime than HBT and FTM as expected. The memory consumption is also low for *FreeS*, whereas for being the small dataset the other two can also perform within the given memory size, Fig. 9(b).

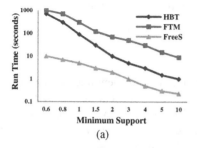 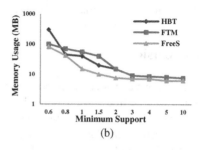

(a)   (b)

Fig. 9. Runtime (a) and memory usage (b) comparison using dataset D1 (a $\log_{10}$ scale is used in Y axis).

The dataset called D2 is generated using high fan-out, $F = 20$ with low number of labels $L = 10$ and a moderate size dataset $T = 10,000$. The rest of the parameters are kept the same. This makes D2 having wider trees than the deep trees. The isomorphic problem is known to occur more commonly when trees have several siblings at same label. This facet of experiment will support the claim that *FreeS* can handle isomorphism more effectively than any other algorithms due to the use of BOCF.

As shown in Fig. 10, *FreeS* consumes much less processing time in comparison to other methods. It happens as *FreeS* does not generate a candidate tree multiple times because of using BOCF that ensures same identity for all isomorphic trees. Therefore, no additional test is required for checking the presence of isomorphism during frequency counting. In contrast, the state-of-the-art algorithms perform a mandatory isomorphism checking which makes them more expensive (Fig. 10a).

Figure 10(b) shows that HBT consumes larger memory space than FTM and *FreeS*, and it becomes worse for smaller support thresholds. As explained before, FTM does not use occurrence list for frequency counting but computes the occurrences of each free tree. Therefore, it saves memory but consumes additional computational time. The usage of occurrence list becomes a pressing concern in terms of memory for large data, especially when the support threshold is low, but allows fast and efficient frequency

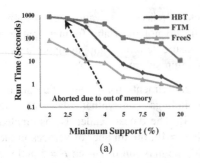
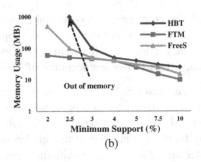

(a)

(b)

Fig. 10. Runtime (a) and memory (b) comparison using dataset D2 (a $\log_{10}$ scale is used in Y axis).

checking. The catching mechanism employed in *FreeS* makes it consume less memory as well as the enumeration strategy does not generate any invalid subtrees, therefore *FreeS* can offer a good trade-off between memory usage and runtime.

6   Conclusion

In this paper, we consider an important problem of mining frequent free subtrees from a collection of free trees. We proposed a computationally efficient algorithm *FreeS* to discover all frequent subtrees in a database of free trees. A novel balanced optimal canonical form is introduced that ensures unique identity of frequent free trees even in presence of isomorphism. Because of this canonical form the isomorphism problem can be handled, that is responsible for computational complexity in this process. Moreover, the proposed tree-structure guided scheme based enumeration enables *FreeS* to reduce the cost for candidate generation by enumerating only valid subtrees. We modified the efficient apriori like occurrence list based frequency counting method that ensures less memory consumption.

Our empirical analyses show *FreeS* is scalable to mine frequent free trees in a large database of free trees with low support thresholds. In future we are planning to extend our algorithm for mining free trees in graph database.

References

1. Chi, Y., Yang, Y., Muntz, R.R.: HybridTreeMiner: an efficient algorithm for mining frequent rooted trees and free trees using canonical forms. In: Proceedings of the 16th International Conference on Scientific and Statistical Database Management, pp. 11–20. IEEE, Santorini (2004)
2. Chi, Y., Yang, Y., Muntz, R.R.: Indexing and mining free trees. In: Third IEEE International Conference on Data Mining, ICDM 2003, pp. 509–512. IEEE (2003)
3. Zhao, P., Yu, J.: Fast frequent free tree mining in graph databases. World Wide Web **11**(1), 71–92 (2008)

4. Rückert, U., Kramer, S.: Frequent free tree discovery in graph data. In: Proceedings of the 2004 ACM Symposium on Applied Computing, pp. 564–570. ACM (2004)

5. Hein, J., Jiang, T., Wang, L., Zhang, K.: On the complexity of comparing evolutionary trees. Discrete Appl. Math. **71**(1), 153–169 (1996)

6. Cui, J.-H., Kim, J., Maggiorini, D., Boussetta, K., Gerla, M.: Aggregated multicast–a comparative study. Cluster Comput. **8**(1), 15–26 (2005)

7. Chi, Y., Muntz, R.R., Nijssen, S., Kok, J.N.: Frequent subtree mining - an overview. Fundam. Inform. **66**(1–2), 161–198 (2004)

8. Chi, Y., Yang, Y., Muntz, R.R.: Canonical forms for labelled trees and their applications in frequent subtree mining. Knowl. Inf. Syst. **8**(2), 203–234 (2005)

9. Valiente, G.: Algorithms on Trees and Graphs. Springer, Heidelberg (2002)

10. Chowdhury, I.J., Nayak, R.: A novel method for finding similarities between unordered trees using matrix data model. In: Lin, X., Manolopoulos, Y., Srivastava, D., Huang, G. (eds.) WISE 2013, Part I. LNCS, vol. 8180, pp. 421–430. Springer, Heidelberg (2013)

11. Chowdhury, I.J., Nayak, R.: BEST: an efficient algorithm for mining frequent unordered embedded subtrees. In: Pham, D.-N., Park, S.-B. (eds.) PRICAI 2014. LNCS, vol. 8862, pp. 459–471. Springer, Heidelberg (2014)

12. Chowdhury, I.J., Nayak, R.: BOSTER: an efficient algorithm for mining frequent unordered induced subtrees. In: Benatallah, B., Bestavros, A., Manolopoulos, Y., Vakali, A., Zhang, Y. (eds.) WISE 2014, Part I. LNCS, vol. 8786, pp. 146–155. Springer, Heidelberg (2014)

13. Ullman, J.D., Aho, A.V., Hopcroft, J.E.: The Design and Analysis of Computer Algorithms. Addison-Wesley, Reading (1974)

14. Zaki, M.J.: Efficiently mining frequent trees in a forest: algorithms and applications. IEEE Trans. Knowl. Data Eng. **17**(8), 1021–1035 (2005)

15. Agrawal, R., Srikant, R.: Fast algorithms for mining association rules in large databases. In: Proceedings of the 20th International Conference on Very Large Data Bases, pp. 487–499. Morgan Kaufmann Publishers Inc. (1994)

Time-Sensitive Topic Derivation in Twitter

Robertus Nugroho[1]([✉]), Weiliang Zhao[1], Jian Yang[1], Cecile Paris[2],
Surya Nepal[2], and Yan Mei[1]

[1] Macquarie University, Sydney, Australia
{robertus.nugroho,yan.mei}@students.mq.edu.au,
{weiliang.zhao,jian.yang}@mq.edu.au
[2] CSIRO, Sydney, Australia
{cecile.paris,surya.nepal}@csiro.au

Abstract. Much research has been concerned with deriving topics from
Twitter and applying the outcomes in a variety of real life applications such as emergency management, business advertisements and cor-
porate/government communication. These activities have used mostly
Twitter content to derive topics. More recently, tweet interactions have
also been considered, leading to better topics. Given the dynamic aspect
of Twitter, we hypothesize that temporal features could further improve
topic derivation on a Twitter collection. In this paper, we first perform
experiments to characterize the temporal features of the interactions in
Twitter. We then propose a time-sensitive topic derivation method. The
proposed method incorporates temporal features when it clusters the
tweets and identifies the representative terms for each topic. Our experimental results show that the inclusion of temporal features into topic
derivation results in a significant improvement for both topic clustering
accuracy and topic coherence comparing to existing baseline methods.

Keywords: Temporal features in twitter · Topic derivation · Joint matrix
factorization

1 Introduction

With about 288 million monthly active users and around 500 million tweets per
day[1], Twitter is one of the most used social media platforms. Topic derivation
from Twitter, to understand what people are talking about, is the foundation
for a wide range of applications such as emergency, social awareness, health
monitoring, and market analysis, and of interest to many organizations [1].

Topic derivation is the process of determining the main topic of every Twitter message (tweet) in a collection (to cluster the tweet based on topics) and
choosing a set of terms to represent each topic [2]. Deriving topics from Twitter
is a challenging task for several reasons: first, tweets are short (140 characters
maximum) and often include informal language (e.g., emoticons, abbreviations)

---

[1] https://about.twitter.com/company, accessed 17 April 2015.

© Springer International Publishing Switzerland 2015
J. Wang et al. (Eds.): WISE 2015, Part I, LNCS 9418, pp. 138–152, 2015.
DOI: 10.1007/978-3-319-26190-4_10

and misspellings, leading to a sparsity problem when approaches only rely on term co-occurrences. Second, the Twitter environment is a highly dynamic one, with topics changing quickly over time.

Existing topic derivation methods based on term co-occurrences, such as LDA [3], PLSA [4] and NMF [5], suffer from the sparsity problem. Some have looked at addressing this problem, e.g., [6–8], by exploiting the relationship between correlated terms. However, they still only use the original tweet content, so that the problem remains. [9] proposed a method to incorporate static external resources to augment the tweet content. None of these approaches considered the information hidden in the interactions amongst posts in the Twitter environment. In their work, [10,11] went beyond terms and exploited content based social features such as hashtag, emoticons, and urls. In our previous work [2,12], we proposed topic derivation models that exploit both complex interaction features and content similarity. The intuition behind the use of interaction features such as *mention*, *reply*, and *retweet* to identify topics is that these features are typically employed to indicate that the posts are part of a conversation, and all posts pertaining to a conversation are likely to be on the same topic. Our experiments showed that, indeed, these models resulted in higher quality topics. To address the dynamic aspect of Twitter, some approaches have exploited temporal features, but only with respect to the tweet content or associated hashtags, e.g., [13–15]. To the best of our knowledge, the temporal features of the posts' *interactions* in Twitter have not been explored for topic derivation. This is what we propose to do in this paper.

While taking conversations into account as in [2,12] can improve topic quality, conversations typically have a time element associated with them. So incorporating a temporal aspect when looking at the interactions might further help topic derivation. For example, two mentions of same users nearly at the same time are more likely to be about the same topic than two mentions of same users within a long time interval. In this paper, we investigate the temporal features of Twitter interactions and propose a topic derivation method that employs these features, building on our previous work [12]. This research is summarized as:

- We discuss the relationships between topics and interaction features (*mention*, *reply* and *retweet*) using a data set obtained by collecting tweets over a month. We found that the *mention* feature is time sensitive with respect to topic assignation.
- We model the time sensitivity of *mentions* as an exponential decay according to the time difference of two tweets with the same mention. The decay parameter is based on an analysis of tweets that include a mention. This time-sensitivity model is then incorporated in the tweet relationship model in order to affect the matrix inter-joint factorization for topic derivation.
- We conducted a comprehensive set of experiments to evaluate our new model with a Twitter dataset covering one-month tweets, using widely accepted evaluation metrics for topic derivation. The results show that the new time-sensitive method results in a significant improvement of the accuracy of tweet clustering and coherence between terms for topic representation comparing with well-known baseline methods and our previous work [12].

The rest of the paper is organized as follows. Section 2 describes an investigation on the temporal features of *mentions, replies*, and *retweets*. Section 3 proposes a topic derivation method that takes these features into account. Section 4 reports on our experiments, with first a discussion of the dataset, the baselines and the evaluation metrics. Related work is provided in Sect. 5, and we conclude in Sect. 6.

2   Temporal Features of Tweet Interactions

Twitter has evolved from a microblogging platform to a medium that also enables people to interact with each other in a conversation-like manner. A user can initiate a conversation by mentioning other users in his/her tweet, and a tweet can be "replied to" by other users, or retweeted to other users. These *mention, reply* and *retweet* features form interactions between users, often related to a particular topic. A *reply* is a clear turn in a discussion between users; a *retweet* resends the message. It is likely that both a message that contains a reply and one that contains a retweet are on the same topic as the original post. Two tweets which mention the same user are also likely to be on the same topic *if* they occur around the same time, but not necessarily otherwise. Time thus plays an important role when attempting to link tweets because they mention the same people. In this section we will analyze the impact of time on user interactions for the same topic based on *mention, reply* and *retweet*.

We investigate users' mention behavior by analyzing tweets in a Twitter dataset to see how time affects the connectivity between tweets. Using the Twitter's streaming API[2], we retrieved all tweets from the top 15 Twitter users in Australia[3] and all the tweets that mention those users during January 12, 2015 until February 12, 2015. Our data set consists of more than 6 million tweets and involves around 800 thousand users. The details of the dataset are shown in Table 1.

Our investigation starts with an analysis of individual user mentions at different level of granularity to see how the mentions are distributed over time. We then look at the topics in the dataset to see if there is a relationship between the mention distribution and the topics. We find that, for all users, when the number of mentions of a specific user rises at a particular time, most of the tweets at that time are on the same topic.

Figure 1 shows the distributions of the tweets that mention *@MrKRudd* in a 3 hr time interval. We can see that there are several fluctuations within different time intervals. We find that each peak in Fig. 1 (an indication of a sharp increase in the number of tweets mentioning *@MrKRudd*) is strongly related to a particular topic. For example, on January 22, 2015 at 7 am (22/7), most of the tweets mentioning *@MrKRudd* were talking about the *"plain packaging act"*. The tweets on January 31, 2015 at 1 pm were about *"Queensland votes"*,

---

[2] https://dev.twitter.com/streaming/overview.

[3] https://followerwonk.com/bio/?q_type=all&l=Australia, accessed January 11, 2015, ordered by number of followers.

Table 1. Top 15 Twitter users in Australia and all related tweets (i.e., tweets that involve these top 15 Twitter users, either by mentioning them, replying to them or retweeting their posts) between Jan 12, 2015 and Feb 12, 2015

Username	# related tweets	# of users involved	# of followers
@CodySimpson	388,970	69,246	7,384,541
@5SOS	2,068,129	258,292	6,619,112
@Calumn5SOS	2,330,628	340,686	5,154,177
@luke_brooks	583,999	56,908	2,242,597
@example	8,464	5,208	2,107,484
@KyrieIrving	46,896	33,311	2,064,137
@BrooksBeau	819,423	95,879	1,932,857
@jascurtissmith	3,318	1,368	1,831,271
@MrKRudd	2,249	1,553	1,524,455
@allisimpson	88,504	20,107	1,418,732
@claireholt	5,413	2,497	1,299,287
@MClarke23	2,442	1,525	1,293,651
@DarrynLyons	1,154	390	1,143,222
@hillsongunited	3,456	2,455	969,020
@imacelebrity	1,675	1,340	894,187
@JordanJansen	10,774	2,512	759,192

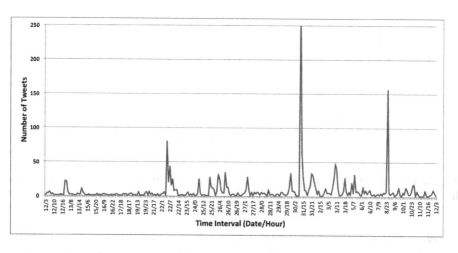

Fig. 1. Tweets mentioning user @MrKRudd with 3 h time interval

and the tweets on February 08, 2015 at 11 pm about *"the end of Kevin Rudd's leadership in February 2012"*.

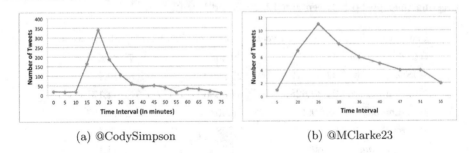

(a) @CodySimpson   (b) @MClarke23

Fig. 2. Tweet distributions of tweets mentioning (a) *@CodySimpson* and (b) *@MClarke23* on 5 min time intervals within 1 h

The rises in the number of tweets with the same mention reaches their peak quickly and then slowly fade away (decay). Figure 2 shows the subset of the distributions of the tweets that mention (a) *@CodySimpson* and (b) *@MClarke23* on 5 min intervals. The specific distributions are different, reaching their peaks and decaying at different rates. What they have in common, however, is that each peak indicates a specific topic. The peak in Fig. 2a is related with the topic: *"Cody's birthday"*; and the peak in Fig. 2b is related with the topic: *"the absence of Michael Clarke on treatment issue"*.

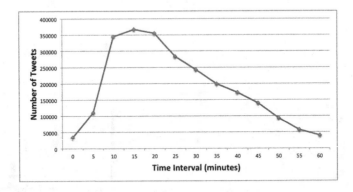

Fig. 3. The sum of all fluctuations in all tweet mention distributions with 5 min time interval

We performed a statistical analysis on all the variations of the tweet distributions, using a 5 min interval. We sum up the number of tweets from all users by choosing the subset of the tweet distributions starting from the closest lowest point before a peak and ending at the lowest point after the peak. Figure 3 shows this sum. We can see from the figure that most of mentions related to a particular topic reach a peak in about 15 min and then gradually fade away. An exponential function is adopted to model the process of fading away. We calculate the half-life of the exponential decay, which is how long the mention frequency decays from its peak to the peak's half value, using the following formula:

$$a = i_{t_{max}/2} - i_{t_{max}} \qquad (1)$$

where $i_{t_{max}}$ is the time when the tweet mention distribution reaches its peak, and $i_{t_{max}/2}$ is the time when the tweet mention distribution reaches half of the peak value after the peak. In Fig. 3, the number of tweets in the highest point (t_{max}) is 367,368, and it is reached after 15 min ($i_{t_{max}}$). Then, $i_{t_{max}/2}$ can be calculated as the time to reach 183,684 after the peak, which is 37 min. So, a for Fig. 3 will be 22 min (1,320 s). This a will be used in the exponential function that models time in the mention behavior in Twitter in the next section.

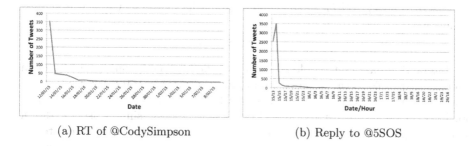

(a) RT of @CodySimpson   (b) Reply to @5SOS

Fig. 4. Tweet distributions of a. retweet to a tweet by (a) *@CodySimpson* and (b) reply to a tweet by *@5SOS* within 1 month period

In contrast to the mention behavior, the topic relationship of a *reply* or a *retweet* with respect to the original tweet is not affected by time. As expected, the analysis of the dataset shows that a *retweet* or a *reply* could occur much after the original tweet and still be on the same topic.

Figure 4a shows the tweet distributions of a retweet to a tweet by *@CodySimpson*: (*"It's the 11th back home in Aus. I m officially 18."*). The tweet was retweeted for 494 times in total, with 354 retweets on the first day, 22 on the third day, and the remaining scattered over time. Irrespective of the time elapsed, the retweets are still on the same topic. Figure 4b shows the tweet distribution of the replies to a tweet by @5SOS (*"Getting lots and lots of ideas for songs! Ready to write a new record!!"*). The total number of replies was 7414 tweets, with a peak on the first day but continuing the following day (with still 291 replies on the next day).

3   Topic Derivation

Our aim is to improve the quality of topic derivation in Twitter. Following our previous work [12], we classify the social interactions present in Twitter messages as interactions based on people and actions. In this paper, we first improve our interaction model to incorporate a time aspect. We then incorporate the new model into a matrix inter-joint factorization process to simultaneously achieve the clustering of the tweets based on topics and the identification of representative terms for each topic.

3.1   Relationship Between Tweets

A tweet is defined as a tuple of $t = \langle P_t, RTP_t, C_t, i_t \rangle$, where $P_t \subset P$ is the union of the author and people mentioned in the tweets, RTP_t the reply and retweet information, $C_t \subset C$ the set of the terms contained by the tweet, and i_t the timestamp of the tweet. We denote a relationship between two tweets t_i and t_j as $R(t_i, t_j)$. A zero value (0) of R means that there is no relation between them, and a higher value indicates the relationship is stronger. The relationship R includes three components: interactions based on people ($po(P_{t_i}, P_{t_j})$), common actions ($act(RTP_{t_i}, RTP_{t_j})$), and content similarity ($sim(C_{t_i}, C_{t_j})$). It is defined as follows:

$$R(t_i, t_j) = po(P_{t_i}, P_{t_j}) + act(RTP_{t_i}, RTP_{t_j}) + sim(C_{t_i}, C_{t_j}) \,. \tag{2}$$

Interaction based on people $po(P_{t_i}, P_{t_j})$ is defined as the number of common *mentioned* people in the tweets t_i and t_j divided by the total number of people *mentioned* in both tweets. As discussed in Sect. 2, time affect the topic behavior in tweet mentions distributions. Tweets that mention similar users within a particular period are more likely to share the same topic. So, for the interactions based on people, we add a temporal factor $f(i_{t_i} - it_j)$. The people-based interaction is calculated as follows:

$$po(P_{t_i}, P_{t_j}) = \frac{|P_{t_i} \cap P_{t_j}|}{|P_{t_i} \cup P_{t_j}|} f(i_{t_i} - it_j) \tag{3}$$

$$where \quad f(i_{t_i} - it_j) = e^{-\frac{1}{a}|it_i - it_j|},$$

$f(i_{t_i} - it_j)$ is the exponential function that models time in the mention behavior in Twitter. Its parameter, a, was defined in the previous section. $f(i_{t_i} - it_j)$ controls the decay rate of the temporal effect.

The interaction based on user actions, denoted as $act(RTP_{t_i}, RTP_{t_j})$, is based on the *retweet* and *reply* relationship between two tweets. As already mentioned, time does not have an effect on these relationships. If tweet A is a *retweet* or *reply* of tweet B (or vice versa), or if both tweets are *replying* to or *retweeting* the same tweet, $act(RTP_{t_i}, RTP_{t_j})$ will be 1 (indicating a strong relationship), otherwise it is 0. We denote a *retweet* or *reply* of tweet t as RTP_t.

$$act(RTP_{t_i}, RTP_{t_j}) = \begin{cases} 1, (RTP_{t_i} = t_j) \; or \; (t_i = RTP_{t_j}) \\ \quad or \; (RTP_{t_i} = RTP_{t_j}) \\ 0, \; otherwise \end{cases} \tag{4}$$

As there are a large number of self-contained tweets (i.e., tweets with no relation to any other tweet), our model for topic derivation also takes content similarity between tweets into account. Before calculating the content similarity, we perform some preprocessing steps to remove all irrelevant terms/characters and stop words. As tweets are short, two tweets sharing at least one (non-stop) word are likely to be on the same topic. For this purpose, $sim(C_{t_i}, C_{t_j})$ denotes the similarity between tweet t_i and t_j, which is measured by *cosine similarity* [16].

$$sim(C_{t_i}, C_{t_j}) = \frac{C_{t_i}.C_{t_j}}{\|C_{t_i}\|\|C_{t_j}\|} . \tag{5}$$

The values of all the relationships among the tweets form a tweet-to-tweet relationship matrix $A \in \mathbb{R}^{m \times m}$, where $a_{ij} = f(R(t_i, t_j))$. $f(R(t_i, t_j))$ is a *sigmoid function* [17] to normalize the value of $R(t_i, t_j)$ for a better relationship distribution.

$$f(R(t_i, t_j)) = \begin{cases} \frac{1}{1+e^{-R(t_i,t_j)}}, R(t_i, t_j) > 0 \\ 0, \; otherwise \end{cases} \tag{6}$$

By incorporating a time factor in the people-based interactions, we obtain a more accurate tweet-to-tweet relationship matrix. This matrix will be used to improve the topic derivation by jointly factorizing it with tweet-to-term matrix, as discussed in the next section.

3.2   Matrix *inter-joint* Factorization for Topic Derivation

We incorporate time into the Non-Negative Matrix inter-joint Factorization (*NMijF*) process described in [12]. We denote the resulting new method as *tNMijF*. Like the method on which it is based, *tNMijF* is an *inter-joint* factorization of a non-negative symmetric matrix $A \in \mathbb{R}^{m \times m}$ and another non-negative matrix $V \in \mathbb{R}^{m \times n}$ within a unified process. In our implementation, matrix A is the new tweet-to-tweet relationship matrix discussed in previous section (which includes a temporal aspect), and V is the tweet-to-term matrix which contains the relationship between tweets and the unique terms appearing in all tweets in the dataset. Each element in V is calculated using the *tf-idf* function described in [18]. We briefly describe the process here. More details can be found in [12].

The tweet-to-tweet matrix A is factorized to the tweet-topic matrix W as a base and W^T as the coefficient matrix. *Within the same process*, the tweet-to-term matrix V is factorized to the shared tweet-topic matrix W and topic-term matrix Y as the coefficient. In this method, matrix A and V share the tweet-topic matrix W. Hence, by implementing *tNMijF*, we can directly retrieve the main topic of a tweet from the tweet-topic matrix W *and* the top-n representative terms for each topic from the topic-term matrix Y within a unified process.

Tweet-to-tweet matrix A is much more dense than the tweet-to-term matrix V. At the best case (all terms are connected), the density of A will be equal to V. Sparsity of V could heavily penalized the quality of topic derivation. So, to handle this problem, the effect of matrix V in the factorization process to retrieve matrix W needs to be reduced. We implement the scale parameter α to control the effect in every iteration to achieve the objective function.

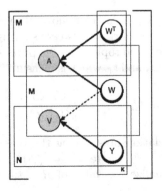

Fig. 5. Graphical Model of *tNMijF*

The inter-joint factorization process in *tNMijF* aims at finding the minimum divergence ($\mathscr{D}$) of $A \approx WW^T$ and $V \approx WY$. The graphical model for *tNMijF* is shown on Fig. 5, with the following objective function ($\mathscr{T}_{tNMijF}$):

$$\mathscr{T}_{tNMijF} = \mathscr{D}(A\|WW^T) + \alpha\mathscr{D}(V\|WY) \qquad (7)$$

$$= \sum_{im} d(a_{im}|(ww^T)_{im}) + \alpha \sum_{mn} d(v_{mn}|(wy)_{mn})$$

where there exists at least one element w and y in matrices W and Y such that $w \geq 0$ and $y \geq 0$, and the scaling parameter α satisfies $0 \leq \alpha \leq 1$.

For each element wise divergence, we employs *Kullback-Leibler divergence*:

$$d(a_{im}|(ww^T)_{im}) = a_{im} \log \frac{a_{im}}{(ww^T)_{im}} - a_{im} + (ww^T)_{im}, and \qquad (8)$$

$$d(v_{mn}|(wy)_{mn}) = v_{mn} \log \frac{v_{mn}}{(wy)_{mn}} - v_{mn} + (wy)_{mn}$$

In each iteration, we apply the following multiplicative update rules to every element in latent matrices W and Y to minimize $\mathscr{T}_{tNMijF}$:

$$\hat{w}_{i,k} = w_{i,k} \frac{(\sum_{m=1}^{M} \frac{a_{i,m}}{(ww^T)_{i,m}} w_{k,m}^T + \alpha \sum_{n=1}^{N} \frac{v_{i,n}}{(wy)_{i,n}} y_{k,n})}{\sum_{m=1}^{M} w_{k,m}^T + \alpha \sum_{n=1}^{N} y_{k,n}},$$

$$and \quad \hat{y}_{k,n} = y_{k,n} \frac{(\sum_{m=1}^{M} \frac{w_{k,m}}{(wy)_{k,m}} w_{k,m})}{\sum_{m=1}^{M} w_{k,m}} \qquad (9)$$

4   Experiments

We now describe our experiments with the new model that is time sensitive. We first present our dataset, followed by the baseline methods and the evaluation metrics we employed. Then, we provide the results with a discussion.

4.1   Dataset

To evaluate our new mothod for topic derivation in Twitter, we employed a data set collected between 03 March 2014 and 07 March 2014 using the Twitter Streaming API. We call this dataset the *TweetMarch*. It includes 729,334 tweets involving 509,713 users all over the world. It contains 12,221 reply tweets, 101,272 retweets, and the rest are self-contained tweets.

We only used tweets in English in the experiments. A pre-processing is employed to remove irrelevant terms or characters (emoticons, punctuations, and terms that less than 3 characters), and stop-words. Then, all terms are lemmatized and all tweets are tokenized. Hashtags are kept unchanged.

Four people manually labeled around 120,000 tweets from the first subset of *TweetMarch* dataset as an evaluation set. From the labeled data, we observe that the *TweetMarch* dataset covers a wide range of topics, from politics and traveling to life entertainment and school activities.

4.2   Evaluation Metrics

For the evaluation purposes, we used several baseline methods:

– *NMijF*. This is our previous model. It takes into account tweet' interactions and employs a non negative inter-joint factorization, but it is not time-sensitive. We use this method as a baseline to see the impact of the temporal features. While we have already shown that *NMijF* improves on the next three baselines, *TNMF*, *LDA* and *NMF*, we still include them for completeness sake.
– *TNMF* [6]. This topic derivation method incorporates a term correlation matrix to improve the quality of the result using matrix factorization techniques.
– *LDA* [3]. The most popular method in topic derivation. It has a "bag of words" assumption and works solely on the content of the document.
– *NMF* [5]. This is the basic method of matrix factorization. It directly factorizes the tweet-to-term matrix into topic-tweet and topic-term matrix.

We conducted the evaluations on both the quality of the clusters and the topics produced by all the methods. The quality of the clusters is measured through their accuracy with respect to our manually labeled classes. We compared the clustering result K from N tweets with an evaluation set of classes C. In particular, three metrics were used in the evaluation on cluster quality [18]: Pairwise *F Measure*, *Purity* and *Normalized Mutual Information (NMI)*.

We used the Pairwise *F-Measure* to measure the accuracy of the clustering result by analyzing the harmonic mean of both precision and recall. In this metrics, precision p is defined as the fraction of pairs of tweets correctly put in the same cluster, and recall r is the fraction of actual pairs of tweets that were identified. The formula of pairwise F-Measure is shown in Eq. 10 below:

$$F = 2 \times \frac{p \times r}{p + r} . \tag{10}$$

Purity is calculated by assigning each cluster in K to class in C, and then counting the number of correctly assigned elements divided by the total of elements in all clusters. A *Purity* value of 1 indicates a perfect clustering, whereas a *Purity* value of 0 means low quality clustering.

$$purity(K, C) = \frac{1}{N} \sum_i \max_j |k_i \cap c_j| . \tag{11}$$

NMI measures the mutual information shared between clusters and classes $I(K; C)$, normalized by the entropy of clusters $H(K)$ and classes $H(C)$. Similar to *Purity*, the value of *NMI* will be ranged between 0 and 1.

$$NMI(K, C) = \frac{I(K; C)}{[H(K) + H(C)]/2} . \tag{12}$$

To evaluate the quality of the representative terms for each topic, we used the *topic coherence*, $Co(k, W)$, for a topic described by its topic-term [19]. It measures the readability of all terms that represent the topic by evaluating the frequency of pair of terms in the same tweet over the original dataset. It is described by the following equation:

$$Co(k, W) = \sum_{m=2}^{M} \sum_{l=1}^{m-1} \log \frac{T(w_m, w_l) + 1}{T(w_l)}, \tag{13}$$

where $w_m, w_l \in W$; $T(*)$ and $T(*, *)$ are document frequency and co-document frequency functions, representing the number of tweets which contain a given term or a pair of two terms respectively; M is the size of the set W of topic-term.

4.3 Results and Discussion

To see the performance of our method on a different number of topics, we used $k = 20, 40,$ and 60 as input for each experiment with every method. We run all methods for 20 times over the dataset and tuned all parameters for the best performance. The average density (non zero element) of the tweet-to-term matrix V is only 0.08 %, which is far below our tweet-to-tweet relationship matrix with 32.64 % density. The scaling parameter $\alpha = 0.1$ was found to be the best for all of the matrix inter-joint factorization processes as the matrix V is very sparse. This α value ensures that the sparsity of V does not heavily penalize the topic-tweet matrix W and still gives good results when factorizing the topic-term matrix Y.

Table 2 shows the results of the pairwise *F-Measure* metrics. It can be seen that the inclusion of time improves both precision and recall in comparison to the baseline methods for all values of k (the number of topics). *tNMijF* consistently provides the best results for both precision and recall, with a positive trend over increasing values of k. Our previous work, *NMijF*, which does not take time into account, also outperforms the other baseline methods. However, as k increases, the improvement in precision and recall lessens. In contrast, the new method proposed in this paper gives a consistent improvement of the precision and recall for all k values.

Table 2. *Precision, Recall and F-Measure for topics* $k = 20, 40, 60$

Method	k=20			k=40			k=60		
	p	r	F-m	p	r	F-m	p	r	F-m
tNMijF	**0.407**	**0.236**	**0.298**	**0.444**	**0.264**	**0.330**	**0.481**	**0.292**	**0.361**
NMijF	0.396	0.218	0.280	0.417	0.227	0.293	0.418	0.227	0.293
TNMF	0.276	0.079	0.123	0.335	0.051	0.088	0.381	0.043	0.078
LDA	0.310	0.084	0.132	0.369	0.057	0.099	0.404	0.047	0.084
NMF	0.271	0.072	0.114	0.336	0.047	0.083	0.405	0.039	0.072

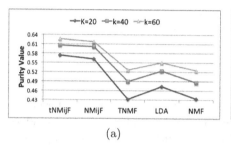
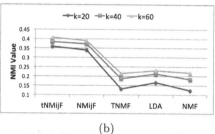

(a)   (b)

Fig. 6. (a) Purity evaluation results and (b) NMI evaluation results

This cluster evaluation is confirmed by other two metrics: *Purity* and *NMI*. Figure 6a shows the evaluation results using the *purity* metrics, and Fig. 6b shows the results of the NMI evaluation. In the purity evaluation, our proposed method *tNMijF* gives about 5 % improvement over our previous work, and 15–30 % over the other baseline methods. For the NMI evaluation, *tNMijF* results in roughly a 5 % improvement compared to *NMijF*, and 90–200 % improvement over the other methods, *TNMF*, *LDA* and *NMF*. We conclude that the introduction of a temporal aspect leads to an obvious improvement over other methods for the accuracy of the topic derivation process.

For the topic coherence evaluation, we use the metric defined in Eq. 7 and take the top-10 terms to represent each topic from the topic-term matrix Y. Figure 7 shows the result of the topic coherence evaluation. We can see that, for a small number of topics (k=20), all methods have quite a good performance. When the number of topics becomes bigger, however, the topic coherence with our new method *tNMijF* reduces only slightly in comparison to the baseline methods which have significant drops. This result shows that *tNMijF* is reliable for different numbers of topics in terms of the topic coherence.

The above results show that introducing a time factor on the interaction features (in particular the *mention*) when performing topic derivation with a non-negative joint matrix factorization process greatly improves the accuracy of tweet clustering and the coherence of topics. This improvement is consistent for any number of derived topics.

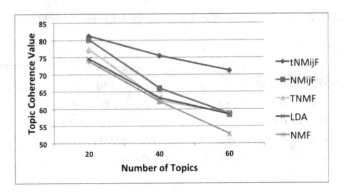

Fig. 7. Topic Coherence

5   Related Work

The short-in-content nature of Twitter presents a challenging problem for deriving the topics of a tweet collection. The very limited length for each tweet renders the frequency of co-occurences between terms extremely low. This sparsity heavily penalizes the performance of the state of the art topic derivation methods such as LDA [3], PLSA [4] and NMF [5], as they generally work solely on content features.

A lot of studies have been conducted to extend those popular methods to handle the sparsity issues. [10] proposed a variant of *labeled-LDA* to work on Twitter environment with the hashtag and other content features as labels for a partially supervised topic learning process. Albakour et al. [9] and Vosecky et al. [11] addressed the problem by expanding the content with the help of external documents collections. However, relying on external documents brings an extra burden when dealing with highly dynamic environments like Twitter. The approach reported in [6,7] exploits the term co-occurrence patterns to improve the topic learning process in a short text environments. Unfortunately, in Twitter environment, the relationship between terms is very sparse and it only provides a small improvement with respect to density in comparison with the original tweet-to-term relationships [2].

To deal with the dynamic nature of the Twitter environment, several methods have been proposed by including temporal features. The proposed method in [11] uses a temporal weight function for the recency sensitivity of the tweet content based on hashtags and urls. [13] proposed a temporal based regularization in NMF method to learn the topics in social media. The study in [14] introduced the content aging theory to mine the emerging topics from Twitter stream. Stilo et al. [15] proposed *Symbolix Aggregate approximation* (SAX) to discretize the temporal series of terms to discover the events from Twitter content. All these studies still focus on contents and overlook the social features available in the Twitter environment. As a result, they still suffer from the sparsity issue.

Different from the topic derivation work that only takes content into account, [12] incorporated the relationships between tweets to deal with the sparsity problem in the Twitter environment and showed improvements in performance. The work presented in this paper builds on this foundation, adding a time dimension to the interactions. To the best of our knowledge, our proposed method is the first one to incorporate temporal features, social interactions and content in a unified model to derive topics from a collection of tweets.

6   Conclusions

In this paper, we investigate the effect of time on user interactions for topic derivation in Twitter. We propose a new topic derivation method that includes this time factor. It can simultaneously achieve the clustering of the tweets based on topics and the identification of the representative terms for each topic. We conducted a set of experiments on a set of tweets collected over a period of one month.

Our results show that incorporating of a time aspect on the interaction features improves the results of the topic derivation process. In particular, the proposed method results in a consistent improvement in the accuracy of the tweet clusterings and topic coherence for different numbers of topics over both well-known baseline methods and our prior method, which was not time-sensitive. Currently, the method works for a static Twitter dataset. We are developing the incremental model of the proposed method to work with the stream based Twitter messages.

Acknowledgment. This work is supported by the Indonesian Directorate General of Higher Education (DGHE), Macquarie University, CSIRO and Australian Research Council Linkage Project (LP120200231).

References

1. Wan, S., Paris, C.: Improving government services with social media feedback. In: Proceedings of the 19th International Conference on Intelligent User Interfaces. IUI 2014, New York, NY, USA, pp. 27–36. ACM (2014)
2. Nugroho, R., Molla-Aliod, D., Yang, J., Paris, C., Nepal, S.: Incorporating tweet relationships into topic derivation. In: Proceedings of the 2015 Conference of the Pacific Association for Computational Linguistics, PACLING (2015)
3. Blei, D., Ng, A., Jordan, M.: Latent dirichlet allocation. J. Mach. Learn. Res. **3**, 993–1022 (2003)
4. Hofmann, T.: Probabilistic latent semantic indexing. In: Proceedings of the 22nd Annual International ACM SIGIR Conference on Research and Development in Information Retrieval, pp. 50–57. ACM (1999)
5. Lee, D., Seung, H.: Algorithms for non-negative matrix factorization. In: Advances in Neural Information Processing Systems, pp. 556–562 (2000)
6. Yan, X., Guo, J., Liu, S., Cheng, X., Wang, Y.: Learning topics in short texts by non-negative matrix factorization on term correlation matrix. In: Proceedings of the SIAM International Conference on Data Mining. SIAM (2013)

7. Yan, X., Guo, J., Lan, Y., Cheng, X.: A biterm topic model for short texts. In: Proceedings of the 22nd International Conference on World Wide Web, International World Wide Web Conferences Steering Committee, pp. 1445–1456 (2013)

8. Hu, Y., John, A., Wang, F., Kambhampati, S.: Et-lda: joint topic modeling for aligning events and their twitter feedback. AAAI **12**, 59–65 (2012)

9. Albakour, M., Macdonald, C., Ounis, I., et al.: On sparsity and drift for effective real-time filtering in microblogs. In: Proceedings of the 22nd ACM International Conference on Information & Knowledge Management, pp. 419–428. ACM (2013)

10. Ramage, D., Dumais, S.T., Liebling, D.J.: Characterizing microblogs with topic models. ICWSM **10**, 1–1 (2010)

11. Vosecky, J., Jiang, D., Leung, K.W.T., Xing, K., Ng, W.: Integrating social and auxiliary semantics for multifaceted topic modeling in twitter. ACM Trans. Internet Technol. (TOIT) **14**, 27 (2014)

12. Nugroho, R., Zhong, Y., Yang, J., Paris, C., Nepal, S.: Matrix inter-joint factorization - a new approach for topic derivation in twitter. In: Proceedings of the 4th IEEE International Congress on Big Data. IEEE Services Computing (2015)

13. Saha, A., Sindhwani, V.: Learning evolving and emerging topics in social media: a dynamic nmf approach with temporal regularization. In: Proceedings of the Fifth ACM International Conference on Web Search and Data Mining, pp. 693–702. ACM (2012)

14. Cataldi, M., Di Caro, L., Schifanella, C.: Emerging topic detection on twitter based on temporal and social terms evaluation. In: Proceedings of the Tenth International Workshop on Multimedia Data Mining, p. 4. ACM (2010)

15. Stilo, G., Velardi, P.: Time makes sense: Event discovery in twitter using temporal similarity. In: Proceedings of the 2014 IEEE/WIC/ACM International Joint Conferences on Web Intelligence (WI) and Intelligent Agent Technologies (IAT), vol. 2, pp. 186–193. IEEE Computer Society (2014)

16. Salton, G.: Automatic Text Processing: The Transformation, Analysis, and Retrieval of Information by Computer. Addison-Wesley, Reading (1989)

17. Von Seggern, D.H.: CRC Standard Curves and Surfaces with Mathematica. CRC Press, Boca Raton (2006)

18. Manning, C., Raghavan, P., Schütze, H.: Introduction to Information Retrieval, vol. 1. Cambridge University Press, Cambridge (2008)

19. Mimno, D., Wallach, H., Talley, E., Leenders, M., McCallum, A.: Optimizing semantic coherence in topic models. In: Proceedings of the Conference on Empirical Methods in Natural Language Processing (EMNLP), Association for Computational Linguistics, pp. 262–272 (2011)

OSCAR: OpenStreetMap Planet at Your Fingertips via OSm Cell ARrangements

Daniel Bahrdt and Stefan Funke[✉]

FMI, Universität Stuttgart, Stuttgart, Germany
{bahrdt,funke}@fmi.uni-stuttgart.de
http://oscar.fmi.uni-stuttgart.de

Abstract. In this paper we show how to preprocess the complete Open-StreetMap (OSM) planet dataset such that efficient (e.g. "search-as-you-type") geo- and *substring* search on all relevant tags of the OSM data is possible. At the core of our OSCAR system lies the construction of a so-called *OSM Cell ARrangement* induced by all regions defined in the raw data. By having our index structure based on this cell arrangement, we can efficiently (with respect to query time and space) incorporate all common set operations with prefix and substring search together with geometric constraints. More importantly, the structure of the induced cell arrangement provides a natural geographic clustering of the results.

Keywords: Spatio-textual search · OpenStreetMap · Planar arrangement

1 Introduction

The goal of the OpenStreetMap (OSM) project is to create a free world map by 'sourcing the crowd'. It was established to provide an alternative to proprietary data, especially when it comes to rendering individual maps, as basis for route planning tools, and for location-based services. Just to provide some numbers: the OSM planet dataset in 2007 contained less than 30 million nodes whereas in 2015 this number has grown to more than 2.6 billion nodes. A limit to this growth is nowhere to be seen due to the demand for a more and more accurate and detailed representation of our environment. There is an obvious need to develop space- and time-efficient algorithms and data structures to handle and access this huge (and rapidly growing) amount of data.

Let us consider a few use cases for text search on geoinformation and see what current commercial (**Google Maps** and **Bing Maps**) and free (**Nominatim** – the official OSM search engine) offerings can do. The following searches were conducted on February, 27th, 2015.

Result Presentation and Clustering. Assume we want to look at all the locations of 'Ace Hardware' (a retailer's cooperative based hardware store). When opening *Google Maps* (http://maps.google.com) you are shown a map of North America,

© Springer International Publishing Switzerland 2015
J. Wang et al. (Eds.): WISE 2015, Part I, LNCS 9418, pp. 153–168, 2015.
DOI: 10.1007/978-3-319-26190-4_11

searching for 'Ace Hardware' displays (all?) locations of stores on this map. There is no indication how many there are, or how they distribute (in numbers) over the different states. Only when zooming out to a world view, oversea stores in Indonesia, India, etc. are shown. But while graphically shown on the map, no clustering with (approximate) hit count is provided. Performing the same search on *Bing Maps* (http://maps.bing.com) starting with a world view for some reason produces no results at all. Only when searching after zooming in, hits are displayed. Strangely, zooming out again, then also yields world-wide hits (only graphically). The same search on *Nominatim* (http://nominatim.openstreetmap. org) produces 10 somewhat arbitrary hits with the map display closely zoomed on the first hit. Zooming out does not display the entirety of hits worldwide or an approximate hit count. We could only have the next 10 (out of how many?) somewhat arbitrary hits be displayed.

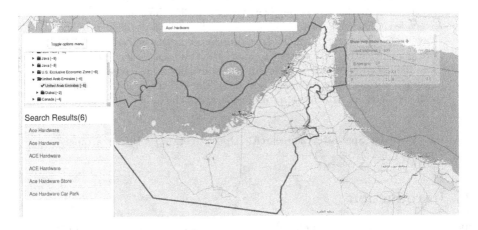

Fig. 1. Zoom-in on the United Arab Emirates for a query 'Ace Hardware'

The web service of our OSCAR system instantaneously provides all results clustered by their geographic location as seen in Fig. 1. Naturally the largest number of hits is inside the United States ($\approx$501), but also several in other countries of the world like Canada, Indonesia, or Mexico. In the current view, the results in the United Arab Emirates are under closer inspection; here, 3 out of 6 hits are in Abu Dhabi, 2 in Dubai, and 1 in Sharjah.

Substring Search. In Germany, many streets named after celebrities bear rather long names. Let's say we want to locate the 'Friedrich-Ludwig-Jahn-Straße' in the village of 'Neuenkirchen', which is often referred to as 'Jahn-Straße' by locals. Searching for 'Jahn-Straße Neuenkirchen' only *Google* and OSCAR produces the correct and unique result, the other search engines did not succeed. OSCAR retrieves the result due to its substring search capability. *Google* probably also considers some natural substrings (in this case breaking the word at the hyphens). However the query 'ahn-Straße Neuenkirchen' fails on Google Maps (no results) whereas OSCAR still succeeds. Also, unknown spelling can sometimes be resolved using substring search, e.g. to match items spelled 'Pfaffenrainweg' or 'Pfaffenreinweg' in the village of Bisingen, OSCAR yields the desired hits with the query 'Pfaffenr inweg Bisingen'. Google and Nominatim require correct spelling (with an 'a'), only Bing apparently employs some fuzzy search.

1.1 OSM Basics

The following basic concepts of the OSM data model are necessary to understand the challenges that arise when processing that data for efficient search. Data in OSM is represented in three forms: 'nodes' (with latitude and longitude), polygonal paths or 'ways' (consisting of one or more nodes) and 'relations' which are compositions thereof (i.e. referencing sets of nodes, ways or other relations). All three forms can be augmented with key-value pairs called tags, which allow to name and classify the data and provide arbitrary additional information.

The road network constitutes a large part of the OSM data, which are naturally modelled as *ways*. Typically, ways are tagged with the type and name of a road. Additional nodes (not necessarily part of a way) can be introduced to denote locations of hotels, restaurants or points-of-interest (POIs) in general.

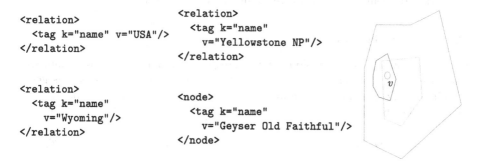

```
<relation>                <relation>
   <tag k="name" v="USA"/>   <tag k="name"
</relation>                      v="Yellowstone NP"/>
                           </relation>

<relation>
   <tag k="name"           <node>
      v="Wyoming"/>           <tag k="name"
</relation>                      v="Geyser Old Faithful"/>
                           </node>
```

Fig. 2. The text information associated with node v should be `Geyser Old Faithful; Yellowstone NP; Wyoming; USA`.

What is also modelled via the OSM constructs node, way, and relation are administrative entities. City, county, state, or country limits are represented as closed polygons consisting of a sequence of ways. An OSM relation groups the ways forming the boundary of such an administrative entity. While administrative entities typically form laminar families (being either contained in each other or disjoint), other entities like regions of transport associations, forests, nature reserves, parks etc. might intersect almost arbitrarily. See Fig. 2 for an example. There, the node v (the POI 'Geyser Old Faithful') is contained within the state of Wyoming and (nested) within the USA. v also belongs to the 'Yellowstone NP' which intersects Wyoming (but is not contained as it partly lies in Montana and Idaho). Observe that typically one does not include the city, state, or country of a hotel or a POI like the Geyser Old Faithful in Fig. 2 in its tags. This information is rather inferred from the containment of the POI in the respective city/state/country polygon regions. In fact this constitutes one of the main challenges when searching e.g. for 'stuttgart hotel' if we do not specify explicitly that we want to find all nodes that contain 'hotel' in its tag set and which are contained in a region whose name tag contains 'stuttgart'. Resolving and unifying the information directly attached to a node and the information inferred by containment will be discussed in detail later on.

Note that there is a considerable number of tags whose values are not really of interest for (sub)string search. For example, many OSM nodes contain in their key-value pairs the creation and revision history of the respective information, which we do not consider. We remove these tags before our processing routines. Amongst the remaining tags we make the values of the keys *name*, *addr**, and *ref* (including their localizations like *name:de* for their German name) *substring searchable* – we also call these the *important tags*. The other key-value pairs can be searched by specifying both key and value, e.g. @highway:motorway. In principle we could make all values (or even keys) substring searchable, but this leads to some odd and unintuitive results. For example, while searching for 'trails' or 'canada', one is probably not interested in OSM data contributed by one of the top contributers with username 'acrosscanadatrails'.

1.2   Problem Formulation

A query consists of an expression E which might contain, intersection, union, set differences as operations and substring and prefix text searches as well as geographic region constraints as operands, hence covering all possible set operations. For example, the query

```
(ibis germany @tourism:hotel $geo:6.767,47.167,13.798,50.401) - budget
```

should return all items which contain the substrings 'ibis' and 'germany' in their (direct or inherited) associated tag set, are of type 'hotel' and lie in the south of Germany (as determined by the keyword specifying ranges for longitude and latitude), but do not contain the word 'budget' in their tag set.

The challenge is to design a data structure allowing for the efficient answering of such queries and presenting the result set geographically clustered.

1.3   Related Work

Apart from the above mentioned search engines that – like OSCAR – work on a planet-sized dataset, search in spatial databases has attracted considerable interest. For example, in [3,7] the authors propose the so-called MHR-tree that allows to retrieve all items within a given query rectangle and (approximately) matching a set of given strings. We want to emphasize, though, that these and several other related results like [8] deal with problem instances which are by orders of magnitudes smaller than OSM planet. For example, [3, 7] only evaluated on *single states* of the US (for comparison: OSCAR works on the whole planet). On the other hand, there is a plethora of indices for text-only string search, e.g. MG4J [2], which can deal with extremely large datasets. They lack the spatial aspect of our search structure, though, both in terms of allowed queries as well as in terms of spatial result clustering.

The available search engines covering the whole world, still have their deficiencies. Google fared pretty well in terms of actual search performance, even though a more structured presentation of the results and substring search would be nice. Current free/OSM-based search engines are still way behind.

Our Contribution. We present OSCAR, a new geo-spatial search engine based on freely available OSM-data that even surpasses Google Maps in some aspects like result presentation. To our knowledge OSCAR is the first framework that can process general query expressions involving unions, intersections, set differences, as well as substring searches and geographic region constraints on the complete OSM planet dataset. Our query scheme is very efficient with respect to both, query times as well as memory footprint such that delopyment on both server as well as mobile platforms is feasible. The main novelty underlying our OSCAR framework is the construction of a cell arrangement induced by the polygonal OSM regions. Showcase implementations both on a server platform as well as for a mobile platform (left out due to page constraints) underline OSCAR's practicability.

Outline. In Sect. 2 we discuss straightforward approaches for a query structure which turn out not to be feasible on the planet scale, though. Hence, in Sect. 3 we propose the concept of cell arrangements for OSM data and based on that our cell-centered index structure. Having illustrated the usage of OSCAR in Sect. 4, we underline the practicability of our approach with extensive benchmarks as well as more details about our client implementations in Sect. 5.

2 Searching OSM-Data: First Attempts

Let us first describe two straightforward approaches to tackle our problem which turn out not to be practical, though.

2.1 Flattening of Tag Information

As the natural semantic of a node is that it inherits all tags of regions it is contained in, a straightforward strategy is to aggregate all information at the respective nodes in a preprocessing step and then build the query data structure on this 'flattened' dataset. The drawback of this approach is somewhat obvious.

Table 1. Blow-up due to flattening of tag information

	Raw	Flattened
California	36.2 MB	2.58 GB
Germany	275 MB	54.5 GB
Europe	1.24 GB	296 GB
Planet	1.84 GB	534 GB

Table 2. Cell arrangement statistics

	Items	Cells	Cell depth		Cell sizes	
			max	avg	max	avg
California	5.05 M	9.29 k	7	3.6	335 k	554.6
Germany	39.2 M	152 k	22	8.2	159 k	269.9
Europe	230 M	790 k	24	7.4	846 k	303.6
Planet	357 M	1.23 M	35	6.8	2.32 M	301.1

In particular for large regions this creates replication of the data associated at all contained nodes. Table 1 compares the space consumption of the flattened representation with the original raw data for our 4 OSM benchmark datasets. More critical, in case of a complex query where result sets of subexpressions of the query expression are to be combined, these operations have to take place on the node level. For example, processing a query 'chapel california', typically all nodes (world-wide) containing 'chapel' in their tag set are determined as intermediate result (or all nodes containing 'california') which here is a huge set. So processing queries at the node level is impractical in terms of query time but even more in terms of space requirements. Our attempts using the flattening approach were only feasible for small datasets and exhibited query times about a factor of 50 slower than OSCAR.

2.2 Geometric Intersection at Query-Time

We can avoid the blow-up in space due to flattening by constructing a text-searchable index on all OSM regions. Querying this structure with the term 'Stuttgart' returns all regions (or the associated node sets) that contain 'Stuttgart' in their tag set. If we have also built a text-searchable index on all OSM nodes, the union of the results of the queries to these two indices consists exactly of all nodes which are directly or indirectly (by inheritance) tagged with 'Stuttgart'. Logical conjunctions can then be implemented via intersection of the respective sets, disjunctions by taking their union. But as for the flattening idea, the problem with this approach is that without further tuning, intermediate results might get huge – in particular compared to the size of the final result of a query.

To avoid dealing with huge node sets as intermediate results one might not return the actual node sets associated with matching regions but their boundary polygon. Then conjunctions or disjunctions of region matches are first executed on the (geometric) polygon level, only at the very end these geometrically determined results are converted to the node level and combined with results from the search structure on the nodes. Here the problem is that the polygon intersection routines can be rather expensive (in particular if the polygons consist of thousands of segments — as it is the case e.g. for boundary polygons of countries). Finally, we have to determine all nodes contained in a polygon which is the result e.g. of a sequence of intersection operations. All these expensive operations have to take place at query time. So while in general this is a reasonable approach, it should be clear that it does not scale well to large datasets like the OSM dataset of the whole planet. In fact our implementation of this approach was more than 100 times slower than OSCAR even for rather simple queries.

3   OSCAR – OSm Cell ARrangements

The key concept behind OSCAR is a so-called *cell arrangement of regions*.

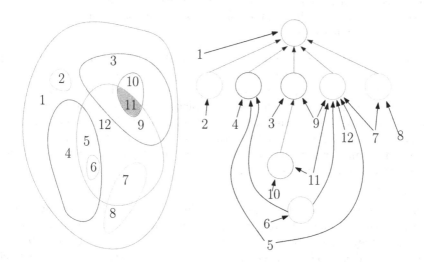

Fig. 3. Left: Cell arrangement – Points in the gray cell 11 have the property of being contained in the brown, purple, orange, and red cell, but not in the others. Right: Corresponding inclusion-DAG (Color figure online).

Consider a set of simple polygons $\mathcal{P}$. Each single polygon $P \in \mathcal{P}$ divides the plane into two regions, the *interior* of the polygon and the *exterior*. The set $\mathcal{P}$ naturally induces a subdivision of the plane into *cells*, see Fig. 3, left, for an example, we call this the *cell arrangement* $\mathcal{A} = \mathcal{A}(\mathcal{P})$ of $\mathcal{P}$. All points contained in a single cell $c \in \mathcal{A}$ have the property that they behave identically with respect to containment in the set of polygons.

The efficiency of our search structure is based on the hope that the number of cells in our cell arrangement $\mathcal{A}$ is considerably smaller than the total number of nodes. Obviously, for n polygons, the cell arrangement can be of arbitrarily high complexity. Even in the simple case where each two polygons intersect at most twice, there might be $\Theta(n^2)$ cells. In practice, though, the number of cells is much lower than the number of nodes, see Table 2. There, we have also listed average and maximum *depth* (i.e. the number of OSM regions a node is contained in) and the number of items which are essentially nodes and (parts of) ways inside a cell that have characterizing tags associated. Note that an item might comprise several nodes. The numbers clearly indicate that processing a query on cell rather than on node level has the potential to lead to drastically reduced processing times.

3.1   A Cell-Arrangement-based Data Structure

With the actual complexity of the cell arrangements of OSM regions in practice being rather small, the basic idea for our query data structure is as follows:

1. create cell arrangement $\mathcal{A}$ induced by OSM regions
2. associate with each cell $c \in \mathcal{A}$ all tags of OSM regions containing c
3. build search structure on 'tagged' cells

A query 'Stuttgart' to this data structure will then return all *cells* containing a tag with substring 'Stuttgart'. If several search terms are specified as part of a complex search query, the set operations take place at the cell level. Only at the very end, a conversion to nodes takes place. There are some details to be filled, though. For example, this approach as such only supports querying for tags associated with OSM regions but ignoring tags that were originally stored only at the nodes. Furthermore, we also incorporate geographical region constraints. These details will be discussed in the following.

Basic Concepts. We call the tags whose values we want to be substring searchable the *important tags*. The remaining, 'unimportant' tags will only be searchable using @key:value expressions – which is much simpler than the substring search and will not be discussed in detail. With each cell $c \in \mathcal{A}$ of our cell arrangement we associate two strings:

– full(c): all important tags of OSM regions containing c
– partial(c): all important tags of OSM nodes or ways contained in c

A region respectively a node shall be matched by q if there exists an important tag matching q. We call a cell $c \in \mathcal{A}$ a *full match* for q if a region spanning c matches q or all nodes in c match q, we call it a *partial match* if only some nodes in c match q.

Construction. We create a text search data structure $\mathcal{D}$ which for a given (single) search string q returns

– full-match(q):={$c \in \mathcal{A} : q \in$ full(c)}
– partial-match(q):={$c \in \mathcal{A} : q \in$ partial(c)}

That is, for a single search string q we can obtain both the set of cells contained in regions with q in the value of their important tags as well as the set of cells which contain some nodes with q contained in the value of their important tags.

There are several ways of implementing $\mathcal{D}$, for example, suffix arrays [4], suffix trees [6], some q-gram-based data structures [5] or even data structures which internally apply compression, e.g. [2]. As it will turn out, that the search structure itself does not dominate the overall space consumption (but rather the associated geoinformation), we refrain from the latter but employ basic suffix arrays. While they incur in their basic form a larger query time than suffix trees, their superior cache locality as well as slightly smaller memory footprint lead to better query times in practice.

Query Language. A query to our cell-arrangement-based data structure needs to be formulated in some form, so we design a simple query language which allows for all standard set operations.

We interpret a sequence of space separated strings as set intersections, '-' denotes a set difference, and '+' a union operation. For example, with the query

```
(california - San\ Diego) (Papa\ John + Papa\ Murphy)
```

we expect as a result all nodes which contain 'California', but not 'San Diego' and 'Papa John' or 'Papa Murphy' in their important tags (the actual intention is to find all places in California where one can have Papa John's or Papa Murphy's pizza excluding the ones in San Diego). In the above example we demand a match of the complete substring 'San Diego' including the space; this is denoted by the backslash preceding the space. OSCAR also supports the specification of geometric ranges (in terms of longitude and latitude) using a special keyword $geo. The result of this keyword is a set of cells that are fully contained in the query rectangle and a set of cells that have a non-empty, partial intersection with it, see the query example in Sect. 1.2.

Query Processing. Let us first describe the basic, cell-centered query routine and later sketch more details. A query expression is recursively processed. So for an expression $E = E_1 \, op \, E_2$ the respective query results for E_1 and E_2 are available as sets R_1^f, R_2^f of full-match cells and sets R_1^p, R_2^p of partial-match cells, $R_i^f \cap R_i^p = \emptyset$ for $i = 1, 2$. Processing an intersection operation (logical AND) is somewhat straightforward. The result set R^f of full-match cells is simply determined as the intersection of R_1^f and R_2^f. The set of partial-match cells is the intersection of partial-match cells for E_1 and E_2 together with all cells which are full-match for E_1 and partial match for E_2 (and vice versa). The other operations work likewise, so we have the following processing rules:

intersection

- $R^f = \{c : c \in R_1^f \cap R_2^f\}$
- $R^p = \{c : c \in R_1^p \cap R_2^f\} \cup \{c : c \in R_2^p \cap R_1^f\}$
  $\cup \{c : c \in R_1^p \cap R_2^p\}$

union

- $R^f = R_1^f \cup R_2^f$
- $R^p = R_1^p \cup R_2^p$

difference

- $R^f = R_1^f - (R_2^f \cup R_2^p)$
- $R^p = R_1^p - R_2^f$

Having recursively processed an expression E we have derived sets R^f and R^p where all nodes in cells $c \in R^f$ for sure match the expression E whereas possibly some of the nodes in each cell $c \in R^p$ match E. Note that some actual full-match cells might be in R^p rather than R^f (in case unions of partial matches yield a whole cell), but this does not affect correctness of the result. Furthermore, a cell $c \in R^p$ might not even contain any node matching E. So in a straightforward implementation, all nodes in cells $c \in R^p$ have to be inspected one by one with respect to E. For most real-world queries this is already very fast since typically only few partial match cells survive, and the inspection of the nodes in these few cells can be done rather quickly. In the worst case, though, during the processing of a query, the number of cells in R_p might grow rapidly without any actual matching node being contained in any cell $c \in R_p$. Inspecting a huge number of such cells yet with empty final result is rather unsatisfactory. In fact, for benchmarking we will deliberately create such queries, which make this straightforward approach break down.

As a remedy we enforce that during the processing of an expression the sets R^p only contain cells c for which a real node $v \in c$ exists matching the expression E. This can be achieved by augmenting the search structure $\mathcal{D}$. Irrespectively whether $\mathcal{D}$ is implemented as a suffix-tree, suffix-array or another index structure, for any (single) string s it refers to the sets of full- and partial-match cells $R^f(s)$ and $R^p(s)$. We augment the sets $R^p(s)$ by creating for each $c \in R^p(s)$ a reference list of all nodes contained in c matching s. Of course, this augmentation has its cost. But being dominated by the geographic location information this additional space is well spent. As result we can quickly inspect the set of potential partial-match cells R^p and prune out cells not containing matching nodes, reducing the size of all R^p sets. To finally obtain all nodes matching the query, it is no more necessary to inspect all nodes in cells of the final R^p but one can simply output the references to nodes that come with the cells in R^p.

Result Presentation. Typically the result of a query contains quite a large number of hits. For example, a query like

Hamburg hotel

not only contains as result hotels in the city of Hamburg, Germany, but also hotels in Hamburg, Iowa or a 'Hotel Hamburg' in the city of Saarbrücken, Germany. The first group of results is arguably the one a typical user would expect and which probably makes up for most of the hit results. We can naturally perform a hierarchical geo-clustering based on the cell representation of the result; the cluster with the largest number of hits typically corresponds to the answer which a user had in mind. Since the clustering also takes place at the cell level of our arrangement, it can be performed quickly.

Inclusion DAG. The inclusion-relation of the regions creating the cell arrangement induces a natural clustering which we use for presenting the result. See Fig. 3, left, for a cell arrangement with 12 cells induced by 8 regions. In Fig. 3, right, we see the induced Directed Acyclic Graph – we call it *Inclusion DAG*. In the inclusion-DAG the *cell nodes* have no incoming edges. Nodes with incoming edges correspond to regions (colored). There is a directed edge (v, w) from a cell/region node v to a region node w in the DAG, iff the cell/region v is fully contained in the region w and there exists no other region $w' \subset w$ in which it is fully contained. For a cell, e.g. cell no. 6, the set of regions it is contained in (here the turquoise, gold, blue, and red) can be easily determined via breadth- or depth-first search in the inclusion DAG starting from the respective node.

Result Clustering. The result of a query expression consists of a set of cells each with a counter for the number of matching nodes inside. By depth- or breadth-first search from all these cells, we can easily determine all regions that contain matching nodes. With a cell being contained in a small number of regions (see Table 2), the exploration of all regions containing result cells is in fact linear in the number of result cells and hence can be performed quite quickly. If we are also interested for every region *how many* matching nodes belong to that region, breadth-first search is not the graph traversal of choice unless the inclusion DAG is in fact a tree. But using depth-first search from each result cell we can propagate its result counter to parent regions without double counting. Again, the cost for each cell is small. Having done this for each result cell, we have faithful counters for all regions. These result counters are exact if we have only matching *nodes*. In case of elements spanning over several cells like *ways*, the counters are only an upper bound, since such ways are counted several times. In practice, the

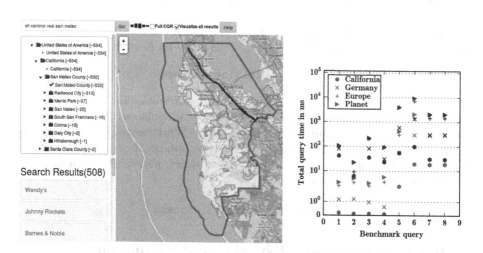

Fig. 4. Left: Result Clustering in our web service. Right: Total query time including set operations and subgraph creation. Hot cache timings are in red, cold cache timings in blue for S-OSCAR (Color figure online).

upper bound already gives a good indication for the result density, though. To present the results in a meaningful manner to the user we simply arrange the results according to the subgraph of the inclusion DAG consisting of regions with matching results, see Fig. 4, left. Here we have searched for 'El Camino Real' and 'San Mateo'; not surprisingly almost all hits are along El Camino Real through San Mateo County. As San Mateo County again is subdivided into the cities of Redwood City, Menlo Park, ... we see the respective clustering on the left hand side of our user interface. By clicking on the cities, the result set can be systematically explored. For the above example of 'Hamburg' and 'Hotel' we get the majority of hits in the Hamburg area and can quickly navigate to that area using our clustered result presentation. Nevertheless, we can also explore the few hotels named Hamburg in Italy, Finnland or Venezuela.

4   Using OSCAR

In this section we want to provide a few examples on how OSCAR can be used to explore the OSM planet dataset. Note that OSCAR is under constant development, so in particular the GUI might change in appearance.

Unspellable Villages. If you want to spend your holidays in the picturesque village of Llanfairpwllgwyngyll in Wales, UK, you might have some trouble remembering the whole village name, but just typing the (maybe memorizable) fragments

```
lanfair gwyngyll
```

make OSCAR lead you to your next holiday destination.

Anyone for dinner? Let's say you are looking for a place to have dinner this evening, preferably Chinese food. The natural query for Stuttgart would be:

```
Stuttgart @cuisine:chinese
```

Somewhat suprisingly OSCAR produces numerous results quite far away from the city of Stuttgart, the reason being that a larger administrative region around Stuttgart (in fact almost half of the state of Baden-Württemberg) bears as name *Regierungsbezirk Stuttgart*, which, of course, also contains the term *Stuttgart* as substring. In this case, we need to insist on an exact, non-substring match for the term *Stuttgart* as follows:

```
"Stuttgart" @cuisine:chinese
```

5   Implementation and Experimental Evaluation

The implementation of OSCAR consists of 4 main modules all written in C++: the *preprocessor P-OSCAR* takes as input OSM data (in its compressed .pbf

Table 3. Hardware specifications of our test environments

	Query/Web server	Preprocessing server
Price	400 €	6,000 €
RAM	8 GiB	256 GiB
CPU	Intel Xeon	2 x Intel Xeon
	E3-1225v3	E5-2630v2

Table 4. Space and time (in hh:mm:ss) for preprocessing

	Data base		Search structure	
	Time	Memory	Time	Memory
California	0:03:46	7.01 G	00:10:11	2.94 GB
Germany	0:25:30	19.0 G	01:21:42	17.8 GB
Europe	3:06:55	119 G	08:20:23	104 GB
Planet	7:24:13	197 G	18:41:08	160 GB

format), constructs the cell arrangement, and based on that the respective *query data structure Q-OSCAR*. P-OSCAR is rather time-intensive, preprocessing OSM planet takes about a day on rather powerful server hardware, see Table 3. The result, Q-OSCAR, is designed to be stored out-of-memory on a harddrive, solid-state disk, or microSD card and has a very small footprint in internal RAM. Q-OSCAR is employed by S-OSCAR – our webserver hosting the search engine.

P-OSCAR, the Preprocessor, and Q-OSCAR, the Data Structure. P-OSCAR runs on the preprocessing server using all available cores. The resulting data structure Q-OSCAR is stored in 3 separate binary files. The *database file* stores all items (nodes, ways, and relations to retrieve in a search), regions and the hierarchy. More than 55 % of the data is made up by geographic information associated with the respective nodes. The *text-search file* contains a suffix-array like data structure which is used to map strings to the respective fully-matched and partial-matched cells. Finally, the *text index file* holds all item sets for all data structures like the set of items of a single region or the items of a particular partial-matched cell. See Table 5 for an overview of the data sizes; *Nominatim* denotes a reduced dataset only storing the tags and items as supported by the official OSM Nominatim search engine (for comparison).

Table 5. Information about the dataset with all tags stored or just a reduced, Nominatim-like set of tags. All tags can be searched by prefix matches, important tags by substring matches. Cell data is the size for the full- and partial matched cell lists excluding the size for the item lists of partial matched cells.

	California		Germany		Europe		Planet	
	Nominatim	all	Nominatim	all	Nominatim	all	Nominatim	all
	Database space consumption							
Total	421 MB	591 MB	2.88 GB	3.63 GB	17.2 GB	22.9 GB	27.7 GB	35.4 GB
	Search data structure space consumption							
Tree	18.8 MB	116 MB	118 MB	235 MB	641 MB	1.89 GB	1.14 GB	3.73 GB
Tree values	103 MB	186 MB	1.03 GB	1.37 GB	6.39 GB	9.07 GB	10.8 GB	15.5 GB
	Index space consumption							
Total	259 MB	465 MB	1.79 GB	2.45 GB	9.62 GB	14.0 GB	16.0 GB	23.5 GB
Cell data	31.6 MB	38.8 MB	433 MB	518 MB	2.76 GB	3.27 GB	4.85 GB	5.67 GB

During the first phase of the preprocessing step regions and items of interest are selected and the hierarchy is built for this set of items and regions. The cell arrangement is represented by the inclusion-DAG we have seen before. Every region has a pointer to an index holding all cells the region is made up of. Furthermore every cell has a pointer to an index holding all the ids of the items in the cell. Vice versa every item knows its cells. To allow for fast set operations all of these lists are lexicographically sorted in ascending order and can be stored in compressed or uncompressed forms as provided by our library from [1].

S-OSCAR, the Web Service. The web service S-OSCAR runs on rather low-end hardware with 8 GB of RAM and a CPU comparable to a Haswell Core i5, see Table 3. Processing of a single query occupies a single core only. S-OSCAR has the precomputed Q-OSCAR data structure on a solid state disk and requires less than 100 MB of RAM. Of course, more RAM improves caching behaviour. The webclient is written in JavaScript using various other freely available libraries. Most of the result data is transferred in binary form, saving transfer time and processing power. For comparison: the default OpenStreetMap search engine Nominatim runs on a server comparable to our *preprocessing server*, having 128 GB of RAM and 12 CPU cores. We have designed Q-OSCAR on purpose to reside in external memory to allow for deployment on less powerful hardware.

5.1   Benchmarks

Timings for P-OSCAR Preprocessing. Preprocessing is split into two separate tasks. We first create a database containing all interesting items (nodes, ways, relations) together with the cell-arrangement and the hierarchy over the regions. Most of the memory is used for a large hash to support fast node-id to location lookups since ways reference nodes only by the id of the node. After that the search structures are created. Table 4 lists the respective measurements.

Query Timings. We designed 3 different search classes. In the first class, we have a very specific query whose result size (in terms of number of cells containing matches as well as result items) does not grow with an increasing dataset (when considering the sequence of datasets Germany – Europe – Planet). For the second class the number of resulting cells and items scales with the dataset size but is rather small. The third class is the worst-case where almost all cells get selected as partial matches and the successive set operation are between such partially matched cells. The respective queries grouped according to their class are listed in Table 6. We expect real-world user queries mostly to be similar to our queries of the first class. The two other query classes are more designed to exhibit the limits of OSCAR. The total time for the queries including both set operations and subgraph creation is shown in Fig. 4, right. *Cold cache* timings describe the case where none of the required data is present in the filesystem cache whereas *hot cache* timings the case where all of the data necessary to answer the query is already in the filesystem cache. For typical queries, the times will be somewhere

Table 6. Three classes of queries: query strings and their result sizes

Query	California Cells	Items	Germany Cells	Items	Europe Cells	Items	Planet Cells	Items
			Local queries					
1 @amenity:restaurant ("stuttgart" + "palo alto")	6	162	156	715	156	715	163	863
2 (@amenity:restaurant "stuttgart") + (@amenity:restaurant "palo alto")	6	162	156	715	156	715	163	863
			Scaling, but small result					
3 @cuisine:italian @cuisine:indian	0	0	49	53	66	71	70	75
4 @waterway:waterfall	58	206	315	489	2251	4471	4009	11800
5 @highway	5612	1335929	102413	9136886	488023	41559791	723255	84984020
			Worst-case					
6 @highway @building	21	40	455	1518	1282	3513	2028	5146
7 @highway @building @area	4	9	33	77	101	273	174	301
8 @highway @building @area @amenity	0	0	3	2	16	12	29	28

inbetween these two extrema, of course closer to the hot cache timings if enough RAM is available. Note that the set operations dominate the total query times. For S-OSCAR and typical queries as we expect them to be issued by a user, query times are only fractions of a second (cold cache). The queries designed to be worst-case for OSCAR might take a few seconds but are not expected to occur too often in practice. Hot cache query times are mostly about one order of magnitude faster, only the queries with huge result sets do not differ much in terms of cold vs hot cache for obvious reasons.

6   Conclusions

We have presented OSCAR, a search engine with the currently most advanced feature set which scales to the complete OpenStreetMap dataset even on desktop commodity hardware. The focus of this work was to get basic search functionality and result clustering as efficient as possible, but of course, there are many challenges and open problems left, e.g. an improved visualization of the clustered results, inclusion of an ontology, ranking of the results, query expression optimization, or improved robustness against poor data quality.

References

1. Bahrdt, D.: Osmfind: Fast textual search on OSM data - on smartphones and servers. In: Proceedings of the 2nd ACM SIGSPATIAL International Workshop on Mobile Geographic Information Systems, MobiGIS 2013, pp. 35–42. ACM, New York (2013)
2. Boldi, P., Vigna, S.: MG4J at TREC 2005. In: Voorhees, E.M., Buckland, L.P. (eds.) Proceedings of the 14th Text REtrieval Conference, TREC (2005)
3. Li, F., Yao, B., Tang, M., Hadjieleftheriou, M.: Spatial approximate string search. IEEE Trans. Knowl. Data Eng. 25(6), 1394–1409 (2013)

4. Manber, U., Myers, G.: Suffix arrays: a new method for on-line string searches. In: Proceedings of the 1st Annual ACM-SIAM Symposium on Discrete Algorithms, SODA (1990)
5. Ukkonen, E.: Approximate string-matching with q-grams and maximal matches. Theor. Comput. Sci. **92**(1), 191–211 (1992)
6. Ukkonen, E.: On-line construction of suffix trees. Algorithmica **14**(3), 249–260 (1995)
7. Yao, B., Li, F., Hadjieleftheriou, M., Hou, K.: Approximate string search in spatial databases. In: Proceedings of the 26th International Conference on Data Engineering, ICDE (2010)
8. Zhong, R., Fan, J., Li, G., Tan, K.-L., Zhou, L.: Location-aware instant search. In: Proceedings of the 21st ACM International Conference on Information and Knowledge Management, CIKM (2012)

Creating Diverse Product Review Summaries: A Graph Approach

Natwar Modani[1]([✉]), Elham Khabiri[2], Harini Srinivasan[2], and James Caverlee[3]

[1] Adobe Research, Bangalore, India
nmodani@adobe.com
[2] IBM Watson Research Lab, Yorktown Heights, NY, USA
{ekhabiri,harini}@us.ibm.com
[3] Texas A&M University, College Station, TX, USA
caverlee@cs.tamu.edu

Abstract. Product reviews play an influential role for the e-commerce websites, as consumers leverage them during the purchase decision process. However, the volume of such reviews can be overwhelming for a web user to comprehend the gist of overall information communicated by other consumers. In this paper, we address the problem of summarizing user contributed product reviews, having certain properties that differentiate them significantly from summarizing of traditional text articles. We propose suitable summarization algorithms that capture useful information with minimum redundancy and maximum information. We present a graph based formulation using a fast and scalable greedy algorithm for the review summarization problem. Our approach provides a rich model that makes certain sentences more *rewarding* based on their properties, in addition to their relation to the other reviews. We evaluate and show that our proposed algorithm outperforms other state-of-the-art summarization algorithms with significance level of 0.01 using automatic evaluation.

Keywords: Text summarization · Graph based modeling · Greedy algorithm · Product reviews

1 Introduction

With the emergence of Web 2.0, user reviews have become prevalent on the e-commerce websites which encourage their consumers to express their opinions and share their experience with others. These reviews are leveraged by potential consumers to learn more and to understand the prevailing opinions about the products/services. This is an important phase of the research performed by users before making purchase decisions. However, quite often, the number of reviews and their length can be overwhelming for a consumer to be able to grasp the opinions expressed by others. Besides, some of the sentences in the reviews may not provide useful information. However, it is useful for the user to get the diverse opinions expressed by the reviewers with an indication of prevalence of

© Springer International Publishing Switzerland 2015
J. Wang et al. (Eds.): WISE 2015, Part I, LNCS 9418, pp. 169–184, 2015.
DOI: 10.1007/978-3-319-26190-4_12

a sentiment expressed by each statement. Therefore, there is a need to filter out non-relevant, non-informative content and present a diversified summary of the relevant content to the consumers. We would like the users to be able to see the comments which correspond to the sentences chosen in the summary. Hence, we choose the extractive summarization paradigm over the abstractive summarization paradigm.

In this paper, we propose an extractive summarizing algorithm for user reviews of products. We model the review summarization as a graph problem, where each sentence is represented as a node, and two node are connected if the corresponding sentences are similar. We associate a reward with each node, which is based on the inherent value of the sentence to the summary. We also associate a weight with each edge based on the degree of similarity between the two nodes. We then use an iterative greedy algorithm to select the sentences to include in summary that provides maximum additional value. The rewards for the sentences connected to the one included in the summary are adjusted. This process is repeated until we find the desired number of sentences for the summary.

Please note that our algorithm does not prescribe any specific formulation of the node reward and costs, as well as, the edge weights. Hence, our algorithm can be easily extended with alternative (and possibly more elaborate) methods for assigning values to these parameters. In fact, the framework is not restricted to text summarization only and is general enough to handle any summarization problem.

We also present available state-of-the-art PageRank based summarization algorithms. We propose different variations of baseline PageRank algorithm where it shows how considering the importance of common terms, and also the weighted version of the graph, will effect on the results. We pick the best variation as one of the possible summarization approaches to compare our result with. Also we compare our results with DivRank [15] and Precedence based Ranking [9] methods which both have set important steps towards considering diversity of the selected sentences as a whole besides taking into account the prestige value of each sentence. We used Rouge as evaluation metric for a small number of products (25), each containing 50 review sentences and we used two automatic evaluation metrics, KL-Divergence and Retention Rate to evaluate the larger dataset of 3,679 products. The problem with Rouge metric is that it needs the ground truth which is very costly to obtain. This is why we applied KL-Divergence and Retention Rate as the two metrics of automatic evaluation for larger dataset.

In summary, our contributions in this paper are the following:

- We propose a graph based formulation of the user review summarization problem which is suitable for the context of social reviews. We provide a fast and scalable greedy algorithm for this problem.
- We also propose some variants of a successful PageRank based summarization algorithm.
- We demonstrate experimentally on real-world dataset that our proposed method outperforms state-of-the-art methods adapted from the literature.

The rest of this paper is organized as follows. We review related literature in Sect. 2 and then formalize and explain proposed graph based algorithm for summarization in Sect. 3. In Sect. 4, we propose some variants of page rank based summarization algorithms. In Sect. 5, we present our experimental results. Finally, we conclude our paper in Sect. 6.

2   Related Work

There has been considerable prior research on summarization, identifying and ranking relevant content and diversifying the ranking results. Summarization approaches could be categorized into two main types: *abstractive* and *extractive*. Abstractive summary [4,5] generates new content from the available input text, while the extractive summary is based on picking the sections of the input text that is more representative for the whole text without any changes to the input. Although abstractive methods would be a desirable way of summarization in general, in the context of review summarization, extractive summary is more appropriate as it can be supported with evidence in the form of the actual user statement. Further, abstractive methods require sufficient number of input texts to construct high quality summary phrases.

Our proposed algorithm is categorized as non-aspect-based, extractive summarization method. Among related work in the context of extractive summarization, machine learning algorithms such as support vector machines (SVM) and regression models have been used to rank the sentences by degree of preference in social communities [8,10]. However, Wu et al. [22] concluded that the combination of graph-based algorithms with length shows better result than SVM regression method. Many of the proposed algorithms are based on graph-based ranking, selecting the top-K sentences as the summaries of the input document(s). Examples include TextRank [16], MEAD [19] and LexRank [20]. Similar to link-based algorithms [12,17], these methods build a graph using the similarity relationships among the sentences in input documents.

Maximum Marginal Relevance (MMR) [3] is another form of diversified ranking, where relevance and diversification are the two optimization objectives. The relevance is defined with respect to a query, and diversification is achieved by incurring penalty for a node which is based on the maximum similarity between the node and the nodes currently in the solution. This method requires a query to measure the relevance and the user to specify the relative importance of the relevance and diversity, because of which this method is suitable for search result diversification but not very suitable for summarization task. Later, DivRank [15] set an important steps towards considering diversity of the selected nodes as a whole besides taking into account the prestige value of a node. Using Vertex Reinforced Random Walks [18] to introduce diversity in the rankings, it outperforms the MMR results. However, [21] argues that DivRank lacks the clarity of the overall objective function. Also it is not clear if it converges and if it does, its stationary state might not be unique. To compensate the mentioned weakness of DivRank, Dragon method was proposed [21], using optimization method with provable near optimal solution that reached higher amount of prestige and diversity values in compare with DivRank. However, there still exists a trade off between the two factors of prestige and diversity (and consequently, the user has to supply a parameter to indicate their relative preference of novelty and prestige) and still the explanation of their method is not easy.

In another direction, [9] proposed Precedence-based algorithm which uses a modified version of the PageRank. They showed that its combination with topic based clustering, LDA [1], results in higher quality summaries as compared to the other state-of-the-art summarization methods such as MEAD and LexRank. The common factor among the state-of-the-art methods introduced above, is that they are all inspired by PageRank algorithm and are based on the concept of convergence in the Markovian Chain matrix. All of these methods need to use a damping factor parameter α that should be specified by the user.

In this paper we propose a parameter free algorithm (beyond preprocessing stage) that selects representative sentences with highest relevance value while retaining diversity for the purpose of summarization. We compare our proposed method with DivRank [15] and Precedence-based algorithm [9] among the above methods, since they offer a clear separation of diversity and prestige which makes them easier to understand, while they have the PageRank algorithm as their main constituent. They are also query free, which makes it compatible for the task of summarization.

3 Reward Based Summarization

We now present a graph model based summarization algorithm that extracts sentences from input text corpus and assigns scores to them. The score indicates the *value of the sentence* from a summarization point of view. Here we focus on the text that is related to the reviews of products. In this context, we contend that a sentence has a high value if it has the following properties:

- It contains informative, relevant concepts about the product. For example, in the reviews related to a cell phone, those containing information about battery life, camera, and call quality are more desirable than the ones about for whom and what occasions this product is purchased for.
- It contains concepts that have received highest emphasis from the community of reviewers. For example, if many consumers have highlighted concerns about the battery life of the product, then a sentence about it should have higher value.
- It contains a diversity of independent concepts, i.e., sentences containing concepts that have not been included in the previously extracted sentences of the current summary are more desirable.

One can associate a notion of cost (say, based on the size of the sentence relative to the space available on the display media) with each sentence. One can order the sentences based on the ratio of their value and cost, and pick the top few sentences to be included in the summary as a natural greedy strategy.

For each product, we break up the reviews of the product into its composing sentences that we consider as our text units. A term is considered *meaningful* if it is either a noun or an adjective in our sentence, and is not a stopword. As a pre-processing step, we retain only those sentences that contain at least one noun and one adjective, so that we have a set of sentences which contains at least one property (adjective) for at least one feature (noun) of a product, and eliminate sentences that do not have any property for a product features, such as: "I bought this speaker for my son's birthday". When a sentence passes this step, we remove any stopwords in the sentence and then stem all the terms inside it, allowing us to consider all different variations of a term to be considered as a single term and leave us with a set of meaningful terms for each sentence. Our goal is to extract a subset of size n from the set of all retained sentences, that contains the most representative sentences. Here, n is the desired number of sentences in the summary.

Please note that we are proposing one pre-processing strategy, but the rest of the steps are not dependent on the specific pre-processing steps. In fact, the rest of the formulation is also flexible and at a meta level. Hence, one may choose alternative ways of defining the costs, the rewards and the edge weights, and the formulation and the algorithm would still be valid.

3.1   Reward Based Algorithm

In this formulation, we represent a collection of sentences, from which to chose the sentences for the summary, as a graph $G = (V, E, W)$. The nodes $v \in V$ represent sentences and the edges $e \in E$ represent a non-zero similarity between them. We associate an initial individual reward score $r_i = r_i^0$ with each node v_i as the number of *meaningful terms* found in its associated sentence (although, the formulation allows it to be more general). The weight $w_{ij} \in W$ associated with an edge $e = (v_i, v_j)$ indicates how much of v_j is known if we know v_i. It is defined as the ratio of *number of common terms between v_i and v_j* to the *number of terms in v_j*. Please note that the graph is directed as the weight is not symmetric.

The *Gain*, or total reward of including a node v_i (i.e., sentence) in the summary at step k, $G_{v_i}^k$, is defined as the weighted summation of the current discounted individual reward values of all the neighbors of v_i.

$$G_{v_i}^k = \sum_{v_j \in N_i} r_j^{k-1} w_{ij} \tag{1}$$

where N_i is the set of neighboring nodes for v_i. When we include node v_i in the summary, the reward score of the neighbor nodes, v_j, will be reduced as: $r_j^k = r_j^{k-1} * (1 - w_{ij})$, i.e., the reward score of each neighbor of a selected node in step k, is its previous reward score multiplied by the amount of its uncaptured similarity with the selected node. This approach prevents the inclusion of similar sentences and tends to pick the sentences with highest degree of diversity in the future steps.

We also associate a cost c_i with each node v_i. The summary selection has a limitation on how much text can be accommodated in the summary. If the summary size is specified in terms of number of sentences, then the cost of the sentence should be taken as 1. On the other hand, if the size of the summary is specified in terms of number of words, the cost c_i of a sentence would taken as the number of words in it. The decision of including a node v_i in the summary is represented by variable d_i, where $d_i = 1$ indicates that the node v_i is included in the summary and $d_i = 0$ indicates that it is not included in the summary. Let the total budget available be B, which implies that the total cost of nodes selected in the summary should not exceed B. We refer to our problem formulation as *Reward Collection Problem*, and call it *RCP* for brevity. The reward obtained by a node v_i from another node v_j varies based on when v_i is selected in the summary. However, the cumulative reward of the summary from the node v_j is independent of order in which the nodes are included in the summary.

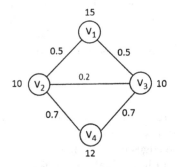

Fig. 1. Example to show the invariance of residual reward in RCP

Let us take a simple example to illustrate this. Consider a simple graph as shown in Fig. 1. The numbers next to the nodes are their rewards and the numbers next to the edges are edge weights. Now consider that the first node to be selected for the summary is v_1. The reward earned by selecting this node is $15 + 10 \times 0.5 + 10 \times 0.5 = 25$. The updated rewards for the nodes v_2 and v_3 are 5 each. Now, suppose we select node v_4 in the summary. The reward earned by selecting this node is $12 + 5 \times 0.7 + 5 \times 0.7 = 19$. The total reward earned by the summary $\{v_1, v_4\}$ is $25 + 19 = 44$. Now consider that the first node to be included in the summary is v_4. The reward earned by v_4 at this stage is $12 + 10 \times 0.7 + 10 \times 0.7 = 26$. The updated rewards for the nodes v_2 and v_3 are 3 each. We then include node v_1 in the summary next. The reward earned by selecting v_1 now is $15 + 3 \times 0.5 + 3 \times 0.5 = 18$. The total reward earned by the summary $\{v_4, v_1\}$ is $26 + 18 = 44$. So while the rewards earned by individual nodes for summary depends on the order in which they are selected, but the total reward of the summary is independent of the order. One can check easily that this property holds in more complicated cases as well.

Algorithm 1. Reward-Based Algorithm

Given: Graph G(V,E,W), reward R, cost C and budget B
Initialize: vertices $U = V$, residual budget $b = B$, solution set $S = \{\}$
while $min(C_{v_i} : v_i \in U) \le b$ and $|U| > 0$ **do**
    for $v_i \in U$ **do**
        $G(v_i) = \sum_{j=1}^{|V|}(r_j * w_{ij})$
    end for
    $v_i = argmax \frac{G(v_i)}{C_{v_i}}$ where $v_i \in U, C_{v_i} \le b$
    $S = S \cup v_i; U = U - v_i; b = b - c_{v_i};$
    for $v_j \in U$ **do**
        $r_j = r_j * (1 - w_{ij})$
    end for
end while

The objective of RCP is to maximize the reward of the summary from the selected nodes. Equivalently, the objective is to minimize the residual rewards for all the nodes. The uncaptured similarity of a node v_j is $(r_j * \prod_{i=1}^{n}(1 - w_{ij} * d_i))$, and hence, the overall total uncaptured similarity is:

$$\sum_{j=1}^{n} r_j * \prod_{i=1}^{n}(1 - w_{ij} * d_i) \qquad (2)$$

Hence, the problem statement of summarization can be written as the following optimization problem, where we would like to minimize the uncaptured value from the sentences.

$$min \sum_{j=1}^{n} r_j * \prod_{i=1}^{n}(1 - w_{ij} * d_i) \qquad s.t. \sum_{i=1}^{n} d_i * c_i \le B \qquad (3)$$

Here B is the budget allowed for the summary (e.g., number of sentences).

Now, we show that the RCP problem (selecting limited number of nodes with maximum total gain or minimum amount of residual values) is an NP-Hard problem. For this, we will reduce the *Budgeted Maximum Coverage Problem* [11] (referred to as BMCP from now on) into an instance of RCP problem. The budgeted maximum coverage problem is defined as follows. A collection of sets $S = \{S_1, S_2, ..., S_m\}$ with associated costs $\{c_i\}_{i=1}^{m}$ is defined over a domain of elements $X = \{x_1, x_2, ..., x_n\}$ with associated weights $\{w_i\}_{i=1}^{n}$. The goal is to find a collection of sets $S' \subseteq S$ such that the total cost of elements in S' does not exceed a given budget and the total weight of elements covered by S' is maximized.

To show the reduction, construct a graph G in the following manner. First, define a vertex corresponding to each element x_i and each set S_j, and define a vertex set V as the collection of all of these vertices. The reward associated with each of the vertex corresponding to the elements x_i corresponds to their weight w_i in the BMCP and the rewards associated with each of the vertices corresponding to the sets S_j is zero. The cost of the vertices corresponding to the sets S_j is assigned as the corresponding costs c_j in the BMCP and the costs of the vertices corresponding to the elements is taken as a large number (say $B + 1$), to ensure that only the sets can be chosen as part of the solution. Now, draw an edge from each of the vertices corresponding to the sets S_j to each of the elements x_i covered by this set, and assign a weight of 1 to these edges. One can easily verify that if one solves this instance of the RCP problem, then it would also be a solution for the corresponding BMCP problem. Hence, *RCP* is also an NP-Hard problem. Therefore, we use a greedy strategy to solve this problem. Algorithm 1 shows how we choose the sentences for the summary. We pick one node at a time, until we exhaust the budget. At each step the node with the highest gain to cost ratio (G_{v_i}/c_i) is selected and the reward values of all the neighbors of the selected node are updated.

4   Summarization Using PageRank Variants

The common factor among the state-of-the-art methods discussed in related work, is that they are all inspired by PageRank algorithm and are based on the concept of convergence in the Markovian Chain matrix. The idea behind using PageRank algorithm is that we interpret the similarity of a node to other nodes as a signal of endorsement by others. If those nodes are themselves similar to other nodes, they will contribute more score to the original node. Such recursive computing of scores will result in higher scores for the sentences that can be considered as the representative sentence of their own clusters. For the original PageRank score of a sentence $PR(s_i)$ we add the score of all the neighbors divided by the number of output links of each of these neighbors. We used 0.85 as our damping factor α.

$$PR(s_i) = \alpha \times \sum_{s_j \in N(s_i)} \frac{PR(s_j)}{outlink(s_j)} + (1 - \alpha) \tag{4}$$

The N_{s_i} is the set of neighbors for the s_i. The $outlink(s_j)$ is the total number of the neighbors for s_j. In the process of constructing the base graph, the edge weights are not considered, i.e., the nodes are simply connected if the number of common terms are greater than a threshold.

Now, we propose some variants of the PageRank based algorithm below. The main idea is to use weights based on various considerations while computing the page rank of the sentences.

PageRank on Weighted Graph (PRW): The first variation is to consider a weighted graph in which the edge weights are defined as a similarity metrics, such as raw number of common terms, normalized number of common terms, Jaccard coefficient, or cosine similarity. For example, if we have two sentences, $S_1 = \{a, b, c\}$ and $S_2 = \{b, c, d\}$. The raw common count will be $|S_1 \cap S_2| = 2$ and the Jaccard coefficient will be $\frac{|S_1 \cap S_2|}{|S_1 \cup S_2|} = \frac{2}{4}$. The Weighted PageRank (PRW) is measured as:

$$PRW(s_i) = \alpha \times \sum_{s_j \in N_{s_i}} PRW(s_j) \frac{W_{out}^{(ji)}}{\sum_{s_k \in N_{s_j}} W_{out}^{(jk)}} + (1 - \alpha) \qquad (5)$$

The N_{s_i} is the set of neighbors for the s_i. $W_{out}^{(ji)}$ is the weight of edges from j to i. Figure 2 shows how the weighted edges affect the score of each node. In this example, the edges are weighted based on the Jaccard Coefficient. Higher similarity in weighted graph results in a larger score for that node. Therefore, having weights will highlight the representative sentence of a cluster with higher resolution.

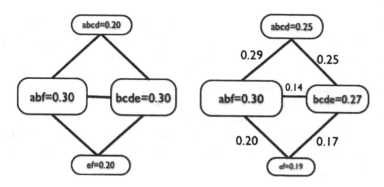

Fig. 2. The left graph does not consider the weight between the nodes. Therefore, both "ef" and "abcd" receive same scores regardless of the extent of similarities to their neighbors. In weighted PR the node with higher similarity to his neighbors receives higher score.

PageRank with *tf-idf* (PRT): The next variant is to consider the importance of the common terms between the nodes. In this variation, more informative common terms will result in a higher score. The *tf-idf* is a good candidate for this purpose. T_{common}^{ij} is the summation of the *tf-idf* scores of all terms that are in common between two nodes i and j.

$$PRT(s_i) = \alpha \times \sum_{s_j \in N(s_i)} \frac{PRT(s_j)}{outlink(s_j)} * T_{common}^{ij} + (1 - \alpha) \qquad (6)$$

PageRank on Weighted Graph with *tf-idf* (PRWT): This is the combination of the two previous variations in which we consider the importance of the common terms, using *tf-idf*, while we apply the PageRank on the weighted graph.

$$PRWT(s_i) = \alpha \times \sum_{s_j \in N_{s_i}} PRWT(s_j) \frac{W_{out}^{(ji)}}{\sum_{s_k \in N_{s_j}} W_{out}^{(jk)}} * T_{common}^{ij}$$
$$+ (1 - \alpha) \qquad (7)$$

We will evaluate the impact of incorporating the weights based on the above suggested variants in the experimental section.

5   Experiments

In this section, we will describe the dataset and discuss the experimental results to evaluate the performance of our proposed summarization algorithm.

5.1   Dataset

The dataset used in the experiments is from Amazon reviews for products of different categories captured by Blitzer *et al.* [2]. From reviews of 20, 423 products provided in [2], we selected those products that had at least 6 reviews. This resulted in 21 different categories with total of 3, 679 products and 29, 884 total reviews. The number of reviews ranges between 6 to 12 and the number of sentences ranges between 11 to 305 sentences per product. We also had 3 volunteer human judges to capture the ground truth summary for review of 25 products, which had nearly 50 review sentences associated with each of them. We asked the judges to select 5 representative sentences each out of the 50 review sentences for each of the products.

5.2   Evaluation Metrics

There are two classes of evaluation methods to evaluate the quality of summary. The first class of evaluation methods measure the quality of the summary by comparing the statistics of the input corpus to the statistics of the generated summary. These methods do not require a ground truth data, and hence are cost effective. Examples of such evaluation methods include KL-divergence and Compression Rate and Retention Rate together. These heuristics are shown to be very useful, for example, Louis *et al.* [14] shows that the automatic evaluation metrics based on information theory such as Jensen-Shannon divergence and KL-Divergence are highly correlated with human judgments. The JS-Divergence is described to be the symmetric version of KL-Divergence.

First, we evaluate the quality using the KL-divergence of the distribution of word in the input corpus and in the summary, factoring in the scores of the sentences selected in the summary. While computing the distribution of the words in the generated summary, we multiply the term frequency by the score assigned by the algorithm. This is done to ensure that if the algorithms pick some terms (as part of the picked sentence) with higher weight, it should be thought of as having higher frequency of that term in the generated summary. Let I be the distribution of terms in input text and S be the distribution of terms in summary. The KL-divergence between these two distributions is given as

$$KL(I||S) = \sum_{x \in Dic} I(x) \log \frac{I(x)}{S(x)} \qquad (8)$$

where x is a term. If KL-divergence is near zero, then it means that the two distributions are highly correlated. Higher value of KL-divergence shows more distance between the distributions of terms.

As discussed in [7], a good summary should be short enough and yet retain the information of the input text as much as possible. Note that there is a trade off between compression rate, defined as $len(Summary)/len(Input)$, and retention rate [6], defined as $info(Summary)/info(Input)$, where information is defined as the number of meaningful terms in a sentence. As the length of the summary is fixed, the compression rate is the same for all the methods, and the retention rate alone is an adequate measure.

The second class of evaluation methods take ground truth summary, and compare the generated summary against the ground truth. We use ROUGE [13], a popular method for evaluating the summary against the ground truth. ROUGE computes the goodness of a generated summary compared to the ground truth or gold-standard summary by computing n-gram recall. We use $n = 1$, as it was shown in [13] that $n = 1$ achieves very good correlation with human judgments. When $n > 1$, the performance is shown to be worse.

Although, ROUGE is used popularly to measure the quality of results, we wish to point out two shortcomings of it. First, ROUGE uses the n-grams from the ground truth, but does not consider synonyms, and hence even if the summary extraction method can factor it in, ROUGE would not rate it better. Second, ROUGE only looks at the recall of the n-grams without considering their importance. For example, if the ground truth summary includes a certain term t_1 as many as 4 times, and another term t_2 only 1 time, for ROUGE, both these terms are equally important. Hence, if two proposed summary both had to include only one of the two terms (say, due to space constraint), picking either would get the same score, which is clearly not desirable. Knowing the drawbacks of the ROUGE, still it is the best possible evaluation metric for the evaluation of the summaries.

5.3   Benchmark Methods

First, we compare the performance of different variations of PageRank based algorithm to see if considering the edge weights and the *tf-idf* of the common terms in the graph improves the results. Table 1 shows that the weighted version of graph in the PageRank based method improves the KL-Divergence. However weighing the common terms by the *tf-idf* results in deterioration of the KL-Divergence. In fact, the weighted *tf-idf* based method performs the worst amongst these methods. We observed similar trend for Retention Rate as well (results not included here), where again, Weighted Page Rank provides slightly better Retention Rate. Hence, in the rest of the experiments, to make the plots more clear, we only include the PRW variation.

For the purpose of remaining experimental evaluation, we compare RCP method with the following algorithms:

Table 1. Comparing KL-Divergence for different variations of PageRank based algorithms. Considering the weighted graph will improve the results.

	PR	PRW	PRT	PRWT
Top 1	**6.656**	6.842	7.300	7.466
Top 2	5.101	**4.965**	6.295	6.295
Top 3	4.682	**4.458**	5.704	5.704
Top 4	4.414	**4.299**	5.034	5.083

1. Random (baseline): This selects k sentences at random from the reviews.
2. TF-IDF: It is computed as the average of *tf-idf* score for all the terms used in a sentence. The purpose of this method is to give higher value to the sentences containing terms that are specific to the target product.
3. LDA+PR: This is the Precedence based algorithm [9] which applies LDA clustering to the sentences first and then sorts each cluster by the PageRank score. This method has already shown to outperform important methods such as MEAD and LexRank as discussed in related work.
4. PRW: This is the weighted version of the PageRank based summarization method. As we have seen above, PRW gives the best results amongst variations of baseline PageRank (PR, PRW, PRT and PRWT).
5. DivRank: This is one of the successful state-of-the-art summarization algorithms. It uses Vertex Reinforced Random Walks to introduce diversity in the rankings, while still taking the prestige value of a node into account.

5.4   Experimental Results

Now, we discuss the results of evaluation using the average values of KL-divergence (Figs. 3 and 4) and Retention Rate (Fig. 5) measures. As shown in Fig. 3, RCP has the lowest KL-divergence for all values of k for the top-k selected sentences. Next, come PRW, DivRank, LDA+PR followed by TFIDF based summarization. The baseline shows the highest value for KL-divergence, which is expected. The improvement in the KL-divergence is more than a factor of 2 for $k = 3$ and $k = 4$ compared to the next best method. We also compared the results statistically and found that RCP outperforms other methods with the significance level of 0.01. The p-value for PRW, DivRank and LDA+PR were, 0.0045, 0.01 and 0.00023, respectively.

Beyond the average value of KL-Divergence, we also want to investigate the spread and variance of the results, so that we can estimate the reliability of our results. We use the 'box-and-whiskers' plot to see the distribution of the observed KL-Divergence values. The box boundaries show the first and third quartiles, and the band (red line) in the middle of the box is the median. The whiskers (the vertical line extending from the box) indicate the extent of remaining data, except the outliers. Figure 4 shows that the variance of the KL-divergence for $k = 4$ is also the lowest for the RCP method, showing that this method is the most reliable one among the competing methods.

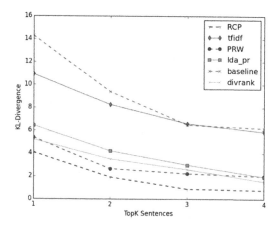

Fig. 3. Average of KL-Divergence for summarization methods (lower KL-Divergence is desirable)

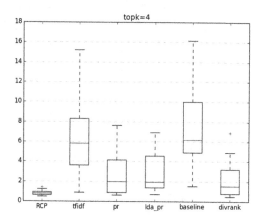

Fig. 4. KL-Divergence for top 4 sentences

Figure 5 shows that the Retention Rate of RCP algorithm is better than other methods, followed by PRW, LDA+PR, DivRank and TFIDF based methods. The baseline, as expected, has shown lowest retention rate.

Now we use ROUGE method to compare the discussed summarization algorithms. Figure 6 shows the ROUGE 1-gram scores for different methods in a 'box-and-whiskers' plot. It suggests that the RCP method gives better results in comparison to the alternative methods. Followed by LDA+PR, PRW and DivRank. The significance level for the outperformance of RCP appeared to be 0.05 for PRW and DivRank and was not significant for the LDA+PR.

One can see that both the non-ground truth based and the ground truth based evaluation methods suggest that RCP algorithm gives best summarization result.

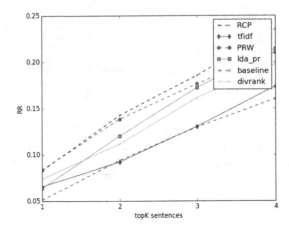

Fig. 5. Retention Rate for different summarization methods (higher value is desirable)

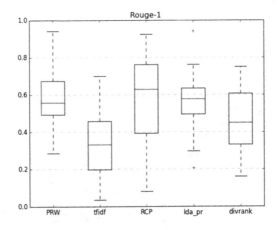

Fig. 6. ROUGE based evaluation using ground truth

6   Conclusion

In this paper we proposed a graph based algorithm for the review summarization problem which is scalable and based on a greedy approach. We found that the proposed algorithm performed better than alternate approaches in terms of emphasis on the diversity of the selected sentences by showing higher retention rate, as well as, following similar language model of the input sentences with lower KL-divergence, compared to the state-of-the-art summarization methods. It also gave best results for ROUGE based evaluation using ground truth.

Our proposed approach is also very flexible and is at a meta level. The approach can not only handle the text summarization, but also can be used in other summarization tasks (e.g., pick a representative set of actors to represent a movie industry). One can choose appropriate cost and reward functions for the nodes (representing the individuals) and the edge weights (representing similarity relations) based on the problem domain. One of the future directions we are exploring is to use more elaborate scheme for assigning the rewards and edge weights based on semantic analysis and sentiment analysis.

References

1. Blei, D.M., Ng, A.Y., Jordan, M.I.: Latent Dirichlet allocation. J. Mach. Learn. Res. **3**, 993–1022 (2003)
2. Blitzer, J., Dredze, M., Pereira, F.: Biographies, bollywood, boomboxes and blenders: domain adaptation for sentiment classification. In: ACL, pp. 187–205 (2007)
3. Carbonell, J., Goldstein, J.: The use of MMR, diversity-based reranking for reordering documents and producing summaries. In: Proceedings of the 21st Annual International ACM SIGIR Conference on Research and Development in Information Retrieval, SIGIR 1998, pp. 335–336. ACM, New York (1998). http://doi.acm.org/10.1145/290941.291025
4. Ganesan, K., Zhai, C., Han, J.: Opinosis: a graph-based approach to abstractive summarization of highly redundant opinions. In: Proceedings of the 23rd International Conference on Computational Linguistics, pp. 340–348. Association for Computational Linguistics (2010)
5. Ganesan, K., Zhai, C., Viegas, E.: Micropinion generation: an unsupervised approach to generating ultra-concise summaries of opinions. In: Proceedings of the 21st International Conference on World Wide Web, pp. 869–878. ACM (2012)
6. Hassel, M.: Evaluation of automatic text summarization, pp. 1–75. Licentiate Thesis, Stockholm, Sweden (2004)
7. Hovy, E., Lin, C.Y.: Automated text summarization in summarist (1999)
8. Hsu, C.F., Khabiri, E., Caverlee, J.: Ranking comments on the social web. In: CSE vol. 4, pp. 90–97 (2009)
9. Khabiri, E., Caverlee, J., Hsu, C.F.: Summarizing user-contributed comments. In: ICWSM (2011)
10. Khabiri, E., Hsu, C.F., Caverlee, J.: Analyzing and predicting community preference of socially generated metadata: a case study on comments in the digg community. In: ICWSM (2009)
11. Khuller, S., Moss, A., Noar, J.: The budgeted maximum coverage problem. Inf. Process. Lett. **70**(1), 39–45 (1999)
12. Kleinberg, J.M.: Hubs, authorities, and communities. ACM Comput. Surv. **31**(4es), December 1999. http://dx.doi.org/10.1145/345966.345982
13. Lin, C.Y.: Rouge: a package for automatic evaluation of summaries. In: Proceedings of the ACL-04 Workshop on Text Summarization Branches Out, pp. 74–81 (2004)
14. Louis, A., Nenkova, A.: Automatically evaluating content selection in summarization without human models. In: Proceedings of the 2009 Conference on Empirical Methods in Natural Language Processing, EMNLP 2009, vol. 1, pp. 306–314. Association for Computational Linguistics, Stroudsburg, PA, USA (2009). http://dl.acm.org/citation.cfm?id=1699510.1699550
15. Mei, Q., Guo, J., Radev, D.: Divrank: the interplay of prestige and diversity in information networks. In: Proceedings of the 16th ACM SIGKDD International Conference on Knowledge Discovery and Data Mining, pp. 1009–1018. ACM (2010)

16. Mihalcea, R.: Language independent extractive summarization. In: Proceedings of the ACL 2005 on Interactive Poster and Demonstration Sessions, pp. 49–52. Association for Computational Linguistics, Stroudsburg (2005). http://dx.doi.org/ 10.3115/1225753.1225766

17. Page, L., Brin, S., Motwani, R., Winograd, T.: The pagerank citation ranking: bringing order to the web. Technical report, Stanford Digital Library Technologies Project (1998). http://citeseer.ist.psu.edu/page98pagerank.html

18. Pemantle, R.: Vertex-reinforced random walk. Probab. Theory Relat. Fields **92**(1), 117–136 (1992)

19. Radev, D.R.: Experiments in single and multidocument summarization using mead. In: First Document Understanding Conference (2001)

20. Radev, D.R.: Lexrank: graph-based centrality as salience in text summarization. J. Artif. Intell. Res. **22**, 457–479 (2004)

21. Tong, H., He, J., Wen, Z., Konuru, R., Lin, C.Y.: Diversified ranking on large graphs: an optimization viewpoint. KDD **11**, 1028–1036 (2011)

22. Wu, J., Xu, B., Li, S.: An unsupervised approach to rank product reviews. In: FSKD, pp. 1769–1772 (2011)

Implicit Trust and Distrust Prediction for Recommender Systems

Xiao Ma, Hongwei Lu, and Zaobin Gan[✉]

School of Computer Science and Technology, Huazhong University of Science
and Technology, Wuhan 430074, People's Republic of China
{cindyma,luhw,zgan}@hust.edu.cn

Abstract. Many trust-aware recommender systems (TARSs) have
explored the value of explicit trust which is specified by users with binary
values. But existing works of TARSs suffer from the problem of the lack
of explicit trust which may not always be available in online social networks. In order to solve this issue, some implicit trust based TARSs
are proposed. However, these methods generally predict implicit trust
scores between users based on the interpersonal aspect, i.e., propagated
trust, similarity obtained by user ratings on co-rated items, while ignore
the personal aspects, i.e., trust bias and the impersonal aspects, i.e.,
local-topology-based features in trust networks, which are also important for implicit trust prediction. In this paper, we attempt to propose a
classification approach to address the trust/distrust prediction problem.
First, we obtain an extensive set of relevant features derived from the
personal aspects, interpersonal aspects and impersonal aspects of trust.
Then a logistic regression model is developed and trained by accommodating these factors, and applied to predict continuous values of users'
trust and distrust. We conduct an empirical study to evaluate the accuracy of the predicted implicit trust. The experimental results on real-
world data sets demonstrate the effectiveness of our proposed model in
employing implicit trust/distrust into existing trust-aware recommendation approaches.

Keywords: Implicit trust/distrust · Logistic regression · Recommender
systems

1 Introduction

Trust has been extensively exploited for improving the prediction accuracy of
the collaborative filtering based recommender systems by alleviating the inherent problems such as data sparsity and cold start [1–3,19,29]. Most trust-aware
recommender systems focus on using the explicit trust issued by users [4–7].
Although having been verified to have high rating prediction coverage and accuracy, existing explicit trust-based TARSs has its own limitation: it is sometimes
time consuming or expensive to get the explicit trust [10]. This is because obtaining the explicit trust needs extra user efforts: users need to specifically point out

© Springer International Publishing Switzerland 2015
J. Wang et al. (Eds.): WISE 2015, Part I, LNCS 9418, pp. 185–199, 2015.
DOI: 10.1007/978-3-319-26190-4_13

their personal opinions on trustees. In addition, explicit trust statements in most practical recommender systems are not available. In order to solve this issue, some implicit trust based recommender systems are proposed, which utilize the inferred trust scores to boost the performance of recommender systems [11,12].

As far as we know, existent implicit trust-based TARSs generally focus on incorporating the inferred trust scores based on the interpersonal aspect of trust, for instance, the propagation of trust [13,14], and similarity obtained by user ratings on co-rated items [8–10]. However, there are other important aspect of trust that should be taken into account, such as the personal aspect of trust, namely, the trust bias. Trust bias is a type of factors extracted from all the trust ratings that one user give or receive, showing his/her dispositional tendency to trust others or to be trusted by others on average (termed as truster bias and trustee bias respectively) [15]. Trust bias is regarded as a very important concept in social science and it is recognized as an integral part of the final trust decision [16]. For instance, some users tend to give relatively high trust ratings more generously than others while some users receive higher trust ratings compared with others. Another important factor has not been considered in implicit trust-based TARSs is the local-topology-based features in trust networks which may affect the establishment of trust links to a large extent [17,18]. Typically, users with higher indegree-trust and lower indegree-distrust are more likely to attract latent trustees.

Another issue of existent TARSs is the ignorance of distrust information. Although distrust is recognized to play an equivalently important role as trust [19], the investigation of utilizing explicit distrust in recommender systems is still in its infancy [20,21]. To the best of our knowledge, few work has attempted to predict distrust for improving recommender systems.

In this paper, we attempt to propose a classification approach to address the trust prediction problem, and incorporating the implicit trust/distrust into existing trust-aware recommender systems to verify the effectiveness of the inferred trust/distrust, as shown in Fig. 1. The main contributions of our work are summarized as follows:

(1) The following aspects of trust are being investigated in our model: two interpersonal features (i.e., propagation of trust and rating-based similarity), two personal features (i.e., truster bias and trustee bias), and four impersonal features (i.e., outdegree-trust, indegree-trust, outdegree-distrust and indegree-distrust).

(2) A logistic regression model is developed and trained by accommodating the aforementioned features, and then applied to predict continuous values of the trust and distrust between users.

(3) Newly generated trust values are then applied into the famous trust-aware recommendation methods (i.e., SocialMF [7], SoReg [22] and RWD [3]). The experimental results on Epinions dataset demonstrate the effectiveness of our proposed model in employing implicit trust into existing trust-aware recommendation approaches when explicit trust is not available.

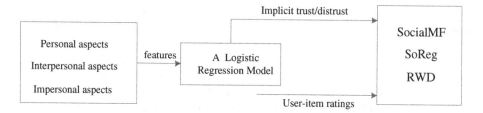

Fig. 1. The framework of our method.

The rest structure of this paper is organized as follows: in Sect. 2, a brief introduction of related work is presented. Section 3 discusses our proposed implicit trust-based recommendation model in detail. Experimental results and analysis are given in Sect. 4, followed by conclusions and future work in Sect. 5.

2   Related Work

Trust has been extensively studied in social science and recommender systems, working as an additional dimension to help model user preference. In this section, we briefly review the related works about trust prediction, explicit trust-based and implicit trust-based recommender systems.

2.1   Trust Prediction

Trust plays an important role in helping online users collect reliable information. Trust prediction is mainly concerned with predicting the unobserved relations between users, finding out who are friends and who are enemies in a social network. Existing trust prediction methods can be roughly divided into two categories: unsupervised trust prediction [8,13,14,23] and supervised trust prediction [26,28].

Most existing unsupervised trust prediction methods are based on some properties of trust to infer unknown trust relations, especially the propagation of trust. Guha et al. [14] propose a trust propagation model which introduces four types of atomic propagations, such as direct propagtation, co-citation propagation, transpose propagation and trust coupling propagation. MoleTrust is a depth-first graph walking algorithm with a tunable trust propagation horizon that allows us to control the distance to which trust is propagated [8]. Zheng et al. [13] propose a trust prediction model based on trust decomposition and matrix factorization, which considers the propagated trust and the similarity of trust values and similarity of trust rating distributions. These methods only consider the interpersonal aspects of trust, i.e., the propagation of trust and similarity, and ignore other aspects that are also important for trust prediction.

On the other hand, there are usually two steps for supervised methods. First, they extract features from available sources to represent each pair of users and consider the existence of trust relations as labels. Second, they train a binary classifier based on the representation with extracted features and labels.

In [28], a taxonomy of trust factors are presented as input features that will be used by learning algorithms to train binary classifiers. They mainly focus on user and interaction factors. Fang et al. [26] consider the multi-aspect of trust and distrust, such as the benevolence, competence, integrity and predictability, and model these factors as features to train the logistic regression models to predict the continuous values of users' trust and distrust.

In addition, some work considers trust relations as positive relations and distrust relations as negative relations, and then trust and dsitrust prediction problem is converted into link prediction in signed networks [17]. In [17], local topology based features based on balance theory are extracted to improve the performance of a logistic regression classifier in signed relation prediction. In our work, we also train a logistic regression model to predict the trust relations between users, but the difference lies in the features being used, that is, we consider all the interpersonal aspects, personal aspects and impersonal aspects of trust in our method.

2.2 Explicit vs Implicit Trust-Based Recommender Systems

In explicit trust-based recommender systems, trust explicitly released by users are utilized to boost performance of recommender systems. Yuan et al. [4] verify the small-world nature of trust networks and demonstrate the effectiveness of incorporating trust networks in trust-aware applications. Jamali et al. [7] introduce SocialMF, which is one of the most popular trust-based recommendation algorithms. The authors incorporate the mechanism of trust propagation into their probabilistic matrix factorization based model, which is a crucial phenomenon in social sciences and increases the rating accuracy to a large extent. In [21], they propose a matrix factorization-based model for recommendation in social rating networks that properly incorporates both explicit trust and distrust relationships to improve the quality of recommendations and mitigate the data sparsity and cold-start users issues.

In contrary, implicit trust-based recommender systems focus on incorporating trust scores inferred from the trust metrics, which are also effective when explicit trust is unavailable. For example, there are Trust Metric module and Similarity Metric module in Massa and Avesani's architecture of trust-aware recommender systems [8]. Therefore, the weights for identifying neighbors can be generated by trust metrics or similar metrics. Yuan et al. [10] construct an implicit trust network, where the implicit trust scores are inferred based on the similarity between users. Guo et al. [11] analyze five existing implicit trust metrics which have been proposed to infer implicit trust from user ratings and conduct an empirical study to explore the ability of trust metrics to distinguish explicit trust from implicit trust and to generate accurate prediction.

3   Our Approach

In this section, we first introduce the notations used in this paper in Sect. 3.1. Then formulate the aspects of trust that influence the establishment of trust in detail in Sect. 3.2. In Sect. 3.3, we present a classification model by incorporating all these influential aspects into a logistic regression model to predict the trust and distrust scores.

3.1 Preliminary

We keep the symbols u,v for two different users, i,j for two different items. Then $r_{ui} \in \{1, 2, 3, 4, 5\}$ represents a rating given by user u on item i. The trust network can be defined as a directed graph $G = \langle U, T \rangle$, where T represents the set of trust relationships. Binary trust relationships (i.e. trust and distrust relationships) are considered in this paper. $\forall u, v \in U, t(u, v) \in \{1, -1\}$, and $t(u, v) = 1$ means user u trusts v, $t(u, v) = -1$ means user u distrusts v.

3.2 Formulations of Aspects

In this section, we present a set of trust factors, including personal aspects, interpersonal aspects and impersonal aspects, that will be used by the machine learning algorithms as input features to train binary classifiers.

- **Personal aspects.** In online social networks, different users tend to show trust to others differently. Some users tend to give relatively higher trust ratings than others, while some users receive higher trust ratings than others, meaning that they are more likely to be trusted. In addition, some users tend to distrust others compared with others. Given two users u and v, in this paper, we explicitly consider two types of trust bias, that is, *trustor bias: Trustor(u)* and *trustee bias: Trustee(v)*. Trustor bias reflects the propensity of a given truster to trust others. It can be calculated as the average of all the trust ratings a user u gives to others [13]. Trustee bias can be treated as another personal property that shows a user's tendency to be trusted. It can be calculated as the average of all the ratings a user v received [13].
- **Interpersonal aspects**
  - **Propagated trust.** An important property of trust that is heavily used in trust-based recommender systems is transitivity. It allows trust to be propagated along paths to reach other users. Based on the transitivity effect, if user u trusts v and v trusts w, it can be inferred that u might also trust w to some extent [23]. It is consistent with real life in which people tend to trust the friend of a friend rather than a stranger. By propagating trust in social networks, we may identify more trusted friends.

    In this paper, we adopt the MoleTrust [8] algorithm to infer the trust value of indirectly connected users. Note that the trust value in the Epinions dataset we use in this paper is 0, 1 or -1, where 0 means no direct trust connections whereas 1 or -1 indicates that a user trusts or distrust another user. As a result, the inferred trust value by the MoleTrust will be 0 or 1, and thus we cannot distinguish trust neighbors in a shorter distance with those in a longer distance. Therefore, we adopt a weighting factor to devalue the inferred trust in a long distance:

$$t_{u,v} = \frac{1}{d} * t'_{u,v} \tag{1}$$

where d represents the shortest distance between user u and v determined by a breadth first search algorithm. As to the distrust value, we use it as a WOT(web of trust) debugger [14,27]. For instance, if user u trusts v completely, v fully trusts w, and u completely distrusts w, then the latter invalidates the propagated trust result (u trusts w). Thus, distrust information can help filter out "false positives".

- **Rating-based similarity.** The homophily effect suggests that similar users have a higher likelihood to establish trust relations [24]. Users with higher similarity are more likely to establish trust relations than those with lower similarity. For example, people with similar tastes about items are more likely to trust each other in product review sites. In this paper, we take into consideration two kinds of similarity, that is, *trust rating-based similarity and product rating-based similarity*. Conventionally, the trust rating-based similarity can be calculated from the common trust ratings that two users give to others [25]. We use Jaccard Similarity Coefficient to measure the similarity from ratings of common trustees that user u and v both have rated respectively:

$$S_T(u,v) = \frac{|N_u \cap N_v|}{|N_u \cup N_v|}, \tag{2}$$

where N_u, N_v represent the trust neighbors user u and v trust respectively, and $S_T(u,v) \in [0,1]$. As to the calculation of product rating-based similarity, the most prevalent approaches are Vector Space Similarity(VSS) and Pearson Corelation Coefficient(PCC). VSS calculates the similarity from ratings of common trustees that user u and v have rated respectively, and PCC takes into account the ratings styles that some users would like give relatively higher ratings to all the others while some may not. Therefore, PCC adds a mean of ratings as follows:

$$PCC(u,v) = \frac{\sum\limits_{i \in \mathcal{I}_{uv}} (r_{ui} - \bar{r}_u)(r_{vi} - \bar{r}_v)}{\sqrt{\sum\limits_{i \in \mathcal{I}_{uv}} (r_{ui} - \bar{r}_u)^2 \sum\limits_{i \in \mathcal{I}_{uv}} (r_{vi} - \bar{r}_v)^2}}, \tag{3}$$

where i belongs to the subset of items that user u and v both have rated, and $\bar{r}_u$ and $\bar{r}_v$ represent the average ratings of u and v respectively. As the range of the PCC is $[-1,1]$, we normalize PCC into $[0,1]$ in applications by $PCC' = (PCC + 1)/2$.

Therefore, the rating-based similarity can be described as follows, where α_1, α_2 are the coefficients to control the contributions of these two factors, and $\alpha_1 + \alpha_2 = 1$.

$$sim(u,v) = \alpha_1 S_T(u,v) + \alpha_2 PCC'(u,v), \tag{4}$$

- **Impersonal aspects.** Due to the availability of trust/distrust links of each user, we specifically identify four kinds of impersonal aspects in our computational model based on the signed degrees of the nodes, which essentially record the aggregated local relations of a node to the rest of the network. As we are interested in predicting the sign of the edge from user u to v, we consider outgoing edges from u and incoming edges to v. Specifically we use $d_{out}^+(u)$ and $d_{out}^-(u)$ to denote the number of outgoing trust and distrust edges from u, respectively. Similarly, we use $d_{in}^+(v)$ and $d_{in}^-(v)$ to denote the number of ingoing trust and distrust edges to v, respectively.

3.3 Trust and Distrust Prediction

After obtaining all the features, we now apply the logistic regression algorithm to predict the implicit trust and distrust between a pair of users u and v. The trust value will be influenced by the set of eight features that we discussed, denoted by

$$f(u,v) = \{trustor(u), trustee(v), t_{u,v}, sim(u,v), d_{out}^+(u), d_{out}^-(u), d_{in}^+(v), d_{in}^-(v)\}.$$
(5)

In practice, users may specify other users as trusted neighbors, that is, $t_{u,v} = 1$. Specifically, the expected probability that trustor u completely trusts the trustee v can be presented as:

$$p^+(u,v) = E(t_{u,v} = 1 | f(u,v)).$$
(6)

We apply the logistic regression to classify trust from distrust, and try to obtain the coefficients of each aspect related with trust $a_m (0 \leq m \leq k)$, in this paper, $k = 8$. Following the approach in [26], the logit of the probability is modeled as a linear combination of $f(u,v)$:

$$logit(p^+(u,v)) = \log(\frac{p^+(u,v)}{1 - p^+(u,v)}) = a_0 + \sum_{m=1}^{k} a_m f_{u,v}^m,$$
(7)

Then the probability $p^+(u,v)$ can be generated by Eq. 8:

$$p^+(u,v) = \frac{1}{1 + e^{-(a_0 + \sum_{m=1}^{k} a_m f_{u,v}^m)}}.$$
(8)

The coefficient $a_m (0 \leq m \leq k)$ are learned from the training data using maximum likelihood estimation, which tries to find the coefficients for which the probability of the observed data is maximized. As to the result, if $p^+(u,v) \geq 0.5$, user u trusts v, else u distrusts v.

4 Experiments

In this section, we conduct experiments on Epinions data sets in order to answer two questions: (1) How is the accuracy of our proposed implicit trust/distrust inference method? (2) Whether the performance of existing trust-aware recommender systems based on the implicit trust predicted by our proposed method will be as effective as employing explicit trust?

Table 1. Statistics of the sampled dataset

	Epinions1	Epinions2
# of user	8053	7124
# of item	21,057	22,129
# of rating	681,754	667,295
# of trust	306,773	28,813
# of distrust	28,813	28,813
trust sparsity	99.48 %	99.89 %
rating sparsity	99.60 %	99.58 %
average rating	4.7649	4.7658

4.1 Dataset Description

Before answering above two questions, we first introduce the dataset we use in our work. Epinions[1] is a consumer review website in which users can express their opinions about items by assigning numerical ratings and writing article reviews. The extended Epinions data set generated by [8] describes the trust (labeled as 1) and distrust(labeled as -1) relations among users and their ratings on other users' articles. Since Epinions is the only available dataset which contains user ratings and trust/distrust links between users, we utilize it to train our proposed logistic regression model for trust and distrust prediction.

In particular, due to limited processing power and memory, we sample a portion of the Epinions dataset. We filter out users who have rated less than 10 items and issued less than 1 trust statement. As the statistics shown in [17], the edge signs in Epinions are overwhelmingly positive. Thus we consider two sampling strategies to tackle with the original dataset. First, we randomly sample 8053 users as well as the user ratings and trust statements. The statistics of the Epinions1 dataset are shown in Table 1. We compute the numbers of trust and distrust relations each user receives and creates, and the distributions are shown in Fig. 2. The distributions for both trust and distrust follow a power-law-like distribution that is typical in social network. Second, we create a balanced dataset with equal numbers of positive and negative edges as well as the user ratings and trust statements as shown in Table 1. Note that the sampled ratings in Epinions dataset are quite skewed, from Fig. 3 we can find that more than 80 % ratings are 5.

4.2 Methodology and Metrics

In our experiments, we perform 5-fold cross validation. In each fold, we use 80 % of the data as the training set and the remaining 20 % as the test set. Then the average performance will be adopted as the final results. The experiments are

---

[1] www.trustlet.org.

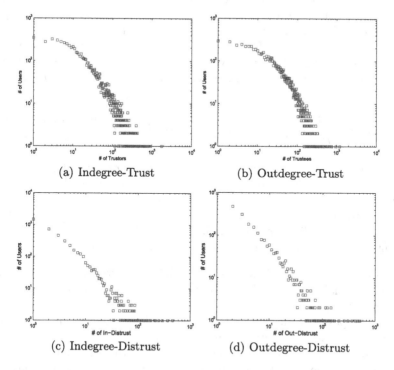

Fig. 2. The Distributions of Indegree and Outdegree of Trust and Distrust Relations of Epinions1.

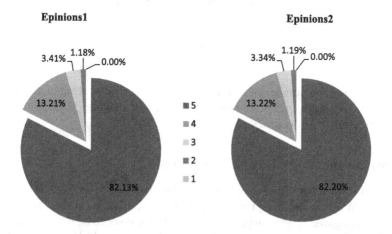

Fig. 3. The rating distributions of Epinions1 and Epinions2 data sets.

conducted in two steps. In step one, by applying the leave-one-out technique, each trust/distrust is iteratively hidden whose value will be predicted by applying our proposed implicit trust inference method until all trust/distrust in the dataset are tested. Since the Epinions1 is quite imbalanced, that is, more positive links exist than negative links. We employ three famous evaluation metrics which are effective for imbalanced dataset: *Precision*, *Recall* and *F1* score, and the closer to 1, the better.

$$Precision = \frac{TP}{TP + FP}, \ Recall = \frac{TP}{TP + FN}, \ F1 = 2 \cdot \frac{Precision \cdot Recall}{Precision + Recall}$$
$$(9)$$

where *TP*, *FP* and *FN* represent true positive, false positive and false negative, respectively. And the *F1* score can be interpreted as a weighted average of the precision and recall.

In the second step, we also apply the leave-one-out technique to evaluate the accuracy of item predictions with the inferred trust and distrust values. Three representative trust-aware approaches are adopted to generate recommendations:

- **SocialMF**, proposed by Jamali and Ester [7], considers the trust information and propagation of trust information into the matrix factorization model for recommender systems.
- **SoReg**, proposed by Ma [22], adds a social regularization term into the matrix factorization model to control friends taste difference.
- **RWD**, proposed by Ma [3], incorporates explicit trust into matrix factorization model and empirically demonstrates that distrust relations among users are as important as the trust relations.

Since the objective of our approach is to test the rating prediction accuracy, we use two standard metrics to measure the accuracy of various models: MAE (Mean Absolute Error) and RMSE (Root Mean Square Error).

$$MAE = \frac{1}{N} \sum_{r=1}^{N} |R - \hat{R}|, \quad RMSE = \sqrt{\frac{1}{N} \sum_{r=1}^{N} (R - \hat{R})^2} \tag{10}$$

where N denotes the number of tested ratings. R is the real rating of an item and $\hat{R}$ is the corresponding predicted rating. In general, smaller RMSE and MAE values indicate better accuracy.

4.3 Evaluating Trust Rating Predictions

We train the data of the sampled dataset by using the regularized logistic regression algorithm[2]. When calculating the rating-based similarity, we set $\alpha_1 = \alpha_2 = 0.5$. As to the propagated trust, the propagation distance is set to be $d = 3$. Intuitively, if d is too small, few indirect trust neighbors can be reached. If d is too large, longer

---

[2] https://github.com/quinnliu/machineLearning.

Table 2. The coefficients of trust aspects.

Coefficients	Epinions1	Epinions2
Trustor bias	4.3012	3.4049
Trustee bias	3.7417	2.9805
Similarity	1.4859	2.5689
Propagated trust	0.5228	0.0304
Trust outdegree	0.0001	0.0028
Distrust outdegree	0.0022	-0.0032
Trust indegree	-0.0002	0.0044
Distrust indegree	-0.0012	-0.0049

propagation path may produce less accurate trust neighbors [8]. The learned coefficients of trust aspects are illustrated in Table 2.

From Table 2 we can find that the trustor bias, trustee bias and similarity are all positively correlated with trust and are more likely to increase the probability of trust, which can be explained that a trustworthy user would further be trusted by more users and users with more similarity are more likely to trust each other. Note that the propagated trust shows more positive correlation with trust in Epinions1 than in Epinions2. It is because Epinions2 are sampled based on the strategy of equal number of trust and distrust relationships, which is inconsistent with the original structure of trust networks as shown in Fig. 2. Thus the propagation property is not as obvious as that in Epinions1. In addition, we can conclude from Table 2 that the impersonal features are less influential in predicting trust values compared with the personal features and interpersonal features. This could be partially explained as that trust are more likely to be generated by personal propensities and the interactions between users.

In this section, we also give out the results of the implicit trust and distrust prediction and investigate which aspects contribute more to the establishment of trust. As shown in Table 3, we predict the trust values based on the learned regression model for four scenarios: "personal", "interpersonal", "impersonal" and "All", where "All" represents the combination of all the three aspects of trust. In both data sets, the personal aspects based method yields the best precision results compared with others, which is consistent with the results in Table 2. The interpersonal aspects based method performs worst in Epinions2. "All" yields the best results in terms of recall and F1 score, as well as comparatively better precision compared with others, which we can conclude that all these three aspects are valuable for the implicit trust predictions. Since the aim of our work is to investigate the effectiveness of incorporating implicit trust/distrust in recommender systems, we don't consider the comparison with other trust inference methods.

Table 3. The prediction results based on different aspects.

Aspects	Epinions1			Epinions2		
	Precision	Recall	F1	Precision	Recall	F1
Personal	**0.9241**	0.9467	0.9353	**0.9641**	0.9467	0.9553
Interpersonal	0.9119	0.9142	0.9130	0.6185	0.6838	0.6495
Impersonal	0.9115	0.9513	0.9310	0.8249	0.9285	0.8736
All	0.9118	**0.9609**	**0.9357**	0.9605	**0.9509**	**0.9557**

4.4 Evaluating Accuracy of Item Rating Predictions

In this section, the newly generated trust values are taken as input to three representative trust/distrust-based recommendation algorithms as shown in Fig. 1. We investigate whether the incorporation of implicit trust could perform as well as the explicit trust in existing recommendation methods. Therefore, we compare our method with the explicit trust/distrust enhanced methods and the conventional matrix factorization model. For MF and SocialMF, we adopt the implementations provided by the MyMediaLite framework[3]. Table 4 illustrates the comparison results on explicit trust and implicit trust. Based on the results we can find that all implicit trust based methods outperforms the MF model. In addition, the results of SocialMF, SoReg and RWD on implicit trust are quite similar to the results of employing explicit trust. Note that the results of RWD on implicit distrust are better than the results on explicit distrust. This can be explained as our implicit trust inference method can better predict the underlying distrust relationships between users, which further demonstrates the effectiveness of our proposed approach.

Table 4. The comparison results based on explicit and implicit trust/distrust.

Methods	Epinions1		Epinions2	
	RMSE	MAE	RMSE	MAE
MF	0.5682	0.3957	0.5649	0.3934
SocialMF-explicit	0.5276	0.2894	0.5189	0.2885
SocialMF-implicit	0.5398	0.2901	0.5332	0.2953
SoReg-explicit	0.5566	0.2981	0.5452	0.2980
SoReg-implicit	0.5694	0.3000	0.5565	0.2934
RWD-explicit	0.5601	0.3102	0.5613	0.3141
RWD-implicit	0.5577	0.2976	0.5596	0.3001

---

[3] www.mymedialite.net.

5   Conclusion and Future Work

In this paper, we employed a logistic regression based classification model to predict the implicit trust/distrust relations between users. Typically, eight aspects of trust are modeled as features to train the classification model. Then the inferred trust/distrust relations are employed into three representative trust-based recommendation approaches to test the accuracy of item predictions. Empirical study was conducted to evaluate the accuracy of the inferred trust and distrust generated by our proposed approach. Real-world data sets based comparison experimental results demonstrated the effectiveness of employing implicit trust into three representative trust-based recommender systems when explicit trust is unavailable.

In the future, we aim to investigate more trust aspects as features to train the logistic regression model, such as the trust dynamicity, context-specific trust and trust rating distribution similarity. We will also evaluate the effectiveness of incorporating implicit trust relations into some other recommendation methods.

Acknowledgment. This research is funded by the National Natural Science Foundation of China under grant No. 61272406 and the Fundamental Research Funds for the Central Universities, HUST: 2013TS101.

References

1. O'Donovan, J., Smyth, B.: Trust in recommender systems. In: Proceedings of the 10th International Conference on Intelligent User Interfaces. ACM, pp. 167–174 (2005)
2. Golbeck, J., Hendler, J.: Filmtrust: movie recommendations using trust in web-based social networks. ACM Trans. Internet Technol. **6**(4), 497–529 (2006)
3. Ma, H., Lyu, M.R., King, I.: Learning to recommend with trust and distrust relationships. In: Proceedings of the 3rd ACM Conference on Recommender Systems. ACM, pp. 189–196 (2009)
4. Yuan, W., Guan, D., Lee, Y.K., et al.: Improved trust-aware recommender system using small-worldness of trust networks. Knowl.-Based Syst. **23**(3), 232–238 (2010)
5. Pitsilis, G., Zhang, X., Wang, W.: Clustering recommenders in collaborative filtering using explicit trust information. In: Wakeman, I., Gudes, E., Jensen, C.D., Crampton, J. (eds.) Trust Management V. IFIP AICT, vol. 358, pp. 82–97. Springer, Heidelberg (2011)
6. Guo, G., Zhang, J., Thalmann, D.: A simple but effective method to incorporate trusted neighbors in recommender systems. In: Masthoff, J., Mobasher, B., Desmarais, M.C., Nkambou, R. (eds.) UMAP 2012. LNCS, vol. 7379, pp. 114–125. Springer, Heidelberg (2012)
7. Jamali, M., Ester, M.: A matrix factorization technique with trust propagation for recommendation in social networks. In: Proceedings of the 4th ACM Conference on Recommender Systems. ACM, pp. 135–142 (2010)
8. Massa, P., Avesani, P.: Trust-aware recommender systems. In: Proceedings of the 1st ACM Conference on Recommender Systems. ACM, pp. 17–24 (2007)

9. Papagelis, M., Plexousakis, D., Kutsuras, T.: Alleviating the sparsity problem of collaborative filtering using trust inferences. In: Herrmann, P., Issarny, V., Shiu, S.C.K. (eds.) iTrust 2005. LNCS, vol. 3477, pp. 224–239. Springer, Heidelberg (2005)

10. Yuan, W., Shu, L., Chao, H.C., et al.: ITARS: trust-aware recommender system using implicit trust networks. IET Commun. **4**(14), 1709–1721 (2010)

11. Guo, G., Zhang, J., Thalmann, D., et al.: From ratings to trust: an empirical study of implicit trust in recommender systems. In: Proceedings of the 29th Annual ACM Symposium on Applied Computing. ACM, pp. 248–253 (2014)

12. Fazeli, S, Loni, B., Bellogin, A., et al.: Implicit vs. explicit trust in social matrix factorization. In: Proceedings of the 8th ACM Conference on Recommender Systems. ACM, pp. 317–320 (2014)

13. Zheng, X., Wang, Y., Orgun, M.A., et al.: Trust prediction with propagation and similarity regularization. In: Proceedings of the 28th AAAI Conference on Artificial Intelligence. ACM (2014)

14. Guha, R., Kumar, R., Raghavan, P., et al.: Propagation of trust and distrust. In: Proceedings of the 13th International Conference on World Wide Web. ACM, pp. 403–412 (2004)

15. Yao, Y., Tong, H., Yan, X., et al.: MATRI: a multi-aspect and transitive trust inference model. In: Proceedings of the 22nd International Conference on World Wide Web. ACM, pp. 1467–1476 (2013)

16. Tversky, A., Kahneman, D.: Judgment under uncertainty: heuristics and biases. Sci. **185**(4157), 1124–1131 (1974)

17. Leskovec, J., Huttenlocher, D., Kleinberg, J.: Predicting positive and negative links in online social networks. In: Proceedings of the 19th International Conference on World Wide Web. ACM, PP. 641–650 (2010)

18. Liben-Nowell, D., Kleinberg, J.: The link-prediction problem for social networks. J. Am. Soc. Inform. Sci. Technol. **58**(7), 1019–1031 (2007)

19. Tang, J., Hu, X., Liu, H.: Is distrust the negation of trust?: the value of distrust in social media. In: Proceedings of the 25th ACM Conference on Hypertext and Social Media. ACM, pp. 148–157 (2014)

20. Victor, P., Verbiest, N., Cornelis, C., et al.: Enhancing the trust-based recommendation process with explicit distrust. ACM Trans. Web (TWEB) **7**(2), 6 (2013)

21. Forsati, R., Mahdavi, M., Shamsfard, M., et al.: Matrix factorization with explicit trust and distrust side information for improved social recommendation. ACM Trans. Inf. Syst. (TOIS) **32**(4), 17 (2014)

22. Ma, H., Zhou, D., Liu, C., et al.: Recommender systems with social regularization. In: Proceedings of the 4th ACM International Conference on Web Search and Data Mining. ACM, pp. 287–296 (2011)

23. Golbeck, J., Hendler, J.A.: Inferring binary trust relationships in web-based social networks. ACM Trans. Internet Technol. (TOIS) **6**(4), 497–529 (2006)

24. Tang, J., Gao, H., Hu, X., et al.: Exploiting homophily effect for trust prediction. In: Proceedings of the 6th ACM International Conference on Web Search and Data Mining. ACM, pp. 53–62 (2013)

25. Jia, D., Zhang, F., Liu, S.: A robust collaborative filtering recommendation algorithm based on multidimensional trust model. J. Softw. **8**(1), 11–18 (2013)

26. Fang, H., Guo, G., Zhang, J.: Multi-faceted trust and distrust prediction for recommender systems. Decis. Support Syst. **71**, 37–47 (2015)

27. Victor, P., Cornelis, C., De Cock, M., et al.: Gradual trust and distrust in recommender systems. Fuzzy Sets Syst. **160**(10), 1367–1382 (2009)

28. Liu, H., Lim, E.P., Lauw, H.W., et al.: Predicting trusts among users of online communities: an epinions case study. In: Proceedings of the 9th ACM Conference on Electronic Commerce. ACM, pp. 310–319 (2008)
29. Ma X., Lu H., Gan Z., Improving recommendation accuracy by combining trust communities and collaborative filtering. In: Proceedings of the 23rd ACM International Conference on Information and Knowledge Management. ACM, pp. 1951–1954 (2014)

Combining Syntactic and Semantic Evidence for Improving Matching over Linked Data Sources

Klitos Christodoulou[✉], Alvaro A.A. Fernandes, and Norman W. Paton

School of Computer Science, University of Manchester,
Oxford Road, Manchester M13 9PL, UK
{christodoulou,alvaro,norm}@cs.man.ac.uk

Abstract. In the context of Linked Data (LD) sources, the ability to traverse links and retrieve further information can be exploited to harvest semantic annotations. Such annotations can, in turn, underpin the inference of semantic correspondences between sources. This paper shows that using semantic annotations as additional evidence of equivalence between schematic representations of LD sources can improve upon the prevalent, purely syntactic approaches. The paper both describes the construction of probabilistic models that yield degrees of belief on the equivalence of the real-world concepts represented by the data and shows how these models are crucial in underpinning a Bayesian approach to assimilating both syntactic evidence (in the form of similarity scores derived by string-based matchers) and semantic evidence (in the form of semantic annotations stemming from LD vocabularies) of equivalence. The paper presents an empirical evaluation of the techniques described. The main finding is confirmation that, with respect to equivalence judgements made by human experts, the use of the contributed techniques incurs significantly fewer discrepancies than purely syntactic approaches.

Keywords: Probabilistic matching · Bayesian updating · Linked data

1 Introduction

The Web of Data (WoD) encourages publishers to make their datasets publicly available. This can lead to a great diversity of publication processes, and inevitably means that resources from the same domain may be described in different ways, using different terminologies. Such heterogeneous representations mean that it can be difficult to identify relationships between published resources, where an understanding of such relationships is useful both for providing an integrated representation of the available data and for linking. Several approaches, from *schema matching* [12] to *ontology alignment* [15], have been

K. Christodoulou—First author has been supported by funding from the UK Engineering and Physical Sciences Research council, whose support we are pleased to acknowledge.

© Springer International Publishing Switzerland 2015
J. Wang et al. (Eds.): WISE 2015, Part I, LNCS 9418, pp. 200–215, 2015.
DOI: 10.1007/978-3-319-26190-4_14

proposed for identifying such candidate relationships (e.g., equivalence). Such techniques typically build on an aggregate of measures of syntactic relationships (such as edit-distance or n-gram intersection) that can be used to hypothesise equivalence. This dependency on syntactic relationships means that decisions tend to suffer from uncertainty. In LD, the fact that resources are described using shared ontologies presents an opportunity to bring together evidence at both the syntactic and semantic levels, i.e., not just names but also semantic annotations that characterise entities at the conceptual level. This paper describes a Bayesian technique for combining syntactic evidence, available in the form of similarity scores computed by string-based matchers, and semantic evidence, available in the form of semantic annotations such as subclass of and equivalent relations that can be formed in, or inferred from, LD ontologies.

Motivating Example. As an example, assume the existence of a LD dataset that describes instance data about music producers, such as, solo artists and groups like "The Beatles". Further, assume that an RDF resource exists for "The Beatles" stating that it is a member of the class *mo:MusicGroup*; *rdf:type(ns1: beatles, mo:MusicGroup)* in the Music Ontology[1]. At the same time, some other music provider models the same instance information about "The Beatles" (potentially under a different URI) stating that a resource for that entity is a member of the class *foaf:Group*; *rdf:type(ns2:beatles, foaf:Group)* using the FOAF vocabulary[2]. Such a scenario is plausible on the WoD since the two resources have been created by different, independent publishers. A system that is interested in merging the two LD datasets needs to deal with such heterogeneity in terminologies by discovering semantic correspondences between the two datasets.

Typically, schema matching techniques utilise knowledge from a formal structure, such as an *ontology description* or a *database schema*, for deriving correspondences of equivalence [2,15]. To make a decision as to the semantic equivalence of the concepts *MusicGroup* and *Group*, assume an approach that applies a set of string-based matchers (such as edit-distance and n-gram) over the local-names of *mo:MusicGroup* and *foaf:Group*. Such algorithms typically use a similarity score from the interval [0,1] as a confidence measure for discovered correspondences. Note that, in addition to their syntactic relationship, a semantic relation exists, stating that *mo:MusicGroup* is subsumed by *foaf:Group*. We suggest that such relations can be used as additional knowledge to improve the decision making of matching techniques beyond the use of syntactic matchers alone.

Summary of Contributions. This paper describes a probabilistic approach for combining evidence from syntactic matchers with semantic annotations modelled as *degrees of belief* on the existence of semantic correspondences of equivalence. The following questions motivated this study. Which semantic annotations can be usefully be taken as additional evidence for the purposes of postulating construct equivalence? How can we reason about syntactic and semantic evidence using probabilistic models? How can we incrementally assimilate different

---

[1] http://purl.org/ontology/mo/.
[2] http://xmlns.com/foaf/0.1/.

kinds of evidence? In seeking solutions to these questions, this paper contributes the following: (a) a methodology that uses kernel density estimation for deriving likelihoods from similarity scores computed by string-based matchers; (b) a methodology for deriving likelihoods from semantic relations (e.g., rdfs: sub-ClassOf, owl:equivalentClass) that are retrieved by dereferencing URIs in LD ontologies; (c) a methodology for aggregating evidence of conceptual construct equivalence from both string-based matchers and semantic annotations; and (d) an empirical evaluation of our approach grounded on the judgements of experts in response to the same kinds of evidence.

The remainder of the paper is structured as follows. Section 2 presents an overview of the developed solution. Section 3 describes the methodology used for deriving probability distributions over similarity scores from string-based matchers, along with a methodology for deriving likelihoods from semantic knowledge defined in LD ontologies. Bayesian updating, as a technique for the incremental assimilation of evidence, is introduced in Sect. 4. Section 5 presents an empirical evaluation of the methodology complemented by a discussion of results. Section 6 reviews related work, and Sect. 7 concludes.

2   Overview of Solution

Given a conceptual description of a *source* and a *target* LD dataset, denoted by S and T, respectively, a *semantic correspondence* of equivalence is a triple $\langle c_S, c_T, P(c_S \equiv c_T | E) \rangle$, where $c_S \in S$ and $c_T \in T$ are constructs (i.e., Classes) from the datasets, and $P(c_S \equiv c_T | E)$ is the conditional probability representing the degree of belief (from now on referred to as dob) in the equivalence ($\equiv$) of the constructs given the pieces of evidence $(e_1, ..., e_n) \in E$. Section 4 describes in detail how to compute the conditional probability using Bayes' theorem. Our approach distinguishes two types of knowledge: (a) *syntactic knowledge*, in the form of strings that are local-names of resources' URIs; and (b) *semantic knowledge*, such as structural relations between entities, either internal to a vocabulary or across different LD vocabularies, e.g., relations such as subclass of and equivalence. Table 1 summarises the types of knowledge construed by our approach as sources of evidence. The set TE is the set of all semantic annotations we consider as evidence, where the subsets EE and NE comprise the assertions that can be construed as *direct* evidence of equivalence and non-equivalence, respectively.

To collect syntactic evidence (represented by the set LE), given two sources, our approach extracts local-names from the URIs of every pair of constructs $\langle c_s, c_t \rangle$ and then derives their pair-wise string-based similarity. Two string-based metrics are used, viz., *edit-distance* (denoted by ed) and *n-gram* (denoted by ng) [15]. Section 3.1 elaborates on how probability distributions can be constructed for each matcher. To collect semantic evidence, our approach dereferences URIs to get access to annotations from the vocabularies that define the resource. For example, the subsumption relation $c_S \sqsubseteq c_T$ is taken as semantic evidence. Section 3.2 elaborates on an approach to constructing probability distributions for each kind of semantic evidence in RDFS/OWL vocabularies.

Table 1. Syntactic and semantic evidence utilised by the technique.

Type		ID	Description	Evidence rule
Syntactic evidence (LE)	-	SLN	similar-local-name	*string similarity*(c_T, c_S)
Semantic evidence (TE)		SU	same-URI	*string equality*(URI_S, URI_T)
	-	SB	subsumed-by	$c_S \sqsubseteq c_T$
	EE	SA	same-as	owl:sameAs(c_S, c_T)
		EC	equivalent-class	owl:equivalentClass(c_S, c_T)
		EM	exact-match	skos:exactMatch(c_S, c_T)
	NE	DF	different-from	owl:differentFrom(c_S, c_T)
		DW	disjoint-with	owl:disjointWith(c_S, c_T)

3 Constructing Likelihoods for Evidence

To assimilate different kinds of evidence, some bootstrapping is needed that will allow the computation of the likelihoods necessary for the calculation of a dob on construct equivalence, as captured by the posterior $P(c_S \equiv c_T | E)$ given both syntactic and semantic evidence. This section describes a principled methodology for constructing probability distributions from similarity scores returned by string-based matchers, as well as a procedure for deriving likelihoods for each type of semantic evidence in Table 1.

3.1 Similarity Scores to Degrees of Belief

We call *syntactic evidence* the likelihoods derived from *similarity scores* produced by string-based matchers. We study the behaviour of each matcher (in our case ed and ng) to derive these likelihoods as follows:

1. From the datasets made available by the Ontology Alignment Evaluation Initiative (OAEI)[3], we observed the available ground truth on whether a pair of local-names, denoted by (n, n'), aligns.
2. We assume the existence of a continuous random variable, X, in the bounded domain [0,1], for the similarity scores returned by each matcher μ, where $\mu \in$ ed, ng. Our objective is to model the behaviour of each matcher in terms of a probability density function (PDF) $f(x)$ over the similarity scores it returns (we refer to them as observations).
3. To empirically approximate $f(x)$ for each matcher we proceed as follows:
   (a) We ran each matcher μ independently over the set of all local-name pairs (n, n') obtained from (1).
   (b) For each pair of local-names, we observed the independent similarity scores returned by the matcher when (n, n') agrees with the ground truth. These are the set of observations $(x_1, ..., x_i)$ from which we estimate $f(x)$ for the equivalent case.

---

[3] http://oaei.ontologymatching.org.

4. The observations $x_1, ..., x_i$ obtained are used as inputs to the non-parametric technique known as kernel density estimation (KDE) (using a Gaussian kernel[4]) [3] whose output is an approximation $\hat{f}(x)$ for both ed and ng for both the equivalent and non-equivalent cases.

We interpret the outcome of applying such a PDF to syntactic evidence as the likelihood of that evidence. More formally, and as an example, $PDF_{\underset{ed}{\equiv}}(\text{ed}(n, n'))$ $= P(\text{ed}(n, n')|c_S \equiv c_T)$, i.e., given a pair of local-names (n, n') the PDF for the ed matcher in the equivalent case $PDF_{\underset{ed}{\equiv}}$ yields the likelihood that the similarity score $\text{ed}(n, n')$ expresses the equivalence of the pair of concepts (c_S, c_T) that (n, n'), resp., denote. Correspondingly, for the non-equivalent case, and for ng in both the equivalent and non-equivalent cases (Fig. 1).

The probability distributions derived by this process are shown in Fig. 2(a) and (b) for ed and in Fig. 2(c) and (d) for ng. The procedure described can be used to study the behaviour of any matcher that returns similarity scores in the interval $[0, 1]$. Note that the PDFs obtained by the method above are *derive-once, apply-many* constructs. Assuming that the sample set used for training remains representative, and given that the behaviour of matchers ed and ng is fixed and deterministic, the PDFs need not be recomputed.

3.2   Semantic Evidence to Degrees of Belief

We call *semantic evidence* the likelihoods derived from *semantic annotations* obtained from the WoD. We first retrieved the semantic annotations summarised in Table 1. The set TE is the set of all such evidence, $TE = \{$SU, SB, SA, EC, EM, DF, DW$\}$. We formed the subsets $EE \subset TE = \{$SA, EC, EM$\}$ and $NE \subset TE = \{$DF, DW$\}$ comprising assertions that can be construed as *direct* evidence of equivalence and non-equivalence, respectively.

To derive probability distributions for semantic evidence, we proceeded as follows:

1. We assume the existence of a Boolean random variable, for each type of semantic evidence in Table 1, with domain $\{true, false\}$.
2. Using the vocabularies available in the Linked Open Vocabularies (LOV)[5] collection.
    (a) We collected and counted pairs of classes and properties that share direct or indirect assertions of equivalence or non-equivalence for all the assertions in TE and NE using SPARQL queries. For example:

---

[4] Gaussian kernel was used due to its mathematical convenience. Note that any kernel other than Gaussian can be applied, however, the shape of the distribution may differ depending on the kernel characteristics.

[5] http://lov.okfn.org/dataset/lov/.

```
SELECT DISTINCT ?elem1 ?elem2
WHERE   {
        {?elem1 a rdfs:Class .} UNION {?elem1 a owl:Class .}
        ?elem1 ?p ?elem2 .
        FILTER (?p = owl:equivalentClass && !isBlank(?elem2))}
```

(b) From the set of pairs derived by the assertions in TE and NE, we counted assertions that can be construed as *evidence* of equivalence or non-equivalence for each pair, grouping such counts by kind of assertion (e.g., subsumed-by (SB), etc.)

3. We used the sets of counts obtained in the previous step to build contingency tables (e.g., see Table 2) from which we can directly derive the probability mass functions (PMFs) for each kind of semantic evidence for both the equivalence and non-equivalent cases.

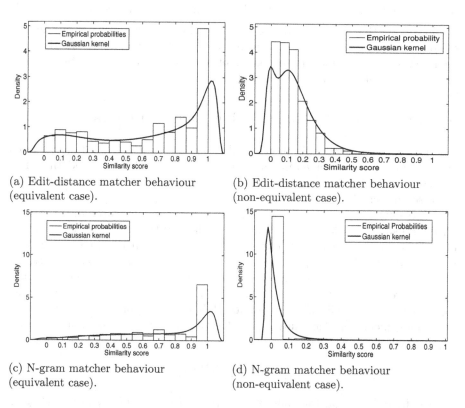

(a) Edit-distance matcher behaviour (equivalent case).

(b) Edit-distance matcher behaviour (non-equivalent case).

(c) N-gram matcher behaviour (equivalent case).

(d) N-gram matcher behaviour (non-equivalent case).

Fig. 1. Illustration of probability distributions for each matcher over $[0, 1]$.

The PMFs obtained through the steps above are also *derive-once, apply-many* constructs, but since the vocabulary collection from which we draw our sample is dynamic, we might wish to view them as *derive-seldom, apply-often.*

Table 2. Example of a contingency table. The likelihood $P(\mathbf{EC}(n, n')|c_S \equiv c_T)$ is estimated by the fraction 305/396.

Contingency table	Semantic evidence		
	EC	¬EC	Total
$c_S \equiv c_T$	305	91	396
$c_S \not\equiv c_T$	0	2552	2552
Total	305	2643	2948

We interpret the outcome of applying such a PMF to semantic evidence as the likelihood of that evidence. More formally, and as an example, $PMF_{\underset{EC}{\equiv}}(\mathsf{EC}(u, u')) = P(\mathsf{EC}(u, u')|c_S \equiv c_T)$, i.e., given the existence of an assertion that a pair of URIs (u, u') have an equivalence relation, the probability mass function for this kind of assertion in the equivalent case $PMF_{\underset{EC}{\equiv}}$ yields the likelihood that the assertion $\mathsf{EC}(u, u')$ expresses the equivalence on the pair of constructs (c_S, c_T) that (u, u'), resp., denote. Correspondingly, for the non-equivalence case and for all other kinds of semantic evidence (e.g., SB, etc.) in both the equivalent and non-equivalent cases.

4   Assimilating Evidence Using Bayesian Updating

The purpose of deriving likelihood models as described in Sect. 3 is to enable the evidence to be combined in a systematic way using Bayesian updating. The procedure for doing so is now described, where the benefits of the procedure are discussed in Sect. 5.

We denote with S and T, resp., the structural summaries (an ontology or a structural summary derived by an approach like [4]) that describe the structure of a *source* and a *target* LD source over which we wish to discover semantic correspondences. Given a pair of constructs $c_S \in S$ and $c_T \in T$ our objective is to derive a dob on the postulated equivalence of a pair of constructs (denoted by H), given pieces of evidence $e_1, ..., e_n \in E$. To reason over our hypothesis, we model it as a conditional probability $P(H|E)$ and apply Bayes' theorem to make judgements on the equivalence of two constructs. In its simplest form, Bayes' theorem states that[6],

$$P(H|E) = \frac{P(E|H)\,P(H)}{P(E)}. \tag{1}$$

Our hypothesis can take one of two states: $P(H) = \{P(c_S \equiv c_T), P(c_S \not\equiv c_T)\}$, i.e., it is a *Boolean hypothesis*. The prior probability, e.g., $P(H) = P(c_S \equiv$

---

[6] Informally, the theorem states that the hypothesis given the evidence (so called posterior) is equal to the ratio between the product of the dob in the evidence given the hypothesis (what we called likelihood in Sect. 3) and the dob in the hypothesis (so called prior) divided by the dob in the evidence.

c_T), is the dob in the absence of any other piece of evidence (we assume a uniform distribution). Thus, for the two possible outcomes our hypothesis can take, $N = 2$, the prior probability that one of the outcomes is observed is given by $1/N$. The probability of the evidence, $P(E)$, can be expressed using the law of total probability [9], i.e., $P(E) = P(E|c_S \equiv c_T) P(c_S \equiv c_T) + P(E|c_S \not\equiv c_T) P(c_S \not\equiv c_T)$. To use Bayes' theorem for deriving a dob on the hypothesis given the available evidence, it is essential to estimate the likelihoods for each evidence: (i.e., $P(E|c_S \equiv c_T)$, and, $P(E|c_S \not\equiv c_T)$). For semantic evidence, the likelihoods are estimated from the contingency tables constructed in Sect. 3.2. For continuous values, like similarity scores, the constructed PDFs for each matcher from Sect. 3.1 are used to estimate the conditional probabilities for the likelihoods. To determine these likelihoods, we integrate the PDF over a finite region $[a, b]$, namely $P(a \leq X \leq b) = \int_a^b f(x)\, dx$, where the density $f(x)$ is computed using KDE with a *Gaussian* kernel.

The idea behind *Bayesian updating* [16], is that once the posterior e.g., $P(c_S \equiv c_T|E)$ is computed for some evidence, $e_1 \in E$, a new piece of evidence $e_2 \in E$, leads us to compute the impact of e_2 by taking the previously computed posterior as the new prior. Given the ability to compute likelihoods for both syntactic and semantic evidence, we can use Bayesian updating to compute a dob on the equivalence of (pairs of constructs in) two structural summaries S and T. To demonstrate this with a concrete example, let $P^{(e_1,\dots,e'_n)}$ denote the dob that results from having assimilated the evidence sequence $(e_1, \dots, e_n)$. The initial prior is therefore denoted by $P^{()}$, and if $(e_1, \dots, e_n)$ is the complete evidence sequence available, then $P^{(e_1,\dots,e'_n)}$ is the final posterior. We proceed as follows:

i. We set the initial prior according to the principle of indifference between the hypothesis that $P(c_S \equiv c_T)$ and its negation, so $P^{()} = 0.5$.

ii. We collect the local-name pairs from the structural summaries S and T.

iii. We run ed on the local-name pairs and, using the probability distributions derived using the methodology described above (Sect. 3.1), compute the likelihoods for each pair and use Bayes' rule to calculate the initial posterior $P^{(ed)}$.

iv. We run ng on the local-name pairs and, using the probability distributions derived using the methodology described above (Sect. 3.1), compute the likelihoods for each pair and use Bayes' rule to calculate the next posterior $P^{(ed,ng)}$. Note that this is the dob given the syntactic evidence alone, which we denote more generally by $P^{(syn)}$.

v. To get access to semantic annotations that span a variety of LD ontologies, we dereference every URI in S and T to collect the available semantic annotations e.g., $SB(c_S \subseteq c_T)$.

vi. Using the methodology described above (Sect. 3.2), we compute, one at a time, the likelihoods for the available semantic evidence, each time using Bayes' rule to calculate the next posterior (e.g., $P^{(ed,ng,SB,\dots)}$), so that once all the available semantic evidence is assimilated, the final posterior, which we denote more generally by $P^{(syn,sem)}$, is the dob on $c_S \equiv c_T$, where, $c_S \in S \wedge c_T \in T$.

Before carrying out the empirical evaluation of this approach using syntactic and semantic evidence described in Sect. 5, we studied analytically, using Bayes's theorem, the effect of each piece of evidence independently. Given a series of initial prior probabilities in the range of $[0, 1]$ and the evidence likelihoods (see Sect. 3) we computed the posterior probabilities given each piece of evidence. Figure 2(a) and (b) show how the posteriors $P(c_s \equiv c_t | ed(c_s, c_t) = s)$, and, $P(c_s \equiv c_t | ng(c_s, c_t) = s)$, resp., are updated when the available evidence is similarity scores computed by the string-based matchers ed and ng. As an example, consider Fig. 2(a), and assume that we are given a prior probability of $x = 0.5$ and a similarity score that is $y < 0.5$, ed will cause the updated posterior probability to fall relatively more. In this case, if the similarity score is $y = 0.2$, the posterior probability drops to $z = 0.2$. In the case of ng, using identical values as previously, the posterior probability drops to $z = 0.36$, which means that ng causes a small decrease in the posterior. In a similar fashion, the independent behaviours of different kinds of semantic evidence have been studied. For example, Fig. 2(c) shows how the posterior is updated when there is direct evidence that a pair of classes stand in a subsumption relationship (i.e., SB). A subsumption relation may indicate that the constructs are more likely to be related than to be disjoint and a low prior is therefore increased. Similarly, Fig. 2(d) shows how the posterior is affected when a pair of constructs stand in an equivalence relation (i.e., EC). This is considered enough evidence to significantly increase a low prior to close to 1; meaning that constructs are more probably equivalent than if that evidence had not been available.

Having observed how different posterior probabilities are updated in the presence of individual pieces of evidence, in Sect. 5 we empirically assess whether the incorporation of semantic evidence from LD ontologies can improve on construct equivalence judgements obtained through syntactic matching alone.

5   Experimental Evaluation

The evaluation of our approach was based on the idea of emulating the construct equivalence judgements produced by human experts in the presence of different kinds of syntactic and semantic evidence[7]. The judgements derived from experts are then compared with the judgements derived by the Bayesian updating approach discussed in Sects. 3 and 4. This section describes an experimental scenario that has a twofold purpose: (a) to compare how well the Bayesian assimilation of syntactic evidence alone performs against the aggregation of syntactic evidence followed by a predefined function, specifically average (AVG) which is commonly used in existing matching systems [2, 15], and (b) to observe empirically whether the incorporation of semantic evidence can improve on construct equivalence judgements obtained through syntactic matching alone.

---

[7] The survey was distributed and completed by 15 human participants all experts in solving data integration tasks, such as schema matching and mapping.

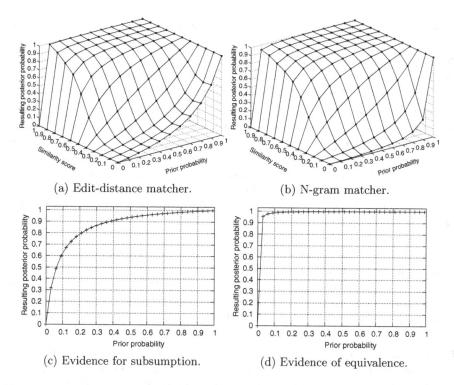

(a) Edit-distance matcher.

(b) N-gram matcher.

(c) Evidence for subsumption.

(d) Evidence of equivalence.

Fig. 2. Effect on the posterior probabilities using particular evidence on different prior probabilities.

5.1 Experimental Setup

To evaluate the application of Bayes's theorem for assimilating different kinds of evidence, the experimental evaluation was grounded on the rational decisions made by *human experts* on data integration and ontology alignment when judging whether a pair of constructs is postulated to be equivalent given both syntactic and semantic evidence as construed in this paper. For the purposes of the experiment, a set of pairs of constructs from different LD ontologies was collected, making sure that different combinations of syntactic and semantic evidence (as in Table 1) were present or absent. To obtain testimonies from the human experts, a survey was designed based on the collected set of pairs of constructs, asking the experts to make judgements on the equivalence of such pairs. Testimonies have been recorded on a discretisation scale [5], as follows: {Definitely equivalent} mapped to a dob of 1.0; {Tending towards being equivalent} mapped to a dob of 0.75; {Do not know} mapped to a dob of 0.5; {Tending towards being not-equivalent} mapped to a dob of 0.25; and {Definitely not-equivalent} mapped to a dob of 0. By observing different pairs of constructs from real ontologies, approximately 40 common combinations of syntactic and semantic evidence have been identified. For each combination, a question was designed to obtain individual testimonies from each responder. Individual testimonies from each question were aggregated using a weighted average, based on the confidence assigned to each item [5]. The aggregated degrees of belief obtained from the survey are treated as an approximation of the experts' confidence on construct equivalence given certain pieces of syntactic and semantic evidence and act as a gold standard.

Datasets. For the purposes of the experiment, the Bayesian technique was evaluated over the class hierarchies of ontologies made available by the OAEI - Conference track, which have been designed independently by different processes but all belonging to the domain of conference organisation. Note also that these ontologies share no semantic relations between them. Since our technique assumes such relations for use as semantic evidence, we made some of these cross-ontology semantic relations explicit using BLOOMS[8]; a system for discovering rdfs:subClassOf and owl:equivalentClass relations between LD ontologies [7]. We note that the contributions reported in this paper are independent of BLOOMS, in that they can be used regardless of the sources of semantic annotations. We found that the LOD cloud at the conceptual level still lacks the abundance of cross-ontology links that, most agree, will one day characterise the SW. We have therefore used BLOOMS to induce some more cross-ontology links in a principled manner. The results reported in this paper consider a single pair of ontologies from the conference track, viz., `ekaw` (denoted by S) and `conference` (denoted by T).

Expectation Matrix. Given a pair of classes from the class hierarchies of the input ontologies and given the available kinds of evidence, both syntactic and semantic, a dob was assigned for each pair on the basis of the experts' testimonies. More formally, we constructed a $n \times m$ structure referred to from now on as the *expectation matrix* and denoted as M_{exp}, where $n = |S|$ and $m = |T|$. The element e_{jk} in the jth row and the kth column of M_{exp} denotes the dob derived from the expert survey between the jth construct in S and the kth construct in T according to the pieces of evidence present or absent.

Evaluation Metric. Let $p_1, p_2, ..., p_n$ be the degrees of belief derived for each pair of classes from the ontologies by either the average aggregated scheme or the Bayesian assimilation, and $a_1, a_2, ..., a_n$ be the corresponding degrees of belief in the expectation matrix just described. We compute the mean-absolute error, $MAE = (|p_1 - a_1| + ... + |p_i - a_i|) \div n$ where $|p_1 - a_1|$ is an *individual error* of a pair and n is the total number of such errors.

5.2   Evaluation Methodology

Traditional matching approaches (e.g., COMA [1]) exploit different pieces of evidence, mostly from string-based matchers, to assess the similarity between constructs in ontologies or in database schemas. Such approaches combine similarity scores computed independently, typically using averages. For this evaluation the antagonist to our Bayesian approach is considered a process that independently runs matchers `ng` and `ed` on the local-names of classes from ontologies S and T,

---

[8] BLOOMS was configured with a high threshold, viz., $> 0.8..$

and produces an average of the similarity scores. The aggregated result of this computation is a matrix M_{avg}. The next step is to measure how close the derived predictions are to the degrees of belief obtained by the experts' testimonies. In doing so, we used MAE as the performance measure since it does not exaggerate the effect of outliers [6]. The result from computing the error between M_{avg} and the expectation matrix M_{exp} is denoted by δ_{avg}.

Similarly, the Bayesian assimilation technique (as described in Sect. 4) was used (instead of an average) to assimilate the evidence computed by the string-based matchers on pairs of local-names. The result of this computation is a matrix M_{syn}, where $n = |S|$ and $m = |T|$. The element e_{jk} in the jth row and the kth column of M_{syn} denotes the posterior probability $P^{(syn)}$ between the jth class in S and the kth class in T according to the syntactic evidence derived from the string-based matchers ed and ng. The next step is to measure how close the predictions from M_{syn} are to the expectation matrix M_{exp}. The result is denoted by δ_{syn}.

To assess whether semantic evidence can improve on construct equivalence judgements that use averaging alone to aggregate syntactic evidence, we first used BLOOMS [7] to make explicit the cross-ontology semantic relations and used this as semantic evidence. In the light of this new evidence, the Bayesian assimilation technique updates the posterior probabilities $P^{(syn)}$ for each pair of classes in M_{syn} accordingly. The result of this process is a new matrix $M_{syn,sem}$ with the same dimensions as M_{syn}, where, the posterior probabilities for the elements e_{jk} reflect both syntactic and semantic evidence, $P^{(syn,sem)}$. Again we denote by $\delta_{syn,sem}$ the error calculated between $M_{syn,sem}$ and the expectation matrix M_{exp}. Finally, to complete the evaluation, the individual absolute errors used for the calculation of δ_{avg}, δ_{syn}, and $\delta_{syn,sem}$ have been examined. The results of the evaluation are now discussed.

5.3   Results and Discussion

Exp. 1: AVG scheme vs. Bayesian Syntactic. The MAE error computed for the average aggregation scheme against the expectation matrix was $\delta_{avg} = 0.1079$ whereas the error as a result of assimilating syntactic evidence using the Bayesian technique was $\delta_{syn} = 0.0698$. The difference of 0.0381 between the two errors can be expressed in percentage terms as 35.32 %. To further understand the difference in errors, we measured the individual *absolute* errors that fall into each of four regions of interest as these are shown in Fig. 3(a). They correspond to the following minimum bounding rectangles, resp., Region 1 lies below the $y = x$ error line where AVG error $>>$ Bayesian error and is the rectangle defined by $y = 0.2$; Region 2 lies above the $y = x$ error line where AVG error $<<$ Bayesian error and is the rectangle defined by $x = 0.2$; Region 3 lies below the $y = x$ error line where AVG error $>$ Bayesian error and is the rectangle defined by $y > 0.2$; and Region 4 lies above the $y = x$ error line where AVG error $<$ Bayesian error and is the rectangle defined by $x > 0.2$. We note that the larger the cardinality of Region 1, the more significant is the impact of using semantic annotations as we propose.

Table 3. AVG scheme vs. Bayesian syntactic.

No.	Region	Count	Perc. (%)
1	$R_{avg>>B_{syn}}$	3833	87.49
2	$R_{avg<<B_{syn}}$	215	4.90
3	$R_{avg>B_{syn}}$	31	0.70
4	$R_{avg<B_{syn}}$	302	6.89

Table 4. AVG scheme vs. Bayesian syntactic & semantic.

No.	Region	Count	Perc. (%)
1	$R_{avg>>B_{syn,sem}}$	125	71.43
2	$R_{avg<<B_{syn,sem}}$	43	24.57
3	$R_{avg>B_{syn,sem}}$	2	1.14
4	$R_{avg<B_{syn,sem}}$	5	2.85

Table 5. Bayesian syntactic vs. Bayesian syntactic & semantic.

No.	Region	Count	Perc. (%)
1	$R_{B_{syn}>>B_{syn,sem}}$	124	89.21
2	$R_{B_{syn}<<B_{syn,sem}}$	9	6.48
3	$R_{B_{syn}>B_{syn,sem}}$	5	3.60
4	$R_{B_{syn}<B_{syn,sem}}$	1	0.72

For the traditional aggregation scheme that produced M_{avg} we counted 3833 matches with individual errors greater than the analogous individual errors derived by the Bayesian technique that produced M_{syn}. The use of Bayesian aggregation significantly outperformed (i.e., has smaller individual errors than) the use of AVG aggregation scheme for 87.49 % of the total. Table 3 summarises the results for each region showing how many individual errors are located in each of the regions of interest in both absolute terms and relative to the total.

Exp. 2: AVG scheme vs. Bayesian Syn. & Sem. To evaluate our hypothesis whether semantic annotations can improve outcomes we compared the aggregated errors denoted by δ_{avg} and $\delta_{syn,sem}$. The mean absolute error $\delta_{syn,sem} = 0.1259$ is lower than $\delta_{avg} = 0.1942$ with a difference of 0.0683 or 35.15 %. Figure 3(b) plots the individual errors for pairs of classes that have some semantic relation between them. We are interested on cases where the individual errors for the Bayesian technique are smaller than the AVG scheme. In particular, the points that lie mostly between 0.1 and 0.3 on the x-axis and below the $y = x$ error line. For 71.43 % of the total matches that have some semantic evidence the Bayesian technique produces results closer to the testimonies, with individual errors that mostly lie in that region. Table 4 summarises the results for each region showing how many individual errors are located in each of the regions of interest in both absolute terms and relative to the total.

Exp. 3: Bayesian Syn. vs. Bayesian Syn. & Sem. Similarly to Exp.2, we compared the aggregated errors denoted by δ_{syn} and $\delta_{syn,sem}$ considering only individual errors that have some semantic evidence. Again in this case $\delta_{syn,sem} = 0.1259$ is closer to the expectation matrix than $\delta_{syn} = 0.2768$ with a difference of 0.1509 or 54.5 %. The results of this experiment are summarised in Table 5.

The points of interest in this experiment are the ones where the individual errors for $B_{syn,sem}$, that considers both syntactic and semantic evidence, are smaller than B_{syn}. For 89.21 % of the total matches discovered, that have some semantic evidence, $B_{syn,sem}$ outperforms the configuration of the Bayesian scheme that utilises syntactic evidence alone, i.e., B_{syn}.

(a)   (b)

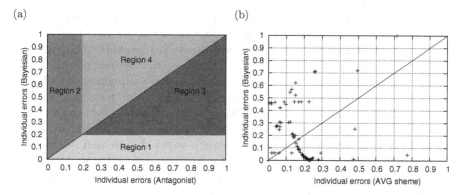

Fig. 3. (a) Shows the regions of interest, (b) Individual errors Bayesian against AVG scheme.

Overall, the experimental evaluation showed that the Bayesian assimilation of syntactic and semantic evidence delivers, in most cases better judgements of construct equivalence than the use of syntactic matchers alone i.e., than the state-of-the-art in matching. The aim of the experiment was to evaluate the Bayesian approach on how well it performed on aggregating different pieces of syntactic and semantic evidence against AVG a common aggregation strategy used in traditional matching approaches. Alignments provided by the OAEI group are tailored towards evaluating approaches that make classification decisions for discovering ontology alignments and are not suitable for judging individual aggregated confidence degrees of belief derived by the approaches. To the best of our knowledge there are no established benchmarks for doing so. Therefore, we consulted human experts for the construction of the baseline used for the evaluation.

6 Related Work

A variety of strategies have been proposed in the literature for solving the problem of combining different pieces of evidence about matches, some examples are: average, weighted average, min, max and sigmoid functions [10]. However, it falls on users to tune or select the appropriate aggregation method manually according to the problem in hand. In contrast, the Bayesian assimilation of evidence technique can be used as an alternative aggregation strategy for assimilating any piece of evidence, complementing typical aggregation strategies used by state-of-the-art schema and ontology matching systems [2,12,15]. When the appropriate probability distributions are made available, the approach presented in this paper can be used as a generic aggregation strategy that is not tied to any specific domain. Sabou et al. [13] presented an ontology matching paradigm that makes use of additional external background knowledge that is made available from ontologies from the Semantic Web. The proposal in our paper makes use of additional semantic annotations from LD ontologies as evidence with the aim of improving the decision making of different matchers that mostly work on syntax. In another note, approaches for discovering semantic relations from ontologies e.g., [14] can be used to provide input to our Bayesian approaches to further improve the accuracy, thus improving the decision making of matching approaches. The uncertainty in the decisions made by different *matchers* has also been observed in [8], where a similarity matrix that describes the outcome of some matcher is modelled as two probability distributions. An alternative statistical analysis is used to model the similarity scores distribution returned by each matcher that uses the parametric beta-distribution to estimate the underlying probability. The proposal in our paper, however, makes no assumptions about the shape or parameters of the underlying distribution, and uses a non-parametric statistical analysis technique, based on kernel density estimation, to approximate the probability distributions for each matcher using the sampled data.

7   Conclusions

The WoD can be seen as vibrant but challenging: vibrant because there are numerous publishers making valuable data sets aware for public use; challenging because of inconsistent practises and terminologies in a setting that is something of a free-for-all. In this context, it is perhaps easier to be a publisher than a consumer. As a result, there is a need for tools and techniques to support effective analysis, linking and integration in the web of data [11]. The challenging environment means: (i) that there are many different sources of evidence on which to build; (ii) that there is a need to make the most of the available evidence; and (iii) that it is not necessarily easy to do (ii). This paper has described a well-founded approach to combining multiple sources of evidence of relevance to matching, namely syntactic matchers and semantic annotations. The findings from the empirical evaluation suggested that the Bayesian aggregation scheme has let to improved decision making of close to 90 % of the total matches when assimilating just syntactic evidence, and the confidence of close to 70 % of the matches that had some semantic evidence has been improved in the light of available semantic evidence. Overall, the suggested approach can be used as a generic methodology for assimilating different kinds of evidence as they become available, or as a method that complements existing aggregation strategies for matching systems.

References

1. Aumueller, D., Do, H.H., Massmann, S., Rahm, E.: Schema and ontology matching with coma++. In: SIGMOD Conference, pp. 906–908 (2005)
2. Bernstein, P., Madhavan, J., Rahm, E.: Generic schema matching, ten years later. Proc. VLDB Endowment **4**(11), 695–701 (2011)
3. Bowman, A.W., Azzalini, A.: Applied Smoothing Techniques for Data Analysis : The Kernel Approach with S-Plus Illustrations: The Kernel Approach with S-Plus Illustrations. OUP, Oxford (1997)
4. Christodoulou, K., Paton, N.W., Fernandes, A.A.A.: Structure inference for linked data sources using clustering. In: EDBT/ICDT Workshops, pp. 60–67 (2013)
5. de Vaus, D.: Surveys in Social Research. Research methods/Sociology. Taylor & Francis (2002)
6. Hyndman, R.J., Koehler, A.B.: Another look at measures of forecast accuracy. Int. J. Forecast. (IJF) **22**(4), 679–688 (2006)
7. Jain, P., Hitzler, P., Sheth, A.P., Verma, K., Yeh, P.Z.: Ontology alignment for linked open data. In: Patel-Schneider, P.F., Pan, Y., Hitzler, P., Mika, P., Zhang, L., Pan, J.Z., Horrocks, I., Glimm, B. (eds.) ISWC 2010, Part I. LNCS, vol. 6496, pp. 402–417. Springer, Heidelberg (2010)
8. Marie, A., Gal, A.: Managing uncertainty in schema matcher ensembles. In: Prade, H., Subrahmanian, V.S. (eds.) SUM 2007. LNCS (LNAI), vol. 4772, pp. 60–73. Springer, Heidelberg (2007)
9. Papoulis, A.: Probability, Random Variables and Stochastic Processes, 3rd edn. McGraw-Hill Companies, New York (1991)
10. Peukert, E., Maßmann, S., König, K.: Comparing similarity combination methods for schema matching. GI Jahrestagung **1**, 692–701 (2010)
11. Polleres, A., Hogan, A., Harth, A., Decker, S.: Can we ever catch up with the web? Semantic Web **1**(1–2), 45–52 (2010)
12. Rahm, E., Bernstein, P.A.: A survey of approaches to automatic schema matching. VLDB J. **10**(4), 334–350 (2001)
13. Sabou, M., d'Aquin, M., Motta, E.: Exploring the semantic web as background knowledge for ontology matching. J. Data Semant. **11**, 156–190 (2008)
14. Sabou, M., d'Aquin, M., Motta, E.: SCARLET: semantiC relAtion discoveRy by harvesting onLinE onTologies. In: Bechhofer, S., Hauswirth, M., Hoffmann, J., Koubarakis, M. (eds.) ESWC 2008. LNCS, vol. 5021, pp. 854–858. Springer, Heidelberg (2008)
15. Shvaiko, P., Euzenat, J.: Ontology matching: state of the art and future challenges. IEEE Trans. Knowl. Data Eng. **25**(1), 158–176 (2013)
16. Spragins, J.: A note on the iterative application of bayes' rule. IEEE Trans. Inf. Theor. **11**(4), 544–549 (2006)

An Efficient Location-Aware Publish/Subscribe Index with Boolean Expressions

Hanhan Jiang[1], Pengpeng Zhao[1]([✉]), Victor S. Sheng[2], Guanfeng Liu[1],
An Liu[1], Jian Wu[1], and Zhiming Cui[1]

[1] School of Computer Science and Technology, Soochow University, Suzhou, China
20145227025@stu.suda.edu.cn,
{ppzhao,gfliu,anliu,jianwu,szzmcui}@suda.edu.cn
[2] Computer Science Department, University of Central Arkansas, Conway, USA
ssheng@uca.edu

Abstract. A location-aware publish/subscribe (pub/sub) system is gaining more and more interest in both industry and academia with the rapid progress of mobile Internet and the rising popularity of smart-phones. Nowadays, with the booming of E-commerce, Object-to-Object (OTO) services are gaining more and more popularity, which results in millions of products with different structured descriptions and locations. To meet this requirement, a pub/sub system should handle subscriptions with location-aware boolean expressions to present users' interests. In this paper, we propose an efficient location-aware pub/sub index for boolean expressions, called RP-trees. RP-trees integrates an R-tree index and a boolean expression index together, can efficiently and simultaneously prune boolean expressions and spatial dimensions. Our experimental results show that RP-trees achieves better performance on both a synthetic dataset and a real-world dataset.

Keywords: Location-aware pub/sub · Boolean expression · OTO

1 Introduction

With the rapid progress of mobile Internet and the growing popularity of smart-phones equipped with GPS, the location-aware pub/sub have recently attracted significant attention from both industrial and academic communities [2,5–7,16, 17]. The subscribers subscribe their interests as subscriptions while a publisher publishes messages as events which are delivered to subscribers whose interests are matched with the messages. Location-aware pub/sub systems can be applied in many real-world domains, such as Groupon, Twitter and Foursquare. Existing location-based pub/sub systems [6,11] can support subscriptions with textual descriptions well. Users on Twitter can register their interests with a textual descriptions and a spatial constraint (e.g.,"Fresh seafood","*distance* $\leqslant$ 3 km"). For each tweet with textual description and location information (e.g.,"Newly stocked seafood: crayfish","31.129025, 120.468226"), the system has to ensure

© Springer International Publishing Switzerland 2015
J. Wang et al. (Eds.): WISE 2015, Part I, LNCS 9418, pp. 216–231, 2015.
DOI: 10.1007/978-3-319-26190-4_15

a timely delivery of textual and spatial matching events to the corresponding subscribers.

However, thanks to the booming of e-commerce and mobile Internet. OTO services are gaining more and more popularity, which bring millions of products with many different attributes, values and geo-locations. This leads to a requirement for structuring descriptions to capture such information. Existing unstructured pub/sub works [2,5,6,17] using textual descriptions cannot accurately represent users interests. Existing structured pub/sub works [4,7,12,13,15] using boolean expressions cannot efficiently process spatial constraints. To meet this requirement, we model a new location-aware structured pub/sub problem in which subscriptions are represented by a combination of boolean expressions and spatial information. To explain this, we have the following example as a working scenario.

We can model services or products in an OTO platform like Groupon[1] as events and their users as subscribers. Users register their interested regions, attributes and values of products as subscriptions. As the Fig. 1 shows, a subscription $\{(B = 3 \wedge A \in (3,2,5) \wedge C \geqslant 2), R_1\}$ is a combination of a boolean expression and a spatial constraint. A product is represented by a list of attribute-value pairs with a spatial point $\{(A = 3 \wedge B = 3 \wedge C = 5), P_1\}$. Users will be notified when a new inserted product matches their subscriptions.

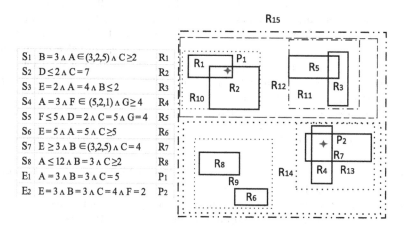

Fig. 1. An example of subscriptions and events

In the following, we present two real-world applications for which a location-aware pub/sub with boolean expression system may benefit.

- City-wide online trading OTO platforms, such as 58 city[2], have a large amount of products with multi-attributes and geo-locations. Location information to

---

[1] http://www.groupon.com.
[2] http://www.58.com.

users is important since users on this online platform tend to trade in a nearby area. A traditional way for users is searching in the system and making a choice from a great number of products. It is obvious that a location-aware pub/sub system may improve the service quality for users. After users register their interests and regions in this system, they can receive a timely notification when a match occurs.

- Group-buying platforms such as LivingSocial[3] and RetailMenot[4] have the location-aware pub/sub gene in nature. Customers in a Group-buying system can register their interests in the form of boolean expressions and a location constraint. When new products are continually being inserted, customers can receive products information timely instead of repeatedly searching the system to find their interested products.

There are several challenges in location-aware boolean expression pub/sub. First, it needs to handle millions of boolean expressions with a large amount of attributes and values. Second, when a spatial constraint is added to subscriptions, its computation cost in both indexing and matching becomes rather larger. Third, its matching process must be efficient enough to ensure a high event arrival rate. Thus, more efficient filtering techniques that support location-based pub/sub with boolean expressions are necessary.

To improve the indexing efficiency, we propose a two-step partitioning R-tree based index structure (called RP-trees since then). In its first step, we partition subscriptions according to their sizes (number of predicates in a subscription). In its second step, we select the most representative attribute of each subscription, and group subscriptions with the same size and the same selected attribute together. To index the spatial constraints, we build an R-tree for each group of subscriptions. Our experiments on large synthetic datasets and real-world datasets show that our method achieves higher performance.

To summarize, we make the following contributions.

- We propose a new problem, i.e. location-aware pub/sub with boolean expressions.
- We propose a new index structure RP-trees,which can efficiently index location-aware boolean expressions.
- We conduct experiments on a synthetic dataset and a real-world dataset to evaluate the efficiency of our proposed index structure RP-trees.

The remaining of this paper is organized as follows. We formalize the problem in Sect. 2. In Sect. 3, we review related works on the pub/sub system. In Sect. 4, we propose two preliminary solutions by extending a well-known existing solution as baselines for comparisons. In Sect. 5, we propose an RP-trees index and present its corresponding index construction algorithm. In Sect. 6, we present our subscription matching algorithm. Extensive experiments results are reported in Sect. 7, and we conclude this paper in Sect. 8.

---

[3] http://www.livingsocial.com.
[4] http://www.retailmenot.com.

2   Problem Formulation

In a location-aware pub/sub system, subscribers register their interests as subscriptions. A subscription includes a boolean expression and a spatial region, denoted by $s= (B, R)$.

A boolean expression is a combination of predicates in Conjunctive Normal Form. A predict contains three elements: an attribute denoted by A, an operator denoted by f_{op} and a value denoted by v. That is, $P_{(A,f_{op},v)}$ denotes a predicate. We can support six operators $(=, \leqslant, \geqslant, <, >, \in)$ in this paper. A boolean expression of a subscription s is modeled as follows:

$$s.B : P_1(x) \wedge P_2(x) \wedge P_3(x) \wedge ... \wedge P_m(x)$$

Thus, a subscription s is modeled over a boolean expression and a spatial constraint as follows:

$$s : \{P_1(x) \wedge P_2(x) \wedge P_3(x) \wedge ... \wedge P_m(x), Region\}$$

An information publisher publishes an event e that contains a collection of value-pairs denoted as $e.V$ and a geo-position denoted as $e.P$. $e.V$ is represented in the form of conjunction of equality predicates and the size of $e.V$ is the number of value-pairs denoted by m. Thus, an event can be denoted as follows:

$$e : \{(A_1 = v_1) \wedge (A_2 = v_2) \wedge (A_3 = v_3) \wedge ... \wedge A_m = v_m, Point\}$$

Definition 1. Predicate Match
A value-pair appears in $e.V$ is denoted by V_n, $V_n.A$ donates the attribute of the value-pair and $V_n.V$ denotes the value. For a predicate P_m appears in a subscription, we said that there is a predicate match if $P_{m(V_n.A,f_{op},V_n.V)} = true$.

Definition 2. Boolean Expression Match
A boolean expression $s.B$ is said to match a collection of value-pairs $e.V$ if each of the predicates in $s.B$ has a match in $e.V$.

Definition 3. Subscription Match
Given a collection of subscriptions $S = \{s_1, s_2...s_m\}$, each subscription has a boolean expression and a spatial constraint denoted by $s.B$ and $s.R$ respectively. For each event e in an event stream with a collection of value-pairs and a point message $e.V$ and $e.P$, we said that $s_i \in S$ match e if $(s_i.R \cap e.P) \neq \varnothing$ and $s_i.B$ matches $e.V$.

Let us revisit the example we mentioned before in Fig. 1. There are 8 subscriptions and 2 events and their locations. For the event $E_1 = \{(A = 3 \wedge B = 3 \wedge C = 5), P_1\}$, the subscription S_1 matches E_1. However, the subscription S_4 doesn't match E_1 as there is no value-pair in E_1 that matches the predicate $G \geqslant 4$. The subscription S_8 is not an answer either, as its region R_8 has no overlap with P_1. That is, the answer of the event E_1 is S_1. Similarly, the answer of E_2 is S_7.

3   Related Works

The pub/sub system with boolean expressions has been studied for many years while the location-aware pub/sub system is a much more hotter issue in both academia and industry. However, existing works on pub/sub with boolean expressions [1,4,12,15] do little for location matches. Recently, there is a structured pub/sub work focuses on location-aware pub/sub with boolean expression over dynamic event streams [7] which mainly deal with the moving features of subscriptions, it adapt a disk-based method to handle the moving feature of events which result in less efficiency in event matching compared to our memory-based method. In contrast, our work is mainly focus on efficiently filter subscriptions with boolean expression and locations. In addition, location-based pub/sub systems are mainly for unstructured information [5,6,8–10,14,16,18]. Traditional pub/sub systems with boolean expressions mainly focus on indexing support to efficiently identify matching subscriptions. Two newly proposed representative indexing methods for boolean expressions are BE-tree [12] and Opindex [4]. It is known that Opindex can achieve better matching performance with much smaller construction cost under different situations.

Opindex is consisted of a two-level index structure [4]. In its first level, subscriptions are grouped together according to selected pivot attributes (i.e., the attributes with minimum frequency) from each subscription. In its second level, subscriptions are divided by the operators ($e.g., \leqslant, =, \geqslant$) of predicates. In each group, predicates are divided by operators. Signature elements are used to map the predicates by a hash function. For each group of subscriptions, there are a corresponding counter arrays to track the matching predicates. The attributes of value-pairs in events are used as keys to grouped subscriptions that are divided by pivot attributes. Given an event, if an attribute in the event is indeed a pivot attribute, then find the corresponding group of predicates and enumerate each value-pair of the event to search predicate lists that are divided by operators.

Opindex is an efficient index for boolean expressions with a large amount of attributes. We can add the predicates for location information into subscriptions. However, the newly added predicates are apparently a burden for Opindex. In the respect of index efficiency, it is a waste to abandon the pruning ability of location information. The existing unstructured pub/sub systems, however, can only support textual descriptions. Thus, it is not accurately enough to represent items with multi-attributes and values. In contrast, our approach can support items with a great amount of boolean expressions with spatial information and achieve a high matching rate.

4   Preliminary Solutions

In this section, we first propose two preliminary solutions by making some extensions for a well-known index Opindex for pub/sub systems to handle the new problem, i.e., location-aware pub/sub with boolean expressions. Note that these two preliminary solutions are used as baselines for evaluating our advanced solution discussed in Sect. 5.

4.1 Extension with Location Predicates

As we know, there exists a well-known index called Opindex for pub/sub systems. Opindex does not contain predicates for locations. If we apply Opindex to location-aware pub/sub systems, we can extend it straightforwardly by just adding predicates into subscriptions to present spatial constraints. For example, for a region ($lat_1 = 51.25144123$, $long_1 = -0.14251263$, $lat_2 = 51.335125445$, $long_2 = 0.12142324$), we can use operators ($\leqslant, \geqslant$) to present region constraints as ($lat_1 \geqslant 51.25144123$, $long_1 \geqslant 0.14251263$, $lat_2 \leqslant 51.335125445$, $long_2 \leqslant 0.12142324$). And then we can straightly utilize Opindex to solve the problem of location-aware pub/sub index with boolean expressions. This extension of Opindex is called Opindex-loc in this paper. Obviously, this method abandons the ability of pruning spatial constraints which leads to high computation cost.

4.2 Extension Using R-tree

R-tree is a well-known index structure to index spatial information, in which can use a minimum bounding rectangular (MBR) to denote spatial regions. The basic idea of this extension is to utilize R-tree as an index for spatial information and organize boolean expressions in the Opindex structure. First, we build an R-tree to index the spatial constraints in subscriptions. Then we organize the subscriptions in the form of the Opindex structure. We append a counter array for each grouped subscription partitioned by a pivot attribute. We initialize the counter array with the sizes of boolean expressions in the subscriptions. Then, for processing event matching, we get the value-pairs of an event to validate the boolean expressions of the subscriptions, for each matched predicate in a subscription, the corresponding element in its counter array is decremented by one. And candidate subscriptions are generated when their corresponding counter elements in counter arrays goes to zero. Then, using the spatial information of the event as an input, we search the R-tree. For each satisfied region, we get its identification to validate the candidates to find the final matched subscriptions to return. Since boolean expressions are first visited to generate candidates, then we called this extension as BF-Opindex.

5 Hybrid Index Structure

As we discussed in Sect. 4.1, both the two mentioned method result in poor performance in event matching since the Opindex-loc abandons the pruning ability of spatial information and the BF-Opindex generates many candidates whose spatial constraints are not satisfied. To improve the performance in event matching, we present a novel partitioning index structure RP-trees, which organizes the regions of subscriptions into disjointed R-trees. In the RP-trees index, each subscription has a representative attribute termed as a pivot attribute denoted by δ_A. The subscriptions are partitioned by two steps. First, the subscriptions are partitioned based on the sizes of subscriptions. And then subscriptions are further partitioned by their pivot attributes. Thus, subscriptions with the same sizes and the same pivot attributes are grouped together using inverted lists. To index the spatial constraints, we build an R-tree using the spatial regions of subscriptions in each list. Given an input event, we search the corresponding R-tree to generate candidates and validate boolean expressions of these candidates. For convenience, we summarize the key notations used in this paper as shown in Table 1.

Table 1. Notation

s	A subscription
e	An event
δ_A	The pivot attribute of a subscription s
$s.R$	The spatial region of a subscription s
$s.B$	The boolean expression of a subscription s
$e.P$	The spatial point of an event e
$e.V$	The value-pairs of an event e
k	The number of predicates of a subscription s
m	The number of attributes in an event
S	The entire subscription dataset
E	The entire events dataset
$L_{(k)}$	List of subscriptions partitioned by k
$L_{(\delta_A)}$	List of subscriptions partitioned by δ_A
$L_{(k,\delta_A)}$	List of subscriptions partitioned by k and δ_A
C	The counter array correspond with $L_{(k,\delta_A)}$
C_i	The counter element in C

5.1 Boolean Expression Index

We index the boolean expressions of subscriptions in two partitioning steps. In the first step, subscriptions are partitioned into disjointed subscription lists based on the sizes of the subscriptions k as follows:

$$S = L_{(k_1)} \cup L_{(k_2)} \cup L_{(k_3)} \cup .... \cup L_{(k_m)}$$

If a boolean expression $s.B$ in a subscription s matches the value-pairs $e.V$ in an event e, then the size of the event m must not less than the size of subscription k. Obviously, if k is larger than m, then, there must be a predicate of any subscription s in $L_{(k)}$ which cannot be matched. According to Definition 2, e is guaranteed not to match with subscriptions in $L_{(k)}$.

In the second step, subscriptions with the same size are further partitioned according to the selected pivot attributes δ_A as follows:

$$L_{(k_i)} = L_{(\delta_{A1})} \cup L_{(\delta_{A2})} \cup L_{(\delta_{A3})} \cup .... \cup L_{(\delta_{Am})}$$

From the Definition 1, we can conclude that if an event e matches a subscription s, then all the attributes in s must appear in e. Obviously, if there is an attribute in s which is not in e, e will be concluded not match s. Thus, given an event e, we only consider the subscriptions whose pivot attributes appears in e. Attributes with less frequency bring more probability to filter subscriptions because attributes with less frequency has less probabilities to appear in subscriptions. Thus, we chose the least frequency attribute in a subscription as pivot attribute.

For each subscription s in $L_{(k,\delta_A)}$, we establish a predicate storage structure based on standard operators $(=,\geqslant,\leqslant)$ of the subscription. Other operators $(<,>)$ are treated similarly. And the collection operator $\in$ is rewritten using the standard operator. For example, $F \in (5,2,1)$ can be rewritten into $F = 5 \vee F = 2 \vee F = 1$. Using a counting-based algorithm, the counter arrays can detect matched subscriptions for an input event. For each subscription s_i in $L_{(k,\delta_A)}$, there is a counter array C to track the number of predicates in $s_i.B$ that has not been matched during an event searching process. Each counter array is initialized as the sizes of its corresponding subscriptions, and the corresponding counter element in counter array is decremented by one whenever a value-pair in $e.V$ of a searching event e matches a predicate in $s_i.B$. Thus, $s_i.B$ matches $e.V$ when the element C_i for S_i in the counter array goes to zero.

The boolean expression index structure for the running example subscriptions in Fig. 1 is shown in Fig. 2. In the first step subscriptions are partitioned by their sizes into three lists L_2, L_3 and L_4. Next, a pivot attribute is selected according to the appearance frequency in the datasets. Thus, A, D, E and G are selected as the pivot attributes respectively. Given an event $E_1 = \{(A = 3 \wedge B = 3 \wedge C = 5), P_1\}$, according to Definition 2, subscriptions in $L_{(4)}, L_{(3,G)}, L_{(2,D)}$ are guaranteed not to match with E_1. Figure 2 also shows the counter arrays and predicates storages structure of S_2, S_5, S_7.

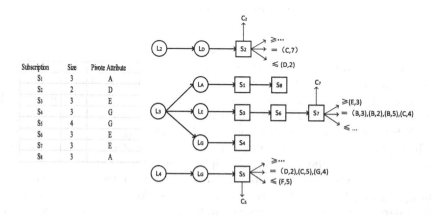

Subscription	Size	Pivote Attribute
S_1	3	A
S_2	2	D
S_3	3	E
S_4	3	G
S_5	4	G
S_6	3	E
S_7	3	E
S_8	3	A

Fig. 2. Boolean expression index structure

5.2   Combine Boolean Expressions and Spatial Index

In the rest of this section, we present the structure of RP-trees. The RP-trees contains three components that are shown in Fig. 3. The first component is a collection of the inverted lists of boolean expressions derived from the two-step partitioning method that have been described above. The second component is a set of corresponding R-trees for each partitioned subscription list $L_{(k,\delta_A)}$. The corresponding R-trees are used to filter the spatial information, and generate candidate subscriptions whose spatial constraints are satisfied by events. The third component is a collection of counter arrays, corresponding to the boolean expression lists $L_{(k,\delta_A)}$.

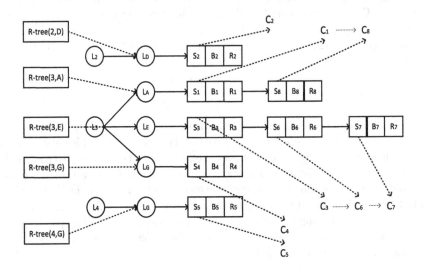

Fig. 3. RP-trees index structure

Algorithm 1 shows an algorithm to insert a new subscription s_i into the RP-trees index. We transfer the collection operator $\in$ into the standard operators as described in Sect. 5.1. Next, we determine the size k_i and the pivot attribute δ_{Ai} of s_i, append a counter element C_i for s_i in counter array C and initial C_i with the size of $s_i.B$. Then the spatial information $s_i.R$ is inserted into the newly built R-tree or an existing R-tree corresponding with $L_{(k,\delta_A)}$.

To explain how a RP-trees index is built, we revisit our running example here. The RP-trees index for the subscriptions in Fig. 1 is shown in Fig. 3. The boolean expressions of the subscriptions are first partitioned and are inserted into lists L_2, L_3 and L_4. And then, each boolean expression of a subscription list is further partitioned into several lists $L_{(2,D)}, L_{(3,A)}, L_{(3,E)}, L_{(3,G)}, L_{(4,G)}$ according to their pivot attributes (i.e., A, E, D and G). For each boolean expression list above, we append an R-tree to index their spatial regions and there are five counter arrays for the eight subscriptions in Fig. 1.

Algorithm 1. *Subscription Insert(s)*

1: Determine the size k of s
2: Get the pivot attribute δ_A from s
3: Get predicates p,and the region R from s
4: **if** $L_{(k)}$ not exists **then**
5:     Define $L_{(k)}$, $L_{(k,\delta_A)}$, $R - tree_{(k,\delta_A)}$, C, Ci
6: **else**
7:     **if** $L_{(k,\delta_A)}$ not exists **then**
8:         Define $L_{(\delta_A)}$, $R - tree_{(k,\delta_A)}$, C, Ci
9:     **else**
10:        **for** each$(R, p) \in s$ **do**
11:            Insert(R) into $R - tree_{(k,\delta_A)}$
12:            Insert (P) into $L_{(k,\delta_A)}$
13:            Initial C_i with k

6 Query Processing

Algorithm 2 provides the process of how RP-trees return matching subscriptions for an input event e. Before the process of searching the RP-trees index, we initialize the set of matching subscriptions R to be empty, and set each value in corresponding counter array to its respective subscription size. First, we calculate the size of e. For each boolean expression list $L_{(k_j)}$, we scan $L_{(k_j)}$ if the size of e is not less than k_j. For each boolean expression list $L_{(k_j,\delta_A)_i}$, we enumerate the attributes A_j from the input event e. If A_j is a pivot attribute, we use a spatial point $e.P$ as a query on the corresponding R-tree to filter out candidate subscriptions. Then, the boolean expressions of the candidate subscriptions are refined by the value-pairs in e. When a predicate stored in $s_j.B$ is matched by a value-pair, the corresponding value in counter array is decremented by one. If the counting value goes to zero, we have a matching subscription for e, which is added to the result set R.

Let us revisit our running example here. Consider the process of the event $E_1 = \{(A = 3 \land B = 3 \land C = 5), P_1\}$ using the RP-trees index in Fig. 3. The size of the event is 3, so the subscription list $L_{(4)}$ is pruned. For the list $L_{(2)}$, there is no pivot attribute in E_1, thus, $L_{(2)}$ is pruned. For the list $L_{(3)}$, only attribute A is contained in E_1. Therefore, only the $R - tree_{(3,A)}$ should be searched for the regions overlapped with P_1. And the region R_1 does overlap with P_1, then, candidate boolean expression $S_1.B$ is scanned, and the value-pairs of E_1 satisfies $S_1.B$. Thus, S_1 is added to the result set R.

7 Experiments

In this section, we report our experimental results of evaluating the performances of our proposed RP-trees index. We compare it with two preliminary methods, which are described in Sect. 4. The implementation of Opindex is implemented according to the algorithms proposed by its authors. All the indexes are memory

Algorithm 2. *Matching Algorithm*

Require: *an event e.*
Ensure: *result set of subscriptions R.*
1: **Initialize** :$R \leftarrow \{\}$
2: Get the size m of e
3: Get the distinct value-pairs $A_j = V_j$ and the point location $e.P$ from e.
4: **for** each $L_{(k_i)}$ in the RP-trees index **do**
5:    **if** $m \geqslant k_i$ **then**
6:       **for** each $L_{(\delta_{A_i})}$ in $L_{(k_i)}$ **do**
7:          **if** $A_j = \delta_{A_i}$ **then**
8:             Search $e.P$ in $RP - tree_{(k_i, \delta_{A_i})}$
9:             **if** $e.P$ overlap s.$R \in RP - tree_{(k_i, \delta_{A_i})}$ **then**
10:                **for** each predicate p in $s.B \in L_{(k_i, \delta_{A_i})}$ **do**
11:                   **if** $A_j = V_j$ satisfies p **then**
12:                      $C_i \leftarrow C_i - 1$
13:                      **if** $C_i = 0$ **then**
14:                         Add the subscription s to the result set R.
15:                         **return** R

---

resident and implemented in Java. We conduct the experiments on a server with 256 G memory, 64 KB L1 cache and 512 KB L2 cache, running centos 5.6.

7.1   Data Generator

To generate the synthetic dataset, we implement a data generator, which can generator attributes, operators and values, uniformly distributed or follows Zipf distribution. Three main operators $(=, \geqslant, \leqslant)$ are supported. we generate 5 k to 20 k distinct attributes, following Zipf distribution from 0.2 to 1.0. We vary the number of subscriptions from 1 M to 10 M to test the scalability. The max size of each generated subscription k varies from 4 to 20 while the size of events m varies from 5 to 20. Default parameters for testing the scalability on a real-world dataset (58 city) are set as follows: The size of each subscription is from 2 to 10, the size of events is from 4 to 12, the number of distinct attributes is set to 5,000, and a zipf law is used in the distribution of attributes with an alpha value 0.8. For the spatial information, we generate 1 M longitudes and latitudes to compose the point locations for an event and a region for a subscription.

Besides the synthetic datasets, we also design a data generator from real-world datasets. The dataset is extracted from 58 city. It is a series products information with its name, price, address and other attributes of the products. Based on the address, we extract the coordinates in the form of longitude and latitude from Baidu API[5]. And then we generate MBRs by selecting a location of a product as the center and extending a random width and height with a upper bound. Similarly, we generate the value of attributes by select the value of the attribute in a product information as mid-value and extending a random value to add to or subtract from the mid-value as a value of predicate in a subscription.

---

[5] http://developer.baidu.com.

In summary, we totally generates 10 M synthetic subscriptions corresponding with 0.1 M events for matching tests. For the real-world dataset, we generate 5 M subscriptions and 0.1 M events from 58 city. Table 2 summarizes the parameters and their settings of the two datasets.

Table 2. Parameters and settings

Parameters	Synthetic dataset	58 city
Number of subscription	2 M, 4 M, 6 M, 8 M, 10 M	1 M, 2 M, 3 M, 4 M, 5 M
Max subscription size	4~20	2~10
Max event size	5 25	4 12
Zipf	0.2, 0.4, 0.6, 0.8, 1.0	Uniformly distributed
Number of distinct attribute	5 k, 10 k, 15 k, 20 k, 25 K	10 K
Node capacity of R-tree	40	40

7.2 Experimental Results

In this section, we will evaluate three indexes on synthetic datasets and real-world datasets. For synthetic datasets we evaluate the performance of three indexes from different perspectives, the memory consumption, a varied number of subscriptions and distinct attributes, varied max size of subscriptions, varied max size of events, varied Zif distributions

Memory consumption: Since all the three indexes are memory-resident, we first evaluate the memory consumption for the three indexes by varying the number of subscriptions. The experimental results are shown in Fig. 4(a). We can see that our RP-trees consumes less memory, comparing with Opindex-loc and BF-Opindex. This is because RP-trees need not to generate signature elements while the other two methods needs signature elements to track the predicates for each subscription.

Matching time on different number of subscriptions: From Fig. 4(b), we can see that the our RP-trees index achieves the best event matching time. This is because partitioning subscriptions using pivot attributes is not as efficient as using the hybrid partitioning of the sizes and the pivot attributes.Furthermore, R-tree with a small amount of regions partitioned by the sizes and the pivot attributes can efficiently generate candidate subscriptions. Compared with Opindex-loc, it avoids a large amount of unnecessary boolean expressions whose spatial constrains are not satisfied to calculate. In summary, our RP-trees can get the least amount of candidates for both spatial information and boolean expressions.

Matching time on different size of subscriptions: The average event matching time on different size of subscriptions for the three indexes are reported in

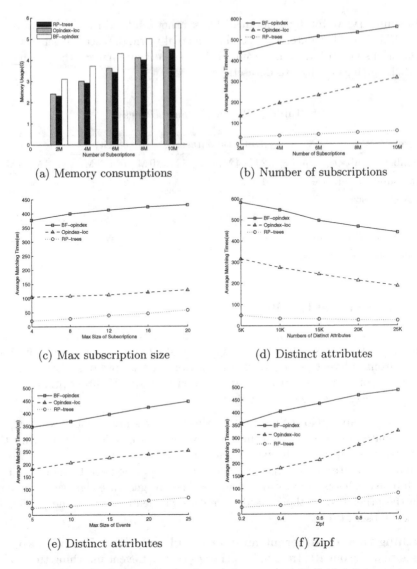

(a) Memory consumptions

(b) Number of subscriptions

(c) Max subscription size

(d) Distinct attributes

(e) Distinct attributes

(f) Zipf

Fig. 4. Evaluation for the three indexes on synthetic dataset

the Fig. 4(c). As we can see from the Fig. 4(c), RP-trees index scales better compared to Opindex-loc and BF-Opindex. The reason is, with the increasing k, the pruning ability for the first partitioning step of RP-trees become more and more powerful since events only visit the subscriptions whose size are not large than that of events.

Matching time on different number of distinct attributes: The number of distinct attributes is an important parameter for the three indexes, because all of the indexes are used pivot attributes to partition subscriptions. An obvious observation from the Fig. 4(d) is that when the number of attributes increases, the matching times of the three indexes decrease because all of the three indexes can generate more narrowed partitions with the increment of the number of attributes. And the matching time of our RP-trees decreases apparently with the increasing number of attributes.

Matching Time under Different sizes of Events: The experimental results are shown in Fig. 4(e). An obvious observation is that our RP-trees scale similarly to Opindex-loc and BF-Opindex. Both RP-trees and the other two indexes are sensitive to the size m of the event. With the size m increases, The collection of candidate subscriptions grows at the same time. There are two reasons for this. First, the pruning ability for the first step partitioning of RP-trees directly decreases since the chance for a larger than m becomes smaller with the increasing of m. Second, events will access more candidate subscriptions partitioned by the second step of partitioning, since there are more attributes in an event with the increasing of m, which increases the probability to access subscriptions partitioned by pivot attributes.

Matching Time under Different Alpha Values of Zipf: We also evaluate the matching time by changing the alpha value from 0.2 to 1.0 in Zipf distribution. The experimental results are shown in Fig. 4(f). An obvious observation is that the matching times for the three indexes all grow with the increasing alpha value of Zipf. Our RP-trees index scales better than both Opindex-loc and BF-Opindex, because of the powerful partitioning ability of RP-trees.

For the real-world dataset from 58 city, we varied the number of the subscription to evaluate the scalability and the memory consumption. We can see from the Fig. 5, our RP-tree achieves the minimum event matching and memory consumption, and with the increases of subscriptions, the event matching time increased nonlinearly. This is because with number of subscriptions increased, RP-trees still pruned large amount of unnecessary subscriptions.

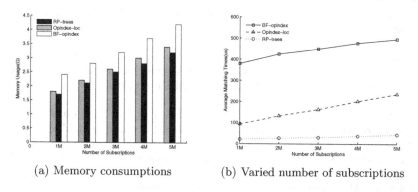

(a) Memory consumptions   (b) Varied number of subscriptions

Fig. 5. Evaluation for the three indexes on 58 dataset

8   Conclusion

In this paper, we tackled the problem of location-based matching in pub/sub for boolean expression which is significant for location-based pub/sub system with a large amount attributes and values. Facing the challenge of efficiently delivering events to the corresponding subscribers, we propose a novel index structure called RP-trees, which using a two-step partitioning method to organize the boolean expression and the state-of-art R-tree to index spatial information. Extensive experiments conducted in both synthetic and real datasets demonstrate the effectiveness of our algorithms. In the future, we plan to support top-k pub/sub matching.

Acknowledgment. This work was partially supported by Chinese NSFC project (61402312, 61170020, 61402311, 61440053), and the US National Science Foundation (IIS-1115417).

References

1. Bittner, S.: Supporting arbitrary boolean subscriptions in distributed publish/subscribe systems. In: Proceedings of the 3rd International Middleware Doctoral Symposium (MDS 2006) (2006)
2. Chen, L., Cong, G., Cao, X., Tan, K.L.: Temporal spatial-keyword top-k publish/subscribe. In: 2015 IEEE 31st International Conference on Data Engineering (ICDE), pp. 255–266 (2015)
3. Cugola, G., Margara, A.: High-performance location-aware publish-subscribe on GPUs. In: Narasimhan, P., Triantafillou, P. (eds.) Middleware 2012. LNCS, vol. 7662, pp. 312–331. Springer, Heidelberg (2012)
4. Zhang, D., Chan, C.Y., Tan, K.L.: An efficient publish/subscribe index for E-commerce databases. Proc. VLDB Endow. **7**(8), 613–624 (2014)
5. Eugster, G.: Location-based publish/subscribe. In: 2013 IEEE 12th International Symposium on Network Computing and Applications, pp. 279–282 (2005)
6. Li, G., Wang, Y., Wang, T.: Location-aware publish/subscribe. IEEE Trans. Knowl. Data Eng. **27**(4), 950–963 (2013)
7. Guo, L., Zhang, D., Li, G., Tan, K.L., Bao, Z.: Location-aware pub/sub system: When continuous moving queries meet dynamic event streams. In: Proceedings of the 2015 ACM SIGMOD International Conference on Management of Data (2015)
8. Huang, Y., Garcia-Molina, H.: Publish/subscribe in a mobile environment. Wireless Netw. **10**(6), 643–652 (2004)
9. Kai Zheng, P.C.F., Zhou, X.: K-nearest neighbor search for fuzzy objects. In: Proceedings of the 2010 ACM SIGMOD International Conference on Management of Data (2010)
10. Lai, S., Wang, G.: P2p streaming media resource location algorithm based on publish/subscribe. Henan Science (2012)
11. Naicken, S.M.: Trusted content-based publish/subscribe trees. University of Sussex (2012)
12. Sadoghi, M., Jacobsen, H.-A.: Be-tree: an index structure to efficiently match boolean expressions over high-dimensional discrete space. In: ACM Conference on Management of Data, pp. 637–648 (2011)

13. Sadoghi, M., Jacobsen, H.A.: Location-based matching in publish/subscribe revisited. In: Proceedings of the Posters and Demo Track (2012)
14. Tam, D., Azimi, R., Jacobsen, H.-A.: Building content-based publish/subscribe systems with distributed hash tables. In: Aberer, K., Koubarakis, M., Kalogeraki, V. (eds.) DBISP2P 2003. LNCS, vol. 2944, pp. 138–152. Springer, Heidelberg (2004)
15. Whang, S.E., Brower, C., Shanmugasundaram, J.: Indexing boolean expressions. Stanford InfoLab $2(1)$, 37–48 (2009)
16. Xiang Wang, Y.Z., Xuemin Line, W.W.: Ap-tree: efficiently support continuous spatial-keyword queries over stream. In: 2015 IEEE 31st International Conference on Data Engineering (ICDE), pp. 1107–1118 (2015)
17. Yu, M., Li, G., Wang, T., Feng, J., Gong, Z.: Efficient filtering algorithms for location-aware publish/subscribe. IEEE Trans. Knowl. Data Eng. $27(4)$, 1 (2015)
18. Zheng, K., Su, H., Zheng, B., Shang, S., Xu, J., Liu, J., Zhou, X.: Interactive top-k spatial keyword queries. IEEE (2015)

EPC Graph Information Service
Enhanced Object Traceability on Unified and Linked EPCIS Events

Jaewook Byun and Daeyoung Kim(✉)

Department of Computer Science, KAIST, Daejeon, South Korea
{bjw0829,kimd}@kaist.ac.kr

Abstract. Object traceability based on Radio Frequency IDentification (RFID), an important capability of the Internet of Things (IoT), has improved our quality of life (e.g. anti-counterfeit). GS1 EPCglobal, a de facto standard in RFID technology, develops EPC Information Service (EPCIS) in order to capture and share standardized events representing various aspects on tag-attached object. However, the current standard is detrimental to the connection of the facts on RFID events, thus yielding query inefficiency, especially for object trace. Also, even the problem becomes worse when the event repositories are distributed. In this paper, we propose the graph-based platform for enhanced object traceability over unified and linked EPCIS events, EPC Graph Information Service. The main contributions of the paper are as follows. Firstly, we propose the graph model on standard EPCIS document called EPC Graph Model, which would be an efficient approach for connecting RFID events in a unified manner. Secondly, we propose the library for the distributed services called EPC Graph Crawler based on fundamental REST APIs for utilizing EPC Graph. By using the library, application developers are able to utilize distributed EPC Graphs as if all the graphs are managed in a centralized platform.

Keywords: EPC Information Service · EPCIS · Object traceability · Graph database · Linked Data · RESTful web service

1 Introduction

Object traceability, an important capability of the Internet of Things (IoT), has improved the quality of our life. For example, we are able to recover lost valuables, detect counterfeit brands and prevent the spread of infectious disease through food [1]. Radio Frequency IDentification (RFID) is one of key solutions since this technology enables everyday-objects to be identifiable and traceable even in indoor-space. GS1 EPCglobal [2], a de facto standard in RFID technology, develops EPC Information Service (EPCIS) [3] to capture and share standardized events representing various aspects on tag-attached object.

Figure 1 shows the motivating scenarios of EPCIS usage in the fields of smart agriculture and automobile. In the smart agriculture scenario, one sellable

© Springer International Publishing Switzerland 2015
J. Wang et al. (Eds.): WISE 2015, Part I, LNCS 9418, pp. 232–246, 2015.
DOI: 10.1007/978-3-319-26190-4_16

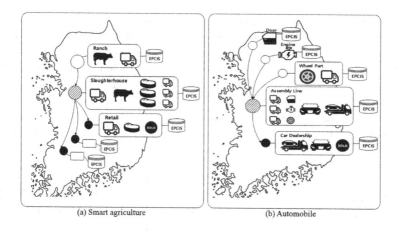

(a) Smart agriculture   (b) Automobile

Fig. 1. The motivating scenarios of EPCIS usage

portion of beef has undergone many business processes (e.g. transportation, transformation and ownership transfer) before it is in retail shop. By utilizing expressiveness of the EPCIS standard, all the processes are visible to customers so that they can enjoy safe foods by avoiding dangerous place-of-origins (e.g. a region where foot-and-mouth, mad cow or avian influenza were rampant). As for the automobile one, various business processes (e.g. transportation, assembly) can be related to one sellable car, and they are visible to decision-maker in an automobile company. Therefore, the decision-maker efficiently enables to recall cars embedding severe faulty parts.

Meanwhile, the current standard is detrimental to the connection of the facts on RFID events, thus yielding query inefficiency when EPCIS is used to trace objects due to event-based data management, diverse event types and different notations on an identical object. For example, a business application, which makes standard EPCIS event called Capturing Application (CA), usually represents the business situation on a beef or a car with the most suitable event type (i.e. one of ObjectEvent, AggregationEvent, TransactionEvent and TransformationEvent) and its corresponding notation (e.g. instance of EPCList, parentID, InputEPCList). However, the another business application, which consumes stored events called Accessing Application (AA), needs to search all the event types over again until there is no additional event found since AA may not know what event type is used for and what object is transformed from or transformed to the queried object in advance. Even this problem becomes worse when the event repositories are distributed. Since it would be expected that unified and linked EPCIS events can resolve the problem, in this paper, we propose the graph-based platform for RFID-related event management, which enhances object traceability.

The main contributions of the paper are as follows. Firstly, we propose the graph model on standard RFID events called EPC Graph Model. This model enables us to see EPCIS event and vocabulary as a unified property graph which consists of vertices of object, location and their relationships regardless of various event types and the different notations above. This would be an efficient approach for connecting RFID events. According to the performance analysis, the unified graph structure enables to enhance query efficiency significantly. Secondly, we propose the library for the distributed services called EPC Graph Crawler based on fundamental REST APIs for utilizing EPC Graph. Each EPC Graph Information Service provides the REST APIs for sharing EPC Graph in a JSON-LD [4] format. Then, by using the crawler, application developers are able to utilize distributed EPC Graphs as if all the graphs are managed in a centralized service.

The remainder of this paper is organized as follows. In Sect. 2, we review related works. Section 3 presents a novel graph model on standard EPCIS document. Then, Sect. 4 shows how the graph can be shared by both EPC Graph REST API for a centralized service and EPC Graph Crawler for distributed services. Based on the aforementioned proposals, Sect. 5 presents a system architecture. Consequently, we evaluate the platform with a performance analysis and a demonstration.

2   Related Work

EPC Information Service (EPCIS) [3] is a standard repository which captures and shares RFID-related events and vocabularies. For global interoperability of the events and the vocabularies, GS1 EPCglobal [2] specifies core event types and standard interfaces. The latest version 1.1 establishes the foundation for improving object traceability with newly added Transformation event and chain of custody. EPC Graph Information Service releases the potential traceability. Also, to be interoperable with existing RFID infrastructure, our capture module is compatible with EPCIS standard v1.1 by leveraging our previous work, Oliot EPCIS [5].

Linked EPCIS [6,7] proposed an ontological model for representing EPCIS events and algorithms for counterfeit detection and pedigree generation. However, the proposed model is mostly syntactical conversion to the ontology so that Linked EPCIS would be also concerned with the query inefficiency we mentioned. Also, Linked EPCIS does not compatible with existing RFID infrastructure, because it does not provide a standard capture interface. Furthermore, in terms of pedigree generation algorithm, Linked EPCIS has the strong assumptions on supply chain (i.e. fixed business vocabulary and business step sequence). This work would be concerned when used in rapidly changing business environments (e.g. logistic and retail). Finally, Linked EPCIS is not suitable for managing a large amount of data since this work uses an in-memory store. Meanwhile, in EPC Graph Information Service, users can explorer EPC Graph without assumptions on supply chain by using REST APIs and EPC Graph Crawler. Also, the service is scalable by leveraging the scalable graph and data storage (i.e. Titan graph database [8], Mongo database [9]).

Turchi et al. [10] designed EPCIS through Linked Data and REST principles. They proved the concept with a small set of standard EPCIS events and that of a standard EPCIS query. However, this work does not fully follow the principles of Linked Data [11]. In EPC Graph Information Service, each URI indicates resources which can be consumed by EPC Graph REST API. Also, the resources (e.g. object and place) are interconnected. Furthermore, EPC Graph REST API returns its results with the standard format, JSON-LD [4].

We believe that our approach can enhance E-pedigree [12], one of EPCglobal components. E-pedigree is a certified record that contains information about each distribution of a prescription drug. Weili Han et al. [13] proposed a master/slave architecture for E-pedigree and an extended pedigree standard for food safety system. However, the existence of master pedigree server called CEPServ results in the demands for additional infrastructure and the additional security issue. It is because managing pedigree information by each business stakeholder (e.g. wholesaler and transportation company) is more secure way in the view of each of them. It is possible that the utilization of EPC Graph Crawler on distributed EPC Graphs can resolve this problem without the additional infrastructure.

3 Graph Modeling on Standard EPCIS Document

In this section, we propose the graph model for standard EPCIS document, EPC Graph Model. A graph type we use is property graph [14]. Briefly, a property graph consists of multiple vertices and edges which have their own unique identifier and a collection of key-value properties. Also, in the property graph, any vertices can have multiple edges. In this section, we present the overall model and how four standard event types and various business vocabularies are converted into an instance of the model.

3.1 EPC Graph Model

Standard EPCIS document contains one or more instance(s) of four core event types and master data (i.e. vocabulary) described in Fig. 2. The figure shows that the four core event types extends EPCIS event, a base event type. Also, each event type has its own semantic. For example, Object Event mainly represents creation, observation and destruction of tag-attached objects. To leverage the expressiveness of various event data types, CAs need to represent the business situation on the objects with the most suitable event type.

Among information types (i.e. *when, where, why* and *what*) which each event data type represents, latter three ones are identified by Uniform Resource Identifier (URI) form. Then, the actual information on the URI is stored as master data. There are 9 vocabulary types in master data as shown in the right side of Fig. 2. On the one hand, four vocabulary types of them are standardized in Core Business Vocabulary (CBV) [15] as standard vocabulary. On the other hand, each business stakeholder can define their own vocabularies as user vocabulary. For example, a business location identified with a Global Location Number (SGLN) code may have detail information such as address, geo-location. Standard EPCIS capture interface can capture the vocabularies and share them with business partners for global consensus.

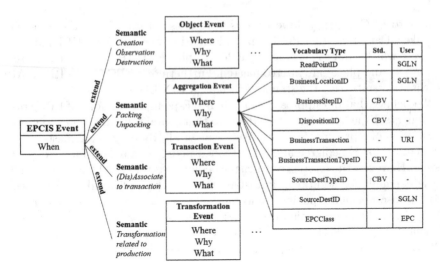

Fig. 2. Graphical view of EPCIS document

Figure 3 depicts EPC Graph Model, which models EPCIS document as a unified property graph. The property graph would contain consistent vertices of object and location and their relationship regardless of various event types and the different notations. In terms of event data, there are two instance types; *Object* instance and *Location* instance where *Object* instance is an instance of *Object* class, and *Location* instance is an instance of *Location* class or *Organization* class. Each instance can be uniquely identified by Electronic Product Code (EPC) [16], a unique identification system of EPCglobal. Also, each of them has key-value properties indicating business situations (e.g. business step and geo-location). Furthermore, any instances are able to have multiple relationships together. For example, an object can have a *isLocatedIn* relationship with locations. Please note that all the relationships are directed ones, and each of them should have its reversed relationship for graph traversal as shown in Relationship Table of Fig. 3. For the simplicity, the model depicts one relationship of any pairs. Lastly, the relationships have key-value properties such as *eventTime* and *xformID*. Meanwhile, vocabularies also share EPC Graph with event data. Each vocabulary can be stored as key-value properties of a vocabulary vertex and queried with its URI or EPC. Therefore, a business application can utilize event data and master data in a unified manner.

3.2 From EPCIS Document to EPC Graph

Based on the proposed EPC Graph Model, an EPCIS document would be converted into an instance of the model. Figure 4 shows how four EPCIS events are transformed to vertices, edges and properties of EPC graph. Since almost every conversions could be self-descriptive and the number of page in the paper is limited, we state some key points for easy understanding.

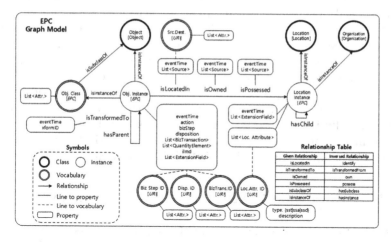

Fig. 3. EPC Graph Model

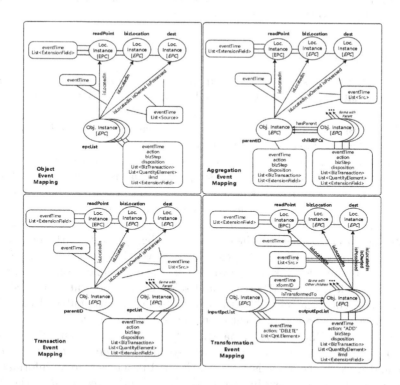

Fig. 4. From EPCIS event to EPC Graph

Firstly, all the properties in vertex are grouped with event time and stored into an external document-based database for achieving scalability. It means that one event would generate one document into the database. The details would be described in Sect. 5. Secondly, in terms of Aggregation Event, a parent object and its child objects share all key-value properties except a list of quantity elements since the quantity elements are related to the child objects. Thirdly, according to the EPCIS specification, a parent object in Transaction Event is just a representative among relevant objects so that they would share all the properties. Finally, input object(s) in Transformation Event would disappear and be transformed to output object(s). Therefore, we semantically label *DELETE* in action field of input object and label *ADD* in that of output object.

4   EPC Graph REST API and EPC Graph Crawler

An EPC Graph is utilizable through EPC Graph REST APIs. These APIs are designed by taking consider both REST principles [17] and Linked Data principles [11]. Return values of the REST APIs are formatted as consistent JSON-LD or JSON, thus leading EPC Graph Crawler easily to aggregate and blend the results of distributed EPC Graph Information Services in order to generate the results equivalent to those of a centralized one.

4.1   EPC Graph REST API

We design REST APIs for a centralized EPC Graph as shown in Table 1. EPC Graph Information Service is based on a Resource-oriented architecture (ROA) so that each resource can be identified by a unique identifier (i.e. EPC of object and URI of business vocabulary), and the resource can be consumed by an appropriate HTTP method in a stateless manner. According to Linked Data principles, the APIs provide useful information on resources (i.e. event/master data, ego networks, siblings and trace history), using the standards (i.e. JSON-LD, JSON Array). Also, results of the APIs include links to other resources, thus leading to discover them. For example, Listing 1 shows an example of the results of *getResource* API. In the result, the resource ID - urn:epc:id:sgtin:0000001.000002.1 - indicates one beef package and the context is explicitly stated in the *epcis* field. Also, users can query a resource for each business vocabulary (e.g. *retail_selling*) since the result includes links to the vocabulary. The example describes that the car was assembled, departed, arrived and sold at each timestamp. It is shown that all the event data related to the car are stored as attributes of the EPC Graph vertex regardless of different event types and notations.

The last five APIs are the fundamental blocks for tracing an EPC Graph. Developer can designate a scope of the APIs to one of *resource, relationship, ego, sibling* and *trace*. If the scope is *resource*, the API provides a detail information on a given resource as shown in Listing 1. If the scope is *relationship*, the API returns all the labels of edges which a given resource has as JSON array. This API would be helpful to utilize the remaining APIs. When the scope is *ego*,

Table 1. EPC Graph REST API

REST API	Parameter or Input Data	Description
POST /Resource	Input Data: Standard EPCIS document	postResource: Reflect vertices and edges in EPCIS document into a graph. The document can be one of event data or master data.
GET /Resource	N/A	getVertexNames: Return existing vertex labels in a graph storage. Return format: JSON Array
GET /Resource/{uri}	Parameters: scope=resource fromTime={Abs|Rel} toTime={Abs|Rel} orderByTime={asc|desc} limit=#of results	getResource: Return attributes of the node relevant to uri. The attributes can be filtered and sorted by time range and orderByTime parameter, respectively. Also, the number of the attributes can be limited. Return format: JSON-LD
GET /Resource/{uri}	Parameters: scope=relationship	getRelationships: Return labels of the edge which uri has. Return format: JSON Array
GET /Resource/{uri}	Parameters: scope=ego fromTime={Abs|Rel} toTime={Abs|Rel} relationship= CSV of edge labels	getEgoNetwork: Return the ego network of the given uri. The network can be filtered by edges having time range with fromTime and toTime. If the relationship parameter is specified, the network only contains the neighbors which have the given relationships. Return format: JSON-LD
GET /Resource/{uri}	Parameters: scope=sibling fromTime={Abs|Rel} toTime={Abs|Rel} relationship= one pair of edge labels	getSiblings: Return neighbors of the given uri as well as neighbors of the neighbors. The network can be filtered by edges having time range with fromTime and toTime. If the relationship parameter is specified, the network only contains the neighbors which have the given relationships. Return format: JSON-LD
GET /Resource/{uri}	Parameters: scope=trace relationship= one edge label	getTraceTree: Return one-directional trace tree of the given uri. Edge label of the tree is limited to given singular relationship. Return format: JSON-LD

the API provides the ego network as JSON-LD. Briefly, ego network means a vertex (i.e. ego) and its one hop networks. This API enables developers to expand the graph one by one. Next one is the case of *sibling*. The functionality of this API is similar to Friend of a Friend (FOAF) [18]. FOAF provides mutual friendship, thus addressing a possibility of interesting social services. This API can be regarded as the extension of FOAF. It gives neighbors of the neighbors to which the given resource is related. We believe that this API would be useful for tracing infectious disease or recalling defective car. Finally, if the scope is *trace*, the API returns a chain of resources which the given resource has the given single relationship with, recursively. For example, if the single relationship is *isTransformedFrom*, the API would return a chain of origin resources until there is no further origin found. We believe that it is useful to automatically trace an origin of any resources.

4.2   EPC Graph Crawler

EPC Graph Crawler aggregates and blends the results of distributed EPC Graph Information Services. When users utilize the EPC Graph Crawler, they initialize

---

Algorithm 1. EPC Graph Crawler - getTraceTree API

```
1: procedure GETTRACETREE
2:     // Distributed Query Invocation and aggregation
3:     U: a URL list of getTraceTree REST APIs for each EPC Graph
4:     R_raw: a JSON-LD list for each query result
5:     for each URL i in U do
6:         Invoke U(i)
7:         Keep the result in R_raw(i)
8:     end for
9:     // Blending: Keep a base result
10:    R_blended = R_raw(0)
11:    Remove R_raw(0)
12:    // Blending: Recursive merge
13:    for each result i in R_raw do
14:        R_blended = recursiveMerge(R_blended, R_raw(i))
15:    end for
16:    // Result: Aggregated and Blended JSON-LD
17:    return R_blended
18: end procedure
19: procedure RECURSIVEMERGE(R_base, R_target)
20:    for each key k in R_raw(i) do
21:        if R_base.isNull(k) then
22:            R_base.put(k, R_target.get(k))
23:        else
24:            R_base.put(k, recursiveMerge(R_base.get(k), R_target.get(k)))
25:        end if
26:    end for
27:    return R_base
28: end procedure
```

---

it along with base URLs of distributed EPC Graph Information Services. If they want more automated initialization, they can leverage class and instance level service registry of EPCglobal infrastructure (i.e. Object Name Service (ONS) [19] and Discovery Service (DS) [20], respectively). The list of APIs which the crawler provides are same as EPC Graph REST APIs, except they are for distributed service. Also, the results of each query are almost equivalent to the results from centralized EPC Graph Information Service, except for *getTraceTree* API.

Listing 1. Example: Query Result of **GET**
/Resource/urn:epc:id:sgtin:0000001.000002.1?scope=resource

```
1  {
2      "@id": "urn:epc:id:sgtin:0000001.000002.1",
3      "@context": {
4          "epcis": "http://gs1.org/docs/epc/epcis_1_1-schema-20140520/EPCglobal-epcis-1_1.xsd"
5      },
6      "1428320011116": {
7          "epcis:action": "ADD",
8          "epcis:disposition":"urn:epcglobal:cbv:disp:sellable_accessible",
9          "epcis:bizStep": "urn:epcglobal:cbv:bizstep:retail_selling",
10         "epcis:ilmd": { "uom": "KGM", "quantity": "0.6" }
11     },
12     "1428194011116": { ... },
13     "1428107611116": { ... },
14     "1428040736591": {
15         "epcis:action": "ADD",
16         "epcis:bizStep": "urn:epcglobal:cbv:bizstep:transforming"
17     }
18 }
```

Except for *getTraceTree* API, blending task is easily achieved since the crawler just needs to merge the results without redundancy. For example, in the case of *getResource* API, the crawler blends the returned JSON-LDs. In terms of event data, the crawler merges all the event data grouped by timestamp into one JSON-LD. If there are redundant event data having same timestamp, EPC Graph Crawler merges sub-attributes in the conflict event data. If sub-attributes also exist in two or more data, the result keeps the sub-attributes of the last EPC Graph, which is the only side-effect. However, the side-effect and the procedure of distributed *getTraceTree* API are more complex than others. Therefore, we present the pseudo code in Algorithm 1 and its side-effect.

The API has the side-effect as shown in Fig. 5. When a *getTraceTree* API for the node *9.1* (i.e. urn:epc:id:sgtin:0000001.000009.1) is invoked, the crawler cannot provide the dashed edges and dotted vertices since the crawler is not able to know that the node *9.1* is connected to other nodes (i.e. *7.1* and *5.1*) in advance. For the complete result equivalent to a centralized version, one solution could be a recursive invocation for each leaf node of the blended result. However, currently it is not available since we expect a performance degradation on the recursive invocation above.

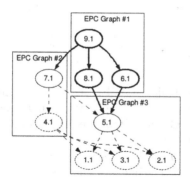

Fig. 5. Side Effect of distributed Trace API

5   System Architecture

Figure 6 depicts the system architecture of EPC Graph Information Service. In the figure, the left part shows how components are organized in a centralized EPC Graph Information Service while the right part presents how applications can utilize distributed EPC Graph Information Services by using the EPC Graph Crawler library.

In terms of Capture Module, we modify Oliot EPCIS [5], our previous work. Firstly, *postResource* API captures event data as well as master data. Then, Capture Service Listener receives the data and sends them to Document Validator. Document Validator validates the data by comparing standard XML Schema of EPCglobal. Then, the validated data are converted to Plain Old Java Object

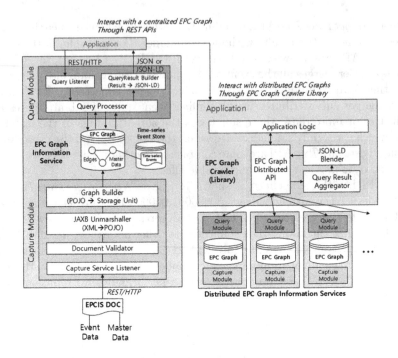

Fig. 6. System Architecture

(POJO) by JAXB Unmarshaller. Finally, Graph Builder converts the POJO into vertices and edges of EPC Graph through the mapping rules described in Sect. 3.

In the Query Module, Query Listener receives a query from users and sends it to Query Processor. Query Processor retrieves the data from EPC Graph and document-based database for REST APIs described in Sect. 4 and sends the data to Query Result Builder. Finally, a client receives the result formatted as JSON-LD or JSON by Query Result Builder.

One important thing of Graph Builder is that event data of each vertex are stored in document-based database (e.g. MongoDB [9]) instead of EPC Graph. This is because, in the property graph, attributes of a vertex are a list of key-value pairs so that the pairs are inappropriate for storing event data grouped by timestamp. Meanwhile, if we utilize document-based database, we can reap benefits from the efficiency of no-sql database and we can efficiently store and compute the vertex attributes with various supported data types and sub-document.

In a distributed environment, EPC Graph Crawler may give a helping hand for application developers. This library described in Sect. 4 enables the developers to utilize distributed EPC Graph like centralized one. This library would save the developers' burden of distributed processing.

6   Evaluation

We evaluate EPC Graph Information Service in this section. Firstly, we conduct a performance analysis. Then, we show the feasibility of our platform with a demonstration.

6.1   Performance Analysis

For the performance analysis, we install EPC Graph Information Service on Intel i7-4770K quad-core 3.50 GHz CPU, 8 GB Ram machine (i.e. Server). The machine runs Linux Mint 17 Cinnamon 64bit operating system, Tomcat 7.0.54 web application server, Titan graph database 0.4.4 over Cassandra 2.1.0 and MongoDB 2.6.4. We adopt nGrinder [22], a performance test platform, introducing a metric of Transaction Per Second (TPS) which measures how many transactions can be dealt with. Another laptop (i.e. Client) runs nGrinder to measure TPS and Average Response Time (ART) of each transaction. There are two experiment as follows.

The purpose of first experiment is to show that externally storing time-series event data into document-based database is more efficient than a situation when EPC Graph Information Service only uses Titan. Figure 7(a,b) shows the trends of how ART and TPS changes against the increasing number of events (i.e. 10, 100, 1000, 10000 and 100000 events) stored as attributes of Titan graphstore and documents of MongoDB database. In this experiment, nGrinder invokes same *getResource* method providing the most recent event of an object. In the case of Titan attribute, the ART increases linearly and the TPS drops with inverse proportional way as the number of events increases. It is because we can leverage various techniques on event data (e.g. index, sharding) and database-side operations (i.e. ordering, limiting) while the case of Titan attributes needs to search all the attribute keys and sort and limit the event data in the server-side. This experiment shows that EPC Graph Information Service is scalable with the number of event data.

The efficiency of *getTraceTree* API is shown in Fig. 7(c). The *getTraceTree* API provides the complete tree containing a resource and any child resources which the resource can reach with a relationship. Therefore, this API enables clients to be free to search of neighbors of neighbors which cannot be known in advance as well as corresponding needless communications with EPC Graph Information Service. We compute ARTs of two cases against the increasing number of tree height (i.e. 10, 100 and 1000). To show the efficiency of *getTraceTree* API, we compute the average response time when tracing the tree step-by-step, then multiplying it with the height. This experiment shows that *getTraceTree* API enhances object traceability. Note that the figure shows logarithm of the ARTs. Also, the latter case does not reflect additional costs to be equivalent to the results of *getTraceTree* API (e.g. finding a neighbor in the result of previous step, merging the intermediate results).

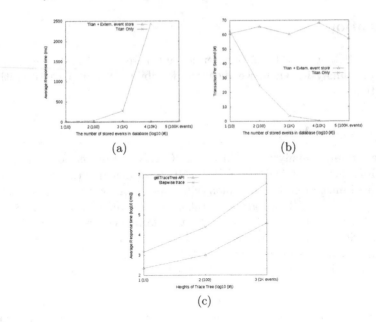

Fig. 7. Performance Analysis: (a,b)- Enhanced ART and TPS of externalized event store, (c)- Efficiency of *getTraceTree* API

6.2   Demonstration - EPC Graph Explorer

We develop EPC Graph Explorer in order to show the feasibility of EPC Graph Information Service. Figure 8 shows a snapshot of EPC Graph Explorer. To initialize a library of EPC Graph Crawler, users need to register base URLs of distributed EPC Graph Information Service. Then, they can explore the visualized EPC Graph implemented by using Prefuse [21]. The evaluation we make is as follows:

- **Enhanced object traceability:** EPC Graph enhances object traceability by representing RFID-related data as consistent graph elements. Also, EPC Graph REST APIs save the burden of clients by imputing it to EPC Graph Information Service. It is because two or more query invocations are originally needed to achieve goals equivalent to some APIs (i.e. sibling and trace) in existing EPCIS.
- **Complexity Alleviation:** When an application developer uses EPC Graph REST APIs instead of the crawler, the development cost increases drastically due to the cost of aggregation and blending distributed results. Furthermore, in the case of existing EPCIS, the cost is aggravated due to the various event types and the different notations.
- **Possibility of graph mining:** As shown in the figure, the visualized graph clearly shows two graph communities: Smart Agriculture and Automobile. We believe that EPC Graph Information Service could provide promising services by reasoning hidden valuable information through various graph mining techniques.

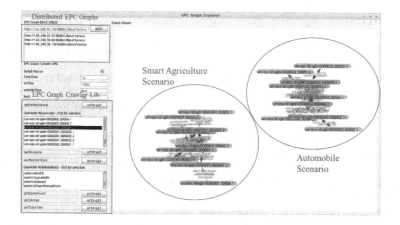

Fig. 8. Demonstration: EPC Graph Explorer

7   Conclusions

In this paper, we propose the graph-based platform for enhanced object trace-ability over unified and linked EPCIS event. The main contributions of the paper are as follows. Firstly, we propose the graph model on standard EPCIS document called EPC Graph, which would be an efficient approach for connecting RFID events in a unified manner. Secondly, we propose the library for distributed EPC Graph Information Service called EPC Graph Crawler based on funda-mental REST APIs for utilizing EPC Graph. By using the library, application developers are able to utilize distributed EPC Graphs as if all the graphs are managed in a centralized platform. We develop EPC Graph Explorer in order to show the feasibility and applicability of our platform. During the development, we identify that our platform enhances object traceability with less complexity.

In future work, we intent to develop an extended version of property graph database and apply it to EPC Graph Information Service. This will enable us not to handle an additional document-based database for efficiently storing time-series events.

Acknowledgement. This research was supported by the ICT R&D program of MSIP/ IITP [R-20150223-000170(R0126-15-1002), Development of agro-livestock cloud and application service for balanced production, transparent distribution and safe con-sumption based on GS1] and also supported by the KUSTAR- KAIST Institute, Korea, under the R&D program supervised by the KAIST.

References

1. Uckelmann, D., Harrison, M., Michahelles, F.: Architecting the Internet of Things. Springer, Heidelberg (2011)
2. Traub, K., et al.: The EPCglobal architecture framework. EPCglobal Ratified specification (2005)
3. EPC Information Service (EPCIS) Version 1.1 Specification, GS1 EPCglobal Standards (2014)
4. Sporny, M., Kellogg, G., Lanthaler, M.: Json-ld 1.0-a json-based serialization for linked data. W3C Working Draft (2013)
5. Byun, J., Kim, D.: Oliot EPCIS: new EPC information service and challenges towards the Internet of Things. In: IEEE RFID 2015, San Diego, USA, 15–17 April 2015
6. Solanki, M., Brewster, C.: Representing supply chain events on the web of data. Detection, Representation, and Exploitation of Events in the Semantic Web, 18 (2013)
7. Solanki, M., Brewster, C.: EPCIS event-based traceability in pharmaceutical supply chains via automated generation of linked pedigrees. In: Mika, P., et al. (eds.) ISWC 2014, Part I. LNCS, vol. 8796, pp. 82–97. Springer, Heidelberg (2014)
8. Titan graph database. http://thinkaurelius.github.io/titan/
9. MongoDB Document-Store NoSQL Database. https://www.mongodb.org
10. Turchi, S., et al.: Designing EPCIS through linked data and REST principles. In: 2012 20th International Conference on Software, Telecommunications and Computer Networks (SoftCOM). IEEE (2012)
11. Berners-Lee, T.: Linked data-design issues (2006)
12. Pedigree Ratified Standard version 1.0, GS1 EPCglobal Standard (2007)
13. Han, W., et al.: The design of an electronic pedigree system for food safety. Inf. Syst. Front. 17(2), 275–287 (2012)
14. Property Graph. https://github.com/tinkerpop/blueprints/wiki/Property-Graph-Model
15. Core Business Vocabulary (CBV) version 1.1, GS1 EPCglobal Standard (2014)
16. Tag Data Standard (TDS) version 1.9, GS1 EPCglobal Standard (2014)
17. Fielding, R.T.: Architectural styles and the design of network-based software architectures. Diss. University of California, Irvine (2000)
18. Brickley, D., Miller, L.: FOAF vocabulary specification 0.98. Namespace document 9 (2012)
19. Object Name Service (ONS) version 2.0.1, GS1 EPCglobal Standard (2013)
20. Evdokimov, S., et al.: Comparison of discovery service architectures for the internet of things. In: 2010 IEEE International Conference on Sensor Networks, Ubiquitous, and Trustworthy Computing (SUTC). IEEE (2010)
21. Heer, J., Card, S.K., Landay, J.A.: Prefuse: a toolkit for interactive information visualization. In: Proceedings of the SIGCHI Conference on Human Factors in Computing Systems. ACM (2005)
22. nGrinder. http://naver.github.io/ngrinder/

A Mashup Architecture with Modeling and Simulation as a Service

Sixuan Wang[✉] and Gabriel Wainer

Department of Systems and Computer Engineering,
Carleton University, 1125 Colonel by Dr., Ottawa, ON K1S5B6, Canada
{swang, gwainer}@sce.carleton.ca

Abstract. Web services have been used in Modeling and Simulation (M&S) for many years, but it is still hard to develop complex M&S applications by using heterogeneous data and varied services. Here we show how to simplify the M&S application development by using mashup technologies. We introduce a novel architecture, the Mashup Architecture with Modeling and Simulation as a Service (MAMSaaS), in order to deploy varied mashup components (e.g. Modeling and Simulation as a Service, existing Web APIs, widgets and operators) and to create M&S mashups for quick application development. We present various tools built to support the architecture and a case study using the mashup architecture.

Keywords: Modeling and simulation as a service · Cloud-based simulation · Web data integration and mashups · Service composition · Web API

1 Introduction

Current Modeling and Simulation (M&S) systems have become more and more complex, involving people different with M&S expertise, and varied M&S resources. In order to simplify the development process of M&S systems, the M&S community has used Web-based Simulation (WBS), Cloud-based Simulation (CBS) and Web Services (WS) technologies for around 20 years. WBS focused on running experiments using the Web and later exposing M&S functions as web services [1]. This method has been successful, and a large number of M&S WSs exist, including SOAP-based WS [2] and RESTful WS [3]. CBS integrates WBS and Cloud Computing. It uses Cloud Computing technologies to reduce costs and make easier to develop M&S systems by exposing M&S resources as Modeling and Simulation as a Services (MSaaS).

Nevertheless, the growth of M&S related WSs and open Web APIs makes it difficult to integrate them for complex applications. In particular, experts from different domains cannot easily use existing M&S related resources for fast application development. With this in mind, we aim to improve the development process of Web information systems for M&S that involve heterogeneous data and services. This process should be such that one could reuse and integrate existing services and resources related to M&S easily. People from different domains should be able to use this process for sharing and combining their works quickly.

In order to do so, we present an architecture based on mashup technologies in the Web 2.0 [4]. Mashups have been widely used in different domains; [5] however, these

© Springer International Publishing Switzerland 2015
J. Wang et al. (Eds.): WISE 2015, Part I, LNCS 9418, pp. 247–261, 2015.
DOI: 10.1007/978-3-319-26190-4_17

technologies have never been used to simplify the process of building M&S applications. Our objective is thus to develop M&S mashup applications to integrate different M&S related WSs and resources. Our novel architecture, named Mashup Architecture with Modeling and Simulation as a Service (MAMSaaS), is a layered architecture to deploy and identify M&S mashup components as well as link and execute mashups for quick M&S application development.

MAMSaaS supports universal identification and development for mashup component (named *Boxes*), which consist of varied M&S resources (MSaaSs, WebAPIs, widgets and operators). In order to link and execute these components, MAMSaaS supports component linking, development and search boxes, then build M&S environments by wiring boxes, and then execute and visualize M&S at run-time.

Following, we will discuss the architecture and the tools we implemented: a Box Development Tool and the MAMSaaS Mashup Platform. In addition, we also present a prototype application integrating a Geographical Information System (GIS), a distributed M&S tool, and visualization services, built as a mashup application.

2   Background

M&S is being applied to almost every aspect of life, and developing of M&S applications has become more and more complex [6]. Varied resources are involved in the development process, such as source systems, models, simulators, experimental frameworks and experiments [7], as well as supported data (e.g. text, file, database) and functions (e.g. data collection, result analysis, visualization) [8].

In order to simplify the application development process, web services have been used in M&S for around 20 years. The basic idea is to expose M&S resources on the Web as services. Web services ease the sharing of M&S resources: for instance, the simulators are located remotely on a server, without worrying about the simulation environment setup and software dependencies. Web services improve data accessibility, interoperability and user experience [9].

In general, web services in M&S include Web-Based Simulation (WBS, which exposes M&S functions as web services) and Cloud-Based Simulation (CBS, which integrates WBS and Cloud Computing). In WBS, the functions of simulators and their simulation environment are exposed as web services [1]. Users can submit their requests (with specified message/parameters) to the simulator through web servers, then simulation experiment runs remotely, and the results are returned to the user. In recent years, numerous WSs in WBS have been developed, which can be categorized into two main frameworks: SOAP-based (e.g. DDSOS [10], SOPM [11], and SASF [12]) and RESTful-based (e.g. RISE [3], RESTful MMVE [13], and RESTful AIS [14]). Numerous simulators (i.e., DEVS/SOA [2] and RISE [3]) implement the DEVS M&S formalism [7] over a WS. RISE is the first and only RESTful distributed simulator to support distributed simulation that supports DEVS and other model formalisms, languages and engines.

In CBS, both web services and Cloud computing are used in M&S. CBS is derived from WBS, using Cloud Computing to manage varied M&S resources and build different simulation environments [15]. The use of web services in CBS has received the name of Modeling and Simulation as a Service (MSaaS). MSaaS is a special form of Software-as-a-Service (SaaS), as it hides the underlying infrastructure, platform and software details from the users. The use of CBS and MSaaS is still in a preliminary stage [16] and little effort has been done to integrate with other services [17], which is an issue because there are many open Web APIs that could be useful for M&S applications (e.g. weather forecast, GIS information, and big data for simulation inputs). They could improve user experience and make richer applications [18].

According to more than 11,000 APIs registered by ProgrammableWeb, REST WSs take 73 % while SOAP-based APIs take 27 % [19]. WADL is a popular language to describe REST web services, and many IT companies describe their REST APIs on the HTML pages, such as Mashape (http://www.mashape.com/explore). SOAP WSs are usually described in WSDL files. For example, WebServiceX (http://www.webservicex. net/ws) has over 70 SOAP WSs using WSDLs. However, there has been no research showing how to integrate these useful Web APIs in the development process of M&S applications. Mashup technologies in Web 2.0 can be used to solve this integration issue and simplify the M&S application development.

Mashups integrate different services from the web, using content from more than one existing source to create a new value-adding application [4]. Mashups integrate heterogeneous data, application logic (exposed as services in general), and UI components (e.g. widgets) [5]. A large number of mashup techniques and tools have been developed in both industrial development and academic research [5]. Many industrial companies have developed their own commercial mashup tools, like IGoogle (http://www.igoogleportal.com), and Yahoo! Pipes (http://www.pipes.yahoo.com). They are based on the visual connection of components of heterogeneous data at the enterprise level, offering Do-It-Yourself (DIY) guidance to meet user requirements [20]. In addition, many academic efforts focus on mashups. Many of them use End-User Programming (EUP), focusing on the composition and integration of web sources for new purposes. Mashroom [21] uses nested relational models and provides mashup operations like merge and filter over tables. In [22], the authors use native language programming in mashup components, linking different logic together.

The fundamental element of current mashup technologies is the *widget*, a small processing unit for performing single purpose task such as fetching, parsing, formatting and visualizing data [23]. For instances, DERI Pipes [24] enables users to build widgets to process data from different sources (e.g. RDF, SPARQL, XML, HTML). In [23], the authors proposed an open mashup platform with linked widgets created freely by users that can be discovered and combined easily. WireCloud [25] is an open source mashup platform provided by the FI-WARE project, which can implement widgets in Java-Script (JS) and HTML5 and build mashups by "wiring" widgets.

However, no one has ever tried to develop such widgets and mashup applications in M&S (as well as integrating available WBS, CBS and Web APIs). Many of the current mashup techniques and tools cannot work directly for M&S because: (1) they are domain specific (i.e., they are useful only for single or limited problems in specific domains); (2) they have been discontinued; (3) they are limited (the widgets do not support MSaaS and different kinds of Web APIs like SOAP and REST). Based on this, we investigated a new mashup method focused on the process of developing M&S applications.

3 The MAMSaaS Architecture

Based on the considerations in Sect. 2, we defined a novel mashup architecture to simplify the process of developing M&S applications, named Mashup Architecture with Modeling and Simulation as a Service (MAMSaaS). Shown in Fig. 1, MAMSaaS is a layered methodology and architecture to create and run M&S mashups.

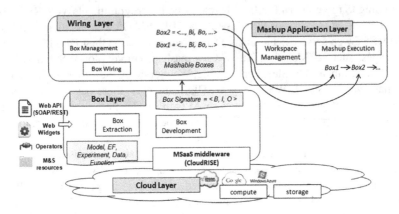

Fig. 1. MAMSaaS architecture

The proposed MAMSaaS architecture has four layers, as follows:

- **Cloud:** it is responsible for supporting Cloud infrastructure and deploying MSaaS. The Cloud infrastructure includes Cloud *compute* units (for building and executing simulation experiments) and Cloud *storage* units (for sharing M&S resources). In addition, this layer is also responsible for deploying user-provided M&S resources as MSaaS in the Cloud by using the *MSaaS middleware* on-demand.
- **Box:** it is responsible for developing mashup components (termed *Boxes*). Boxes can have different categories (e.g. MSaaS from the Cloud Layer, existing open APIs, widgets, operators). Each box is identified by a uniform box signature, and has its own function for handling input messages and has its own visual form.
- **Wiring:** it is responsible for connecting boxes into a mashup. Boxes can be linked with each other by their inputs/outputs that are identified in their box signatures. A same box can be reused and re-wired in different mashups for new purposes.
- **Mashup Application:** it is responsible to select and wire boxes in workspaces, and run applications. They can add box, wires and visualize the results at runtime.

Here we will discuss the top three layers: Box, Wiring, and Mashup Application layers, and discuss how to develop boxes and build M&S mashups. For the details about the Cloud layer, the reader can refer to [26], in which we introduced the CloudRISE middleware, which implements the concept of MSaaS (exposing all kinds of M&S resources as services). CloudRISE uses a resource-oriented design via RESTful WSs in which M&S resources are identified through URIs in the Cloud.

4 M&S Mashup Methodology

Box, wiring, and M&S mashups make up the complete set of an M&S mashup application. In this section, we will discuss details of these three concepts.

4.1 Box – Developing Mashup Components

Boxes represent mashup components used for M&S. They are mini applications for M&S-related scenarios (e.g. fetching a model, reproducing an experiment, visualizing results, etc). Boxes receive heterogeneous data from varied services. A Box modularizes specific functions and sends data to others; they can be shared or published on the Web. There are four basic types: *MSaaS, WebAPI, Widget,* and *Operator* Boxes.

- **MSaaS Box:** it uses MSaaS services from CloudRISE to handle input data, and then outputs results. CloudRISE works as a repository interface to expose M&S resources as MSaaS. There are six main types: simulations, supporting functions, models, supporting data, semantic data and instances. The ones related to the M&S mashup are simulations, functions, and models. Users can manage resources and control the lifecycle of the execution of simulations and functions by using the HTTP methods GET/PUT/POST/DELETE to corresponding URIs. MSaaS Boxes simplify the execution of experiments by combining several MSaaS into one box. For example, an MSaaS simulation box can create a new experiment configuration file, create a new experiment using that file, start the simulation and check its status; when it finishes, it sends the simulation results to the output.
- **WebAPI Box:** it calls existing open Web APIs. It exposes the function following a WS principle (RESTful/SOAP-based). RESTful Web APIs are usually described in WADLs or HTML pages, while SOAP-based Web API is described in WSDLs.
- **Widget Box:** it is a lightweight web application that shows the data on web browsers. They provide a visual representation for particular data. They can be reused for web development or other mashup platforms. For example, http://www.100widgets.com provides different widgets, supporting varied type of data, e.g. forms, diagrams, tables, maps, photos and videos.
- **Operator Box:** it takes input data from other boxes and it generates the output based on a customized process. The reason why we need operators is to address the inconsistencies between boxes (i.e. boxes with similar ports that cannot connect directly). Operators can be viewed as a converter between boxes. It can be a filter, aggregator, splitter, or adapter. For example, if one port of a box is *full name*, and one port in another box is *given name*, an operator can be a splitter that extracts *given name* from *full name*, so these boxes can be connected.

Though boxes have different types, they are managed in a similar way. Each box is packaged in a separated archive file, so it can be developed, downloaded and installed on different servers. Each box package has three parts, as follows:

- Box Signature: To manage the varied boxes, we provide a uniform structure,

Box Signature (B) $= \;<B_x, I, O>$ (**Definition 1**)

$B_x = <B_n, B_t, B_d, B_s, B_a, B_p, B_m>$ *is the general information of the box,*
$I = \{p\}$ *is a set of input ports of the box,*
$O = \{p\}$ *is a set of output ports of the box,*
$p = < p_n,\; p_t,\; p_d>$ *includes port name, type, and description.*

The Box Signature identifies each box with its basic information (B_x), input ports (I) and output ports (O). The basic information includes its name B_n, type B_t (e.g. MSaaS, WebAPI, Widget, Operator), description B_d, subtype B_s (e.g. simulation MSaaS, RESTful WebAPI), author B_a, path B_p (the URI of a related WS), and method B_m (the method name of a related WS). A box can have multiple input and output ports, which are used to connect the boxes in a mashup. Each input or output port include a port name, type (the message type in the port) and description (a text that describes the port). For instance, given a WebAPI RESTful WS to forecast the weather, its name is *WeatherForecast*, type is *WebAPI*, description is *return weather forecast information*, subtype is *SOAP*, author is *JohnDoe*, path is http://www.webservicex.net/weatherforcat, and method is *GetWeather*. This WebAPI Box has one input port (with *CityName* as its name, *xsd:string* as its type and *city name* as its description), and one output port (with *weather* as its name, *xsd:string* as its type and *weather forecast in 5 days* as its description).

Box Signatures can be described in XML files that can be provided by users or constructed automatically from existing sources.

- Box Function: Each box has a function to response the input events. Boxes are event-driven. When an input event comes, it triggers a function in the Box; then, it sends data through output ports. Different types of boxes have different functions. For the MSaaS Box, it combines multiple MSaaSs related to a same experiment into a single box. For example, for the box of executing a new simulation experiment, box function will: (1) construct a new experiment.xml; (2) create a new experiment; (3) start the simulation; (4) check the result; (5) get result and send it to the output ports. For the WebAPI Box, the function executes as defined in the Web API (e.g. SOAP WS, RESTful WS). For the Widget Box, the function tells how to handle the input data for visualizing in web browsers. For the Operator Box, the function is the action to be executed (e.g. splitting, combing, and data conversion).
- Box View: Each box can also have a view in web browsers. Boxes can have HTML/CSS files for visualization purposes. For boxes of MSaaS and WebAPIs, their views could be either their signatures or the execution status. For Widget Boxes, they can reuse existing HTML/CSS files in existing widgets. Users can also customize these files to change the view how the data will show.

4.2   Box Wiring – Linking Boxes

A key feature of the boxes is that they can be connected to each other, which is called *Box Wiring*. It is for composing different boxes through inputs and outputs.

Box Wiring (W) $= < \{B\}, \{B_{x\bullet1}, B_{y\bulletO}\} >$ (**Definition 2**)

$B = \{B_x, B_y...\}$ *is a set of boxes,*
$\{B_{x\cdot1}, B_{y\cdot0}\}$ *is a set of connections between boxes*

A Box Wiring is a combination of boxes and connections among them. Each Wiring (*W*) contains a set of boxes $\{B_x, B_y...\}$ and connections $\{B_{x.1}, B_{y.0}\}$. For instance, $B_{x.1}, B_{y.0}$ means the output port *O* of Box B_y can be linked to the input port *I* of Box B_x. Boxes notify about their changes via events on their output ports; other boxes can consume these events via input ports. By wiring boxes, users can reuse them in multiple M&S scenarios without understanding the internal details.

4.3   M&S Mashup – Building Mashup Applications

The boxes and wiring mechanism can be used to build *M&S mashups*. They are used to build a composite M&S application by selecting and wiring boxes. It is based on data flow and event-based mechanism among the boxes using a visual representation.

Mashup (M) $= < \{B\}, \{W\}, U >$ (**Definition 3**)

$B = \{B_x, B_y...\}$ *is a set of boxes,*
W *is a set of Box Wirings,*
U *is a user workspace for this mashup.*

An M&S Mashup consists of a set of Boxes (*B*), Box Wirings (*W*), and a User Workspace (*U*). Users can have different workspaces. In any workspace, users select boxes and wire boxes through their input/output ports. After that, the mashup application is ready. Users can run and visualize it.

Figure 2 shows an example of an M&S Mashup. It is made up of six boxes (*B1* to *B6*), which interoperate with each other by exchanging data. Consider we have MSaaS Boxes *B1* and *B2*, Web API Boxes *B3* and *B4*, Operator Box *B5*, and Widget Box *B6*. Users can build a mashup application by wiring these boxes as shown in the right part of Fig. 2. At run-time, the data generated in B1 will pass to B3 and trigger B3's function, and then B3 will output its data to B2. Similar actions happen in other boxes. Users can see the mashup as defined in each Box View of boxes on web browsers.

5   MAMSaaS Implementation

We have developed different tools supporting MAMSaaS and the M&S Mashup methodology. Here we discuss the Box Development Tool (used to extract and develop boxes) and the Mashup Platform (used to wire boxes and run M&S mashups).

5.1   Box Development Tool

The Box Development Tool is used to develop boxes rapidly. It can load and save Box Signatures in XML for the different kinds of Boxes. It can also extract

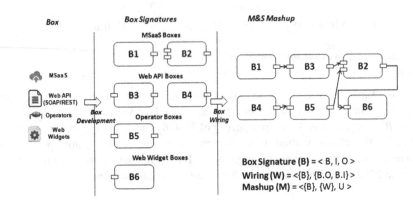

Fig. 2. Box/Wiring/Mashup example

Box Signatures from existing files (e.g. MSaaS Experiment Frameworks, WSDL for SOAP-based WebAPI, and WADL for REST-based WebAPI). In addition, it can generate the configuration XML file that is used in the box package. This tool can also suggest users with similar existing boxes.

The Box Development Tool was developed in Java using SWT (Standard Widget Toolkit), which is a graphical widget toolkit. The Box Development Tool follows the MVC design pattern. Figure 3 shows its class diagram, which consists of three groups:

- The *Data* classes manage the data contained in the *BoxSignatures*. *BoxSignatures* keeps all the signatures in a list. *Signature* keeps the information of a box. *Operation* contains the information of each port. Each *operation* has one *parameter* or *ComplexParameter*. *Parameter* ports have basic information (name/type/description) and *ComplexParameter* uses complex data types (e.g. user-defined XSD).

- The *Logic* classes are used to define the logic of the boxes. They use *Signature* as a bridge. They can extract information from other files to construct a *Signature*, and convert it to other files. *SimulationFramewrokXMLLoader* and *FunctionFrameworkXMLLoader* load Experimental Frameworks from CloudRISE, extracting information from it and saving it as a *Signature* in Data. Similarly, *WADLReader* and *WSDLReader* load and parse the description file of Web APIs. *SignatureXMLHandler* loads and saves the box signature XML files. *Signature 2ConfigurationXMLConverter* converts a *Signature* into a box configuration file.

- The *View* classes are used to build User Interfaces (UI). It supports user-friendly UI to control the process of generating Boxes. The default UI is *MainShell*. It has two menus: *BoxSignatureUI* for extracting and loading a *Signature*, and *BoxDevelopmentUI* for saving a *Signature* and developing box packages.

As discussed before, each box is packaged in an archive file. This file contains a Box Signature, a Box Function and a Box View. In our tool, Boxes are developed using current web technologies (XML, JS, and HTML/CSS). This archive file is an extended version of the widget package used in WireCloud [25], which will be discussed in the next

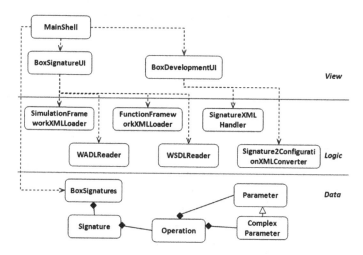

Fig. 3. Class diagram of box development tool

section. The Box Signature is a XML configuration file. The Box Function is a JS file that defines the actions for input events, while the Box View contains HTML/CSS files to show the data in web browsers. For each box, we generate the configuration XML file. For the JS and HTML/CSS files, it can suggest users with similar boxes, in order to reuse them. In particular, the JS function triggers functions for input events by reusing the WireCloud's API *MashupPlatform.wiring. registerCallback (inputName, callback);* and when the function finishes, it outputs message by reusing the WireCloud's API *MashupPlatform.wiring.pushEvent(outputName, data).* Inside the JS function, it can execute SOAP or RESTful WS.

Figure 4 shows an UI example of the Box Development Tool after extracting the Box Signature from an Experimental Framework in the cloud. Users can select the existing MSaaS file, load it, and then the tool parses the file and extracts the information needed for its Box Signature. Later, users can modify this signature, and save it in the format used by the M&S Mashup Platform.

5.2   M&S Mashup Platform

The M&S Mashup Platform has a wiring editor that allows users rapidly building M&S mashups. Users choose a type of box, drag and drop appropriate boxes into a workspace, and then connect the output of a box to the input of another one. After that, the mashup is ready and users can visualize the mashup at run-time.

The M&S Mashup Platform is an extended version of WireCloud [25], an open source mashup platform. It supports widgets uploading and wiring, user workspace management, and mashup execution. However, WireCloud does not support different type of Boxes (in particular the MSaaS Box and WebAPI Box). WireCloud is a general-purpose mashup platform, but not for M&S mashups (which should manage different boxes with different handling processes). The M&S mashup platform we

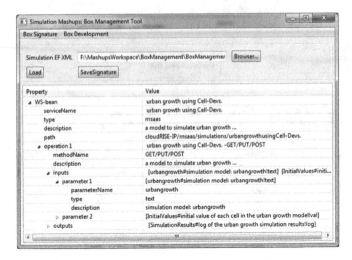

Fig. 4. Extracting box signature from MSaaS file.

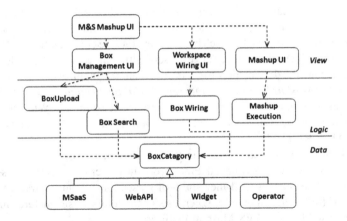

Fig. 5. Class diagram of M&S mashup platform (powered by WireCloud)

developed extended WireCloud to support boxes for M&S mashup, which are MSaaS, WebAPIs, widgets, and operators. Figure 5 shows its class diagram, it consists of three groups of classes:

- The *Data* classes extend the *Category* in WireCloud to *BoxCategory*, in which we added two new types, which are *MSaaS* and *WebAPI*. We reuse most features of *widget* and *operator* provided by WireCloud. Each box has three elements: signature, view, and function.
- The *Logic* classes extend the uploading and searching logic of WireCloud to support all types of Boxes. In *BoxUpload*, we changed the package format as archive file, and modified the uploading and parsing logic. In *BoxSearch*, we changed the databases of resources and searching logic. We extended the *BoxWiring* mechanism and *MashupExecution* mechanism in WireCloud to support all types of boxes.

- The *UI* classes change the UI of WireCloud. The overall UI has been modified in *M&S Mashup UI*. For *Box Management UI*, we changed the box uploading and searching pages. For *Workspace UI*, we modified functions with *Add Box* button to add boxes into user workspace, *Wire Box* button that drags and drops boxes in the wiring editor, *My Box* to select from available boxes. For *Mashup UI*, we reuse the mashup executing engine of WireCloud for box execution.

6 Case Study: A Land Use Modeling Application

In this section, we show how to use MAMSaaS and our tools to build and visualize an M&S mashup application from scratch. The case study focuses on building a Land Use M&S mashup using GIS and other related M&S resources. GIS allows managing, analyzing, and displaying geographically referenced information, and it has been studied for several years in M&S. However, developing M&S applications using GIS is still a complex process [27]. GIS M&S requires many M&S related resources including experts from different domains.

Changes for Land Use have drawn much attention in urban planning, engineering, urban economics, and related fields. Land use can also affect the development of transportation, population, and land distribution. Our mashup application includes:

- **Environmental modeling:** we built a Cell-DEVS environmental model to simulate the land use scenarios.
- **Data collection:** we need a function for generating initial data files from GIS to be used as inputs to the Cell-DEVS model.
- **Cloud-based simulation:** we execute simulation experiments in the Cloud.
- **Results analysis:** it is a function to analyze simulation results, e.g. parsing, converting, statistical analysis.
- **Visualize results:** it is a widget in web browsers to visualize the simulation result in a vivid way.
- **WebAPIs:** they are existing web APIs useful for the GIS M&S. For instances, to better predict the landuse tread, web API to forecast weather for the studied GIS area is needed; to know the zip code information, a web API to search zip code based on geographic information is helpful.

The Box Development Tool can help users develop boxes for the above M&S resources. The developed boxes for this case study are as follows:

- **MSaaS Boxes** use CloudRISE. They contain *Landuse model simulation* (a simulation experiment for the landuse model), *GIS_data_collection* (a function experiment for collecting data), and *GIS_KML_analysis* (a function experiment to parse simulation results to KML file). For each MSaaS Box, its signature is extracted from the corresponding configuration files; its function combines multiple MSaaS in a same experiment framework; and its view shows its execution status.
- **WebAPI Boxes** call existing open Web APIs. They contain *City_weather* (a SOAP WS to forecast weather) and *GeoIP_to_address* (a RESTful WS to get an address from a GIS). For each WebAPI Box, its signature is extracted from the

existing WS description file or HTML files; the function calls the Web API through HTTP request/response; and the view shows its output message.

- **Widget Boxes** show input data in web browsers. They contain *Input_box* (allowing users to input messages), *Input_show* (showing input messages), *KML_viewer* (viewing KML files in Google Map), and *Wikipedia* (getting Wikipedia information). For each Widget Box, its signature is provided by users; the function analyzes input data; and the view visualizes input data.
- **Operator Boxes** handle inconsistencies between boxes. They contain *Zip_to_log* (extracting Log files from an archive), and *Geo_spliter* (splitting Geo information into coordinates). For each Operator Box, its signature is provided by users; the function converts the input data and outputs it; and there is no view for operators.

After the boxes are developed, user can upload them into the M&S Mashup Platform. Now it is time to wire these boxes into a mashup. Figure 6 shows the box-wiring page in the M&S Mashup Platform. In the wiring editor, users can drag and drop boxes from different types, wire the boxes with their input and output ports. There are three user inputs: *model* (the land use model in CloudRISE); *GIS Tiff* (the user-selected area in a GIS dataset) and *Geo Info* (the global geographical references of GIS dataset). We can wire the output of *GIS Tiff* as input of *GIS Data collection*, to send *GIS Tiff* and extract an initialization file for the *Land Use Simulation*. Similarly, we can wire *model* and *GIS data collection* to *Land Use Simulation* to receive inputs of the model and initial files. Then, it can run the simulation. After that, we can wire the *Land Use Simulation* with the Operator Box *get log file from zip* to extract the log file and then wire this to *Results Parsing*, so a KML file can be generated, which wires to the Widget Box *Google Map* for visualizing purposes. In another path, the *Geo Info* wires to Operator Box *Split Geo Info* to get the coordinates of the area under study, then the coordinates link to the WebAPI Box *get IP to Address* to get *zipcode* and *city address*. After that, the *zipcode* wires to WebAPI Box *City weather* for getting the weather forecast (shown in the Widget Box *ShowInput*); and *city address* wires to Widget Box *Wikipedia* to get wiki information.

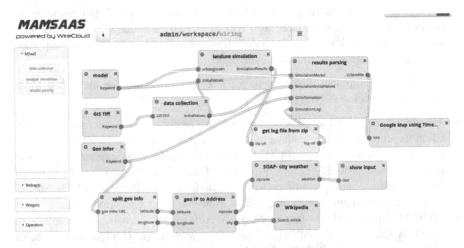

Fig. 6. Boxes wiring for GIS M&S mashup.

After the boxes have been wired, a new M&S mashup application is ready. Users can go back to the workspace and execute it. After setting three inputs (i.e. *Model URL; GIS Tiff URL;* and *Geo infor URL*), the M&S mashup will run. The messages follow the wired boxes flow, and each box runs its function when receiving input events and shows the corresponding visualization view in the workspace. Figure 7 shows its execution view in Google Chrome. We can see that the MSaaS Boxes of data collection, model simulation, and results parsing have been executed successfully, and a KML with simulation results was generated and shown in the Google Map with a timeline control (in which user can choose time to predict the land use population). In addition, this mashup also shows the information of city address, weather forecast in following 5 days and Wikipedia with *North Carolina*. From this case study, we can see that anyone with basic web development knowledge (like HTML/CSS/JS) can easily develop boxes. For developing M&S mashup, no specific knowledge is required. User can simply select boxes and wire them as mashup.

Fig. 7. Executing the GIS M&S mashup application

7 Conclusion

We presented a new method for building M&S mashups for fast application development by integrating heterogeneous data and services. We introduced a novel architecture, named the Mashup Architecture with Modeling and Simulation as a Service (MAMSaaS), a layered architecture to deploy and identify M&S mashup component as well as link and execute mashups for quick M&S application development. M&S Mashup components are called boxes, which consists of MSaaSs, Web APIs, widgets and operators. Each box has its own structure, function and view. M&S Mashup is created through box wiring mechanism. We developed tools for developing boxes, wiring boxes and run mashups. We presented a prototype with GIS M&S mashup application, which has shown that the proposed method using boxes and wiring can create and run simulation mashup in an easy and rapid way. The future work includes developing more M&S mashup applications using the proposed architecture and tools. This is an ongoing project, in order to further test its simplicity for developing M&S applications, we are going to let graduate students in our M&S course to develop their own boxes and mashups using the proposed tools. Another work is to study more on the operator boxes to handle semantic issues between boxes.

References

1. Byrne, J., Heavey, C., Byrne, P.J.: A review of web-based simulation and supporting tools. Simul. Model. Pract. Theory **18**(3), 253–276 (2010)
2. Mittal, S., Risco-Martin, J.L.: Netcentric System of Systems Engineering with DEVS Unified Process: A Book in System of Systems Engineering. CRC/Taylor Francis, Boca Raton (2013)
3. Al-Zoubi, K., Wainer, G.: RISE: a general simulation interoperability middleware container. J. Parallel Distrib. Comput. **73**(5), 580–594 (2013)
4. Balasubramaniam, S., Lewis, G.A., Simanta, S., Smith, D.B.: Situated software: concepts, motivation, technology, and the future. IEEE Softw. **25**(6), 50–55 (2008)
5. Gebhardt, H., Gaedke, M., Daniel, F., Soi, S., Casati, F., Iglesias, C.A., Wilson, S.: From mashups to telco mashups: a survey. IEEE Internet Comput. **3**, 70–76 (2012)
6. Papelis, Y., Madhavan, P.: Modeling human behavior. In: Sokolowski, J.A., Banks, C.M. (eds.) Modeling and Simulation Fundamentals: Theoretical Underpinnings and Practical Domains, pp. 271–324. Wiley, Hoboken (2010)
7. Zeigler, B.P., Praehofer, H., Kim, T.G.: Theory of Modeling and Simulation: Integrating Discrete Event and Continuous Complex Dynamic Systems. Academic Press, San Diego (2000)
8. Skoogh, A., Perera, T., Johansson, B.: Input data management in simulation–Industrial practices and future trends. Simul. Model. Pract. Theory **29**, 181–192 (2012)
9. Fortmann-Roe, S.: Insight maker: a general-purpose tool for web-based modeling and simulation. Simul. Model. Pract. Theory **47**, 28–45 (2014)
10. Tsai, W.T., Fan, C., Chen, Y., Paul, R: DDSOS: a dynamic distributed service-oriented simulation framework1. In: Proceedings of the 39th Annual Symposium on Simulation. Washington, D.C. (2006)
11. Brebner, P: Service-oriented performance modeling the MULE enterprise service bus (ESB) loan broker application. In: Proceedings of 35th Euromicro conference on Software Engineering and Advanced Applications. Patras, Greece (2009)
12. Smit, M., Stroulia, E.: Simulating service-oriented systems: a survey and the services-aware simulation framework. IEEE Trans. Serv. Comput. **6**(4), 443–456 (2013)
13. Lopes, C.V., Debeauvais, T., Valadares, A: Restful massively multi-user virtual environments: a feasibility study. In: 2012 IEEE International Games Innovation Conference (IGIC), Rochester, NY (2012)
14. Arroqui, M., Mateos, C., Machado, C., Zunino, A.: RESTful Web Services improve the efficiency of data transfer of a whole-farm simulator accessed by Android smartphones. Comput. Electron. Agric. **87**, 14–18 (2012)
15. Cayirci, E: Modeling and simulation as a cloud service: a survey. In: Proceedings of the 2013 Winter Simulation Conference, Savannah, GA (2013)

16. Garg, S.K., Versteeg, S., Buyya, R: SMICloud: a framework for comparing and ranking cloud services. In: Fourth International Conference on Utility and Cloud Computing, Melbourne, Australia (2011)
17. Taylor, S.J.E, Khan, A, Morse, K, Tolk, A, Yilmaz, L, Zander, J: Grand challenges on the theory of modeling and simulation. In: Proceedings of the Symposium on Theory of Modeling and Simulation, San Diego, CA (2013)
18. Jung, W., Kim, S.I., Kim, H.S: Ontology modeling for REST open APIs and web service mash-up method. In: 2013 International Conference on Information Networking (ICOIN), Huket, Thailand (2013)
19. Siriwardena, P.: Advanced API Security. Springer, Heidelberg (2014)
20. Lizcano, D., Soriano, J., Reyes, M., Hierro, J.J.: A user-centric approach for developing and deploying service front-ends in the future internet of services. Int. J. Web Grid Serv. 5(2), 155–191 (2009)
21. Wang, G., Yang, S., Han, Y: Mashroom: end-user mashup programming using nested tables. In: Proceedings of the 18th World Wide Web Conference. Madrid, Spain (2009)
22. Aghaee, S., Pautasso, C.: End-user programming for web mashups. In: Harth, A., Koch, N. (eds.) ICWE 2011. LNCS, vol. 7059, pp. 347–351. Springer, Heidelberg (2012)
23. Trinh, T.D., Wetz, P., Do, B.L., Anjomshoaa, A., Kiesling, E., Tjoa, A.M: Open linked widgets mashup platform. In: 2014 ESWC, Crete, Greece (2014)
24. Le Phuoc, D., Polleres, A., Morbidoni, C., Hauswirth, M., Tummarello, G: Rapid semantic web mashup development through semantic web pipes. In: Proceedings of the 18th World Wide Web Conference. Madrid, Spain (2009)
25. Zahariadis, T., Papadakis, A., Alvarez, F., Gonzalez, J., Lopez, F., Facca, F., Al-Hazmi, Y: FIWARE lab: managing resources and services in a cloud federation supporting future internet applications. In: 2014 UCC. London, UK (2014)
26. Wang, S., Wainer, G: Semantic selection for model composition using SAMSaaS. In: Proceedings of Symposium on Theory of Modeling & Simulation. Alexandria, VA (2015)
27. Wang, S., Wainer, G.: Web-based simulation using Cell-DEVS modeling and GIS visualization. Model. Simul.-Based Syst. Eng. Handb. 3, 425–467 (2014)

Temporal PageRank on Social Networks

Weishu Hu[1]([✉]), Haitao Zou[2], and Zhiguo Gong[3]

[1] School of Science, Jiangnan University, No. 1800 Lihu Avenue, Wuxi, Jiangsu,
People's Republic of China
huweiss@outlook.com
[2] School of Computer Science and Engineering, Jiangsu University of Science
and Technology, No. 2 Mengxi Road, Zhenjiang, Jiangsu, People's Republic of China
nkroben@outlook.com
[3] Department of Computer and Information Science, University of Macau Avenida
da Universidade, Taipa, Macau, China
fstzgg@umac.mo

Abstract. Social network has been a widely accepted way for people to
communicate and interact online. However, few of existing works studied temporal dimension in assessing the authority of nodes on social
networks. In this paper, a novel Temporal PageRank (T-PR) algorithm
is proposed for analyzing the authority of nodes. Three temporal factors
are adopted to personalize PageRank, which favors the nodes that are
more important to people. They are Built-up Time-length Factor (BTF),
Frequency Factor (FF), and Similarity Factor (SF). The experiments on
a real data set demonstrate T-PR algorithm provides the best ranking
results over recent competitor methods.

Keywords: Temporal ranking · PageRank · Link analysis · Search
engine · Time-weighted ranking

1 Introduction

Over the past decade, the social networks have evolved from being an information
source to a center of the world for commercial and social role. Web search engines
are important to help users to find the most useful resources for their specific
interest, which is to bring the most relevant web pages to the top ranked list for
a given query. Most search engines include a ranking algorithm that computes a
page's authority based on either the link structures of the web, e.g., PageRank
[25] and HITS [17], or mining of users' web histories, e.g., BrowseRank [19],
Traffic-weighted Ranking [22], and BookRank [9]. All of these algorithms produce
rankings for different uses. But these algorithms may be biased against more
recent pages [13,24], while such pages have less time to accumulate in-links to
contribute to their link-based ranking, and less chance to be involved into a web
history.

However, an important factor that is not considered by these algorithms is
temporal information which is critical to user's interest in other users. Social

© Springer International Publishing Switzerland 2015
J. Wang et al. (Eds.): WISE 2015, Part I, LNCS 9418, pp. 262–276, 2015.
DOI: 10.1007/978-3-319-26190-4_18

networks are drastically different from traditional web, which are dynamic inter-action environments. Quality users in the past may not be quality users now or in the future. In this paper, we study search from the temporal dimension, which is important due to the following reasons:

- Users are often interested in the recent or active user. Except for the users are celebrities or stars, most users on the network change constantly. New friends are added; ideally, outdated friends are deleted. However, in practice many outdated links are not deleted. This fact prevents the ranking algorithms from retrieving the update results.
- Existing Web page evaluation algorithms basically favor pages that have many in-links. Thus, older users are favored because they tend to accumulate more in-links due to longer existence. In contrast, new users that are high quality will not be ranked high.

We believe that dealing with the problems related to the temporal dimension is of great importance to future developments of rank technology in social net-works. In this paper, we investigate the value of incorporating temporal aspects into ranking of users. Three temporal factors are considered: Built-up Time-length Factor (BTF), Frequency Factor (FF), and Similarity Factor (SF), which capture the intuitive notion that a user with recent updates, special time occur-rence, or trend in revision is potentially more important to users. The hypoth-esizes of this paper are (1) the longer interval between registration time of user and creation time of link is more trustworthy; (2) trustworthy users are often active and regularly adding/deleting links, and seldom perform majority opera-tions at once; (3) recent link is more important than old link.

Trust network is a directed graph, which has explicit links to express one node trusts/distrusts other nodes. In a trust network, there exists two important times (the registration time of node and the creation time of trust link). Individual users are represented by nodes with registration time, having the relationship "User X trusts User Y on Time t" resulting in an edge directed from User X's node to User Y's on Time t. Everything has its cause; there is no absolutely independent behavior without a cause. Under the theory of sociology, i.e., trust increases over time in relationships [28], which means the trust from one node to another node should satisfy the constraints of time. In some social networks such as Facebook and Twitter, an explicit link implies that two nodes are very close for their frequent communication. However, in a trust network, two nodes may be connected but the link may be untrustworthy. More importantly, a trustworthy link in trust network is the situation where two connected nodes have proper time interval. If two users are trustworthy in terms of their time intervals, then their trust links are more trustworthy. For instance, users have trust relations as Fig. 1, where $t_A < t_B < t_C < t_D < t_E$ and $t_1 < t_2 < t_3 < t_4 < t_5 < t_6 < t_7 < t_8$. It is intuitive that the trust link from B to A on t_1 is more trustworthy than that from C to E on t_5. Because E's registration time is later than C's, old node seldom actively trusts new node in general. In contrast, B's registration time is later than A, so B trusts A firstly and then A trusts B as a feedback, which

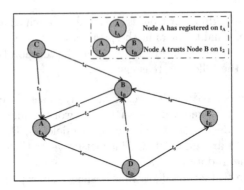

Fig. 1. An example of trust network

is trustworthy. D is a suspicious node used to promote E by connecting other trustworthy nodes such as A and B.

The contribution of this paper include: (1) a novel temporal method (T-PR) which builds upon PageRank and captures how the structure of the interaction network changes over; (2) a new approach for aggregating scores from each of the temporal features over which T-PR is running, the first based on build-up time-length, the second based on frequency, and the third based on similarity; (3) an extensive experiment on a real world data set to evaluate the performance of our proposed method T-PR. Results show that the T-PR outperforms both the state-of-the-art ranking models, and can improve the performance of user recommendation.

The rest of this paper is organized as follows. We briefly describe the PageRank algorithm and review some related work in Sect. 2. In Sect. 3, we define measures of temporal factors, and then describe how to unify them in a Temporal PageRank (T-PR). In Sect. 4, we show and discuss the results of experiments. Finally we make a conclusion and present some future researches in Sect. 5.

2   Related Work

2.1   Brief Review of PageRank

The basic definition of PageRank [25] can be stated as follows. Given a directed graph $G = (V, E)$, while V is the set of nodes and $E \subset V \times V$ is the set of links/edges, without multiple edges. If $u \in V$ has a link to $v \in V$, it implies that u implicitly confers some importance to v. Let $pr(u)$ be the PageRank score of u and $w(u, v)$ be the proportion of importance propagated from u to v, which is normally set to $1/|d^o(u)|$, where $|d^o(u)|$ is the out-degree of u in G. Therefore, link $(u, v) \in E$ confers $pr(u)/|d^o(u)|$ units of rank to v. The PageRank score can be calculated by the following equation:

$$\forall v \in V, pr^{(\phi+1)}(v) = \sum_{u \in d^i(v)} w(u, v) pr^{(\phi)}(u) = \sum_{u \in d^i(v)} \frac{pr^{(\phi)}(u)}{|d^o(u)|} \tag{1}$$

where $d^i(v)$ represents the set of nodes linking to v. The total amount of score conferred on v is the summation of the score of each u divided by its out-degree.

2.2 Other Related Work

Since the PageRank [25] and HITS [17] algorithms were published, there are a large number of papers on improvements , variations, and speed-up of the algorithms have proposed in [4,6,16,23]. These works are still within the framework of the original algorithms, and do not consider the temporal aspect.

There are also some studies that concentrate on incorporate temporal factor into the PageRank algorithm. [2] proposes a modified PageRank by weighting pages with an exponential decay function according to their age, which is a limited attempt to tackle the problem without evaluation. TimedPageRank (TimedPR) [30] is to weight each citation by an exponential decay function according to publication time of the papers, which cannot be directly applied to social networks. Scientific papers have static information fixed a publication time. Since articles cannot be deleted, their citation counts are monotonically increasing. By contrast, social networks can be modified by adding/deleting links. And [3] presents Time-Aware Authority Ranking (T-Rank), by weighting the page transition and random jump probabilities. Their results may not be representative since each domain usually has its own pattern. [7,10,15,21] are enhanced PageRank with time. However, these approaches are not suitable for social networks; user's behavior on social networks is very different from web pages. In summary, all the researchers of the above studies on (time-)weighted PageRank show their methods are better than the standard PageRank, but neglect a high correlation of various PageRank variants and other social network measures such as build-up time-length, frequency, and similarity. None of the studies, however, has combined time information from both the nature of social networks and the characteristic of users via the "Temporal" PageRank described in this paper.

Recently, several researchers, such as [5,8,20,26,29], have applied machine learning techniques to train the ranking model using queries previously and relevance information on retrieved results derived from user's browser behavior, to improve ranking quality. The success of these "learning to rank" approaches depends on both query and document information, which is very difficult to tune the parameter of theses machine learning techniques.

3 Temporal PageRank

In this section, we first describe quantitative measure of three temporal factors. To evaluate the usefulness of these factors, we then present how to incorporate all three factors into the personalized PageRank algorithm.

3.1 Temporal Aspects

Social network is continually changing, i.e., users are registered, modified, or deleted over time. Based on the hypothesizes: (1) the longer interval between registration time of user and creation time of link is more trustworthy; (2) trustworthy users are often active and regularly adding/deleting links, and seldom perform majority operations at once; (3) recent link is more important than old link. Let us consider a social network as a directed graph $G = (V, E, T)$, where the nodes represent users with registration time, the edges represent the trust links with creation time. For $\forall u, v \in V$, $t_u, t_v \in T$ is the registration time of u, v respectively, and $\exists (u, v) \in E$, $t_{uv} \in T$ is the creation time of link from u to v.

Built-Up Time-Length Factor. We define built-up time-length factor to describe the interval between registration time of node and creation time of link. A longer time-length is considered to be more trustworthy than a shorter one. Time-length is used to describe the length of the time that something continues or exists.

Let t_{uv} - t_u denote the built-up time-length of u adds a link to v, obviously $t_u \leq t_{uv}$. The Built-up Time-length Factor of (u, v) $BTF(u, v)$, is simply defined in a exponential time-scale as:

$$
BTF(u, v) = \begin{cases}
1 - e^{-\frac{(t_{uv} - t_u)}{2\sigma^2}} & \text{if } t_u \geq t_v, \\
1 - e^{-\frac{(t_{uv} - t_{vu})}{2\sigma^2}} & \text{if } t_u < t_v \text{ and } \exists t_{uv} \geq t_{vu}, \\
0 & \text{otherwise.}
\end{cases} \tag{2}
$$

Note that $\sigma = 14$ means fortnight, which is the best empirical setting. This built-up time-length factor is bound by $[0, 1)$. Given the observation period t_{uv} through t_u, the built-up time-length factor of u has value 0 if the registration time of node u is the same as the creation time of link (u, v), increasing to 1 for existing interval between t_{uv} and t_u / t_{vu}. If $t_u < t_v$ and $\exists t_{uv} > t_{vu}$, which mean u is older than v and (u, v) is a feedback of (v, u), then $BTF(u, v) \geq 0$.

Frequency Factor. Let $t_i(u)$ is the i-th timestamp of node u adding links, $N_{t_i(u)}$ is the count of links added by node u on time t_i. n_u is the number of timestamps when u adds links. $\overline{N_{t(u)}} = \frac{\sum_{i=1}^{n} N_{t_i(u)}}{n_u}$ is the average of $N_{t(u)}$, and the variance is $var(u) = \frac{\sum_{i=1}^{n} (N_{t_i(u)} - \overline{N_{t(u)}})^2}{n_u}$. n_u is bigger means u is more active. $var(u)$ is smaller means u is more trustworthy, otherwise, $var(u)$ is bigger means u is less trustworthy. In Fig. 2, node D is more trustworthy than other nodes, since node D is an active user and adds trust links regularly, which satisfies the $var(D)$ is smallest and n is biggest. Frequency factor (FF) can be computed by Eq. 3.

$$
FF(u, v) = e^{-\frac{var(u) + N_{t_{uv}(u)}}{n_u \cdot \overline{N_{t(u)}}}} \tag{3}
$$

The frequency factor of (u, v) is the proportion of trustworthiness of u adding a link to v. This frequency factor is bound by $(1, 0)$.

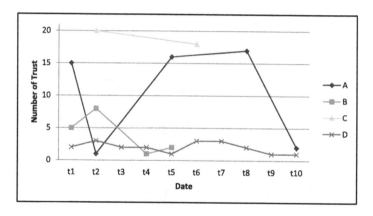

Fig. 2. An example of diverse frequency for adding trust

Similarity Factor. The Built-up Time-length and Frequency factors only consider the temporal factors of individual users. We hypothesize that the pattern of user adds trust link in other causes may also have significance in estimating a user's importance. A user that adds trust links in accordance with a special or good reason (e.g., interest, personality, etc.) reflects trustworthiness in itself, which has been verified experimentally in [12]. However, considering only similarity between all users without examining their ephemeral of behavior may be biased and unfair to those users who have added trust links recently. During a period of time, one user tends to show interest in similar things, such as focusing on similar users. So it is acceptable that an old user trusts a new user because they are similar during a period of time.

A similarity factor describes how similar change the behavior of a user is to others in the network, based on the hypothesis that authoritative users would behave in a similar way [1]. Here, can simply be the SimRank similarity function used in [14]. The similarity factor of u and v on time t is thus considered as the similarity of their in-links and out-links before t_{uv}. Similarity Factor (SF) can be defined as:

$$SF(u, v) = Sim_{t_{uv}}(u, v)e^{-\gamma(t - t_{uv})} \tag{4}$$

A key parameter of the above formula is γ which denotes how fast the value of similarity factor decreases over time. If γ is too small, then the similarity factor will have a long effect. Especially, if $\gamma = 0$ then similarity factor does not decrease with time. On the other hand, if γ is too large, then similarity factor has very short effect, getting quickly to 0. Intuitively, γ should be set according to the timescale of the users to reflect how fast real rankings change. Moreover, it should also be related to the number of trust links added per unit of time (e.g., year or month) of the users. In our experiments, we investigate the impact of various values for γ. In our data set, we find $\gamma = 0.05$ gets the best performance.

And $Sim_{t_{uv}}(u, v)$ is computed by Eq. 5. Let $d^o_{t_{uv}}(u)$ denotes the set of all outgoing neighbors from node u before t_{uv}, and likewise, $d^i_{t_{uv}}(u)$ denotes the set of all incoming neighbors to node u before t_{uv}. $|d^o_{t_{uv}}(u)|$ and $|d^i_{t_{uv}}(u)|$ denote the number of nodes in $d^o_{t_{uv}}(u)$ and $d^i_{t_{uv}}(u)$ respectively.

$$Sim_{t_{uv}}(u,v) = \frac{\lambda \cdot \theta}{|d^i_{t_{uv}}(u)||d^i_{t_{uv}}(v)|} \sum_{p \in d^i_{t_{uv}}(u)} \sum_{q \in d^i_{t_{uv}}(v)} Sim_{t_{pq}}(p,q)$$
$$+ \frac{\lambda \cdot (1-\theta)}{|d^o_{t_{uv}}(u)||d^o_{t_{uv}}(v)|} \sum_{p \in d^o_{t_{uv}}(u)} \sum_{q \in d^o_{t_{uv}}(v)} Sim_{t_{pq}}(p,q) \tag{5}$$

Here, λ is a decay factor between 0 and 1, $\theta \in [0,1]$ is a parameter, which determining the importance of in-links and out-links in the graph. In this paper, we take $\lambda = 0.8$ refers to [14], and $\theta = 0.5$, which is the best empirical setting.

3.2   Temporal Ranking

In this section, we describe a new ranking function that incorporates the built-up time-length, frequency, and similarity factors into PageRank. Let $w_t(u,v)$ be the weight for a transition from node u to node v on time t, $d^i_t(v)$ denotes the set of all incoming neighbors to node v before t. Then, the PageRank formula in Eq. 1 can be rewritten in a temporal form (T-PR) as follows:

$$\forall v \in V, tpr^{(\phi+1)}(v) = \sum_{u \in d^i_t(v)} w_t(u,v)tpr^{(\phi)}(u) \tag{6}$$

In the original PageRank, $w_t(u,v)$ is set to $1/|d^o(u)|$. Some studies [18,27] have proposed biased PageRank. In this paper, our purpose is primarily to investigate the contribution of temporal information (i.e., built-up time-length, frequency, and similarity) to node ranking by weighting in node transitions, and subsequently to reduce the effect of bias against new-born nodes by weighting nodes with only link information.

The time-dependent interactions among users clearly have a fundamental impact on their rankings. For example, a series of recent trust links is likely to be more important in determining the current ranking of a set of users than a series of trust links among the same set of users that occurred far in the past. Intuitively, the importance of trusts link for ranking users decays over time: the older a trust link is, the less important is its result. Thus, a good ranking for todays users should give more importance to recent trust links.

The importance of relationships is usually captured by assigning weights to network edges. Thus, based on this intuition, we propose the use of temporal edge weights to reflect the fact that the importance of trust links decays with time. In particular, we will consider an exponentially decaying weight, controlled by a parameter that determines how fast importance decays over time. Consider two nodes u and v. Moreover, let $w_t(u,v)$ denote the weight of the directed edge from u to v on time $t \geq 0$. We define the edge weight $w_t(u,v)$ as follows:

$$w_t(u,v) = \frac{\alpha BTF(u,v) + \beta FF(u,v) + (1-\alpha-\beta)SF(u,v)}{\sum_{s \in d^o_t(u)} \alpha BTF(u,s) + \beta FF(u,s) + (1-\alpha-\beta)SF(u,s)} \tag{7}$$

Note that $\alpha \in [0,1]$ is a parameter used to weigh the importance of BTF and $\beta \in [0,1]$ is a parameter used to weigh the importance of FF. In this paper, we take $\alpha = \beta = 0.3$, which is the best empirical setting.

4 Evaluation Methodology

We implemented five ranking algorithms: PR [25], TWPR [21], T-Rank [3], TimedPR [30], and T-PR (new) for comparison of the quality of ranking results.

4.1 Experimental Data Set

As the source of web data, we selected a portion of the set of trust network from Epinions.com. To study the evolution of trust network, we downloaded them until April 6th, 2010. The data include the user ID with registration date and the trust link from one user to another user with creation date. Because the creation date of trust link is unavailable before January 1st, 2001, our experimental data only include the users having trust link after January 1st, 2001. Table 1 shows the various classes of statistics about Epinions data. Figure 3 describes the evolution of registered nodes and trust links per half a year. Figure 4 displays the count of users with different in-degrees on April 6th, 2010.

Table 1. Statistical information of Epinions trust network

Epinions	Number	From	To	Total days	Avg
Nodes	47109	25-Jun-1999	23-Mar-2010	3768	12.502
Trusts	258966	01-Jan-2001	02-Apr-2010	3359	77.096
Trusts/Nodes:	5.497	Nodes trust:	33875	Nodes trusted:	29807

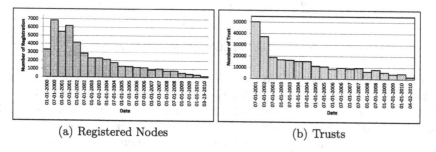

(a) Registered Nodes   (b) Trusts

Fig. 3. The evolution of Epinions trust network

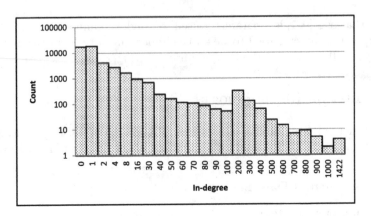

Fig. 4. Distribution of in-degree for Epinions

4.2 Evaluation Methods

In this section, we detail the three measures used in this paper. The $OSim$ and $KSim$ metrics, proposed by Haveliwala [11], measure the similarity of any two ranked lists l_1 and l_2, each of size k. $OSim(l_1, l_2)$ determines the degree of overlap between the top-k nodes of two rankings l_1 and l_2.

$$OSim(l_1, l_2) = \frac{|l_1 \cap l_2|}{k} \tag{8}$$

where l_1 and l_2 are the lists of top-k nodes.

$KSim(l_1, l_2)$ determines the degree of agreement in which the relative ordering of the top-k nodes of two lists l_1 and l_2, have the same relative order in both rankings. Consider two lists l_1 and l_2 of top-k rankings. Let U be the union of nodes contained in both lists and define l_1' as the extension of l_1 to add the elements $U - l_1$ at l_1''s end. Similarly, l_2' is also defined as the extension of l_2. We can define $KSim$ as follows:

$$KSim(l_1, l_2) = \frac{|(u, v) : l_1', l_2' \text{ agree on order of } (u, v) \text{ and } u \neq v|}{|U| \times (|U| - 1)} \tag{9}$$

where the numerator denotes the number of pairwise agreements of elements between l_1' and l_2'.

The final measure we used to evaluate the performance of these 5 algorithms is the Mean Reciprocal Rank (MRR). Given an ordered list of predicted nodes (these nodes are in a descending order according to their ranking values) and the "actual node" that is trustworthy, the reciprocal rank is calculated as $1/\rho$ where ρ is the position of the "actual node" in this ordered list, and it is set to 0 if the "actual node" is not in this list. The MRR value of an algorithm that is closer to 1 denotes a better performance.

4.3   Results and Discussions

We conducted experiments on our crawled data to study the node authority assessment of the PR, TWPR, T-Rank, TimedPR, and T-PR (new) methods. In the following, we first discuss the time evolution of authoritativeness. Then, we describe results of a user study to evaluate the quality of ranked results.

The first set of experiments measure the comparison of different scores of the nodes, which compares the ranking of nodes using the different scores against that produced by the PageRank [25] on January 1st, 2010. Figure 5 shows the comparison of results. Note that we have scaled the scores by multiplying with 1000. The distribution of T-PR is steepest due to the consideration of the effect of 3 temporal factors. One common trend we observe is that nodes with less PR have low score, and those with high PR have high score. This shows that the ranking determined by the T-PR conform to the perception that more popular nodes have more trustworthy.

For each timestamp, we first computed an authoritative score for each node according to a ranking method, and then sorted them from highest to lowest to produce a ranked list. To investigate the time evolution effect (i.e., how the authoritativeness changes over time), we compared the lists of top authoritative nodes obtained from those 5 ranking methods for several consecutive half a year. Figure 6 shows the evolution of top-500 authorities in terms of OSim and KSim. Note that the observation period starts from January 1st, 2002 until April 2nd, 2010. Hence, each point on the graph is the similarity between the ranking at that time and that of the previous half a year.

As shown in the Fig. 6, T-PR yields OSim values of 0.819-0.845 (average 0.831) and KSim values of 0.822-0.851 (average 0.837). This implies that PR produces nearly the same rankings over the entire observation period on the experimental data set. In contrast, the consecutive rankings of TWPR, T-Rank, TimedPR, and T-PR have lower similarities because of their temporal approach. We also see that T-PR produces more similar rankings than TWPR, T-Rank

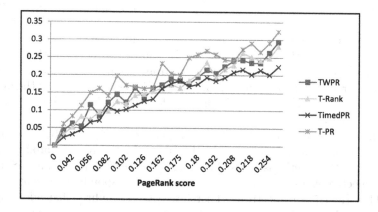

Fig. 5. Comparison of different scores with PR for Epinions

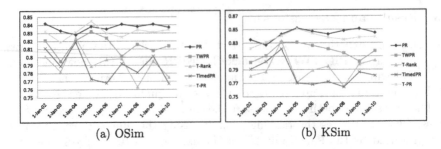

(a) OSim   (b) KSim

Fig. 6. Similarities between lists of top-500 authoritative nodes produced by the same ranking methods for consecutive years

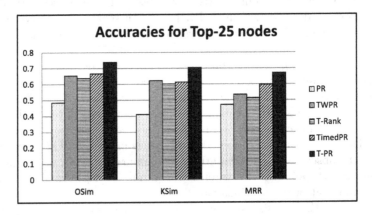

Fig. 7. Average OSim, KSim and MRR values for top-25 nodes

and TimedPR, respectively. This phenomenon may come from the effect of the BTF and FF factors in T-PR that allow some authoritative nodes in the past to be ranked in the top authorities.

We used these 5 algorithms (PR, TWPR, T-Rank, TimedPR, and T-PR) to obtain the sets of the top-25 and top-50 most highly-ranked nodes for the most popular nodes we chose from the data set. Then for each of the most popular nodes, we compared these sets of the top-25 and top-50 nodes with the sets of the top-25 and top-50 most popular nodes from Epinions.com respectively by using the aforementioned evaluation measures. Figures 7 and 8 show the average OSim, KSim and MRR values for the top-25 and top-50 rankings of these 5 different algorithms.

As depicted in Figs. 7 and 8, T-PR outperforms PR, TWPR, T-Rank and TimedPR in all OSim, KSim and MRR values. In the top-25 ranking, T-PR outperforms other methods by 10.8 %, 13.1 % and 12.5 % for OSim, KSim and MRR respectively. In the top-50 ranking, the corresponding percentages are 8.9 %, 11.4 %, 11.7 % respectively. Therefore, we can confirm that adopting the built-up time-length, frequency, and similarity of a node adding trust link can improve the accuracy of nodes ranking.

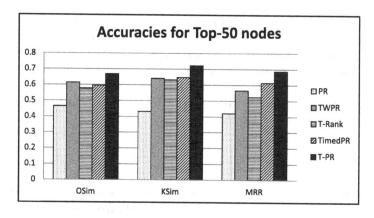

Fig. 8. Average OSim, KSim and MRR values for top-50 nodes

Table 2. Top-20 users' different scores in Epinions

Name	Registration date	PR	TWPR	T-Rank	TimedPR	T-PR
jo.com	02/23/2000	0.285	0.325	0.313	0.297	0.341
dkozin	10/04/1999	0.256	0.293	0.267	0.281	0.323
mkaresh	10/16/1999	0.262	0.301	0.255	0.266	0.305
Freak369	02/05/2000	0.277	0.256	0.261	0.243	0.298
Bryan_Carey	12/21/1999	0.214	0.272	0.234	0.258	0.292
three_ster	06/26/2000	0.226	0.207	0.210	0.213	0.286
shoplmart	11/26/2001	0.265	0.197	0.187	0.221	0.281
dlstewart	05/31/2002	0.244	0.211	0.196	0.206	0.264
Howard_Creech	08/16/1999	0.229	0.174	0.182	0.195	0.249
ChrisJarmick	07/19/2000	0.212	0.203	0.173	0.188	0.243
popsrocks	08/25/2002	0.221	0.199	0.164	0.175	0.233
mrkstvns	09/09/1999	0.208	0.215	0.158	0.191	0.210
yusakugo	04/14/2000	0.258	0.178	0.152	0.162	0.202
melissasrn	02/03/2001	0.211	0.164	0.171	0.153	0.194
surferdude7	01/08/2000	0.188	0.193	0.163	0.147	0.186
AliventiAsylum	08/11/2000	0.174	0.161	0.144	0.168	0.181
jeremy1456	03/17/2001	0.166	0.186	0.155	0.156	0.168
marytara	04/02/2001	0.171	0.157	0.149	0.144	0.157
captaind	01/19/2003	0.175	0.151	0.133	0.132	0.152
yakkowarner	03/01/2004	0.149	0.146	0.143	0.135	0.141

4.4   Case Study

In the final experiment, we want to compare the scores computed by 5 different algorithms including PR, TWPR, T-Rank, TimedPR, and T-PR(new) for

Top-20 trustworthy users from Epinions.com. We select Top-20 reviewers for the most popular authors overall obtained from Epinions.com. In Table 2, we can find that our method produces a better result than others. The T-PR scores conform to the manual ranking. All scores in Table 2 is multiplied by 1000. From the Table 2, we find the PR, TWPR, T-Rank, and TimedPR values fluctuate erratically, and our method get better result than others, which proves Eq. 6 is an effective improvement of PageRank.

5   Conclusion

In this paper, we propose a novel Temporal PageRank algorithm on social networks. Three temporal factors (built-up time-length, frequency, and similarity) are used to improve PageRank so that it favor the nodes that were registered for a longer time, more frequently, and recent trust link weighed more than others. We conduct extensive experiments to evaluate the performance of T-PR. Evaluation results show that the proposed T-PR outperforms other timed PageRank algorithms significantly. Moreover, our T-PR algorithm provides most similar rankings to human beings' preference. In our future work, we are interested in studying the effect of the time factor using other decay functions. Since social networks are multi-relational networks, we will incorporate more relations to improve the ranking of Temporal PageRank.

Acknowledgments. This work was partially sponsored by Grant FDCT/106/2012/ A3, FDCT/116/2013/A3 from Fund of Science and Technology Development of Macau Government and MYRG105-FST13-GZG from University of Macau Research Committee.

References

1. Anagnostopoulos, A., Kumar, R., Mahdian, M.: Influence and correlation in social networks. In: Proceedings of the 14th ACM SIGKDD International Conference on Knowledge Discovery and Data Mining, pp. 7–15. ACM (2008)
2. Baeza-Yates, R., Saint-Jean, F., Castillo, C.D.: Web structure, dynamics and page quality. In: Laender, A.H.F., Oliveira, A.L. (eds.) SPIRE 2002. LNCS, vol. 2476, pp. 117–130. Springer, Heidelberg (2002)
3. Berberich, K., Vazirgiannis, M., Weikum, G.: Time-aware authority ranking. Internet Math. **2**(3), 301–332 (2005)
4. Borodin, A., Roberts, G.O., Rosenthal, J.S., Tsaparas, P.: Finding authorities and hubs from link structures on the world wide web. In: Proceedings of the 10th International Conference on World Wide Web, pp. 415–429. ACM (2001)
5. Burges, C., Shaked, T., Renshaw, E., Lazier, A., Deeds, M., Hamilton, N., Hullender, G.: Learning to rank using gradient descent. In: Proceedings of the 22nd International Conference on Machine learning, pp. 89–96. ACM (2005)
6. Cho, J., Roy, S.: Impact of search engines on page popularity. In: Proceedings of the 13th International Conference on World Wide Web, pp. 20–29. ACM (2004)

7. Fiala, D.: Time-aware PageRank for bibliographic networks. J. Informetrics **6**(3), 370–388 (2012)
8. Geng, X., Liu, T.Y., Qin, T., Arnold, A., Li, H., Shum, H.Y.: Query dependent ranking using k-nearest neighbor. In: Proceedings of the 31st aNnual International ACM SIGIR Conference on Research and Development in Information Retrieval, pp. 115–122. ACM (2008)
9. Gonçalves, B., Meiss, M.R., Ramasco, J.J., Flammini, A., Menczer, F.: Remembering what we like: toward an agent-based model of web traffic (2009). arXiv preprint arXiv:0901.3839
10. Guo, Y.Z., Ramamohanarao, K., Park, L.A.: Personalized pagerank for web page prediction based on access time-length and frequency. In: IEEE/WIC/ACM International Conference on Web Intelligence, pp. 687–690. IEEE (2007)
11. Haveliwala, T.H.: Topic-sensitive PageRank: a context-sensitive ranking algorithm for web search. IEEE Trans. Knowl. Data Eng. **15**(4), 784–796 (2003)
12. Hu, W., Gong, Z.: Assessing the credibility of nodes on multiple-relational social networks. In: Benatallah, B., Bestavros, A., Manolopoulos, Y., Vakali, A., Zhang, Y. (eds.) WISE 2014, Part II. LNCS, vol. 8787, pp. 62–77. Springer, Heidelberg (2014)
13. Ieong, S., Mishra, N., Sadikov, E., Zhang, L.: Domain bias in web search. In: Proceedings of the Fifth ACM International Conference on Web Search and Data Mining, pp. 413–422. ACM (2012)
14. Jeh, G., Widom, J.: Simrank: a measure of structural-context similarity. In: KDD, pp. 538–543 (2002)
15. Júnior, P.S.P., Gonçalves, M.A., Laender, A.H., Salles, T., Figueiredo, D.: Time-aware ranking in sport social networks. J. Inf. Data Manag. **3**(3), 195 (2012)
16. Kamvar, S.D., Haveliwala, T.H., Manning, C.D., Golub, G.H.: Extrapolation methods for accelerating pagerank computations. In: Proceedings of the 12th International Conference on World Wide Web, pp. 261–270. ACM (2003)
17. Kleinberg, J.M.: Authoritative sources in a hyperlinked environment. J. ACM **46**(5), 604–632 (1999)
18. Kohlschütter, C., Chirita, P.A., Nejdl, W.: Utility analysis for topically biased pagerank. In: Proceedings of the 16th International Conference on World Wide Web, WWW 2007, pp. 1211–1212. ACM, New York (2007). http://doi.acm.org/10.1145/1242572.1242770
19. Liu, Y., Gao, B., Liu, T.Y., Zhang, Y., Ma, Z., He, S., Li, H.: Browserank: letting web users vote for page importance. In: Proceedings of the 31st Annual International ACM SIGIR Conference on Research and Development in Information Retrieval, pp. 451–458. ACM (2008)
20. Liu, Y., Liu, T.Y., Gao, B., Ma, Z., Li, H.: A framework to compute page importance based on user behaviors. Inf. Retrieval **13**(1), 22–45 (2010)
21. Manaskasemsak, B., Rungsawang, A., Yamana, H.: Time-weighted web authoritative ranking. Inf. Retrieval **14**(2), 133–157 (2011)
22. Meiss, M.R., Menczer, F., Fortunato, S., Flammini, A., Vespignani, A.: Ranking web sites with real user traffic. In: Proceedings of the 2008 International Conference on Web Search and Data Mining, pp. 65–76. ACM (2008)
23. Ntoulas, A., Cho, J., Olston, C.: What's new on the web? The evolution of the web from a search engine perspective. In: Proceedings of the 13th International Conference on World Wide Web, pp. 1–12. ACM (2004)
24. Otterbacher, J., Hemphill, L., Dekker, E.: Helpful to you is useful to me: the use and interpretation of social voting. Proc. Am. Soc. Inf. Sci. Technol. **48**(1), 1–10 (2011)

25. Page, L., Brin, S., Motwani, R., Winograd, T.: The pagerank citation ranking: bringing order to the web. Technical report, Stanford InfoLab, November 1999. http://ilpubs.stanford.edu:8090/422/
26. Richardson, M., Prakash, A., Brill, E.: Beyond pagerank: machine learning for static ranking. In: Proceedings of the 15th International Conference on World Wide Web, pp. 707–715. ACM (2006)
27. Rungsawang, A., Puntumapon, K., Manaskasemsak, B.: Un-biasing the link farm effect in pagerank computation. In: 2007 21st International Conference on Advanced Information Networking and Applications, AINA 2007, pp. 924–931. IEEE (2007)
28. Vanneste, B., Puranam, P., Kretschmer, T.: Trust over time in exchange relationships: meta-analysis and theory. Strat. Manag.J. (2013). Forthcoming
29. Xu, J., Li, H.: Adarank: a boosting algorithm for information retrieval. In: Proceedings of the 30th Annual International ACM SIGIR Conference on Research and Development in Information Retrieval, pp. 391–398. ACM (2007)
30. Yu, P.S., Li, X., Liu, B.: Adding the temporal dimension to search-a case study in publication search. In: 2005 Proceedings of the 2005 IEEE/WIC/ACM International Conference on Web Intelligence, pp. 543–549. IEEE (2005)

Improving Marketing Interactions
by Mining Sequences

Ritwik Sinha[1], Sanket Mehta[1(✉)], Tapan Bohra[2], and Adit Krishnan[3]

[1] Big Data Intelligence Lab, Adobe Research, Bangalore, India
{risinha,samehta}@adobe.com
[2] Georgia Institute of Technology, Atlanta, GA, USA
tbohra3@gatech.edu
[3] IIT Madras, Chennai, India
adit@cse.iitm.ac.in

Abstract. The advent of digital marketing has revolutionized how a marketer reaches the organization's customers. Since each interaction with the customer is recorded today, the marketer can do a better job of measuring the effectiveness of marketing efforts. With multi-channel marketing data, comes a new set of challenges; those of measuring the effect of individual channels, understanding synergistic effects, and finally leveraging the information stored in the sequence of marketing activities. While there is some work addressing the first two challenges, we aim to shed light on the last question. The combinatorial explosion in the number of possible marketing sequences requires a systematic approach to address this problem. We propose an approach based on sequence mining to identify marketing touch sequences that are most likely to lead to a stated marketing goal. Our approach provides a rapid way of creating marketing campaigns with the highest chance of success. We test our proposed approach on a real world dataset of a retail chain (with web visits, digital marketing channels, email data, instore and online purchase data). We compare against baseline approaches, and observe interesting insights in the real data.

Keywords: Sequence mining · Association rules · Digital marketing · Marketing attribution

1 Introduction

Marketing is the process of communicating the value of products or services to customers. Marketing campaigns carried out through a range of channels play a paramount role in this process. Hence, the task of understanding the value of individual campaigns and marketing channels is important to the marketer. The area of Marketing Attribution is devoted to understanding the true value of different marketing mediums in a multivariate fashion. With the advent of

---

All authors contributed equally to this work.

© Springer International Publishing Switzerland 2015
J. Wang et al. (Eds.): WISE 2015, Part I, LNCS 9418, pp. 277–292, 2015.
DOI: 10.1007/978-3-319-26190-4_19

digital marketing, much of today's marketing data is recorded in a variety of databases (web logs, transaction databases, email marketing databases, social media sites, search marketing, and display networks). The area of marketing attribution has three aspects. First, algorithmically computing attribution to campaigns or marketing channels. Second, using this information for optimal marketing spend allocation. Finally, all algorithmic attribution needs the entire interaction history between an organization and customers; this data can be mined to design improved targeting strategies. This last aspect is what our work is related to.

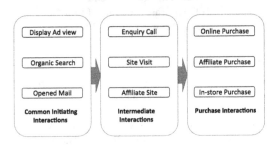

Fig. 1. A customer can follow many paths to a purchase. This schematic shows possible interactions a customer can have at different stages of the purchase process.

Marketing automation refers to a solution that is designed for marketing organization to more effectively manage multi-channel marketing. Many organizations that provide marketing automation solutions claim to reach the customers with the *right message* at the *right time* through the *right channel* [1]. But, most such solutions are not algorithmic in nature, rather, they depend on a marketer's understanding of the business. Hence, this is an area where significant advances can be made by designing data driven algorithmic solutions. We aim to address the question of identifying which channel to use at which point of a customer's journey.

An important source of information about a customer's preferences is stored in the sequence of interactions the customer has with the organization. Figure 1 is a schematic of how a customer may interact with different channels at different stages of the purchase and complete the purchase on different mediums. To test if the sequence of touches is indeed important, we conducted a hypothesis test on a large database of sequences from a major office and home supplies retailer. The hypothesis test, comparing future purchase propensity for certain observed sequences against complements of these sequences, was found to be highly statistically significant (details in Sect. 5). This indicates that the sequence of marketing activities plays an important role in determining marketing effectiveness. This motivates the need for our approach to find high impact marketing sequences which leads us to the primary contributions of our work. In addition to finding evidence for the importance of marketing sequences in real data, we propose an extension of sequential mining techniques to identify sequences that are most likely to lead to a pre-specified marketing goal. We show significant improvements over baseline targeting schemes. Finally, we build a tool that lets the end-user define a goal and retrieve the marketing sequences that have the highest likelihood of achieve the goal.

This paper is organized as follows. In Sect. 2, we describe how our work fits into existing literature. After that, we describe the methodology of our approach and algorithms in Sect. 3. Next, we describe the data on which we applied our work in Sect. 4. In Sect. 5 we describe some of the interesting results we obtained. Finally, we finish with our conclusions and future work in Sect. 6.

2   Related Work

Measuring the effect of different marketing activities has long been an important problem. This problem is as old as the famous quote by John Wanamaker, "Half the money I spend on advertising is wasted; the trouble is I don't know which half". Much of the effort to address this question has been based on Market Mix Models (MMM). MMMs work on time series data aggregated at the marketing channel level. The goal is to find relationships between the time series of revenues, with those of spends for different marketing channels. For example, temporal analysis of marketing channel data is performed in [11,13]. However, with a significant shift of marketing to digital mediums, it is today possible to evaluate the value of advertising at an individual customer level. This aids to draw better causal interpretations of the role played by each advertisement.

Using individual level data for credit assignment to marketing channels falls under the area of Multi-Channel Marketing Attribution. In [2], the authors propose a Hidden Markov Model based approach to model a customer's journey across different states. While theoretically appealing, their model requires a large number of parameters to be estimated from the data. An attribution model aimed at estimating the incremental effect of individual channels has been proposed in [16]. A game theoretic formulation of the marketing attribution model aimed at estimating the causal effects is studied in [7]. While some of these approaches address the question of interaction between multiple channels, none of them look at the effect that the sequence of marketing activities has, on the potential for purchase.

The area of Association Rule Learning has been developed to find interesting relationships between variables in large databases. The literature of this field may be classified into two broad categories. In the first category, association rules are used as an exploratory tool [4], while in the other group they are used as a predictive tool [14]. The problem of mining a large collection of transaction baskets was introduced in [3]. While the authors address the question of finding patterns within transactions, the sequence the items appear in is not considered. In [10], the problem of finding sequential patterns within transactions is addressed where sequence in which items are purchased is considered. As [3,10] are concerned with intra-transactions, they assume that there is no repetition of items in the basket. The question of mining sequential patterns in large databases of transactions has been addressed in [4]. In this work, the authors look beyond situations where the entities may be considered a set (like a basket), but consider a temporal ordering of the items, thus leading to a sequence. In [4,17] a horizontal representation of the database is used for mining sequential patterns while in [8,9] a vertical representation of the database is used. A vertical representation leads to faster algorithms to computing frequencies compared to horizontal representations.

In this work, we have applied algorithms for fast computation of frequencies of different marketing touch sequences. In particular, we have explored SPADE [17], CM-SPADE, CM-ClaSP [8], and VMSP [9]. In our experience, the VMSP algorithm leads to fast computations, and has the beneficial property that it extracts maximal sequences thus avoiding short sequences which are obvious and do not reveal much information, for instance search followed by web visit. These are eliminated by maximality of our frequent sequences. In the next section, we provide the details of our work.

3   Method

Here we provide the foundation of our approach. First, we describe the statistical hypotheses tests that are conducted to show that the sequence of marketing activities is important, when it comes to informing what happens in the future. Next, we describe the exploratory part of our work, that is, how we compute the frequency of sequences, and their corresponding confidence for certain objectives.

3.1   Importance of Sequences

The whole premise of our work hinges on the notion that the sequence in which a set of marketing activities takes place is important to the future relationship an organization has with a customer. We test this hypothesis in a staged manner. In the first step, we conduct a hypothesis test for the importance of the interaction of marketing channels.

Interaction Effects Between Marketing Touches. Let x_{ij} denote the indicator variable of whether customer i $(i = 1, \cdots, N)$ has had an interaction of type j $(j = 1, \cdots, J)$ with the organization. Also, let y_i denote whether this customer made a purchase. Then, consider the logistic regression

$$\log \left(\frac{P(y_i = 1 | \mathbf{x}_i)}{P(y_i = 0 | \mathbf{x}_i)} \right) = \alpha + \sum_{j=1}^{J} \beta_j x_{ij} + \sum_{j=1}^{J} \sum_{k=1}^{J} \beta_{jk} x_{ij} \times x_{ik}.$$

Where $\mathbf{x}_i = (x_{i1}, \cdots, x_{iJ})$. Then the Wald test [5] statistic for the parameter β_{jk} may be used to test the hypothesis that the jth and k^{th} channels have an interactive effect. A positive coefficient estimate of the interaction term denotes that the presence of both channels leads to a larger increase in the log-odds ratio of purchase, than just the sum of the effects of the two channels. Such a finding will show that channels have synergistic effects. With J marketing channels, we will be testing $\binom{J}{2}$ different hypothesis. Since we will be conducting multiple hypothesis tests, we use the Bonferroni correction [6] to control for it with a family wise error rate of 0.05.

Importance of Marketing Touch Sequence. If the above hypothesis is established, one may further be interested in understanding if, it is not just the presence of both channels j and k, but that their appearance in a particular order is also important. To test this hypothesis, we performed a formal evaluation.

Before we describe our approach, let's introduce some notation. Let Γ be the set of possible interactions $\{C_1, C_2, \cdots, C_J\}$ (for example, opened email, viewed display ad, or clicked paid search). Define a marketing interaction sequence of a customer as the chronologically ordered list of interactions. We denote a sequence S_i of length i_k by $< I_1, I_2, \cdots, I_{i_k} >$, where $I_j \in \Gamma$ for all $j = 1, \cdots, i_k$. Further, the sequence S_l denoted by $< I_1^l, I_2^l, \cdots, I_{l_k}^l >$ is *contained in* sequence S_m denoted by $< I_1^m, I_2^m, \cdots, I_{m_k}^m >$ if $I_1^l = I_{m_1}^m, I_2^l = I_{m_2}^m, \cdots, I_{l_k}^l = I_{m_{l_k}}^m$ for $m_1 < m_2 < \cdots < m_{l_k} \leq m_k$. For example, the sequence $S_a = < S, A >$ is contained in the sequence $S_b = < S, W, A, G >$, however it is not contained in the sequence $S_c = < S, T, U, D >$.

For a sequence $S_i = < I_1, I_2, \cdots, I_{i_k} >$, define its complement $\Omega(S_i^c)$ as the set of all sequences $S_m = < I_1^m, I_2^m, \cdots, I_{m_k}^m >$ such that all singleton sequences $< I_j^m >$ are contained in S_i, for all $j = 1, \cdots, m_k$, but, S_i is not contained in S_m. In other words, the set of sequences that contain all the individual marketing channels, but not in the exact same order. For example, if the sequence whose effect is being tested is $S_d = < W, A, S >$, then $\Omega(S_d^c)$ has the five sequences $< W, S, A >$, $< A, S, W >$, $< A, S, W >$, $< S, W, A >$, and $< S, A, W >$.

To test the effect of a sequence, we performed the following exercise. For a sequence S_i, we can compute how often the sequence ends with a purchase, denoted by $p_S = P(S_i \rightarrow 1)$ (S_i leads to a purchase). Similarly, define $p_{S^c} = P(\Omega(S_i^c) \rightarrow 1)$ (for sequences in the complement of S_i, how often do they lead to a purchase). The hypothesis $H_0 : p_S = p_{S^c}$ may be tested using the asymptotic normality test [12]. To ensure we do not bias our hypothesis tests, we selected a number of sequences ensuring sufficient presence in the data, but without looking at its association with purchase. As in the earlier section, we will use the Bonferroni correction to control for multiple tests with a family wise error rate of 0.05. Further, since we are working with large sample sizes, traditional statistical significance may be associated with small effects, hence we also consider the effect size given by $(p_S - p_{S^c})$ when deciding whether there is an effect of business importance.

3.2 Confidence and Support

Before we describe our approach, let's define a few terms. Let D denote $\{S_1, S_2, \cdots, S_n\}$, as a database of user activity sequences. A sequence S_l *supports* a sequence S_m if sequence S_m *is contained in* sequence S_l. The *support for a sequence* is defined as the fraction of total sequences which support this sequence. For example, in Table 1, the sequence $< P, A, W >$ has a support of 3/6.

We define a *frequent sequence* as a sequence that has support greater than or equal to a user-specified minimum support (*minsup*). A *frequent maximal sequence* is one which is not strictly contained in any another frequent sequence. In Table 1, if *minsup* = 0.2, the sequence $< S, A >$ fails to meet the minimum support. The sequence $< P, A >$ has a support of 4/6. However, $< P, A >$ is not a frequent maximal sequence because it is contained in $< P, A, W >$ which is also a frequent sequence.

Table 1. Some example user activity sequences, each letter denotes a different marketing interaction.

Customer ID	User activity sequence
1	$O \to W \to A \to D \to 1$
2	$D \to P \to A \to M \to W \to 1$
3	$D \to P \to A \to 1$
4	$P \to M \to A \to O \to D \to W \to 0$
5	$S \to W \to A \to G \to 1$
6	$O \to P \to M \to A \to M \to W \to 0$

(1 – Purchase, 0 – Non Purchase)

Note that some sequences may end with a purchase (call them purchase sequences). In other words, $S_i = < I_1, I_2, \cdots, I_{i_k} >$, such that, I_{i_k} is a purchase event (a sequence cannot have a purchase in its interior, if they do, such sequences are broken into separate sequences, each with at most one purchase event). A purchase may have additional attributes like revenue, product type, mode of purchase, and so on. Let us denote a mapping of purchase events to classes of *objectives*, $F(I_{i_k}) \in \{O_1, \cdots, O_K\}$. An objective is a marketing goal that the marketer may envision for customer sequences, for example, one objective could be purchase of product type "furniture", with a "high" revenue that happen "online". Also, for a purchase sequence S_i, define the function $antecedent(S_i) = < I_1, I_2, \cdots, I_{i_k-1} >$, the sequence S_i without the purchase event.

Let the database of sequences $D(O)$ denote all observed sequences that satisfy objective O, for example, all sequences ending in high revenue online furniture purchases. Algorithm 1 outlines the steps to compute the support and confidence of sequences. In the first step, the VMSP sub-routine [9] is called on the sequence database $D(O)$, this finds all frequent maximal sequences with a specified minimum support. All such sequences end with the objective O, next take the antecedent of the sequence, and compute the support for this sequence in the entire data (D). The ratio of the two frequencies gives us the confidence of the sequence leading to the objective of interest. This approach lets us find the best sequences likely to lead to a certain objective. It also gives a measure of the chance of this objective being satisfied by the sequence.

Algorithm 1. Mining High Confidence sequences

Require: Two database of sequences D and $D(O)$
1: Supp $\leftarrow$ VMSP($D(O)$,*minsup*,*maxlen*)
2: Count $\leftarrow$ empty dictionary
3: **for** seq in Supp **do**
4:    antseq $\leftarrow$ *antecedent*(seq)
5:    **for** x in D **do**
6:       **if** antseq is contained in x **then**
7:          Count[seq]++
8:       **end if**
9:    **end for**
10: **end for**
11: Conf $\leftarrow$ empty dictionary
12: **for** seq in Supp **do**
13:    Conf[seq] = Supp[seq] / Count[seq]
14: **end for**

Table 2. Details of the marketing interactions along with frequencies (Frequency is the total count of sequences where that particular channel is reflected).

Channel	Symbol	Source	Frequency	Percentage
Direct Web Visit	T	Web	9,088,525	36.28
Display Ad View	d	Display Ad	4,167,055	16.64
Display Ad Clicked	D	Web	218,604	0.87
Opened Email	o	Mail	13,955,334	55.71
Clicked Email	O	Web	1,699,826	6.79
Social	S	Web	10,537	0.04
Paid Search	P	Web	682,837	2.73
Organic Search	G	Web	1,933,748	7.72
Owned Ads	A	Web	1,408,392	5.62
Money Saving Sites	M	Web	295,668	1.18
Instore	I	Instore	16,770,764	66.95
Online	E	Web	4,422,140	17.65

4   Data Description

While our approach can be reasonably easily applied to any marketing interaction data, in this section, we provide details of the data-set on which we applied our approach.

We had access to the marketing (email and display), web interaction, as well as online and in-store purchase data for a large office and home supplies retailer. The data ranged over a 100 day period during the summer of 2013. The data contained about 54 Million interactions with 18 Million customers. For marketing activities to be valuable, it is important for organizations to stitch customer data across different sources. The data from different sources like email marketing, display marketing, web analytics and transaction databases were stitched to have a single key with the use of email addresses, loyalty cards, and web cookies. While there has been some work on probabilistically stitched customer data [15], we assumed that the data was correctly stitched in a rule based fashion.

Table 2 provides details of the channels from which we had data, and the relative frequencies of them. We had the following interactions: Direct Web Visit (typing of url or bookmark visit), Display Ad View, Display Ad Click, Opened Email, Clicked Email, Visit from a Social Site, Click on a Paid Search, Click on an Organic Search, Click on an Owned Ad (ads shown on affiliate sites), and Click on a link on a Money Saving Site (bargain hunting sites). Lastly, we had two kinds of purchase events possible, either an In-Store purchase or an Online purchase. There were more in-store purchases than online purchases in this dataset. Among the interactions, email opens, direct web visits, and display ad views were the more common interaction channels.

Table 3. Distribution of the sequences across multiple-channels. The diagonal value denotes the percentage of sequences containing that channel. The off-diagonal value denote the percentage of sequences containing the two channels corresponding to the row and column. Please refer to Table 2 for what the row and column names stand for.

	Channels										
	T	d	D	o	O	S	P	G	A	M	I	E
T	36.284	11.415	0.632	22.208	3.095	0.031	1.305	3.698	3.980	0.725	15.313	14.147
d		16.636	0.854	9.536	1.948	0.018	0.953	2.557	1.714	0.476	7.427	6.838
D			0.873	0.487	0.141	0.003	0.060	0.170	0.276	0.076	0.312	0.434
o				55.714	5.669	0.028	1.220	3.728	3.402	0.675	37.207	5.442
O					6.786	0.009	0.237	0.951	0.734	0.191	4.156	0.747
S						0.042	0.003	0.009	0.009	0.004	0.023	0.008
P							2.726	0.523	0.352	0.075	0.998	1.29
G								7.720	0.983	0.243	2.868	3.445
A									5.623	0.258	2.239	2.315
M										1.180	0.503	0.420
I											66.954	0.000
E												17.654

Since our problem looks at sequences, we explore the interaction between channels and the importance of sequences, we looked at how often multiple channels appear in the same sequence. Table 3 describes these interactions. We see that there is significant cross-channel interactions. Additionally, we find in our data, 75 % of the customers had at interactions with at least 2 channels, 30 % of them interacted with at least 3 channels, and 24 % of them interact with 4 or more channels.

We define a sequence as a chronologically ordered collection of interactions for an individual. Thus, the entire sequence for the individual can be $< A, B, C, P_1, D, E, P_2, F, G, H >$ (alphabets A–H denote different interaction channels, while P_1 and P_2 denote purchase events), such an sequence is broken into the following 3 sequences $< A, B, C, P_1 >$, $< D, E, P_2 >$, and $< F, G, H >$. The first two are purchase sequences, but the last one is a non-purchase sequence.

Table 4. Sequences distribution across Product Category and Mode of Purchase.

Product category	Purchase mode	Total sequences	Unique sequences
Furniture	Online	115,987	34,998
	Instore	265,662	46,758
Machines	Online	901,906	214,633
	Instore	3,769,764	467,098
Electronics	Online	272,500	83,443
	Instore	1,597,079	231,740
Home	Online	2,275,956	485,701
	Instore	5,248,325	638,993
Services A	Online	0	0
	Instore	1,304,517	106,761
Office	Online	736,024	184,215
	Instore	3,837,277	480,918
Services B	Online	5,542	2,593
	Instore	228,657	34,964
Store	Online	114,225	49,697
	Instore	519,483	81,429
Non purchase		3,855,469	277,762
Total		25,048,373	808,254

Breaking an individual's entire sequence by purchase events was necessary because we wanted to profile sequences by various attributes of the purchase. We had additional information for purchase interactions, in particular, the product purchased, the revenue of the purchase and the mode of purchase. Table 4 shows the distribution of purchases from the different product categories. In all, there were about 25 Million sequences considered in our analysis. Each of which was one of 808, 254 unique sequences. This table also provides the distribution for different product categories and mode of purchase. Apart from "Services A" and "Services B" being only (or mostly) sold in-store, all other product categories have a large number of sequences.

Figure 2 shows the sequence for three customers. The x-axis of the plot is the date of interaction, on the y-axis, we have the sequence number. In the first panel, we see a customer who may be dubbed an "impulsive customer". This

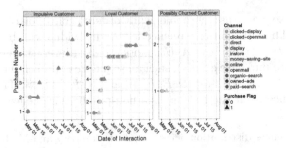

Fig. 2. The marketing interaction sequences for 3 customers. The x-axis has the date of interaction, on the y-axis are the resulting sequences from the customer, after breaking them by purchases. The three panels denote three different kinds of customers. The color and symbol denote the type of interaction and whether the interaction included a purchase or not (Color figure online).

customer had 7 purchases (denoted by seven rows of data in the panel), in each case making the purchases soon after the interaction. The second customer may be called a "loyal customer" because of the large number of interactions with the organization, as well as large number of purchases. The last customer is possibly a churned customer, after a single purchase. This customer's last sequence is a non-purchase sequence. Apart from such customers there were also customers who had no purchases over the entire 100 day period.

5   Results

In this section, we describe our results. We first provide the findings of the hypotheses tests that prove the importance of marketing sequences, next we describe some exploratory data analysis. After that, we describe some results contrasting high confidence sequences for different marketing objectives.

Table 5. Test of the importance of interactions between channels. Only selected interaction coefficients are displayed.

Variable interaction	Estimate	Std. error	z value	Pr(> |z|)
Direct Web Visit × Display Ad	2.85	0.02	125.37	0.00
Direct Web Visit × Openmail	1.61	0.04	45.79	0.00
Display Ad × Paid Search	0.87	0.04	20.95	0.00
Display Ad × Organic Search	0.83	0.03	31.56	0.00
Openmail × Clicked Openmail	1.34	0.04	35.71	0.00
Openmail × Owned Ad	0.71	0.03	22.71	0.00

5.1 Importance of Marketing Sequence

As described in Sect. 3, we first test two hypotheses. The first looks at the importance of interaction between channels. Table 5 shows selected coefficients from the logistic regression, along with the standard error, z–values and p–values. Of the 45 interaction effects in the model, 28 interactions are statistically significant at the Bonferroni corrected level of 0.001 (0.05/45), also 20 of these are positive. Thus, in our data, there is a strong evidence that interaction between marketing channels is important to encourage a positive likelihood of purchase.

In the second hypothesis, we test the effect a sequence has on the likelihood of a future purchase. Table 6 displays the results from a selection of this analysis. Of the 219 sequences tested against their complements (please refer to Sect. 3 for the definition), we found 212 of them to have significantly different purchase propensities (at the Bonferroni corrected level of 0.0002). Also, as displayed in the table, the magnitude of this difference was often large. In fact, for 208 of these hypotheses, the effect size was 1 % or more. The effect may have been either positive or negative, but since we did not select the sequences to test based on their propensity of purchase, this is to be expected. Also note that the effect sizes were often large.

Table 6. Test showing the confidence of sequences and their complements. Please refer to Table 2 for the sequence name legend.

Sequence	Sequence confidence (%)	Complement confidence (%)	P-value	Difference (%)
dodT	83.41	79.19	0.00	4.22
TdT	85.89	85.26	0.00	0.63
Todo	80.95	85.14	0.00	−4.19
dGdT	81.55	86.27	0.00	−4.72
odoT	82.94	77.53	0.00	5.41
GdT	82.16	89.78	0.00	−7.62

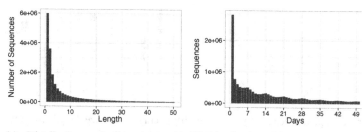

(a) Distribution of Sequences by Length of Sequence.

(b) Distribution of Sequences by Duration of Sequence.

Fig. 3. Number of sequences plotted against the length of the sequence (in number of touches or duration in days).

5.2   Exploratory Data Analysis

Our data contains a number of interesting facets on which we performed exploratory data analysis. In particular, in addition to the sequence data, we had the timing of interactions, the types of products purchased (if any), the mode of purchase (online or in-store) as well as the revenue of the purchase.

We first looked at the distribution of the number of sequences by the number of touches in the sequence (Fig. 3(a)). As expected, we see a highly positively skewed distribution, with the largest number of sequences being short (of length 1 or 2), but there is still a significant number of sequences longer than 2. When looking at the duration of these sequences (Fig. 3(b)), we again see a positive skew, but with a large proportion of sequences lasting a week or more. We also see an interesting seasonal pattern, with a periodicity of a week (due to a weekly activity cycle).

Next, we explored how the length of the sequence associates with the mode of purchase (Fig. 4(a)) and the revenue of the purchase (Fig. 4(b)). We see that the online sales tend to have a larger number of touches, across all product categories. We also see that the higher revenue sequences require a higher number of touches, indicating that customers take longer to decide on higher revenue purchases. As before, we again see that "Services A" and "Services B" purchases require fewer touches.

To explore the effect of duration further, we looked at the purchase sequences by different product categories (Fig. 5(a)) and at the average revenue of the purchase against the duration of the purchase (Fig. 5(b)). We see that different products take different number of days for the purchase to be completed. We also see that, on the average, the shortest (in terms of days) sequences lead to the highest revenues. On further exploration, we found a lot of high revenue business purchases (in product category "Services A" and "Services B") happening rapidly. The weekly recurring pattern is more apparent when plotting the average revenue against duration of the sequence.

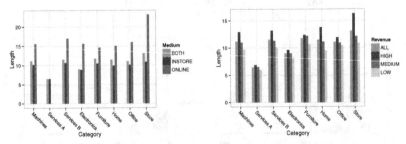

(a) Average sequence length by Purchase Mode and Categories.

(b) Average sequence length by Revenue Levels and Categories.

Fig. 4. Plot of Sequence length by Category, Revenue Level and Purchase Mode.

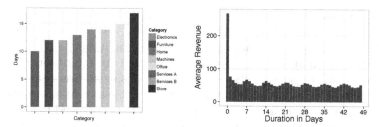

(a) Average conversion time (in days) by product category.

(b) Total Revenue by Duration of sequence.

Fig. 5. Duration of the sequence plotted against the product category and Average Revenue of the transaction.

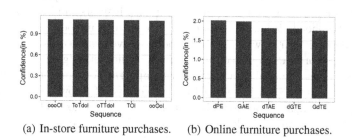

(a) In-store furniture purchases.

(b) Online furniture purchases.

Fig. 6. Comparing Furniture buys across online and in-store modes across all revenue categories. The x-axis has the sequence names and the height of the bar denotes the confidence of the sequence. Please refer to Table 2 for the sequence name legend. We present the top five sequences, in terms of their confidence toward the marketing goal.

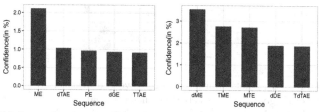

(a) Low revenue category electronics purchases.

(b) High revenue category electronics purchases.

Fig. 7. Online Electronic purchases across low and high revenue categories. Please refer to Table 2 for the sequence name legend.

5.3   Confidence and Support

The primary goal of our work is to give the marketing practitioner the ability to find the best sequences for a given objective. Let's say a marketer wants to know the kind of sequences that have the highest confidence for furniture purchases that happen online, we provide a tool for the marketer to identify these sequences readily. Figure 6(b) shows the most promising sequences for such purchases. The marketer can additionally contrast the in-store furniture purchases (Fig. 6(a)) against the online furniture purchases (Fig. 6(b)), and find vastly different sequences with the highest confidence. We see that display followed by paid search leads to the highest confidence for an online furniture purchase, where as the high confidence sequences for in-store furniture sales tend to be more influenced by emails and direct web visits.

When we compare high revenue and low online Electronic purchases (Fig. 7(b) and (a)), we see that the presence of a display ad prior to the money saving sites visit can leads to higher confidence in the high category purchases. This could be an indication for the marketer to target customers with display ads, to try to get them to money saving sites. Many similar such insights can be gleaned by the marketer when he explores marketing sequences using our tool.

Figure 8 provides a view of the demo tool we have designed help the marketer identify top performing sequences for a particular marketing goal. At the top, the marketer specifies the goal of a particular campaign. For example, in this view, the marketer is interested in selling furniture online of a low revenue category. The best marketing sequence for this goal is Display ads followed by Paid Search. At the bottom are the confidence and support for the five sequences with the highest confidences for the specified marketing goal. It is easy for the marketer to use this tool to rapidly create marketing campaigns for any given marketing goal. It is also possible to contrast sequences for different marketing goals.

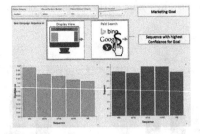

Fig. 8. A tool created to help marketer to retrieve the best sequence of interactions based on a marketing goal. The marketer specifies the goal, the best sequence to target for this goal is identified. At the bottom, on the left are the confidences of the different sequences, and on the right are their corresponding supports.

5.4   Comparison to Baseline

We compare the results of our proposed approach to two baseline approaches. The first is the random strategy. The random strategy is the one where we

Table 7. Comparison of high confidence sequences, in their confidence to lead to orders of "Machines". The last two columns provide multiplicative lifts compared to two baseline strategies.

Marketing sequence	Confidence (%)	Lift over random	Lift over 100th sequence
odMAE	4.00	3.33	2.01
dodGE	3.20	2.67	1.61
MPE	3.17	2.64	1.59
GdE	2.76	2.30	1.38
GTE	2.69	2.24	1.35
TTAE	2.60	2.17	1.31

consider the confidence of a randomly selected sequence to lead to a certain marketing goal. The second comparison is the following. For a given marketing goal, we look at the ranked list of sequences, ranked by their confidence to achieve the given marketing objective. We then compare the highest ranked sequences with the hundredth ranked sequence. Table 7 provides these details. The last two columns of the table provides the lift achieved by using the highest ranked sequences as a multiplicative factor compared to the two baselines. We see that the top six sequences have between 2 and 3.3 times increase over the random approach. When we compare them to the 100th most successful sequence, we see lifts of 1.3 to 2 times.

6   Conclusions

In this work, we have shown on a large retail data-set that the sequence of marketing touches has a significant effect on the propensity of achieving a stated marketing goal. Next, we propose a novel application of sequence mining to marketing touch data. While association rule learning has been applied to the market basket problem in the realm of marketing, this is, to our knowledge, the only application of sequence mining to marketing interaction data. In doing so, we look beyond the individual effect of marketing channels toward combinations of marketing channels in the form of sequences. Our approach aids better measurement of marketing effectiveness, and aids in better targeting. We also provide a tool to the marketer which can be used to rapidly mine this data to find high confidence sequences for a particular marketing goal.

Our approach has a number of advantages, we were able to mine large amounts of data in a scalable fashion. We handle in excess of 25 Million marketing interaction sequences without difficulty, a size that is likely to be bigger than most organizations' customer base. We also consider all possible interactions between marketing channels and capture the most important of these. We show the marketing lift achieved by our approach over random targeting.

There are a number of possible extensions to our work. While this paper talks about finding an optimal sequence to launch a particular campaign, it does not consider the attributes of an individual customer. We also aim to overlay sequence mining with the stage of a customer in the purchase life cycle. This will aid us in understanding if certain channels play a more important role in different stages of a customer's purchase cycle. We also aim to explore the effect that channels like television have on sequences. Such marketing channels are harder to capture in sequence mining because confirmed touch information is not available.

References

1. Gartner names salesforce.com a leader in the CRM customer engagement center magic quadrant. sforce.co/1z5Lesd. Accessed: 22 July 2014
2. Abhishek, V., Fader, P., Hosanagar, K.: Media exposure through the funnel: a model of multi-stage attribution (2012). SSRN 2158421
3. Agrawal, R., Imieliński, T., Swami, A.: Mining association rules between sets of items in large databases. ACM SIGMOD Rec. **22**, 207–216 (1993). ACM
4. Agrawal, R., Srikant, R.: Mining sequential patterns. In: Proceedings of the Eleventh International Conference on Data Engineering, pp. 3–14. IEEE (1995)
5. Agresti, A., Kateri, M.: Categorical Data Analysis. Springer, Heidelberg (2011)
6. Bland, J.M., Altman, D.G.: Multiple significance tests: the bonferroni method. BMJ **310**(6973), 170 (1995)
7. Dalessandro, B., Perlich, C., Stitelman, O., Provost, F.: Causally motivated attribution for online advertising. In: Proceedings of the Sixth International Workshop on Data Mining for Online Advertising and Internet Economy, p. 7. ACM (2012)
8. Fournier-Viger, P., Gomariz, A., Campos, M., Thomas, R.: Fast vertical mining of sequential patterns using co-occurrence information. In: Tseng, V.S., Ho, T.B., Zhou, Z.-H., Chen, A.L.P., Kao, H.-Y. (eds.) PAKDD 2014, Part I. LNCS, vol. 8443, pp. 40–52. Springer, Heidelberg (2014)
9. Fournier-Viger, P., Wu, C.-W., Gomariz, A., Tseng, V.S.: VMSP: efficient vertical mining of maximal sequential patterns. In: Sokolova, M., van Beek, P. (eds.) Canadian AI 2014. LNCS, vol. 8436, pp. 83–94. Springer, Heidelberg (2014)
10. Kamakura, W.A.: Sequential market basket analysis. Mark. Lett. **23**(3), 505–516 (2012)
11. Naik, P.A., Raman, K., Winer, R.S.: Planning marketing-mix strategies in the presence of interaction effects. Mark. Sci. **24**(1), 25–34 (2005)
12. Newcombe, R.G.: Interval estimation for the difference between independent proportions: comparison of eleven methods. Stat. Med. **17**(8), 873–890 (1998)
13. Ramaswamy, V., DeSarbo, W.S., Reibstein, D.J., Robinson, W.T.: An empirical pooling approach for estimating marketing mix elasticities with pims data. Mark. Sci. **12**(1), 103–124 (1993)
14. Rudin, C., Letham, B., Salleb-Aouissi, A., Kogan, E., Madigan, D.: Sequential event prediction with association rules. In: Conference on Learning Theory (2011)
15. Saha Roy, R., Sinha, R., Chhaya, N., Saini, S.: Probabilistic deduplication of anonymous web traffic. In: Proceedings of the 24th International Conference on World Wide Web Companion, pp. 103–104. International World Wide Web Conferences Steering Committee (2015)
16. Sinha, R., Saini, S.K., Natarajan, A.: Estimating the incremental effects of interactions for marketing attribution. In: 2014 International Conference on Behavior, Economic and Social Computing (BESC), pp. 1–6. IEEE (2014)
17. Zaki, M.J.: SPADE: an efficient algorithm for mining frequent sequences. Mach. Learn. **42**(1–2), 31–60 (2001)

A Majority of Wrongs Doesn't Make It Right - On Crowdsourcing Quality for Skewed Domain Tasks

Kinda El Maarry[1(✉)], Ulrich Güntzer[2], and Wolf-Tilo Balke[1]

[1] IFIS, TU Braunschweig, Brunswick, Germany
{elmaarry,balke}@ifis.cs.tu-bs.de
[2] Inst. F. Informatik, Universität Tübingen, Tübingen, Germany
ulrich.guentzer@informatik.uni-tuebingen.de

Abstract. Today, crowdsourcing has emerged as a promising paradigm for annotating, structuring, and managing Web data. Still, as long as the problem of the crowd workers' trustworthiness in terms of result quality is not essentially solved, all these efforts remain doubtful. Therefore, in this paper we look at today's dominant quality assurance techniques and investigate how they cope with Web data, i.e. typical long-tail distributions, making it easy for strategic spammers to guess the prevalent answers and thus to go undetected. We provide a thorough theoretical analysis, quantifying the success of different methods on such skewed domains by means of test theory and show their individual weaknesses. Exploiting our case study analysis, we propose a simple privacy-preserving, task-agnostic model to improve test reliability, while actually decreasing overhead costs for quality assurance. Finally, we show the stability of our method for even higher numbers of spammers in controlled crowdsourcing experiments.

Keywords: Crowdsourcing · Fraud detection · Result quality · Quality control

1 Introduction

In recent years, the paradigm of crowdsourcing has been discussed in many computer science disciplines, ranging from simple tasks like: content annotation [1] or IR evaluation [2], to complex tasks like: crowd-enabled database retrieval [3]. In fact, the integration of human assessments for intelligent steering of algorithms, promises a high benefit, especially in many typical Web data processing tasks, where state of the art algorithms are still lacking (e.g., text translation, image annotation, opinion mining, or sentiment analysis). In any case, for crowdsourcing to be practical, the main threat of *fraudulent workers affecting the overall result quality* has to be managed: Since spammers are easily attracted to cheat crowdsourcing systems due to the highly distributed, virtual, and anonymous nature of crowdsourcing, task providers are forced to apply strict quality control measures to ensure their exclusion, as their contributions jeopardize the entire quality of the output.

Task-aware control measures include *gold questions*, i.e. a set of test questions whose answers are already known, which are currently one of the widely applied

© Springer International Publishing Switzerland 2015
J. Wang et al. (Eds.): WISE 2015, Part I, LNCS 9418, pp. 293–308, 2015.
DOI: 10.1007/978-3-319-26190-4_20

quality safeguards (although their applicability has been doubted in many typical crowdsourcing scenarios, see e.g. [4]). Simply put, gold questions are randomly injected in each task to test and monitor the reliability of workers. Such questions could be either (1) blindly injected, where the workers don't know they're answering a gold question, in which case, gold questions are more like a quality control detection mechanism, or (2) openly injected, where the workers get instant feedback when they answer a question incorrectly, in which case, gold questions are more like a teaching and future failure prevention mechanism that explains to workers why they failed. In addition, *reputation-based systems* are widely employed to track the worker's history and the requestor's satisfaction in association with each worker. Yet, computing workers' reputation in distributed settings (in particular across the boundaries of several systems) still poses a real world challenge.

In contrast, *redundancy* is a task-oblivious mechanism for quality control, where the aggregation of results from multiple workers for the same task is ultimately chosen as the final answer, e.g., by using averages or performing a majority vote. Such task-oblivious measures have the valuable advantage of being applicable for all kinds of tasks (even opinion-based tasks and tasks where the notion of correctness depends on consensual agreement) and without any knowledge about the worker. In that sense, redundancy forms a light-weight and privacy preserving quality control, where the aggregation techniques allow to measure each worker's quality as a function of his/her agreement with the crowd. However, besides incurring higher costs, redundancy-based mechanisms have also been shown to have their limitations, especially for higher ratios of spammers, see e.g., [5].

Apart from such general limitations, all of the above control measures drastically fail for crowdsourcing tasks whose *answer sets are intrinsically skewed*, i.e. there is a predominant answer, which can be exploited as a difficult-to-detect-default-answer by spammers. Especially in Web data, this kind of (long-tailed) skew that is favoring certain: entity types, image content, or data properties is well known. In practice, individual crowd-sourcing tasks, or human intelligence tasks (HITs), requiring the crowd to search for patterns or rare occurrences in Web data sets, usually follow a power law, namely a Zipfian distribution.

Example 1 (Adult website classifier). A study on quality management in Amazon Mechanical Turk (for details see [6]), attempted to train an adult website classifier based on data points labeled by the crowd. In reality, 85 % of the websites are suitable for general audiences, while only 15 % are actual porn. The crowd was asked to label a website as G when suitable for general audiences and as P when comprising adult content. The results reported that strategic spammers (i.e., workers submitting always the prevalent answer/the highest class prior) indeed only showed an error rate of 15 %. Yet, honest workers sometimes exhibited even higher error rates.

Obviously for the above example, all common measures, whether it's gold questions, majority votes, or reputation-based systems, would fail in identifying strategic spammers, because they're seemingly doing a reasonably good job. Thus, as soon as spammers realize a skewness in some task, they can easily cheat the system by providing only the prevalent answers and thus render the final results useless.

To better understand the actual problem, we turn to classical measures from test theory: *sensitivity* and *specificity*, and illustrate how they individually influence a worker selection decision. We then argue how in datasets, whose answer set are highly skewed (i.e. in a binary setup, one class label occurs seldom, e.g. porn label as in Example 1), we need only focus on the specificity measure, rather than the sensitivity as well. Building on this insight, we propose double/triple testing model, which aims at boosting a worker's specificity, which in turn improves the positive predictive value, i.e. the ability to avoid incorrect answers submitted by strategic spammers who would always submit the frequent class label. Unlike majority voting, the double/triple testing model only seeks a second and/or a third opinion when the first response belongs to the frequent class. Using a controlled environment for simulations on crowdsourcing formulated tasks, provide evidence for both the cost efficiency and the accuracy of double/triple testing relative to traditional aggregation techniques based on redundancy.

2   Related Work

Crowdsourcing provides an inexpensive and easy to set up solution for companies having typical digital business problems needing intelligent or perceptual input like e.g., Web resource tagging [7], completing missing data [8], sentiment analysis [9], text translation [10], information extraction [4], etc.

A lot of research has focused on the quality problem in crowdsourcing and consequently different approaches have emerged. The most commonly used approaches aim at controlling the quality by: (1) reliably identifying and excluding spammers, (2) identifying low quality results and cleverly integrating the results to produce higher resulting quality, (3) a combination of both, or (4) providing the workers with incentives to demotivate them from cheating the system.

Gold questions are a typical example of the *first approach* to quality control. Questions, whose answers are already known, are added to a crowdsourcing task. Failing to correctly answer a certain percentage of gold questions raises a flag, and the worker is identified as a spammer to be excluded from the workforce. Here, the workforce is basically filtered, and only those workers satisfying a certain quality threshold are allowed to work on a given task, all others are simply discarded.

Under the *second approach* falls the family of aggregation methods. Widely employed aggregation control measures include those relying on redundancy: by assigning the same task to several workers, the correct answer can be identified through aggregation, e.g. majority voting. Nevertheless, it suffers from severe limitations, see e.g. [5]. Different variations of weighted aggregation methods include: Dawid and Skene's work [11], where an expectation maximization (EM) algorithm is used to consider the responses' quality based on the individual workers. The responses' quality is then used to compute a weighted aggregation. When applied to skewed domain tasks, EM would fail, as it's unable to identify the strategic spammers who supply high prevalent answers [6]. With such error rates in focus, other approaches alike emerged, such as: a Bayesian version of the expectation maximization algorithm approach [12], a probabilistic approach taking into account both the worker's skill and the difficulty of the task at hand [13]. A more elaborate algorithm was introduced in [6] for building an adult web classifier (see Example 1). It aimed at separating the unrecoverable error rates from the recoverable bias by generating a scalar score. This score would represent the inherent quality of each worker. Moreover, unlike EM, it takes into consideration the uncertainty of the strategic spammers' prevalent answers. However, the algorithm accurately functions with approximately 5 labels per question and for 20 or 30 labels from each worker. Fewer labels per question would be dramatic in terms of quality. Our approach requires only two labels per question and there's no restriction on the number of labels acquired from a worker.

Reputation based systems is a good representative for the *third approach*, where the focus is on eliminating unethical workers throughout longer time scales. This is attainable either through observing constant workers' performance via a reputation-based system (based on a reputation model [13, 14], on feedback and overall satisfaction [15], or on deterministic approaches [16], etc.) With such systems in place, a worker's reputation score can be used either as a threshold, which either allows or denies him/her access to a task (first approach), or can be used as a weight when aggregating the results (second approach). Except, computing a reliable aggregated reputation scores for workers still poses a real challenge.

Lastly, the *fourth approach* comprises the different motivational incentives that are used on crowdsourcing platform, which can be divided into intrinsic and extrinsic incentives [17]. Intrinsic motivation is related to the task, where the workers are motivated to work on the task without expecting anything in return because it's: interesting, a hobby, or for a good cause, e.g. ByMyEyes – an iOS application that is designed to help the blind with daily challenges –. Extrinsic motivation is related to the return-yielding-mechanisms, where workers are motivated to work on the task expecting: monetary compensation, fame, status, recognition, etc. Investigations on using money as an incentive have shown it to be quite tricky, where low paid jobs yield sloppy work, and high paid jobs attract unethical workers [18].

Our proposed model falls under the second approach, where we investigate a way to combine the opinions of different independent workers to decrease the number of false negatives, consequently improving the positive predictive value i.e. the resulting quality. Accordingly, our evaluations will be set against the most commonly used aggregation technique on crowdsourcing platforms: Majority Voting.

3   Case Study: Insights from Medical Test Theory

Since many of the crowdsourcing tasks rely on binary decisions (e.g., recognizing whether a certain object is contained in some image or classifying a business document or news article), the classic measures from medical test theory, in particular, sensitivity and specificity [19] apply.

In this section, we delve into medical test theory and closely examine these measures to determine the factors influencing the design of a high performance test. Building on the insight's drawn from this case study, we abstract these factors and transform them to our crowdsourcing setup, where the individual workers become the individual tests, whose performance should be boosted. In particular, we focus on skewed datasets: in medical test theory, one example could be the skewness of people diagnosed with AIDS, while in crowd sourcing, this would be tasks based on zipfian web datasets.

3.1 Sensitivity, Specificity and Prevalence

The classic measures of medical test theory: sensitivity and specificity are used to measure the performance of a binary classification test (e.g. Is the patient diagnosed with AIDS?) reflecting the true positive rate and the true negative rate respectively (see Definition 1). The prevalence on the other hand, defines the percentage of population shown to have the tested classification condition (i.e. percentage of people found to have the disease).

Definition 1 – Sensitivity and Specificity: sensitivity σ and specificity τ of some test T are defined as:

$$Senstivity\ \sigma(T) := \frac{|true\ positives|}{|positively\ tested|}$$

$$Specificity\ \tau(T) := \frac{|true\ negatives|}{|negatively\ tested|}$$

Where *positively tested* refers to those being predicted as positive and *negatively tested* refers to those being predicted as negative.

Generally, the usefulness of a test [21] can be measured by the positive predictive value $ppv(T)$: the probability that a positive answer (i.e. have AIDS) is indeed correct, and the negative predictive value $npv(T)$: the probability that a negative answer (i.e. doesn't have AIDS) result is correct. By considering the probability distribution for all possible outcomes for a tested condition (see Fig. 1), we can define both positive and negative predictive values as follows:

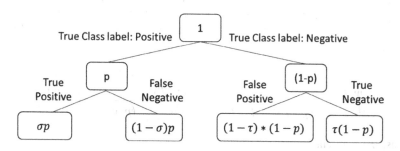

Fig. 1. Testing error diagram with prevalence p, sensitivity σ, and specificity τ

Definition 2 – Positive and negative predictive values: using Bayes theorem, we can directly derive from Fig. 1 the following formulas for both positive (*ppv*) and negative predictive values (*npv*) for some tested condition T:

$$ppv(T) := \frac{\sigma p}{\sigma p + (1 - \tau)(1 - p)} = \frac{1}{1 + \frac{(1-\tau)}{\sigma}\frac{(1-p)}{p}}$$

$$\approx \frac{1}{1 + \frac{1-\tau}{p}}, \textit{for } \sigma \textit{ close to } 1 \textit{ and } p \textit{ close to } 0$$

$$npv(T) := \frac{\tau(1 - p)}{\tau(1 - p) + (1 - \sigma)p} = \frac{1}{1 + \frac{(1-\sigma)p}{\tau(1-p)}}$$

$$\approx \frac{1}{1 + (1 - \sigma)p}, \textit{for } \tau \textit{ close to } 1 \textit{and } p \textit{ close to } 0$$

The approximation shows that the positive predictive value depends mostly on τ and p, whereas the negative predictive values strongly depends on σ and p. For high values of σ and τ however, a small prevalence would lead to very bad positive predictive values, see Example 2.

Example 2. Given a tested condition T exhibiting both high sensitivity and specificity $\sigma(T) = \tau(T) = 0.99$. Assume however, that the tested condition (e.g. is a product review negative?) only shows a prevalence of $p = 0.001$. Then the positive predictive value is $ppv(T) = 0.09$, i.e. only 9 %.

On the other hand, the respective negative predictive value is almost perfect at 99.99 % and can be safely ignored.

To see how a test can be influenced by improvements of sensitivity and specificity, let us again consider a prevalence of $p = 0.001$ and change either measure while fixing the other. Table 1 shows the respective increase in positive prediction values. As can be observed, even extreme improvements in sensitivity σ doesn't lead to improvements beyond 1 % for the positive predictive values. In contrast, increasing values of τ quickly leads to acceptable positive prediction values. Thus, our focus should be fixed on the *specificity* (Table 2).

Lemma 1 – Designing Reliable Tests based on Specificity:

Let $ppv(T)$ be a function of p, σ and τ. Looking at the partial derivatives with respect to σ and τ analytically proves why the design of a reliable test should focus on specificity.

Proof:

$$\frac{\partial ppv(T)}{\partial \tau} = \frac{1}{(ppv(T))^2}\frac{1-p}{p}\frac{1}{\sigma} \text{ and } \frac{\partial ppv(T)}{\partial \sigma} = \frac{1}{(ppv(T))^2}\frac{1-p}{p}\frac{1-\tau}{\sigma^2}$$

Thus, the quotient:

$$\frac{\partial ppv(T)}{\partial \tau} \Big/ \frac{\partial ppv(T)}{\partial \sigma} = \frac{\sigma}{1-\tau}$$

$$\approx \frac{1}{1-\tau}, \textit{for } \sigma \textit{ close to } 1 \qquad \blacksquare$$

Table 1. Positive prediction values for a test T with p = 0.001 and $\tau = 0.9$

σ	0.9	0.95	0.99	0.999	0.9999
$ppv(T)$	0.0089	0.009	0.0098	0.0099	0.00991

Table 2. Positive prediction values for a test T with p = 0.001 and $\sigma = 0.9$

τ	0.9	0.95	0.99	0.999	0.9999
$ppv(T)$	0.0089	0.0177	0.0826	0.4739	0.9

It is easy to see that improvements in σ will yield almost no result, however even small improvements in τ may directly change $ppv(T)$ drastically. Accordingly, to design a reliable test, one must focus on boosting the specificity, which in turn improves the positive predictive value.

4   Integrating Reliable Answers from Crowd Workers

Our analysis in the above case study revealed that solely focusing on the specificity would lead to significant improvements to a test's performance. In this section, we adapt our findings to crowdsourcing, and design a model for an efficient workers' answer selection scheme.

4.1   From Medical Test Theory to Crowd-Sourcing

For crowdsourcing tasks that are formulated based on skewed datasets (e.g. is there a cat in the picture?), strategic spammers can gain highly accurate results by simply always giving the frequent class label (i.e. there is no cat). Since the performance of a model for a reliable crowd worker's answer selection depends mainly on τ, boosting answers from users with high specificity is key. Still, strategic spammer would always appear to have high specificity by giving the frequent class label. On the other hand, their sensitivity becomes nearly non-existent, since they hardly ever give the less frequent class label (i.e. there is a cat).

Upon focusing on the specificity, the measure's bound for both the honest workers and the strategic spammers differ: for the honest workers, the specificity is bounded by their skill levels, where areas for the strategic spammers, it is bounded by the prevalence, that is, the skewness of the dataset. Since the prevalence is low, strategic spammers will tend to be highly accurate. At which point, redundancy-based approach like Majority Votes fail. Accordingly, as shown in Fig. 1, an efficient model's focus should be on the false positives (i.e. frequent class label responses). And even though false negatives are also problematic, the prevalence's imbalance adds in most of the weight and the big mass within the false positive branch, leaving the false negative branch unworthy to focus on.

4.2   Double Testing Model: Asking for a Second Opinion

Following our analysis' insights, we propose a light weight redundancy-based model, which procures redundant opinions to decrease the number of false positives, which is responsible for low positive predictive values. Namely: the double testing model, which seeks a second independent opinion.

The double testing model is inherently more cost and time efficient than traditional redundancy-based quality control measures. In double testing, a redundant opinion is only retrieved when the first opinion lies in the frequent class label. If the second opinion, coincides with the first opinion, the frequent class label is accepted. Otherwise, the non-frequent class label, as dictated by the second opinion, ends up as the accepted label. Simply put, if the rare label is given at least once, it is accepted. Such a redundancy-based model has the advantages of being privacy preserving, where no information about the worker is required, and task agnostic, where no information about the task is needed.

Next, we examine the false positives frequency and the quality of positive predictive value after double testing is applied.

Example 3. Given two independent workers w_i, where $(i = 1, 2)$ with sensitivity σ_i and specificity τ_i. Let p be the prevalence of the positive answer. A task q is given to worker w_1 and is denoted as q_1. Upon seeking a second opinion from worker w_2, it's denoted as $q1, 2$.

As illustrated in the tree representation in Fig. 2, we can formally then define the combined sensitivity and specificity as follows:

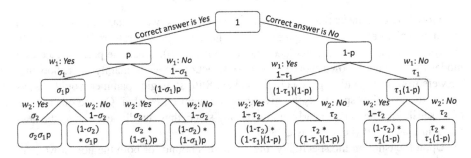

Fig. 2. Double testing (seeking second opinion) tree diagram

Lemma 2 – Combined Sensitivity $\sigma_{1,2}$

$$\sigma_{1,2} = \sigma_1.\sigma_2$$

In fact, $\sigma_{1,2}$ has a lower value than σ_1, but only by a factor of σ_2 (which has a value close to 1 in general). Accordingly, $\sigma_{1,2}$ can be also estimated as follows

$$\sigma_{1,2} \geq (\min\{\sigma_1, \sigma_2\})^2$$

Lemma 3 – Combined Specificity $\tau_{1,2}$:

$$\tau_{1,2} = \tau_1 + \tau_2 - \tau_1.\tau_2$$

Where:

$$(1 - \tau_2)(1 - \tau_1)(1 - p) = (1 - \tau_{1,2})(1 - p)$$

$$\Leftrightarrow$$

$$(1 - \tau_2)(1 - \tau_1) = (1 - \tau_{1,2})$$

$$\Leftrightarrow$$

$$1 - \tau_2 - \tau_1 + \tau_1 * \tau_2 \Leftrightarrow 1 - \tau_{1,2}$$

$$\Leftrightarrow$$

$$\tau_{1,2} = \tau_1 + \tau_2 - \tau_1 * \tau_2 \quad \blacksquare$$

$$\tau_{1,2} \geq max\{\tau_1, \tau_2\}$$

Where:

$$\tau_{1,2} = \tau_1 + \tau_2 - \tau_1 * \tau_2$$
$$\tau_1 + \tau_2 * (1 - \tau_1) \geq \tau_1$$
$$Since \ (1 - \tau_1) \geq 0$$
$$By \ symmetry : \tau_{1,2} \geq \tau_2$$

The frequency of false positives for the combined question $q_{1,2}$, namely $(1 - \tau_2) * (1 - \tau_1)(1 - p)$, massively decreases relative to the frequency of false positives of q_1, namely $(1 - \tau_1)(1 - p)$. Since the frequency of false positives is responsible for extremely low $ppv(T)$ values, $ppv(T)_{1,2}$ will be accordingly much better than either $ppv(T)_1$ or $ppv(T)_2$, regardless of the high σ and τ values.

To further analyze the difference between $ppv(T)_{1,2}$ and $ppv(T)_1$, we compare the false positives and true positives of w_1 with the false and true positives of the combined answers.

Lemma 4 Let $ppv(T)_{1,2}$ be the combined positive predictive value, accordingly

$$ppv(T)_{1,2} = \frac{1}{1 + (1 - \tau_2)/\sigma_2 * (1/ppv(T)_1^{-1})}$$

Proof:

$$ppv(T)_1 = \frac{\sigma_1 p}{\sigma_1 p + (1 - \tau_1)(1 - p)} = \frac{1}{1 + (1 - \tau_1)/\sigma_1 * (1 - p)/p} \qquad (1)$$

$$ppv(T)_{1,2} = \frac{\sigma_2 \sigma_1 p}{\sigma_2 \sigma_1 p + (1 - \tau_2)(1 - \tau_1)(1 - p)}$$

$$= \frac{1}{1 + (1 - \tau_2)/_{\sigma_2} * (1 - \tau_1)/_{\sigma_1} * (1 - p)/_p} \qquad (2)$$

From (1), we directly get:

$$\frac{1}{ppv(T)_1} - 1 = \frac{1 - \tau_1}{\sigma_1} * \frac{1 - p}{p} \qquad (3)$$

By Substituting (3) in (2), we get:

$$ppv(T)_{1,2} = \frac{1}{1 + (1 - \tau_2)/_{\sigma_2} * (1/_{ppv(T)_1} - 1)} \qquad \blacksquare$$

Accordingly, the improvement of $ppv(T)_{1,2}$ over $ppv(T)_1$ can be described by the $((1 - \tau\_2))/\sigma\_2$ factor. So even if $ppv(T)_1$ is very bad, e.g. $ppv(T)_1 \geq 0.1$, $ppv(T)_{1,2}$ will be much better. For both σ_2 and $\tau_2 \geq 0.9$, substituting in the above definition will give a combined predictive value bigger than 50 %. Similarly, if σ_2 and $\tau_2 \geq 0.99$, double testing gives a $ppv(T)_{1,2}$ of more than 90 %, despite of either the prevalence or $ppv(T)_1$ being very low.

4.3   Triple Testing: Asking for a Third Opinion

Sometimes it might be necessary to ask for a third worker's opinion. The associativity will hold, if all three workers w_1, w_2 and w_3 are independent.

Definition 5 – Triple combined Sensitivity $\sigma_{1,2,3}$ and Specificty $\tau_{1,2,3}$:

$$\sigma_{1,2,3} = \sigma_3 * \sigma_{1,2} = \sigma_3 * \sigma_2 * \sigma_1$$
$$1 - \tau_{1,2,3} = (1 - \tau_3) * (1 - \tau_{1,2}) = (1 - \tau_3) * (1 - \tau_2) * (1 - \tau_1)$$

Accordingly, a second and/or a third opinion can be sought, whenever the initial response resides in the frequent class of the answer set. Triple testing, as will be shown in the evaluation section, is more cost efficient than a majority vote of three workers and outperforms it in terms of final resulting quality.

In triple testing a third opinion is sought when w_1 responds with the frequent answer. As soon as one of the other two workers respond with the non-frequent class label, their answer is taken as the final label. Otherwise, the frequent label is accepted.

4.4   Statistical Independence Assumption

So far we've assumed statistical independence, that is, the errors committed by the workers are random. Even if there exists a systematic bias in the submitted answers

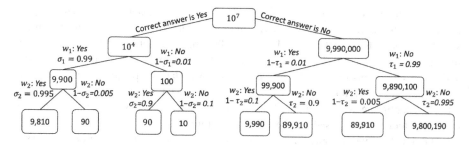

Fig. 3. Second opinion example tree diagram

(i.e. the frequency of errors depends on the time of the day or on the workers' cultural background), we can beat this bias by employing a heterogeneous set of workers, coming from different parts of the world and having different education levels, etc.

However, if the workers' error rate depends on the intrinsic difficulty of a question, then the error probability for worker w_2 increases given that worker w_1 already failed this task. In the example below, we try to assess the consequences of this effect.

Example 4. In order to see how $ppv(T)$ is affected when seeking a second opinion, we use the tree diagram in Fig. 3. Assume $p = 10^{-3}$ and start with a population $N = 10^7$. Accordingly:

$$ppv_{1,2} = \frac{9810}{9810 + 9990} = 0.495$$

As seen above, both $\sigma_{1,2}$ and $\tau_{1,2}$ aren't bad at all. The combined positive predictive value $ppv_{1,2}$ has improved when compared to ppv_1. However, it's much less than in the case of independence. This is to be expected, as the dependency leads to much more false positives. As soon as worker w_1 produces a false positive, which might be due to the question's difficulty, worker w_2 will with a higher probability reproduce the same mistake. The more false positives there are the lower the predictive value $ppv_{1,2}$ will be.

To face that, a third, a fourth and up to n-opinions could be asked for. Another proposition would be to categorize the gold questions according to their difficulty level as perceived by the workers. In previous work of ours (see [22, 23]), we employed the Rasch Model from psychometrics to compute the questions' difficulty, which is perceived differently by the heterogeneous set of workers, who have varying skill levels. Knowing the questions' difficulty level beforehand, allows for aligned questions assignment to workers having the appropriate skill levels.

5 Evaluation

In this section, we evaluate the accuracy and cost efficiency of both double and triple testing and compare them to the most commonly used redundancy quality control technique: majority voting.

We simulated the crowdsourcing tasks and conducted extensive synthetic experiments, which on one hand allows for high numbers of validation runs to ensure statistically significant results, and on the other hand enables a flexible and deterministic parameter control: namely, the percentage of spammers within a workforce and the exact composition of the answer set, in particular, the percentage by which the crowdsourcing tasks' answer set is skewed. Since for easy tasks, honest workers' error rates are negligible when compared to the error introduced by spammers, throughout our experiments we focused on the error of spammers.

The HITS were designed to comprise 20 binary classification problems. As a default, 85 % of the correct problem solutions belong to one class label, and 15 % belong to the other class (unless stated otherwise). As a strategy for exploiting the system, a spammer in our experiments always gives the frequent answer and thus has an average error rate of only 15 %. All of the following results are averaged over 1000 binary decision problems assigned to a workforce comprised of 100 workers.

5.1 Accuracy of Double and Triple Testing

We start by looking into the efficiency of our model in terms of the resulting quality, and compare their results to majority voting of three workers. We examine how that quality changes under varying parameters: (1) different percentage of spammers making up the entire workforce (100 workers) and (2) percentage of skewness of the crowdsourcing tasks' answer set.

Impact of the Percentage of Spammers. Figure 4, illustrates the quality achieved through majority voting, double and triple testing. The worst quality in terms of average correctness for our evaluation environment lies at about 85 %, because constantly responding with the prevalent answer will be 85 % of the time correct. Figure 4 shows that both double and triple testing always perform better than majority voting, regardless of the percentage of spammers in the system. In fact, their performance even gets better, when the percentage of spammers in the workforce increases: assuming an existence of 80 % spammers, majority voting's resulting quality resides at 86.5 %, while double testing and triple testing scores 90.4 % and 92.5 % respectively. Note that, employing gold questions, which only target the low prevalent class, would immediately discard about 80 % of the workforce at this point. With triple testing, the curve becomes bigger than when a second opinion is sought by the double testing, indicating improved resulting quality.

Impact of Skewness Percentage of Crowdsourcing Task's Answer set. To illustrate the danger of skewed datasets, we changed the degree of skewness while setting the percentage of spammers to 80 % (i.e. high). Figure 5 shows that the more a crowdsourcing task's answer set is skewed, the higher the achieved quality is. This is especially true, when the number of strategic spammers who submit the prevalent answer increases. The quality attained by the majority vote is always bounded by the answer set skewness, i.e. at 95 % skewness, a quality of 95 % can be attained, which corresponds to the prevalent class. Both double and triple testing can achieve higher quality values, scoring at 95 % skewness resulting quality corresponding to 96 % and

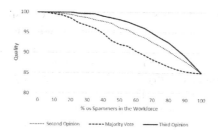

Fig. 4. Impact of the percentage of spammers

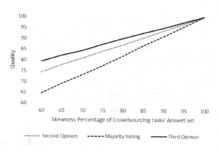

Fig. 5. Impact of skewness percentage of crowdsourcing tasks' answer set

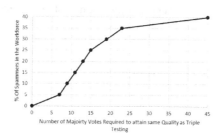

Fig. 6. Number of Majority Votes requires for comparable Resulting Quality

97 % respectively. Of course, the difference in performance is more significant with less skewed answer sets, where the majority vote can't attain high resulting quality anymore. That is, as the skewness decreases, the prevalent class shrinks in size, and with it the high quality that can be attained by the strategic spammers. For instance, at 70 % skewness, majority voting scores a quality of 73 %, while double and triple testing score 81 % and 84 % respectively.

5.2 Comparable Resulting Quality with Majority Voting

In the previous set of experiments, we compared the resulting quality achieved by majority voting of three workers and that of both double and triple testing. Now, we test how many workers' vote are needed for the majority voting to reach comparable resulting quality as that of the triple testing.

Our experiments show that as soon as the percentage of spammers exceeds 40 %, even a majority vote over the entire taskforce doesn't add up to the targeted quality. At lower percentage of spammers however, we can monitor the cost associated with reaching comparable results. As seen previously in Fig. 4, at the existence of 35 % spammers in the workforce, majority voting's resulting quality resided at around 96 %, while triple testing was at 99.55 %. To reach the same resulting quality, a majority of 23 votes is required. This means much higher costs for redundancy based on majority voting.

5.3   Crowdsourcing Cost

Finally, we measure both double and triple testing in terms of overhead cost of quality assurance, as incurred when submitting the tasks on a crowdsourcing platform. A task is made up of 20 binary classification problems. Each of which costs 0.05$. That is, a task costs 1$. Crowdsourcing 5 tasks with a total of 100 classification problems would always incur 15$ when using majority voting, regardless of the underlying composition of the workforce i.e. percentage of spammers.

In case of double and triple testing, the costs varies as the percentage of spammers in the workforce change. As shown in Fig. 7, as the percentage of spammers increase, the cost gradually increases. Double and triple testing has an upper bound of 10$ and 15$ respectively, at which point the worst case scenario materializes: the entire crowd is made up of spammers, and every classification problem needs to be asked respectively twice or thrice.

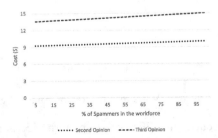

Fig. 7. Impact of the percentage of spammers on the Crowdsourcing costs

At the other end of the spectrum, the lower bound would be equivalent to the cost of all the tasks being submitted once to the crowd, i.e. 5$ for both the double and triple testing. The cost instantly increases after that point, as spammers are gradually introduced into the workforce, and accordingly for every high prevalent answer that is submitted, a second and/or a third opinion is sought.

In a nutshell, both double and triple testing will always incur lower costs than the traditional majority voting scheme.

6 Conclusion

In this paper, we focus on crowdsourcing tasks involving skewed data, in particular, the typically long-tailed Web data and the quality control challenge they pose: detecting strategic spammers who exploit the intrinsic skewness of the answer sets to get a simple, yet hard to detect spamming strategy: they always respond with the prevalent answer. To get a deeper understanding of the problem and a guideline for designing effective countermeasures against spammers, we turned to test theory, namely, to the classical measures: sensitivity and specificity (Fig. 6).

A thorough analytical analysis revealed a surprising result. A redundancy-based quality measure's performance, even when facing low prevalence, can be significantly improved by only focusing on the measure's specificity rather than its sensitivity: in brief, getting a few more redundant opinions has more impact on the overall result quality than designing a better and more reliable test. In fact, we proved that enhancing the specificity, drastically improves the positive predictive value of the test. We then proposed double and triple testing model, which seeks a second and/or a third opinion whenever the crowdsourcing task's response is in the prevalent class.

To verify and test the quality and cost efficiency of our model, we conducted extensive controlled crowdsourcing experiments, which allowed for flexible parameter control. Our results show that both double and triple testing outperform majority voting in terms of quality and cost, even if the percentage of spammers in the crowd increases. Thus, our simple, yet effective design also features a highly robust behavior for Web information systems.

References

1. Sorokin, A., Forsyth, D.: Utility data annotation with amazon mechanical turk. In: 2008 IEEE Computer Society Conference on Computer Vision and Pattern Recognition Workshops, CVPR Workshops (2008)
2. Kazai, G., Craswell, N., Yilmaz, E., Tahaghoghi, S.M.: An analysis of systematic judging errors in information retrieval. In: Proceedings of the 21st ACM International Conference on Information and Knowledge Management - CIKM 2012, p. 105 (2012)
3. Selke, J., Lofi, C., Balke, W.-T.: Pushing the boundaries of crowd-enabled databases with query-driven schema expansion. In: 38th International Conference on Very Large Data Bases (VLDB), pp. 538–549 (2012)
4. Lofi, C., Selke, J., Balke, W.-T.: Information extraction meets crowdsourcing: a promising couple. Datenbank-Spektrum 12(1), 109–120 (2012)
5. Kuncheva, L.I., Whitaker, C.J., Shipp, C.A., Duin, R.P.W.: Limits on the majority vote accuracy in classifier fusion. Pattern Anal. Appl. PAA 6(1), 22–31 (2003)
6. Ipeirotis, P.G., Provost, F., Wang, J.: Quality management on amazon mechanical turk. In: Proceedings of the 2nd Human Computation Workshop (HCOMP), pp. 0–3 (2010)
7. Finin, T., Murnane, W., Karandikar, A., Keller, N., Martineau, J., Dredze, M.: Annotating named entities in twitter data with crowdsourcing. In: Proceedings of the NAACL HLT 2010 Workshop on Creating Speech and Language Data with Amazon's Mechanical Turk, vol. 2010, pp. 80–88, Hune (2010)

8. Lofi, C., El Maarry, K., Balke, W.-T.: Skyline queries in crowd-enabled databases. In: International Conference on Extending Database Technology (EDBT), March 2013

9. Kouloumpis, E., Wilson, T., Moore, J.: Twitter sentiment analysis: the good the bad and the OMG! In: Fifth International AAAI Conference on Weblogs and Social Media, pp. 538–541 (2011)

10. Callison-Burch, C.: Fast, cheap, and creative: evaluating translation quality using amazon's mechanical turk. In: Proceedings of the 2009 Conference on Empirical Methods in Natural Language Processing EMNLP 2009, vol. 1, issue 1, pp. 286–295 (2009)

11. Dawid, A.P., Skene, A.M.: Maximum likelihood estimation of observer error-rates using the EM algorithm. J. Roy. Stat. Soc. Ser. C Appl. Stat. JSTOR 28(1), 20–28 (1979)

12. Raykar, V.C., Yu, S., Zhao, L.H., Valadez, G.H., Florin, C., Bogoni, L., Moy, L.: Learning from crowds. J. Mach. Learn. Res. 11, 1297–1322 (2010). JMLR.org

13. Whitehill, J., Ruvolo, P., Wu, T., Bergsma, J., Movellan, J.: Whose vote should count more: optimal integration of labels from labelers of unknown expertise. Adv. Neural Inf. Process. Syst. 22(1), 1–9 (2009)

14. El Maarry, K., Balke, W.-T., Cho, H., Hwang, S., Baba, Y.: Skill ontology-based model for quality assurance in crowdsourcing. In: UnCrowd 2014: DASFAA Workshop on Uncertain and Crowdsourced Data, Bali, Indonesia (2014)

15. Ignjatovic, A., Foo, N., Lee, C.T.L.C.T.: An analytic approach to reputation ranking of participants in online transactions. In: IEEE/WIC/ACM International Conferences on Web Intelligence and Intelligent Agent Technology-Workshops, vol. 1 (2008)

16. Noorian, Z., Ulieru, M.: The state of the art in trust and reputation systems: a framework for comparison. J. Theor. Appl. Eelectron. Commer. Res. 5(2), 97–117 (2010)

17. Hossain, M.: Users' motivation to participate in online crowdsourcing platforms. In: ICIMTR 2012 – 2012 International Conference on Innovation, Management and Technology Research, pp. 310–315 (2012)

18. Kazai, G.: In search of quality in crowdsourcing for search engine evaluation. SIGIR Forum 44(2), 165–176 (2011)

19. Altman, D.G., Bland, J.M.: Diagnostic tests. 1: sensitivity and specificity. Br. Med. J. 308 (6943), 1552 (1994). (Clinical research edition)

20. Loong, T.-W.: Understanding sensitivity and specificity with the right side of the brain. Br. Med. J. 327(7417), 716–719 (2003). (BMJ Publishing Group Ltd.)

21. Altman, D.G., Bland, J.M.: Diagnostic tests 2: predictive values. Br. Med. J. 309(6947), 102 (1994). (Clinical research edition)

22. El Maarry, K., Balke, W.-T.: Retaining rough diamonds: towards a fairer elimination of low-skilled workers. In: Renz, M., Shahabi, C., Zhou, X., Cheema, M.A. (eds.) DASFAA 2015. LNCS, vol. 9050, pp. 169–185. Springer, Heidelberg (2015)

23. El Maarry, K., Güntzer, U., Balke, W.-T.: Realizing impact sourcing by adaptive gold questions: a socially responsible measure for workers' trustworthiness. In: Dong, X.L., Yu, X., Li, J., Sun, Y. (eds.). LNCS, vol. 9098, pp. 17–29. Springer, Heidelberg (2015)

Modelling Processes of Big Data Analytics

Verena Kantere[✉] and Maxim Filatov

University of Geneva, Geneva, Switzerland
{verena.kantere,maxim.filatov}@unige.ch

Abstract. Analytics tasks in scientific and industrial environments are to be performed in some order that, as a whole, represent the rationale of a specific process on the data. The challenge to process the data is, beyond there mere size, their dispersion and the variety of their formats. The data analysis may include a range of tasks to be executed on a range of query engines, which are created by various users, such as business analysts, engineers, end-users etc. The users, depending on their role and expertise, may need or care for a different level of abstraction with respect to the execution of the individual tasks and overall process. Therefore, a system for Big Data analytics should enable the expression of tasks in an abstract manner, adaptable to the user role, interest and expertise. In this work we discuss the modelling of Big Data Analytics. We propose a novel representation model for analytics tasks and overall processes, that encapsulates their declaration, but, also, their execution semantics. The model allows for the definition of analytics processes with a varying level of abstraction, adaptable to the user role. Our motivation derives from real use cases.

1 Introduction

The analysis of Big Data is a core and critical part in multifarious domains of science and industry. Such analysis needs to be performed on a range of data stores, both traditional and modern, on data sources that are heterogeneous in their schemas and formats, and on a diversity of query engines.

The users that need to perform such data analysis may have several roles, like, business analysts, engineers, end-users, scientists etc. Users with different roles may need different aspects of information deduced from the data. Therefore, the various users need to perform a variety of tasks, like simple or complex data queries, data mining, algorithmic processing, text retrieval, data annotation, etc. Moreover, they may need to perform such tasks in different scheduling schemes, for example short or long-running queries in combinations with a one-time or a continuous output. Finally, the users may differ in their expertise with respect to their data management skills, as well as on their interest in implementation specifics. Thus, a system for Big Data analytics should enable the expression of simple tasks, as well as combinations of tasks, in a manner that describes the application logic of the tasks and is adaptable to the user role, interest and expertise.

© Springer International Publishing Switzerland 2015
J. Wang et al. (Eds.): WISE 2015, Part I, LNCS 9418, pp. 309–322, 2015.
DOI: 10.1007/978-3-319-26190-4_21

To fulfil the above requirements we propose a novel workflow model for the expression of analytics tasks on Big Data. The proposed model allows for the expression of the application logic while abstracting the execution details of tasks and the details on the data formats and sources. The model enables the separation of task dependencies from task functionality, as well as the adaptation of the level of description of execution semantics, i.e. the execution rationale. In this way, the model can be easily, i.e. intuitively and in a straightforward manner, used by any type of user with any level of data management expertise and interest in the implementation. Therefore, using the proposed model, a user is not only able to express a variety of application logics for Big Data analytics, but also to set her degree of control on the execution of the workflow. This means that the model enables the user to express specific execution semantics for parts of the workflow and leave the execution semantics of other parts abstract. The latter are decided by the analytics system at the processing time of the workflow.

This work is part of the ASAP[1] research project that develops a dynamic open-source execution framework for scalable data analytics. Our motivation derives from the requirements of applications in the ASAP project. In the rest of the paper, Sect. 2 defines the workflow, Sect. 3 describes the enhancement of the workflow with detailed execution semantics, Sect. 4 discusses related work and Sect. 5 concludes the paper.

2 Workflow Definition

The goal of the workflow is to enable the expression of the logical definition of user applications, which include *data processing*, i.e. *data accessing* and *computation*, as well as *dependencies* between instances of data processing. Computation may refer to algebraic computation or to more elaborate, algorithmic computation. The workflow models such applications as a graph. The vertices in the graph represent application logic and the edges represent the flow of data. Application logic includes (a) the analysis of data, and (b) the modification of data. Edges are directed and connect the vertices that produce and consume data. The rationale for adopting a graph model for the definition of a workflow is that the latter can enable the expression of application logic in an intuitive manner.

There are three types of vertices in a workflow, namely *root* vertices, *sink* vertices and *plain* vertices. The root vertices have only outgoing edges and they represent entry points of the application logic. Figure 1 explains the notation in all the figures that represent workflows and workflow parts. We require that each workflow has at least one root vertex. The sink vertices have only incoming edges and they represent final points of the application logic. We do not require that each workflow has one or more sink vertices. The vertices that are not of type root or sink, are plain vertices, which means that they have both incoming and outgoing edges. For applications that include many phases of data modifications or analysis, we expect that most vertices in respective workflows are plain, as

---

[1] http://www.asap-fp7.eu.

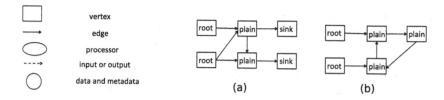

Fig. 1. Notation for workflows

Fig. 2. Workflow examples

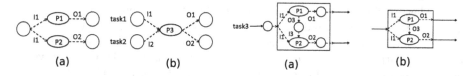

Fig. 3. Task examples

Fig. 4. A vertex with multiple tasks

they represent points in the application logic where data are both produced and consumed. Workflows that do not have sink vertices are those that express an application logic of continuous execution. It is easy to see that workflows without sink vertices are graphs with cycles. Figure 2a shows a workflow with two root and two sink vertices. Figure 2b shows a workflow with no sink vertices, and, therefore, a cycle. Nevertheless, a workflow may comprise both acyclic sub-graphs and sub-graphs with cycles. A trivial case of such a workflow is one that expresses the logic of continuous querying that also outputs processed data, e.g. some final results to be archived. The formal definition of a workflow is the following:

Definition 1. *A workflow is a directed graph $G = (\mathcal{V}, \mathcal{E})$ where $\mathcal{V} = \mathcal{V}_r \cup \mathcal{V}_s \cup \mathcal{V}_p$ is a set that consists of three sets of vertices, the root $\mathcal{V}_r$, sink $\mathcal{V}_s$ and plain $\mathcal{V}_p$ vertices. The three sets do not overlap, i.e. $\mathcal{V}_r \cap \mathcal{V}_s \cap \mathcal{V}_p = \emptyset$, and there should be at least one root vertex, i.e. $\mathcal{V}_r \neq \emptyset$. Also, $\mathcal{E} = \{E_1, \ldots, E_m\}$ is a set of edges. An edge $E \in \mathcal{E}$ is an ordered pair of vertices, i.e. $E = \{(V_i, V_j)|V_i, V_j \in \mathcal{V}\}$.*

Vertices and edges of workflows have properties. The properties of a vertex are related to tasks of the application part represented by this vertex, as well as corresponding metadata. The properties of an edge are related to data flow, and respective metadata, represented by this edge.

2.1   Vertices

Each vertex in a workflow represents one or more tasks of data processing. Each task T is a set of *inputs, outputs* and a *processor*. An input inputs data to a processor; the latter represents the core of the data processing of the task, and, furthermore, an output outputs data generated from the processor. Therefore, inputs and outputs are related to descriptions of data and respective metadata. Figure 3 shows task examples. Figure 3a shows two tasks that have a shared input and one output each. Figure 3b shows a task with two inputs and two outputs.

Definition 2. *A vertex $V \in \mathcal{V}$ corresponds to non-empty set of tasks $\mathcal{T} \neq \emptyset$ such that each task $T \in \mathcal{T}$ is a set of inputs $\mathcal{I}$, outputs $\mathcal{O}$ and a processor P, i.e. $T = \{\mathcal{I}, \mathcal{O}, P\}$. Each input $I \in \mathcal{I}$ and output $O \in \mathcal{O}$ is a pair of data D and metadata descriptors M, i.e. $I = (D_I, M_I)$ and $O = (D_O, M_O)$.*

As defined, a vertex may represent one or multiple tasks of the application logic. These tasks may share or not inputs, but they do not share processors and outputs. The inputs and outputs of the tasks of a vertex can be related to incoming and outgoing edges of this vertex, but they do not identify with edges: inputs and outputs represent consumption and production of data, respectively, and edges represent flow of data. Similarly, vertices do not identify with processors. This semantic differentiation is necessary in order to allow the management of the dependencies in the workflow through graph manipulation, separately from the management of data processing and computation in the workflow. Hence, it is easy to see that a vertex of any type, root, sink, or plain, may consist of tasks with non-empty sets of inputs and outputs, since the latter do not imply the existence of incoming or outgoing edges, respectively. The incoming and outgoing edges of vertices are related in a 1-1 fashion with inputs and outputs, respectively, of vertices. Therefore, if $\mathcal{E}_I$ and $\mathcal{E}_O$ are the sets of incoming and outgoing edges, respectively of a vertex V, and $\mathcal{T}$ is the corresponding set of tasks, then $|\mathcal{E}_I| \subset |\cup_{T.\mathcal{I}}|$, $\forall T \in \mathcal{T}$ and $|\mathcal{E}_O| \subset |\cup_{T.\mathcal{O}}|$, $\forall T \in \mathcal{T}$.

Figure 4 shows an example of a vertex with two tasks. (Figure 4a is the detailed representation and Fig. 4b is a simplified representation where the cycles that represent data are omitted). The tasks share input $I1$ and each has one output, $O1$ and $O2$ that each are an input to an edge outgoing from this vertex. The input $I1$ is the output of an edge incoming to this vertex. Also, one of the tasks has one more output, $O3$, which is an additional input, $I3$, of the other task. The input/output $O3/I3$ is not related with any edge, meaning that these data are not input to tasks that correspond to any dependent vertex. Figure 5 shows a vertex with one task, which creates a histogram of the input data. The task outputs the histogram and additional statistics. These two outputs are separated and are input to two different edges to feed two different tasks. The histogram is further processed and the statistics are logged.

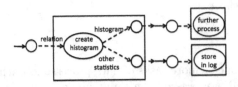

Fig. 5. A vertex with two outputs

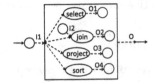

Fig. 6. A vertex for a SQL query

A vertex needs to correspond to at least one task, but it can also correspond to more than one task of the application. Such tasks may or may not adhere to any sort of relation, e.g. concerning associations or similarities of their inputs, outputs or processors. Nevertheless, the reason why the proposed model allows the definition of vertices with multiple tasks, is to enable the user to express such associations or similarities. Therefore, the definition of a vertex so that it consists of multiple tasks, enables the definition of workflows that are intuitive with respect to the rationale of the application logic.

Figure 6 shows a vertex that represents a SQL query. The vertex includes separate tasks for different parts of the SQL query. All the tasks share the input data, and one of them, the task that represents a *join*, has an additional input. Each task has an output. Note that the output data of this vertex, which is the input to the outgoing edge, is actually the output data after executing the whole group of tasks represented by this vertex. Notably, the output of the vertex can be any of the $O1$, $O2$, $O3$, $O4$, depending on the execution plan of this group of tasks. Allowing the user to define vertices with multiple tasks, enables her to represent in a vertex part of the application logic that she considers to be, conceptually, one unified complex task, without requiring her to define at the same time the way that this complex task should be executed, i.e. the execution semantics of it.

The processors in tasks realise the application logic, which is, as mentioned earlier, the analysis or the modification of data. Section 2.4 gives details on the concept of the processor and discusses proposed instantiations.

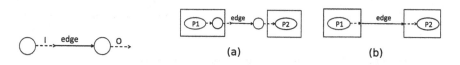

Fig. 7. An edge   Fig. 8. Edge connecting two vertices

2.2   Edges

Each edge in a workflow corresponds to a pair of an input I and an output O of the same data D. As mentioned, the I and the O of an edge correspond in a 1-1 fashion to an I and an O, respectively of a task. The data D are accompanied by metadata M, which can be different for the input and the output of the same edge. Figure 7 depicts the input and the output of an edge. Figure 8 shows an edge that connect two vertices, with one task each. The output of one task becomes the input of the other, via the dependency created by the edge connecting the two vertices. (Figure 8a shows the detailed representation of this example, and Fig. 8b shows the simplified representation of this example, where the cycles representing the data are omitted.) Formally:

Definition 3. *An edge* $E = (V_i, V_j)$, $V_i, V_j \in \mathcal{V}$, *in the workflow corresponds to a pair of an input and an output* (I, O). *Input* I *is a pair of data* D *and some metadata* M_I, *i.e.* $I = (D, M_I)$, *and output* O *is a pair of data* D *and some metadata* M_O, *i.e.* $O = (D, M_O)$. *Input* I *is equivalent with an output* O' *of a task that corresponds to vertex* V_i, *i.e.* $\exists T \in V_i.\mathcal{T}, \exists O' \in V_i.T.\mathcal{O}$ *such that* $I \equiv O'$. *Also, output* O *is equivalent with an input* I' *of a task that corresponds to vertex* V_j, *i.e.* $\exists T \in V_j.\mathcal{T}, \exists I' \in V_j.T.\mathcal{I}$ *such that* $O \equiv I'$.

Hence, an edge defines the flow of data from one vertex to another according to some metadata that describe production and consumption information for these data. The production and consumption information can be the same or different and are related to (a) the data flow (b) the data persistence (c) the data reliability. Other types of metadata may be added in future work. In general, such metadata can be any information that plays a role in determining the execution plan of the workflow.

2.3   Data

The data D of inputs and outputs of edges, as well as of inputs and outputs of tasks consists of information on the data source where these data reside, as well as information on the data *unit*. Formally:

Definition 4. *The data* D *of an input* $I = (D, M_I)$ *or an output* $O = (D, M_O)$ *is a set* $D = \{S, u, \mathcal{A}\}$, *where* S *is the data source, u is the basic data unit and* $\mathcal{A}$ *includes additional information. The data source is a pair* $S = (n, t)$ *of the name n and the type t of the source. The unit u takes values from a constraint domain* $\mathcal{D}$, *which includes the names of the basic units for known types of data sources.*

The type t of a data source can be one of the well known ones, e.g. 'relational', 'rdf', 'xml', 'key-value' etc. The unit for each type is unique and pre-specified; e.g. the unit of the relational type is the 'tuple', the unit for the 'rdf' type is the 'triple' and the unit for the 'key-value' type is the 'pair'. Moreover, data may include the description of additional information, such as relation and attributes names, as well as schema information (e.g. primary and foreign key constraints) and information on the respective processing engine, (e.g. engines of NoSQL databases, relational DBMSs etc.).

2.4   Processors

The tasks included in vertices take as input data and metadata, process the data using a processor and output some data and metadata. Each processor can have an abstract definition and several implementations, i.e. one or more implementations per platform. For example a processor that implements a 'join' for two data inputs, has an abstract definition, and can be implemented for a relational DBMS and a NoSQL database. In order for a processor to be used on a specific platform, it is required that this processor is implemented for the specific platform. The same holds for processors that perform more complex operations, such as algorithmic computation. A processor definition includes restrictions on the type and number of inputs and specifies the number and type of outputs. Defined and implemented processors form a library from which a user can select processors to describe tasks. Users can define their own processors and provide respective implementations, in which input and output data can be in the form of raw bytes/records/key-value pairs etc.

In the following we give examples of the definition of basic processors, namely the *select*, *calc* and *join*:

$$O(select, I) = \{r \mid r \in I \ \wedge \ SelectPredicate(r)\}$$
$$O(calc, I) = \{r \cup \{attr : value\} \mid r \in I \ \wedge \ value := CalcExpression(r)\}$$
$$O(join, I_1, I_2) = \{t \cup s \mid t \in I_1 \ \wedge \ s \in I_2 \ \wedge \ JoinPredicate(t \cup s)\}$$

The input and output data of a processor are accompanied by metadata that describe their type, format and other characteristics. The metadata defined for each processor have a generic tree format (JSON, XML etc.). In order to allow for extensibility, the first levels of the meta-data tree are predefined; yet, users can add their ad-hoc subtrees to define customized processors.

3   Workflow Execution

A workflow represents the dependencies among processing tasks that analyse or modify data, as well as the input and output data of these tasks, together with respective metadata. The defined workflow structure allows for the user to depict the application logic in mind in a straightforward and intuitive manner. This is achieved with two design choices for the workflow structure: (a) the semantic abstraction and separation of the description of processing tasks from the dependencies of processing tasks, and (b) the association of vertices with one or multiple tasks.

The first choice enables the user to describe the application logic and the processing units in the application independently, allowing for easy changes and updates of the workflow structure, as well as a modular and gradual definition of a workflow. This choice also allows the user to be agnostic with respect to the execution semantics of the dependencies between tasks. This execution semantics is determined based on the combination of input and output data and metadata of edges, and will be discussed in the following.

The second choice enables the user to depict in the workflow structure the semantic dependencies of processing units with the depiction of edges and vertices, allowing her to be agnostic on the execution semantics of the set of processors that correspond to a single vertex. Therefore, a user can define a vertex with multiple processors, which, as a group, define a complex operation on the data. Such an operation may be a traditional way of data querying, for example, a vertex may correspond to a SQL query of the Select-Project-Join form; or the

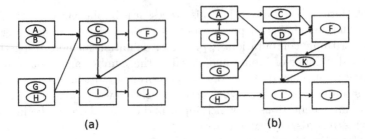

Fig. 9. The original and analysed version of a workflow

operation may be a processing module that comprises simple and complex computation units, like algorithms, for example a data mining algorithm and some sorting of the output data.

3.1 The Analysed Workflow

The workflow structure alleviates from the user the burden of determining any execution semantics for the application logic. The execution semantics of the workflow includes the execution of tasks of vertices and the execution of input-output dependencies of edges. The determination of the execution semantics of vertices and edges leads to an execution plan of the workflow. We refer to this plan as the *analysed* workflow. The latter is actually an enhancement of the initial workflow with more vertices, and substitution of vertices and/or edges in the initial workflow with others.

More specifically, in the analysed workflow, an edge with different input and output metadata, may be replaced with two edges and a new vertex; the new vertex corresponds to a new task that takes the data and metadata of the input of the initial edge and produces the data and metadata of the output of the initial edge. In other words, since the data of the input and the output of an edge are equivalent, this task changes only the metadata. Such vertices are *associative*, as they encompass associative tasks. Also, a vertex that includes multiple tasks, in the original workflow, is replaced, in the analysed workflow, with a set of new vertices that each includes one task of the original vertex. The new vertices may or may not be connected with new edges.

Figure 9 shows an example with the original and the analysed version of a workflow. The original workflow in Fig. 9a has 9 tasks, 3 vertices with two tasks each and 3 vertices with 1 task each. The analysed workflow has more vertices: each one of the 3 vertices with two tasks are replaced with two vertices with 1 task each; also, the edge connecting vertices with tasks F and I is replaced with two more edges and an associative vertex, which includes the associative task K. The analysed version shows that the execution of the vertex with tasks A, B is planned as: first, execution of task B; second, the output of B is input to task A; and third, execution of A. The analysed version also shows that the dependency of tasks C, D on tasks A, B in the original version, means that the output of task A is input to both tasks C and D, which are executed in parallel. Furthermore, the dependency of task F from tasks C, D in the original version, is executed with the input of the outputs of both C and D to task F.

Definition 5. *An associative vertex V_a corresponds to an associative task $T_a = \{I_a, O_a, P_a\}$, where $I_a = (D, M_I)$ and $O_a = (D, M_O)$.*

An associative vertex together with a pair of new edges replaces an edge in the initial workflow. Such new edges are also called associative. We call this triple the *associative triple* of an edge.

Definition 6. *An associative triple $A = (V_a, E_a, E'_a)$ of an edge $E = (V_i, V_j)$ is a set that consists of an associative vertex V_a and a pair of associative edges E_a, E'_a, where $E_a = (V_i, V_a)$ and $E'_a = (V_a, V_j)$. If E, E_a and E'_a correspond to (I, O), (I_a, O_a) and (I'_a, O'_a), respectively, then it holds that $I \equiv I_a \wedge O_a \equiv I'_a \wedge O'_a \equiv O$.*

Essentially, the eventual replacement of edges with associative triples realises the execution semantics of the replaced edge, by creating, through the processor of the associative vertex, an explicit execution plan of the dependency represented by the replaced edge. In Sect. 3.2 we discuss the types of processors for associative vertices. Furthermore, a vertex that corresponds to multiple tasks is replaced with an *associative subgraph* that contains a set of new vertices. The latter correspond to the tasks of the initial vertex: each new vertex corresponds to one task; vertices may correspond 1-1 to tasks, but it can be the case that two or more vertices correspond to the same task[2]. The incoming edges of the initial vertex may have to be replicated, since they may correspond to the input of more than one tasks. The outgoing edges, however, remain the same, as each corresponds to the output of one task. The replacing subgraph may also contain new edges that connect the replacing vertices. Such edges represent task dependencies related to their execution semantics, which are not originally specified by the user.

Definition 7. *An associative subgraph $G_a(V_a, \mathcal{E}_a)$ of a vertex V consists of a set of new vertices V_a and a set of new edges $\mathcal{E}_a$. If V corresponds to a set of tasks T, then it holds that $\forall V \in V_a, \exists T \in T$ such that V corresponds to T, and $\cup_{V.T,} \equiv T$.*

Hence, the analysed workflow is the initial workflow where some edges and vertices are replaced by associative triples and subgraphs, respectively.

Definition 8. *An analysed workflow is a directed graph $G^A(V^A, \mathcal{E}^A)$, where $V^A = V \cup V_{new} - V_{rep}$ and $\mathcal{E}^A = \mathcal{E} \cup \mathcal{E}_{new} - \mathcal{E}_{rep}$. The set $V_{rep} \subseteq V$ includes the replaced vertices and the set $\mathcal{E}_{rep} \subseteq \mathcal{E}$ includes the replaced edges in a*

---

[2] Replication of tasks using many associative vertices that correspond to the same task of an original vertex may be needed for optimisation of workflow execution.

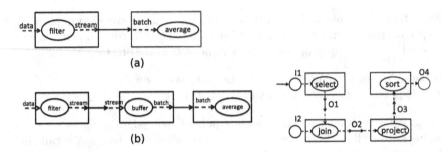

Fig. 10. Example of a scheduling task

Fig. 11. Example of an analysed work-flow for a vertex that represents a SQL query

workflow $G(\mathcal{V}, \mathcal{E})$. *It holds that* $\forall V \in \mathcal{V}_{rep} \;\exists\; G_a(\mathcal{V}_a, \mathcal{E}_a)$ *and* $\forall E \in \mathcal{E}_{rep} \;\exists\; A = (V_a, E_a, E'_a)$. *Also, it holds that* $\cup_{\mathcal{V}_a, V_a} \equiv \mathcal{V}_{new}$, *where* $V_a \in \cup_{\mathcal{G}_a}$ *and* $V_a \in \cup_A$; *and* $\cup_{\mathcal{E}_a, E_a, E'_a} \equiv \mathcal{E}_{new}$, *where* $E_a, E'_a \in \cup_A$ *and* $\mathcal{E}_a \in \cup_{\mathcal{G}_a}$.

The analysed workflow represents the execution semantics of the application logic represented by the initial workflow.

3.2   Execution Semantics of Edges

The input and output of an edge describe data, and information concerning the production and consumption, respectively, of these data. This information, as discussed in Sect. 2.2 can be related to properties concerning the flow, the persistence and the reliability of data. New properties can be added.

The qualitative or quantitative difference in the values of the same property for the input and output of an edge implies some sort of compatibility or incompatibility between the tasks in the connected vertices, and creates constraints for the execution of the dependency represented by the edge. Therefore, the property values of the input and output metadata require appropriate execution semantics of the edge. An associative triple replaces, in the analysed form of the initial workflow., this edge. The associative triple has a vertex that corresponds to a new task, which realises the execution semantics of the input and output metadata of the initial edge. This is an associative task and includes an associative processor. An associative processor is dedicated to a specific associative task. These tasks are categorized as follows:

Scheduling Tasks. These tasks realise data flow patterns, by scheduling the propagation of data from the production to the consumption vertex:

- Sequential data propagation: The data are propagated from input to output in a sequential manner. For example, the trivial case of an edge that has $I = O = (D, \{`stream', rate = 1\,\mathrm{ms}\})$, necessitates a task that realises the execution semantics of a pipeline.

– Concurrent data propagation: The data are propagated from input to output in a concurrent manner. For example, the trivial case of an edge that has $I = O = (D, \{'batch', size = 10\,MB\})$, necessitates a task that realises the execution semantics of a buffer of size that equals $10\,MB$.

Figure 10 shows the associative triple that substitutes an edge that connects two vertices with a filtering task and a task that computes an average value. The triple includes a new edge that connects the filtering task with the associative task, and a new edge that connect the associative task with the *average* task. The associative task performs buffering of the streaming input data, so that they can be input as a batch in the *average* task. The data, but also the flow metadata of the input and output for both new edges are the same: the left edge has as input and output streaming data, whereas the right edge has as input and output batch data.

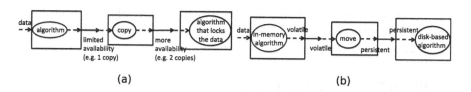

(a)   (b)

Fig. 12. Example of availability tasks

Availability Tasks. These tasks realise patterns of data persistence, by replicating and moving data in order to change their availability as formed by the production vertex and guarantee their availability as requested by the consumption vertex. Such tasks may realise degrees of:

– Data replication: The produced data are replicated in order to be available for consumption. The replication may be performed for several reasons. In case of produced data that are volatile, i.e. they are stored in memory, they are replicated in permanent storage; in case of produced data that have some degree of persistence in permanent storage, they are replicated in the same storage in order to increase their availability in terms of time; in case of produced data that are totally persistent, they are replicated in the same or other storage in order to increase availability in terms of processing throughput. Figure 12a shows the associative triple that replaces the edge connecting two algorithmic tasks. The associative task makes a copy of the data output by the first algorithm, so that one copy is available for other tasks to read, while one copy is locked by the dependent algorithmic task.

– Data movement: The produced data are moved in order to be available for consumption. The movement may be performed for several reasons. Persistent data may be moved to other storage in order to be closer to the processing in the consumption vertex and, therefore, increase availability in terms of processing throughput. Also, persistent data may be moved in order to reduce the risk of unavailability due to source or engine failures. Figure 12b shows the associative triple replacing the edge connecting an in-memory and a disk-based algorithmic task. The associative task moves the data from the main memory to the disk, so that the dependent algorithm can access them.

Cleaning Tasks. These tasks realise patterns of data reliability, by controlling and mending the quality of the data as formed by the production vertex and guarantee their reliability as requested by the consumption vertex[3]:

- Data checking: The produced data are checked for the correctness and completeness in order to decide if the reliability requested by the processing in the consumption vertex can be guaranteed.
- Data mending: The produced data are processed in order to be corrected and/or completed, so that the reliability requested by the processing in the consumption vertex can be guaranteed.

Each task category corresponds to a library of processors that implement specific instances of this category. Such a library contains actually code for processors, and it can be extended with new processors.

3.3 Execution Semantics of Vertices

A vertex in the initial workflow corresponds to a set of tasks that are defined, by the user, to be performed as a unit in the application logic. The degree of detail, in terms of processing, of such a set, depends on the nature of the application, the user role, the user experience and knowledge in creating workflows etc. This means that different users may describe the same part of an application logic in different granularity, and, consequently, by defining a different set of vertices with a small or big number of tasks each. For example, such an application logic may refer to a SQL query, with a *Select-Project-Join-Sort* form. If the user is a business analyst, then she may define one single vertex that corresponds to all four tasks, i.e. *Select, Project, Join, Sort*, as in Fig. 6, or, if the user is a programmer, she may define one vertex per task, as in Fig. 11. In the last case, the user has to also define a set of edges that connect the four vertices. This set of edges, (which may be empty, in case of parallel execution), in essence, indicates how the user perceives the execution semantics of the tasks. Therefore, the user may impose the execution semantics for a set of tasks, or she may be agnostic to this. In the last case, the execution semantics of these tasks should be defined in the analysed workflow. This is achieved by replacing the initial vertices with associative subgraphs, that define the respective execution semantics. In the example of the SQL query above, the vertex corresponding to the four tasks is replaced by a subgraph that contains one vertex per task. This subgraph is actually a query execution plan, e.g. created by a DBMS optimiser. The plan structure depends on the tasks and on the type of the execution engine.

---

[3] Note that such tasks may involve also human interaction and may be performed online or offline.

4 Related Work

Artemis [1] queries multiple heterogeneous data sources using ontologies and metadata, and integrates metadata in terms of semantics. The proposed workflow model enables the creation and execution of associative tasks that process and integrate intermediate results.

HFMS [2] builds on top of previous work on multi-engine execution optimization [3]. Their study is more focused on optimization and execution across multiple engines. The design of flows in HFMS is agnostic to a physical implementation. HFMS handles flows as DAGs (i.e. Directed Acyclic Graphs) encoded in xLM, a proprietary language for expressing data flows. The proposed workflow model aims at modularity of workflow manipulation, expressibility of application logic, and adaptability to the user interests and role, goals that are out of the scope of the HFMS flow model definition.

Pegasus [4] is another workflow management system that allows users to easily express multi-step computational tasks. The workflow description is separated from the description of the execution environment [5,6]. This allows the system to perform optimizations at 'compile time' and/or at 'runtime'. A drawback of this approach is that the executing workflow be different than what the user anticipated when she submitted the workflow. As a result, in Pegasus a lot of effort is devoted toward developing a monitoring and debugging system that can connect the two different workflow representations in a way that makes sense to the user. The proposed workflow model overcomes such problems, by separating the definition of the dependancies and the processing tasks in the application logic. In this way, the user controls the detail of execution semantics she describes.

Taverna [7] is an open source domain-independent workflow management system, which includes a suite of tools used to design and execute scientific workflows. Research in [8] is focused on the issue of the analysis of data from heterogeneous and 'incompatible' sources. While Taverna includes tools for the composition and enactment of bioinformatics workflows, the composition of workflows is done through a graphical user interface and does not provide sophisticated methods for their efficient execution.

Apache Tez [9] is an extensible framework for building high performance batch and interactive data processing applications. Tez models data processing as a DAG. The task design is based on inputs and outputs that exist in pairs. In the proposed model inputs and outputs of dependent tasks are not connected. Our model dictates that inputs and outputs of edges, and not of tasks, represent the dependencies between tasks and realise the relation between their inputs/outputs. In this way, tasks are inherently independent, allowing modular manipulation of tasks and task groups, as well as separate manipulation of task execution and dependencies in the application logic.

The Stratosphere project [10] tackles the challenge of executing workflows with the PACT programming model, based on the Nephele execution engine [11]. This approach introduces the notion of workflows in cloud-based systems, but the solution is not mature enough to give the necessary efficiency or full-fledged capabilities of adaptive execution.

5   Conclusion

This paper discusses the modelling of processes that perform Big Data analytics. It proposes a novel workflow model for the expression of such analytics tasks. The model enables the separation of task dependencies from task functionality. Employing the model, a user is able to express a variety of application logics and to set her degree of control on the execution of the workflow. The originally defined workflow is augmented with associative tasks and task dependencies that specify missing execution semantics. The analysed form of the workflow is ready for execution. Ongoing and future work focuses on defining methods for the manipulation of the workflow structure in order to optimise its execution semantics.

Acknowledgment. The research leading to these results has received funding from the European Union Seventh Framework Programme (FP7/2007-2013) under Grant Agreement n° 619706 ASAP.

References

1. Tuchinda, R., Thakkar, S., Gil, Y., Deelman, E.: Artemis: integrating scientific data on the grid. In: IAAA, pp. 25–29 (2004)
2. Simitsis, A., Wilkinson, K., Dayal, U., Hsu, M.: HFMS: managing the lifecycle and complexity of hybrid analytic data flows. In: ICDE 2013, pp. 1174–1185 (2013)
3. Simitsis, A., Wilkinson, K., Castellanos, M., Dayal, U.: Optimizing analytic data flows for multiple execution engines. In: SIGMOD 2012, pp. 829–840 (2012)
4. Pegasus. http://pegasus.isi.edu/
5. Deelman, E., Vahi, K., Juve, G., Rynge, M., Callaghan, S., Maechling, P.J., Mayani, R., Chen, W., da Silva, R.F., Livny, M., Wenger, K.: Pegasus: a workflow management system for science automation. Future Gener. Comput. Syst. **46**, 17–35 (2015)
6. Malawski, M., Juve, G., Deelman, E., Nabrzyski, J.: Cost- and deadline-constrained provisioning for scientific workflow ensembles in IaaS clouds. In: SC 2012, pp. 22:1–22:11 (2012)
7. Oinn, T., Addis, M., Ferris, J., Marvin, D., Carver, T., Pocock, M.R., Wipat, A.: Taverna: a tool for the composition and enactment of bioinformatics workflows. Bioinform. **20**, 3045–3054 (2004)
8. Wolstencroft, K., Haines, R., Fellows, D., Williams, A.R., Withers, D., Owen, S., Soiland-Reyes, S., Dunlop, I., Nenadic, A., Fisher, P., Bhagat, J., Belhajjame, K., Bacall, F., Hardisty, A., de la Hidalga, A.N., Vargas, M.P.B., Sufi, S., Goble, C.A.: The taverna workflow suite: designing and executing workflows of web services on the desktop, web or in the cloud. Nucleic Acids Res. **41**, 557–561 (2013)
9. Apache tez. http://hortonworks.com/hadoop/tez/
10. Alexandrov, A., Bergmann, R., Ewen, S., Freytag, J.-C., Hueske, F., Heise, A., Kao, O., Leich, M., Leser, U., Markl, V., Naumann, F., Peters, M., Rheinländer, A., Sax, M.J., Schelter, S., Höger, M., Tzoumas, K., Warneke, D.: The stratosphere platform for big data analytics. VLDB J. **23**(6), 939–964 (2014)
11. Battré, D., Ewen, S., Hueske, F., Kao, O., Markl, V., Warneke, D.: Nephele/pacts: a programming model and execution framework for web-scale analytical processing. In: SoCC 2010, pp. 119–130 (2010)

WISEngineering: Achieving Scalability and Extensibility in Massive Online Learning

Xiang Fu[1](✉), Tyler Befferman[1], Jennie Chiu[2], and M.D. Burghardt[1]

[1] Hofstra University, Hempstead, NY 11549, USA
{Xiang.Fu,M.D.Burghardt}@hofstra.edu, tbeffe1@pride.hofstra.edu
[2] University of Virginia, Charlottesville, VA 22904, USA
jlchiu@virginia.edu

Abstract. Massive Open Online Courses (MOOCs) have raised many unique challenges to online learning platforms. For example, the low teacher-student ratio in MOOCs often means lack of feedback to students and poor learning experiences. We present WISEngineering, a MOOCs platform that provides a rich set of features for overcoming these challenges. The system embraces social media for fostering student reflection. Its automated grading system adopts an open-architecture and uses stack generalization to blend multiple machine learning algorithms. A Zookeeper based computing cluster runs behind auto-grading and provides instant feedback. A behavior tracking system collects user behavior and can be later used for learning outcome analysis. We report the design and implementation details of WISEngineering, and present the design decisions that allow the system to achieve performance, scalability and extensibility in massive online learning.

Keywords: Online learning platform · Automated grading · Web application · Scalability · Extensibility

1 Introduction

For thousands of years, humankind has been trying to lower the cost of education, for making it more accessible. Massive open online courses (MOOCs) [18] are the latest attempt. MOOCs have great potential to revolutionize how people learn and how people teach. Using MOOCs in engineering classes, however, faces several challenges. Distance learning lacks face to face social interaction and larger class size can often contribute negatively to learning outcomes [1]. These problems are magnified in engineering education when learners need to be deeply engaged in hands-on environments and the feedback from peer learners and teachers is important.

We present WISEngineering [3,4], a distributed and web-based MOOCs platform that intends to address the above challenges. WISEngineering embraces social media computing for encouraging learner engagement. Its mobile portal, running as a Google Chrome application, provides easy access in a hands-on

© Springer International Publishing Switzerland 2015
J. Wang et al. (Eds.): WISE 2015, Part I, LNCS 9418, pp. 323–337, 2015.
DOI: 10.1007/978-3-319-26190-4_22

lab environment using 7 inch Android tablets. In particular, WISEngineering offers features that are available in a typical social media website, for learners to scaffold engineering design.

Automated and instant feedback is the key to providing quality learning experiences in WISEngineering. The system adopts an open architecture that blends a variety of automated grading algorithms and modules. It can be trained by providing manual grading samples, and be further calibrated at run time. A learning outcome system is built upon the automated grading system. All questions are tagged with learning outcome goals, and a weighted sum formula can be defined to take into account various aspects of a learning process. A reporting system is available for performing comparative study of learning outcomes of any selected learner(s).

WISEngineering adopts component based software engineering [13], and it builds the features set by integrating components from a number of open source traditional and MOOCs web platforms. For example, its course and user management system is centered around the WISE system from UC Berkeley [2,22]. Its automated grading module uses the EASE module from edX [5]. Its mobile portal uses the responsible style template from Twitter [17]. Its instant user feedback module is built upon Apache HDFS and Zookeeper [9]. The video processing uses Google cloud and it is optimized by a local load balancer.

To construct a heterogeneous software system like WISEngineering is challenging. An open and extensible architecture is required for accommodating a wide variety of web and mobile application technologies and languages in one general framework. Caution has to be exercised when wrapping up the components, while at the same time, it has to provide authentication/security, atomic transaction, synchronization, and aggregation of data.

Efficiency and scalability are the key requirements of WISEngineering. For instance, to support hands-on engineering experiments, the system has to process large quantity of video data. For another example, to provide instant feedback to users, multiple auto-graders have to run alive, with each consuming large amount of RAM resources. The system has to address these issues by leveraging parallel and distributed computing techniques.

This paper reports the rich feature set as well as the design trade-offs and implementation details of WISEngineering. Section 2 introduces the system features of WISEngineering. Section 3 presents general architectural decisions of the system. Section 4 discusses how extensibility is accomplished in design. Section 5 addresses system performance and scalability. Section 6 discusses related work, and Section 7 concludes.

2   System Features

To better understand the architectural and design decisions in developing WISEngineering, we present a number of its important system features. A working copy of the system is available at [3]. Section 2.1 first presents the Web-based Inquiry Science Environment (WISE) system from the University of California,

Berkeley [2], upon which WISEngineering is built. Then, the rest of the section discusses the new features.

2.1 Existing Features Provided by WISE

WISE [2], like other educational platforms such as Blackboard [11] and Moodle [14], provides core functions such as user registration, curriculum development and manual grading. The WISE environment has been developed based on extensive research in classroom instruction. It uses rich online interactive plug-ins (such as PhET [25]) which are used by students to experiment with scientific concepts in hands-on exploration.

WISE provides a hosting platform for many science and engineering educational initiatives, such as the WISE Guys & Gals project (WGG) [3]. WGG introduces middle school age youth to innovative and engaging blended STEM based engineering design activities. Each activity is framed to expose youth to an engineering discipline (e.g. Mechanical, Electrical, and Civil).

WISE provides authoring tools for developing curriculum materials. In WISE, each activity is structured as a tree view of learning steps, where a step can either be an HTML page that presents a scientific concept, or an assessment step that collects a student's feedback and reflection. WISE supports typical assessment approaches such as short answer, multiple-choice, match and sequence, and discussion. Assessment steps such as multiple-choice can be automatically graded by WISE, however, at this moment, short answer questions still have to be manually graded by teachers.

2.2 WISEngineering Mobile Portal

Powered by WISE technologies [2], the WISEngineering system has incorporated a number of important system features for meeting the challenge of massive online learning. The first addition is a mobile portal for students.

The mobile portal is a Google Chrome Web application that ports the major functions of WISE to tablets. Students no longer have to walk between their computers and workbench for data entry and analysis. For example, students can use data plotting and tabulating tools on tablets to analyze and visualize data on the spot. In particular, WISEngineering embraces social media computing to engage students in reflection and collaboration. Tools such as the design journal and design wall are used to share ideas and designs. The mobile tools allow students to take live pictures and videos, which puts demands on the system to process large video uploads efficiently.

2.3 Automated Grading

WISEngineering uses an open architecture to embrace automated grading technologies such as the edX EASE engine [19]. This section presents the user interface of the system. Later, Sect. 4.2 presents the details of design and implementation.

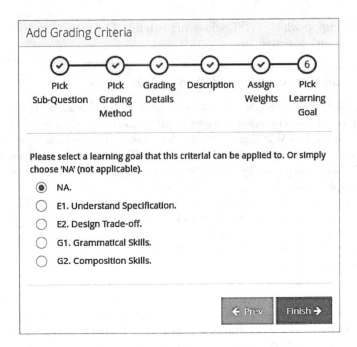

Fig. 1. Grading Criteria

Learning Goals and Grading Criteria. WISEngineering supports outcome-based education [24]. Before the curriculum is developed, a set of learning outcome goals can be entered into the system. Later, they are associated with assessment questions in activities. As an example, Fig. 1 presents the user interface for defining a grading criteria. A curriculum developer can choose to use the question as an indicator for selected learning outcome goals.

Training. Each automated grading criteria has to be trained and calibrated. At any moment, the system keeps two sets of data: one training set and one calibration set. Each training/calibration sample consists of three parts: a student response, the grade assigned by a human grader, and a grade assigned by the AI grader (in calibration data only). The training set is used to generate the grading model, while the calibration set is used to measure the quality of the model.

It is recommended that for each grading level, at least ten samples are entered for the grading engine to function correctly. WISEngineering provides tools for creating training samples, and student response samples can be retrieved directly from the database of the WISE system. Figure 2 displays the training statistics of a grading criteria. It shows the progress of data entry (of samples), and the current quality/precision of the model. At the bottom of the page, it also displays the details of those mismatched records in calibration set.

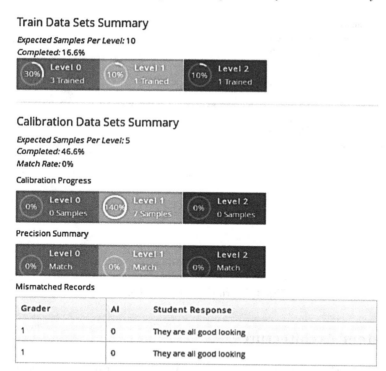

Fig. 2. Training Grading Criteria

Grading. After a model is generated for all criteria, a project can be graded. An automated grading script runs on each server overnight. The script loads each grading model, performs the grading, calculates the weighted sum of the grade, and pushes the grade into the database of the original WISE system. Students can view their grade and the automatic hint/response generated by the AI grader the second day. The default automated grading module is *not* instant because model generation and loading is expensive in both time and space. We present the details of the instant feedback system in Sect. 5.2.

2.4 Learning Outcome Analysis and Reporting System

The system periodically generates learning outcome reports for all groups of learners. Figure 3 shows a part of one sample report. Student activity data are generated by aggregating data from the WISE database. Learning outcomes are computed using the weighted sum formula associated with each question, which serves as an indicator of learning goals. Histograms of learning outcomes and the achievement of each individual learning outcome by each club can be generated and included in the report. The reporting function allows the curriculum developer to perform comparative study of learning outcomes on selected learners over the time.

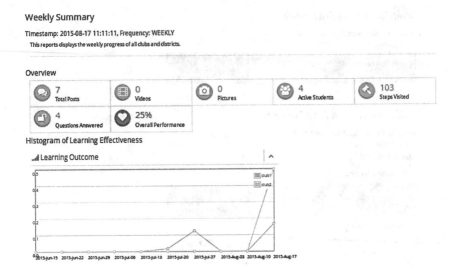

Fig. 3. Learning Outcome Report

3   System Architecture

In addition to those introduced in Sect. 2, WISEngineering has provided many other features, such as a user avatar system, a user behavior tracking system, a video processing cluster, and an instant feedback cluster. To include these heterogeneous software components, while at the same time achieving performance and scalability requires careful design decisions in its software architecture.

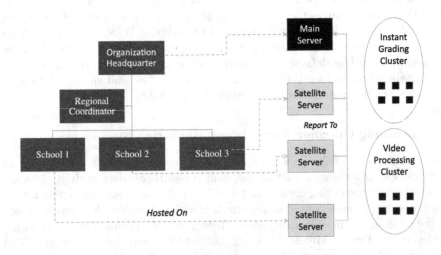

Fig. 4. WISEngineering Network Topology

3.1 Network Topology

The deployment of WISEngineering requires multiple loosely coupled servers and several clusters. Figure 4 displays the WISEngineering network topology.

To all users of the system, the main access point of WISEngineering is the *main server*. It has the identical software stack as all other satellite servers. The only difference is that the main server has a load balancer distributing the incoming traffic, and it is equipped with a report aggregation system.

WISEngineering has the ability to support large-scale educational activities nationwide (e.g., [3]). Such activities are usually organized by national organizations with a hierarchical structure. The main server records the information and utilizes this information to generate aggregated report. For example, to retrieve the number of active students in one particular region, the main server will first query the organizational structure and submit data retrieval requests to the related satellite servers for aggregation.

All servers are supported by two back-end clusters. The video processing cluster accepts video uploads from mobile devices, pre-processes/compresses video files, and stores the media file in the Google Cloud. The design of the instant grading cluster will be described in details in Sect. 5.2.

3.2 Software Stack

Figure 5 displays the software component stack at each satellite server. It addresses the challenge of integrating a wide variety of components and web technologies in one general framework.

The left side of Fig. 5 shows the structure of the original WISE system [2], upon which WISEngineering is built. WISE adopts a typical three tier structure. At the bottom, sits the data tier. It consists of two MySQL databases that store user and session information. Student responses and teacher feedback are also

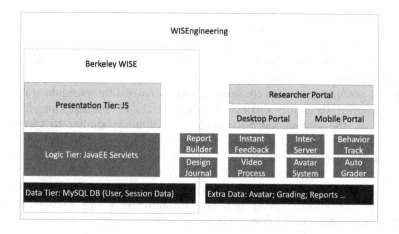

Fig. 5. WISEngineering Software Stack

stored there. The logic layer is implemented as a collection of JavaEE servlets. Then the presentation layer (customized for desktop browsers) renders the data generated by the logic layer using a combination of JQuery and CSS.

The main design goal of WISEngineering is to keep all the original functions of WISE, and extend it with other major functions such as automated grading and a mobile portal. This is achieved using a similar three-tier structure, and making minimal changes to the original WISE source code.

As shown on the right of Fig. 5, the extra WISEngineering components are also structured in three layers. Several more MySQL databases are added to the system to store information for automated grading system, design journal/wall, and a user avatar system. Then an extra logic layer is added, using a variety of platforms such as PHP and Python, depending on the components being integrated. At the presentation layer, three portals are provided: a desktop portal which enriches the original WISE system, a mobile portal which uses the Bootstrap [17] responsive templates for mobile devices, and a special desktop portal for educational researchers to train automated grading system and generate system reports.

Notice that some WISEngineering components, such as the report generator, do have to interact with the data tier of the original WISE environment. The system is designed in a way so that tight coupling is minimized.

4   Achieving Extensibility

This section presents a number of design decisions we made to achieve extensibility in WISEngineering. When new components are "plugged" into WISEngineering, we use a service-oriented approach to wrap-up the standard interface of a component so that it can be invoked by others.

4.1   Inter-Server Communication

In many cases, satellite servers have to exchange information with the main server. One typical example is to generate aggregated reports. For instance, at the main server, when a user requests the total number of video posts in a specific date range, the data has to be retrieved from all satellite servers and then aggregated by the main server.

Report generation is time consuming, mostly caused by huge join statements in SQL queries. To simply retrieve data via an HTTP request can often time out. Instead, the system has to provide a mechanism similar to restful web services [20]. The detailed communication protocol is described below.

We assume $E(k, m)$ is an encryption operation which uses key k to encrypt message m and it is always true that $E^{-1}(k, E(k, m)) = m$. Let the requester be R (e.g., the main server) and the server be S (e.g., a satellite server). If R has access to an operation P on S, there is a common secret $s_{R,P,S}$ shared between R and S. The interaction follows the steps below:

1. $R \rightarrow S$: request for operation P.

2. $S \rightarrow R$: (id, n). Here id is a unique service request identifier and n is a nonce (random number). In the database of S, a new entry is established for service request id where the field of operation_result is left blank. A timeout process is started simultaneously to kill the request operation if it is timed out. At time out, the database entry is removed as well, to avoid denial of service attack.

3. $R \rightarrow S$: ($E(s_{R,P,S}, (\text{id}, \text{n}))$ to prove that R has the access. Here, the use of nonce n is to avoid replay attack.

4. S verifies the access right of R using E^{-1} and starts the operation P. When P completes, it writes the data into operation_result.

5. R will periodically check the status of request id, until the data is available or time out.

On each server, an administrator application is provided for managing keys. We use this lighter weight authentication protocol, instead of Kerberos [12], to avoid failure of single point.

4.2   Open Architecture of Automated Grading

The extensibility of automated grading module is a similar but separate problem. In this case, we would like to make the system extensible, in the sense that, new machine learning algorithms can be added to the system in the future, to further improve the precision of grading.

We use an open architecture and stack generalization algorithm to repeatedly train and blend the results of a collection of automated grading algorithms. Details are given below.

Automated grading is essentially a machine learning problem. Take edX EASE [19] as an example. To train EASE, a user has to provide a training set of samples, where each is a pair of string (student answer) and a number (trainer assigned score). EASE, based on the number of samples (whether greater than 5), takes one of the regression or classification approaches. Using sklearn [21], a machine learning package in Python, EASE builds a classifier for each training set. Then, the classifier (also called the model), can be used for grading, i.e., it generates a numeric score for any string input.

The vector of features extracted for a model and the training algorithm used usually determine the quality of a classifier. For example, the feature set of EASE includes the bag of words (n-grams), length feature (counts of words, punctuation etc.), the number of spelling errors, and the number of grammar errors. This is suitable for essay grading, but may not be ideal for grading short answer questions in a specific technical context in WISEngineering. For another example, the sklearn package provides many different learning algorithms such as support vector machine, nearest neighbors, and Gaussian Processes. A great number of factors can determine the quality of auto-grading. Our framework tries to use one more level of training to find an optimized combination of auto-grading algorithms. We formalize our algorithm below.

Let Σ be the English alphabet, 2^Σ is the domain of input. Let $\mathcal{N}$ be the target range of scores. A classifier $\mathcal{C}$ is a function from 2^Σ to $\mathcal{N}$. Let $2^\mathcal{C}$ be the domain of all classifiers. A training algorithm $\mathcal{T}$ is modeled as $\mathcal{T} : 2^{2^\Sigma \times \mathcal{N}} \rightarrow 2^\mathcal{C}$, which given a training set T, generates a classifier.

Our framework is a 2-level application of the stack generalization by D. Wolpert [27]. The input of the algorithm is a collection of auto-grading (training) algorithms $\mathcal{T}_1, ..., \mathcal{T}_n$. They will be the "ensemble" training algorithms at level 0. Let $\theta = \{e_1, ..., e_k\}$ be the training samples, where each sample is a tuple $e_i = (s_i, n_i)$ where s_i is a string and n_i is the numeric score. Given a training sample, let input(e_i) be its s_i and let score(e_i) be its n_i. At level 1, we use k-NN as a selection algorithm (written as $\mathcal{K}$) that selects the result generated by level 0 algorithms. The training process works as follows:

1. Partition: build a set of partitions over θ. For each partition p_i, it is constructed by picking one training sample from θ. More formally each p_i is a tuple $(p_{i,1}, p_{i,2})$ where $p_{i,2}$ is a singleton element $\{e_i\}$, and $p_{i,1} = \theta - p_{i,2}$.
2. Train level 1 selector $\mathcal{K}$: the classifier generated by $\mathcal{K}$ will be a mapping from $\mathcal{N}^n \rightarrow [1, n]$ (where n is the number of level 0 training algorithms). Intuitively, given a vector of numeric scores produced by all level 0 trainers, the level 1 selector chooses the result produced by one of the level 0 trainers.

   Now we discuss how $\mathcal{K}$ is trained. For each partition p_i, use its $p_{i,1}$ as the training set for each level 0 $\mathcal{T}_j$, we obtain a classifier. Then using the level 1 classifier, we compute the output of $\mathcal{T}_j$ on $p_{i,2}$. Given n level 0 trainers, we have a vector of n output scores, and then we use score($p_{i,2}$) as the ground fact for training. Officially, one training sample of $\mathcal{K}$, derived from partition p_i is defined as below: Let $\mathcal{C}_j$ be the classifier generated by $\mathcal{T}_j$ on $p_{i,1}$, i.e., $\mathcal{C}_j = \mathcal{T}_j(p_{i,1})$. We have a training vector:

$$((\mathcal{C}_1(\text{input}(p_{i,2})), ..., \mathcal{C}_n(\text{input}(p_{i,2}))), \text{score}(p_{i,2}))$$

Now given any question text q, let $\mathcal{C}_\mathcal{K}$ be the classifier generated by the above algorithm. Let $(v_1, ..., v_n)$ be the vector of results generated by level 0 classifiers. Feed the vector to $\mathcal{C}_\mathcal{K}$ we get an index number between 1 and n, let it be i, then the corresponding output is v_i.

In summary, the two level stack generalization algorithm [27] allows the system to extend with an arbitrary number of auto-grading algorithms.

5   Design for Scalability and Performance

System performance and scalability are the key to user satisfaction. This section presents the related implementation details.

5.1   Video Processing

WISEngineering uses a loosely distributed server cluster for handling video upload. The nodes of this cluster do not even have to belong to the same local area network (LAN). Load balancing is achieved using a simple scheme.

The main server keeps track of a list of servers capable of handling video uploads. Each video processing server provides a set of web servlets, implemented using PHP, for uploading video files. Once a file arrives, a video processing servlet compresses the file, and uploads it to the Google/Youtube cloud service. All video processing servers use the same Google/Youtube account. When a user requests a video upload, e.g., in creating a design journal post, an AJAX request is sent to the main server, to ask for a random decision on the video server to use. This random video server is then used for upload and processing. Clearly, this scheme is easily scalable by adding more processing servers to the pool.

5.2   Instant Feedback System

The preliminary automated grading system introduced in Sect. 4.2 is not appropriate to serve as an instant feedback system. To grade a question, the grader has to load a pre-compiled classifier model, which takes at least 30 seconds on a typical server. The grading itself, however, only takes less than a fraction of a second.

One naive solution is to make the grader running as a service/daemon process in OS, and responding to grading request. This solution does not work for WISEngineering because there are many graders and each consumes at least 10 MB of RAM. One single server cannot host them all. On the other hand, grading requests (due to how class sections are run) usually come in bursts. That is, in a short period of time, the majority of grading requests can be for a small subset of questions. We need a flexible way to create a large number of graders and put down unused graders for saving system resources. This solution has to be scalable to add new hardware resources.

General Architecture. The entire instant grading/feedback system (IFS), as shown in Fig. 6, can be deployed in a local area network (LAN) different from the main server. However, all cluster nodes of IFS have to be located in the same LAN, for the convenience of setting up HDFS and Zookeeper services.

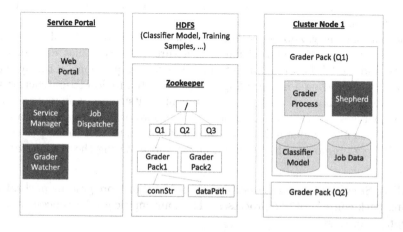

Fig. 6. HDFS/Zookeeper Cluster for Instant Feedback

The only externally visible component of IFS is its web portal. When a student user submits to WISEngineering, an AJAX request is sent to IFS web portal for instant feedback. Once the question is graded, a callback Javascript function is invoked at the client side to display the auto-grader feedback.

Running behind the web portal is the entire IFS cluster. Figure 6 displays its structure. IFS utilizes two Apache cloud frameworks: the HDFS distributed file system and the Zookeeper synchronization service [9]. We use Zookeeper as both a centralized directory service of all of our auto-grader services and a synchronization service for providing atomic and mutual exclusive access of shared data.

IFS Service Portal. The IFS web server and several Java applications (such as IFS Service Manager and Job Dispatcher) are hosted in a physical server called "Service Portal" (as shown on the left of Fig. 6).

The Service Manager is a daemon process. It examines the request/performance statistics of the IFS web portal periodically and decides to start new grading services or decommission unused ones. As the directory information of all grading services are contained in Zookeeper, the Service Manager has write access on the Zookeeper nodes.

The Job Dispatcher, once receiving a grading request (the question ID and a student response) from the WISEngineering main server, generates a new service ticket ID, and locates an available grading service by querying the directory information in Zookeeper. The Grader Watcher service is also a daemon process. It periodically checks the status of all auto-graders by sending them heartbeat messages.

HDFS and Zookeeper. The IFS cluster relies on HDFS and Zookeeper. HDFS serves as a shared file system among all computing nodes. It stores the classifier models for grading, training data, and other pertinent information.

Zookeeper is used for synchronizing the requester processes (such as the Job Dispatcher) and the service processes (such as the auto-graders) that are distributed on different servers.

More importantly, Zookeeper stores a central service directory of all services. As shown in Fig. 6, the information is structured as a tree that embodies the location and model data related to the graders for all questions in the system.

For example, when a new grading job is submitted to Job Dispatcher. It will first create a job node in Zookeeper (documenting its status, unique ticket ID). It then searches for an available grader in the pool of graders and updates its status. When the grader finishes its job, it will update the Zookeeper central directory (e.g., removing the job information and updating the grader status).

Grading Services. For each question, there is one or more grading packs. Each grading pack consists of two processes: (1) an automatic grader process, and (2) a shepherd process.

The grader process is a daemon process. Depending on the grading module it uses, it can be implemented using different languages (e.g., Python). The main body of the process basically runs as an infinite loop, waiting for requests from network sockets, which are sent from the shepherd process. When a request signal arrives, it reads the details of the request, which is prepared by the Shepherd process in advance.

The atomicity/mutual exclusive access of shared local files and classifier models are guaranteed by the Shepherd process (and Zookeeper). When the grader process finishes processing, it writes the output to a local pipe. Then the Shepard process reads the output and transfers it to HDFS, and then updates Zookeeper to wake up the waiting requester.

In summary, the use of HDFS and Zookeeper makes IFS a highly scalable, reliable, and robust system that can handle in-burst requests with reasonable consumption of hardware resources.

6 Related Work

Traditional web based educational platforms, such as Blackboard [11] and Moodle [14] provide standard course management functions, however, lack the support for massive online learning. Emerging platforms such as edX [5] and OpenMooc [16], embrace social media, mobile portal and cloud based video processing services such as YouTube.

WISEngineering has the above features developed in parallel with the aforementioned platforms and it has several unique features. First of all, WISEngineering serves a special sector of massive online learning - large scale implementation of a same curriculum with the support of a national organization (e.g., running Engineering educational activities through Boys & Girls Clubs [3]). In WISEngineering, the same curriculum could be potentially used in hundreds of classrooms in distant geographical regions. These classes are usually small, in contrast to huge classes hosted on edX, which could have thousands of students in one class section. The application context of WISEngineering raises many interesting problems. For example, for the same curriculum, why does one region have better learning outcomes than another? This demands that WISEngineering provides a well-defined learning outcome structure and correlate it with assessment questions and user behavior. It also determines the loosely coupled network topology because organization units may be hosted on different servers. This further leads to the need to aggregate reports from satellite servers and hence the remote information exchange protocol among servers.

WISEngineering uses but goes one step beyond the edX EASE automated grading module [19]. It adopts the stack generalization algorithm [27] for an open architecture to blend multiple machine learning algorithms for grading. Instant feedback system is built upon the auto-graders, and can handle the grading requests in burst. It leverages the highly robust Zookeeper cluster [9] for synchronizing and coordinating distributed auto-grader processes. This is in line with the practice of big cloud systems such as the Eclipse Communication Framework [10] and the Apache HBase [8].

Design for performance and scalability is always one of the greatest challenges faced by web application developers [26]. The REST (Representational State Transfer) pattern [6,7], allowing redundancy and caching, has been followed for decades to scale up services. The design of WISEngineering adopts most of its principles (e.g., the stateless request property), for serving client requests by randomly selecting cluster nodes at run time. WISEngineering also tries to maximize the interoperability of components by standardizing their communication interface, following the principles outlined in [15]. In general, the distributed architecture allows WISEngineering to scale up easily by adding more hardware resources. Its performance can be further tuned using Software Performance Engineering [23].

7   Conclusion

This paper has presented WISEngineering, a novel web application addressing the needs of massive online learning. The system provides many interesting features such as a powerful and extensible automated grading and instant feedback system, upon which a learning outcome analysis and reporting module is constructed. To achieve extensibility, an open software architecture is adopted to integrate heterogeneous software components. To improve service performance and scalability, back-end cloud clusters are used and interoperability is achieved using a simple service invocation protocol. The WISEngineering system is a promising platform that supports both learning and the research on learning. Future directions include investigating data mining analysis techniques that associate user behavior data with learning outcomes.

Acknowledgments. This work is partially supported by the National Science Foundation Grants DRL-1422436 and DRL-1253523. The instant grading service cluster is hosted by Hofstra Big Data Lab, funded by grant ESD CFA 29409.

References

1. Barwick, D.: Does Class Size Matter? Inside Higher Education. http://www.insidehighered.com/views/2007/12/06/barwick
2. UC Berkeley. Berkley WISE System. https://wise.berkeley.edu/
3. Hofstra STEM Research Center. Research and Development: Advances in Wise Guys & Gals - Boys and Girls as WISEngineering STEM Learners. http://www.hofstra.edu/Academics/Colleges/SEAS/CTL/wise/research-development.html
4. Hofstra STEM Research Center. WISEngineering Web Portal. http://wgg1.hofstra.edu
5. edX Inc., OpenEdx educational platform. https://open.edx.org/
6. Fielding, R.T., Taylor, R.N.: Principled design of the modern web architecture. ACM Trans. Internet Technol. **2**, 115–150 (2002)
7. Fielding, R.T.: Architectural Styles and the Design of Network-based Software Architectures. PhD thesis, University of California, Irvine (2000)
8. Apache Foundation. Apache HBase. http://hbase.apache.org/

9. Apache Foundation. Apache Zookeeper. http://zookeeper.apache.org/

10. The Eclipse Foundation. Eclipse Communication Framework Project Home. http://www.eclipse.org/ecf/

11. Blackboard In.c. Blackboard Learning Platform. http://www.blackboard.com

12. Kohl, J., Neuman, C.: The Kerberos Network Authentication Service (V5). http://tools.ietf.org/html/rfc1510

13. McIlroy, C.: Mass produced software components. In: Software Engineering: Report of a conference sponsored by the NATO Science Committee, Garmisch, October 1968

14. Moodle.org. Moodle: Modular Object-Oriented Dynamic Learning Environment. http://www.moodle.org

15. Nelson, D.: Next gen web architecture for the cloud era. In: SATURN 2013 Software Architecture Conference (2013)

16. openmooc.org. OpenMooc: A fully open source MOOC solution. https://openmooc.org/

17. Otto, M.: Say Hello to Bootstrap 2.0. https://blog.twitter.com/2012/say-hello-to-bootstrap-2

18. Pappano, L.: The Year of MOOC. http://www.nytimes.com/2012/11/04/education/edlife/massive-open-online-courses-are-multiplying-at-a-rapid-pace.html?pagewanted=all&_r=0

19. Paruchuri, V., Huang, D., Jarvis, J., Tauber, J., Aune, N., Kern, J.: EASE Auto-Grading Module. https://github.com/edx/ease

20. Richardson, L., Ruby, S.: RESTful Web Services. O'Reilly (2007)

21. scikit-learn developers. Scikit-learn: machine learning in python. http://scikit-learn.org/stable/

22. Slotta, J., Linn, M.: WISE Science: Web-based Inquiry in the Classroom. Teachers College Press, New York (2009)

23. Smith, C., Williams, L.: Building responsive and scalable web applications. In: 26th International Computer Measurement Group Conference, pp. 127–138 (2000)

24. Spady, W.: Outcome Based Education: Critical Issues and Answers. American Association of Schol Administrators, Arlington Virginia (1994)

25. The PhET Team. PhET: Interactive Simulation for Science and Math. http://phet.colorado.edu/

26. Williams, L., Smith, C.: Web application scalability: a model-based approach. In: Software Engineering Research and Performance Engineering Services, pp. 215–226 (2004)

27. Wolpert, H.D.: Stacked generalization. Neural Networks 5, 241–259 (1992)

A Non-parametric Approach to the Multi-channel Attribution Problem

Meghanath Macha Yadagiri[✉], Shiv Kumar Saini,
and Ritwik Sinha

Adobe Research, Bangalore, India
{mmacha,shsaini,risinha}@adobe.com

Abstract. Multi-channel marketing attribution modeling is a two-stage process. First, the value of exposure from different marketing channel needs to be estimated. Next, the total surplus achieved needs to be assigned to individual marketing channels by using the exposure effects from the first stage. There has been limited work in exploring possible choices and effects of determining the value of exposure to different marketing channels in the first stage. This paper proposes novel non-parametric and semi-parametric approaches to estimate the value function and compares it with other natural choices. We build a simulation engine that captures important behavioral phenomenon known to affect a customer's purchase decision; and compare the performance of five attribution approaches in their ability to closely approximate the known ground truth. Our proposed method works well when marketing channels have high levels of synergy. We apply the proposed approaches on two real-world datasets and present the results.

Keywords: Multi-channel attribution · Shapley Value · Simulation study

1 Introduction

Digital marketing has given a marketer greater access to customers as well as ability to observe each customer's interaction with different marketing channels. Access to the digital marketing data has allowed a marketer to assign each conversion event to various marketing channels. This is called the multi-channel marketing attribution (attribution) problem. The attribution modelling is a two stage process. In the first stage, the value of exposure from different marketing channel needs to be estimated. In the second stage, the total surplus achieved needs to be assigned to individual marketing channels by using the exposure effects from the first stage. There has been little work in trying to understand the possible choices and effects of determining the value of exposure to different marketing channels in the first stage. Similarly, there is a lack of literature on how to distribute the conversions that would have occurred without exposure to any marketing channel. The goal of this article is to fill these gaps in the literature on attribution modelling.

© Springer International Publishing Switzerland 2015
J. Wang et al. (Eds.): WISE 2015, Part I, LNCS 9418, pp. 338–352, 2015.
DOI: 10.1007/978-3-319-26190-4_23

The data available for estimating an attribution model consists of a series of marketing touches that may or may not result in a conversion event. An attribution model first estimates the effect of exposure (exposure effect) to the marketing channel(s) on a conversion event using observational data. The exposure effect is captured by modelling the effect of a subset of the marketing channels on the likelihood of the conversion event. The exposure effect can be highly non-linear due to, among other reasons, the synergies between the marketing channel. The non-linearities in the exposure effect explored in [5,7,9] are captured by using the non-linear parametric models such as a logistic regression model. However, there is no justification for using a particular model. The answer to the attribution problem is highly dependent on the exposure effect model. We evaluate the parametric models used to estimate the exposure effect in the past using a simulation engine. The popular parametric models do not perform well, especially, when the level of synergies between the marketing channels are high. We propose a novel non-parametric approach to estimate the exposure effect. The proposed model performs well even when the synergies between the marketing channels are high.

The exposure effect model can be used in the second stage to divide the conversion events to each marketing channel. We use the Shapley Value approach to assign the conversion events that occurred due to the observed marketing touches. The Shapely Value approach used here is similar to the approach in [5]. This is called marginal attribution because only the increase in the conversions due to marketing touches are assigned. At times, the goal of attribution modelling is to assign all of the conversion events. The past literature on attribution modelling do not provide any answer other than the rule based approaches such as the first touch and last touch attribution models. We propose a Nash Bargaining [3] based approach to assign the left-over return to each marketing channel. The Nash Bargaining solution is a cooperative game theoretic concept to divide surplus in a bargaining games where players have outside options. We use the marginal attribution to each channel as the outside option of the channel. We apply the proposed models using two real-world data sets. The models generate consistent results across different real world datasets.

The rest of the paper is organized in the following manner. In Sect. 2, we present the related work in the field of Shapley Value and Algorithmic Attribution. In Sect. 3, we formally define the marketing attribution. Section 4 presents our proposed methodology and its advantages. In Sect. 5, we present our simulation study and compare the proposed models with the existing parametric models. In Sect. 6, we present our results on two datasets. We conclude the paper and discuss directions for future work in Sect. 7.

2   Related Work

As noted earlier, there is no past work in evaluating the models that estimate the exposure effect. In this section, we describe the work that touches upon the other aspects of the attribution model.

Algorithmic Attribution. There have been several efforts to solve the problem of marketing attribution in an algorithmic fashion. One of the first models was proposed by [7]. In this paper, the authors propose two data-driven attribution models depending on estimated measure of channel relevance. The models lack a clear interpretation of fairness while assigning credit to the marketing channels. The authors in [1] model the customer journey as a funneling process leveraging the concepts of Hidden Markov Model to relate the stages of a customer to their conversion behavior. Different advertising campaigns may have varied notions of customer stages making such a model difficult to deploy in real-world scenarios. Similarly, [11] models the customer's journey as a Markov process and provides methods to estimate each channel's effective contribution and prediction of conversion rate. In [9], the authors present a model based on counterfactual analysis to solve the attribution problem and estimates the incremental effect of each marketing channel.

The only game-theoretic approach for marginal marketing attribution was proposed in [5]. In this paper, the authors initially recommend properties of a good attribution model. Further, the authors frame the problem of attribution as a causal estimation problem and then propose two approximate methods based on co-operative game theory. However, the proposed models are parametric and usually cannot be simplified. Further, there is no comparison of the two proposed models to evaluate which model performs better. Also, the paper does not highlight the difficulties of deploying a game-theoretic framework to real world settings. Additionally, the paper is restricted to marginal marketing attribution - assigning the surplus produced by customers to the marketing channels. The methods and evaluations presented in this paper are motivated by the limitations of the current literature.

3   Problem Definition

Problem Setting. Let $\Upsilon = \{U_1, U_2, ..., U_n\}$ be n users targeted by the marketer using k marketing channels $\Omega = \{C_1, C_2, ..., C_k\}^1$. We define *return* from a user as a measure of response to marketing activity undertaken by the marketer. Some examples of return may include purchase, revenue, page views etc. The *left-over* for a user is defined as the *return* that would have occurred without exposure to any of the k marketing channels. Also, *surplus* for a user is defined as the total *return* minus the *left-over*. Hence, *return = surplus + left-over*. Given that the marketer has generated a return R, surplus S and left-over L, we would like to algorithmically assign $\Psi = (a_1, a_2, ..., a_k)$,to the k marketing channels, where $\sum_j a_j = S$ and $\Pi = (l_1, l_2, ..., l_k)$ where $\sum_j l_j = L$ to each of the channels that the marketer has used. For this purpose, we model the problem at a customer level by considering various marketing channels through which each customer in Υ has been targeted.

---

[1] We use i as a counter for elements (users) in Υ as well as elements (users) in Υ depending on the context, j is also used in a similar fashion for the set Ω.

Customer Level Definition. Let $\mathbf{E_i} = (e_{i1}, e_{i2}, ..., e_{ik})$ be the binary vector of a user i who has been exposed to the k available marketing channels. Here, $e_{ij} = 1$ if and only if the user i has been exposed to a channel j and is 0 otherwise. Let $\tau = (b_1, b_2, ..., b_n)$, where $\sum_i b_i = R$ be the total *return* generated by the marketer on all the users targeted by the k available marketing channels. For all the users in Υ, we would like to come up with $\rho = (s_1, s_2, ..., s_n)$ the surplus produced by each customer. We achieve this by computing s_i^j, the surplus produced by marketing to customer i due to channel j in the matrix $S_{att} = [s]_{ij}$. In S_{att}, the row-sum is the total surplus produced by each customer i, $\sum_j s_i^j = s_i, s_i \in \rho$. The column-sum is the attributed surplus to a marketing channel j, $\sum_i s_i^j = a_j, a_j \in \mathbf{\Psi}$ and $\sum_j \sum_i s_i^j = S$ where S is the total surplus achieved by all marketing efforts. Further we use the customer surplus estimations to calculate $(l_1, l_2, ..., l_k)$, the $left - over$ vector $\mathbf{\Pi}$.

4   Approach

To solve the problem defined in the above section, we first estimate the surplus of each customer. Next, we model the multi-touch attribution problem as a coalitional game[2] at a customer level and then apply Shapley Value to determine the surplus that needs to be assigned to each channel. In the general marketing attribution problem where multiple channels are involved, some of the channels may influence a user more than others, as a result, may possess different bargaining power, hence the concept of *fairness* of distribution of the final gains is an essential property which Shapley Value captures. Also, the measure of channel importance in an attribution model is the channel's expected marginal impact on conversion, where the expectation is taken over the possible orderings of the channels in Ω which is well addressed by Shapley Value. Finally, we propose a method to assign the *left-over* return to the marketing channels.

Customer-Level Approach. We model the attribution problem by defining coalitional games (f, Ω_i) at a customer level. Where Ω_i are the various channels a customer i has been exposed to, $\Omega_i \subseteq \Omega$, $|\Omega_i| = k_i$ and the value of the characteristic function f is defined to be an estimate of the surplus produced from the customer. Further, we use the concept of Shapley Value to distribute the gains to the all the channels in Ω_i that were collaboratively involved in influencing the customer to generate return. However, Shapley Value is not readily applied to a problem, one needs to define the value function that satisfy certain criterion[3] [8]. Also, the theory assumes that it is possible to assess the expected gain for every possible co-operation, that is, for all 2^{k_i} possible combinations. Hence, one would need to estimate the expected surplus of a user when he has been

---

[2] A coalitional game denoted by (f,N) is defined by a characteristic function f and total number of players in the game N, where f maps subsets of players to real numbers: $f : P(N) \rightarrow R$ with $f(\emptyset) = 0$, where $\emptyset$ denotes the empty set and $P(N)$ is the power set of the N players.

[3] One key criterion is that the characteristic function f should satisfy $f(\emptyset) = 0$.

exposed to all the possible combinations of channels in Ω_i. A marketer may not necessarily have information of the behavior of the user when exposed to various combinations of marketing channels which makes the direct application of Shapley value to real-world highly impractical. To address the issues that stem from the definition of Shapley value, we initially provide non-parametric and semi-parametric approaches to estimate the expected surplus from a customer.

Existing Methods. In [5], the authors propose to estimate the expected surplus from a customer using a fully parametric approach. The models proposed in the paper leverage popular binary classification models like (1) Logistic Regression (2) Elastic-Net Regularised Logistic Regression. Another parametric binary classification model one could use to estimate the surplus is Random Forest [4]. That brings to our first contribution in this paper, we propose using non-parametric and semi-parametric models to estimate the surplus for each customer.

Initial Setting. Firstly, for a given data set, if the total number of marketing channels is k, a maximum of 2^k possible combinations (excluding repetitions) of marketing channels could be used by the marketer to target the users. Hence, a maximum of 2^k channel combinations could be observed in the dataset. For each such combination of marketing channels s, we define $f(s)$ as

$$f(s) = \frac{Purchases(s)}{Purchases(s) + Non-purchases(s)} \tag{1}$$

This function is the frequentist estimate of the conditional probability of purchase given exposure to the channels in s. Typically, in a dataset, all the 2^k channel combinations are not observed. Also, $f(\emptyset)$, where $\emptyset$ is the null set may not be calculated by the above formulation since the marketer may not have information about the users who were not targeted through any of the marketing channels and have converted. Hence, we propose the below methods to estimate the $f(s)$ of unobserved channel combination including $f(\emptyset)$. The proposed estimates are further used to form a coalitional game at a customer level.

Table 1. Notation for Existing(E) and Proposed(P) methods

Notation	Existing estimation	Notation	Proposed estimation
E1	Logistic regression	P1	(1), (2) and (3)
E2	Elastic net regularised logistic regression	P2	(1) and Logistic regression
E3	Random forests	P3	(1) and Random forest

Non-parametric Model. Let P_Ω be the power set of k marketing channels in Ω and by using the initial setting, say, we have information about a subset of the all possible combinations. Let this set be $S_{obs} \subseteq P_\Omega$. We approximate the value of $f(s), \forall s \in P_\Omega \setminus S_{obs}$ in the following manner
(a) For each $f(s)$ to be estimated, $s \in P_\Omega \setminus S_{obs}$, consider all the subsets of the

combination of channels s (P_s, the power set of combination s) that belong to S_{obs}. Let this set be T_{obs}. Hence $T_{obs} = P_s \cap S_{obs}$

(b) The value of $f(s)$ is given by averaging over all the elements in T_{obs}

$$f(s) = \frac{1}{|T_{obs}|} \sum_{j \in T_{obs}} f(j), \forall s \in P_\Omega \setminus S_{obs} \qquad (2)$$

We use the above equation to estimate $f(s)$, $\forall s \in P_\Omega \setminus S_{obs}$. Note that the above definition may also not always estimate $f(\emptyset)$ if a marketer does not have information about customers who have not been exposed to any channels and have converted. To tackle such instances, we provide a formulation for estimation of $f(\emptyset)$.

Estimation of $f(\emptyset)$. In our formulation of f, both the effect of channels and $\emptyset$ (no-channels) are inherently captured. We split $f(s) = f(s + \emptyset)$ to separate out both the effects. $f(s) = f(s + \emptyset) = t(s) + f(\emptyset)$. Now, we assume that t is linear. Let Ω^* be the set of all non-overlapping cover sets[4] of the set comprising elements of s. For each such covering set K in Ω^*, $f(s) = f(s + \emptyset) = \sum_{p \in K} t(p) + f(\emptyset)$. If $|K|$ is the cardinality of the set K, then the equation could be re-written as

$$f(s) = f(s + \emptyset) = \sum_{p \in K} t(p) + f(\emptyset) + |K|f(\emptyset) - |K|f(\emptyset)$$

$$= \sum_{p \in K} f(p) - (|K| - 1)f(\emptyset)$$

$$\Rightarrow f(\emptyset) = \frac{1}{|K| - 1} \left[\sum_{p \in K} f(p) - f(s) \right] \qquad (3)$$

We average the estimates of $f(\emptyset)$ for all covering sets K $\in \Omega^*$ to compute the final estimate of $f(\emptyset)$. The estimation steps (1), (2) and (3) are combined denoted as P1 throughout this paper.

Semi-parametric Model. Alternatively, to estimate the value of $f(s)$ $\forall s \in P_\Omega \setminus S_{obs}$, one could use probabilistic estimates from a binary classification algorithm. We train the classification algorithms in the following manner. (a) *Predictive variables*: For each customer, the feature vector is equal to E_i, as defined in Sect. 3. (b) *Response variable*: For each customer i, the response variable is assigned 1 if the user i produced return to the marketer and 0 otherwise. The probability estimates from the classification models are interpreted as the likelihood of a customer to provide some return to the marketer given the customer has been exposed to a particular set of marketing touches. This is exactly what we are estimating in a non-parametric fashion in Eq. (2) and hence could be replaced by the probability estimates from the binary classification models. Thus, unlike a fully parametric model as proposed in [5,7], we suggest to use a

---

[4] Non-overlapping cover set: Given a set of elements $\Theta = \{1,2,...,n\}$, $\Delta = \{U_1, U_2, .., U_k\}$ is a non-overlapping cover set of Θ if $U_1 \cup U_2 \cup .. \cup U_k = \Theta$ and $U_i \cap U_j = \emptyset$, $\forall$ i,j in Δ

combination of Eq. (1) and the probabilistic estimates from the binary classification models. The two models used in this paper for the parametric estimates are Random forest [4] and a logistic regression [6]. Also, the probability estimate of a customer to produce return given that he has been exposed to no marketing channels is equated to be $f(\emptyset)$. In a logistic regression, this leads to the effect of the intercept term towards the conditional probability. For future use, we denote the estimation step (1) combined with logistic regression as $P2$ and (1) combined with Random forest as $P3$. Throughout this paper, we denote various attribution models using the notation defined in Table 1.

Attribution of Surplus. After calculating the surplus estimates for all the possible 2^k subsets of the marketing channels, the next step is to assign the surplus to the marketing channels. This is achieved in the following manner.

- Consider all the users Υ in the given data set who have made a purchase. Let this set of users be Λ, $\Lambda \subseteq \Upsilon$.
- For each customer in Λ, consider the channels the customer has been exposed to. Let this set of channels be $E \subseteq \Omega$.
- Frame a coalitional game (g, E), $g(s) = f(s) - f(\emptyset)$ where $f(s)\ \forall s \in$ power set of E is calculated using one of the estimation models described above.

Observe that $g(\emptyset) = 0$ and g satisfies all the properties of the value function of Shapley value [8]. Hence, using the concept of Shapley, the total gain $g(E)$, interpreted to be the fractional contribution the marketing channels in s has made to the surplus generated by the marketer is distributed to all the marketing channels involved. The channel-level attributions (pay-offs) for each marketing channel in E for the customer in Λ according to Shapley is given by

$$s_i^j(g) = \sum_{T \subseteq E \setminus \{j\}} \frac{|T|!\,(|E| - |T| - 1)!}{|E|!}(g(T \cup \{i\}) - g(T)) \qquad (4)$$

The s_i^j calculated here is plugged into the matrix S_{att} defined in Sect. 3. For the users who belong to $\Upsilon \setminus \Lambda$, s_i^j is 0, since there was no *return* from these users due to the exposure of various marketing activities. The aggregated channel attribution for all the k channels is calculated by cumulating the customer-level channel attributions obtained from (3). If n be the total users under consideration, then the aggregated channel attributions for each channel in Ω is given by,

$$\sum_i s_i^j = a_j, j = 1, 2, ..., k \qquad (5)$$

and $a_j \in \Psi$ as defined in Sect. 3.

Observe that instead of forming coalitional games (g, E) at a customer level, we could group all the customers who have been exposed to a particular set of the marketing channels. Once the grouping is done, we could form coalitional games for each set rather than at an customer-level. Such a set-level formulation would hugely reduce the number of games formed and is hence computationally more efficient.

Attribution of Left-Over. After attributing the surplus, our aim now is to attribute the *left-over*. The total *left-over* to be attributed to the k channels is obtained by using a_j computed through Eq. (4) and the left-over is computed by $L = \sum_i b_i - \sum_j a_j = R - S$, b_i and a_j are defined in Sect. 3. Next, we assign channel-wise left-over attributions using a_j by $l_j = \frac{a_j}{\sum_j a_j} * L$.

This formulation is inspired by the concept of Nash bargaining solution. Nash bargaining is a bargaining problem modeled with an outside option. If the bargaining collapses (if there is no co-operation among the players) each players gets the outside option. In our case, the players, as above, are the k marketing channels and the outside option is the surplus attributed to each channel. Since we are interested in assigning the *left-over* return to the k marketing channels, it is assumed that the *left-over* return was generated due to no co-operation among the marketing channels. Once we have computed $\Psi = (a_1, a_2, ..., a_k)$ and $\Pi = (l_1, l_2, ..., l_k)$, the final attributions would be given by the sum $(a_1 + l_1, a_2 + l_2, ..., a_k + l_k)$.

5   Simulation Framework

All attribution models aim to formalize an answer to the credit assignment problem, but a natural question to ask is which of these is better, or which of these gives a more correct answer. Here lies one of the biggest problems of all Attribution models; the inability to evaluate different attribution models. No academic work has thus far explored this question. Here, we try to formalize an approach to evaluating multiple attribution models. Our approach includes simulating data with many of the behavioral characteristics that marketers and customer display when interacting with each other.

Here are the major aspects of our simulation set-up. Given n customers, we first simulate the touches they experience. The touches can be from different marketing channels. Next, these marketing touches lead to some propensity of the customers to conduct a transaction with the marketing organization. Finally, the goal of the simulation framework is to compare an estimation method with the true values of the simulation. At the end, we describe the parameter values we used for the simulation set-up.

5.1   Parameters of Simulation

Number of Touches. For each of n customer, we simulation the number of interactions they will experience with the marketer. This is simulated to be a Poisson random variable. Let's say that the ith customer has M_i number of interactions with the marketer, we assume that $M_i - 1 \sim Poisson(\lambda)$. In other words, $P(M_i = k + 1) = \lambda^k \frac{\exp - \lambda}{k!}$, $k = 0, 1, 2, \cdots, \infty$. Note that since we are adding 1 to a Poisson random variable because we would not see a customer in a dataset unless the individual had at least one interaction with the marketer.

Type of Interaction. Once an interaction has taken place, we need to assign the type of interaction. This is achieved by sampling the interaction from a all possible interactions with a Multinomial distribution. Assuming there are k types of interactions, let $(\pi_1, \cdots, \pi_k)$ denote the multinomial probabilities for each type of interaction ($\sum_1^k \pi_i = 1$). Define $(Z_{i1}, \cdots, Z_{ik})$, where Z_{ij} is the number of interactions of type j for the ith customer. Then, $P(Z_{i1} = z_1, \cdots, Z_{ik} = z_k) = \frac{M_i!}{z_1! \cdots z_k!} \pi^{z_1} \cdots \pi_k^{z_k}$, x'where $\sum z_j = M_i$. Since all the attribution models we study only consider whether or not a channel has been activated, and not the number of touches of a particular channel, we define $X_{ij} = 1_{Z_{ij} > 0}$ (where 1 denotes the indicator function).

Probability of Event. Given a set of touches of different types, we need to next simulate whether the interactions $(X_{i1}, \cdots, X_{ik})$ (denoted by $\mathbf{X_i}$) will lead to a conversion event resulting in returns for the marketer. Let Y_i denote the binary event of whether the ith customer converts. We model the distribution of Y_i as

$$\log\left(\frac{P(Y_i = 1|\mathbf{X}_i)}{P(Y_i = 0|\mathbf{X}_i)}\right) = \alpha + \sum_{j=1}^k \beta_j X_{ij} + \sum_{j=1}^k \sum_{l=1}^k \beta_{jl} X_{ij} \times X_{il}.$$

Here α is the intercept term, this term can be controlled to have different proportion of purchasers in the population. Note here that β_j is the main effect, that is, the increase or decrease in the log-odds ratio if the jth channel is added to the media-mix of an individual customer. The term β_{jl} is the interaction between the jth and l^{th} channels, in other words, the effect on the log-odds ratio when both channels are simultaneously active.

5.2   Evaluation of Truth and Comparison

For comparing the proposed models, we initially introduce the notion of *true* attributions of each marketing channel. This is done by equating $f(s)$, as defined in Sect. 4 to the sigmoid function with the assumed main channel effects β_j and interaction β_{jl}. Note that the same sigmoid function is used to generate the propensity of each customer in the simulated dataset. Hence, we believe a perfect attribution model would capture the marketing effects in a similar manner. Consider s be a set of marketing channels which belong to P_s, the power set of k marketing channels in Ω. If $s = (X_1, \cdots, X_k)$, $f(s)$ is given by

$$f(s) = \alpha + \sum_{j=1}^k \beta_j X_j + \sum_{j=1}^k \sum_{l=1}^k \beta_{jl} X_j \times X_l$$

We simulate a total of three datasets by varying the interaction effect β_{jl} between the channels - Low (interaction factor $\beta_{jl} = 0.01$), Medium ($\beta_{jl} = 0.05$) and High ($\beta_{jl} = 0.1$). We assume the number of marketing channels to be fixed to 7 in all our experiments. We provide a comparison of the proposed models with the existing parametric models (defined in Table 1). We omit the model E1 (which leverages sigmoid function) from the comparison since we use sigmoid function to calculate the propensity of each customer in the simulated dataset. In each dataset, we generate 1 Million users. Next, we randomly sample 10,000 users and run the proposed models - iteratively for 100 times on each dataset. For each instance in the iteration, we compare the channel attributions of all the *five* models (Proposed and Existing omitting E1) with the true attributions. For comparison, we use two metrics to calculate the proximity of the models to the truth. Firstly, we calculate the Root Mean Square Error (RMSE) [2] from the true attributions for each model. This error gives a measure of closeness between the true and estimated channel level attribution.

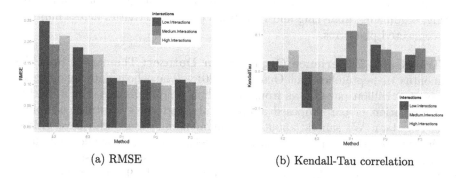

(a) RMSE   (b) Kendall-Tau correlation

Fig. 1. Comparison of RMSE (Lower is better) and Kendall-tau (Higher is better) from true attributions - Proposed models vs Existing models. (Refer Table 1 for notation)

Next, for each iteration, we compute the Kendall-tau rank correlation coefficient [10] between the estimated attribution values to channels from the models mentioned in Table 1 to the true attribution values based on the ground truth. For each dataset, we average the RMSE and Kendall-tau rank correlation over all the iterations. The results are presented in Fig. 1. The two accuracy measures supplement each other. RMSE captures the big differences in the attributed values and may be insensitive to the smaller differences. Also, RMSE can be agnostic towards the ranking of the marketing channels. Whereas, Kendall-tau rank correlation captures the ranking of the marketing channels and hence is more sensitive towards the smaller differences in the attributed values supplementing RMSE.

Firstly, in the existing methods, we notice that Random Forests (E3) performs better than Elastic net regularized LR (E2)[5] in terms of RMSE, but are worse off in the Kendall-tau rank correlation. Hence, a marketer interested in strategizing the preference order of his marketing channel spends, would be better off if he does not use a parametric model E2 for the estimation phase. The proposed semi-parametric models (P2, P3) outperform the non-parametric models (Kendall-Tau) in case of lower interactions between the marketing channels, whereas, the non-parametric models outperform the semi-parametric models in case of medium and high interactions. Elastic-net regularized LR performs well in the case of higher interactions between the marketing channels. Such a simulation study could be further extended to help the marketer decide the best attribution model amongst the various existing models. The analysis performed also introduces a notion of evaluation framework to compare the performance of different attribution models.

6   Empirical Results

Data Description. We apply our approach to two web analytics datasets. The data was collected using Adobe Analytics, an industry-leading solution for collecting, organizing, analyzing and reporting customer activities across multiple web-connected platforms.

(a) **Travel and Experience Organisation Dataset:** The data is from the months of September and October of 2013. The whole data amounts to about 2 Billion page views from 26 Million unique visitors. For a quicker examination of our proposed approach, we perform a stratified sampling of our data. We sample about 1.5 million unique users who have visited the web property during the last two weeks of the data window, of these visitors, about 300,000 of them have made a purchase in the assumed timeframe. The users could be targeted through one of the 9 marketing channels described in Table 2. For each user in the data, we have information about the various marketing channels the user has been exposed to and the purchase, revenue generated by the user as a result of these marketing interactions. In this dataset, we calculate both the order and revenue attributions of the channels. This dataset is referred to as Dataset-1 in figures.

(b) **E-commerce Retailer Dataset:** The next data set we considered belonged to a large e-commerce retailer. The data ranged over a 100 day period during the summer of 2013. The data contained about 54 Million interactions with 18 Million customers. For the purpose of evaluating our proposed framework, we performed a stratified sampling on our data. The sampled data comprised of 400,000 unique users of which about 200,000 have made a purchase. The users were targeted by the marketer through a variety of 10 marketing channels. We omit the tabular description for this dataset due to space constraints. In this dataset, for each user, we have information about the marketing channels the user has been exposed to and if a purchase has been made by the user, the platform the purchase has been made ($"Instore(I)"$, $"Online(E)"$) and the type of product that has been purchased. Leveraging this information, we find channel attributions specific to product category and purchase medium. This dataset is referred to as Dataset-2 in figures.

Table 2. Dataset-1 : Definition of various marketing channels along with the frequency of their occurences in purchases as well as the whole data

interaction	Definition	Total (%)	Orders (%)
direct	User directly navigating to site	33.8	27.5
display_ad	User clicking on a display ad	0.5	0.3
email	A click on an email from the advertizer	1.4	1.7
other_owned	A click from otherowned web properties	12.5	24
other_website	Clicks from other websites not owned by the advertizer	10	8.8
social_media	User navigates from a social media site	4.3	2
search	Clicks on organic search	31.5	28
search_ad	Clicks on search ad	5	5.2
travelagents	A visit from a travel agents site	1	2.5

Exploratory Data Analysis. Before applying our approach, we conducted exploratory data analysis in both datasets. We perform our analysis on all the users (purchasers and non-purchasers) and specifically on the converting (purchasing) customers since these are the users who have produced return for the marketer. Table 2 provides definition of each marketing channels and has details about the number of times each of the marketing channels was exposed to users (Dataset-1). We also analyzed the number of marketing touches the customers in our dataset were exposed to. Due to space constraints we only provide a detailed analysis of Dataset- 1. From Table 2, we observe that *direct* and *search* are the marketing channels that all the users and also specifically purchasers are most exposed to. Whereas *display_ad* ,*travelagents* and *social_media* are the least occurring marketing channels. An accurate attribution model should gauge the incremental and interactive effect of each marketing channel without biasing for the frequency of occurrence of the channels. Also, from the exploratory analysis, we observe that more than 80 % of the non-purchasers and more than 50 % of the purchasers have only one marketing interaction. Traditional attribution techniques such as First touch and Last touch would attribute all the generated return from the purchasers to these marketing channels without considering the users that have not made a purchase. An ideal attribution model should find the true attribution of each marketing channel by contrasting the purchases with non-purchases.

Estimation. We calculate the surplus estimations using the proposed models in the Approach section. Our assumption of linearity of t and the approximation of the value of $f(\emptyset)$ was validated by noting that the parametric estimate of $f(\emptyset)$ using a logistic regression and Random forest was fairly consistent with the estimated $f(\emptyset)$ using Eq. (3). For the second dataset, along with the channel-level attributions for the whole dataset, we calculate channel attributions specific to each product category and purchase media and for this purpose, we train different models for each of the product categories and purchase medium. Further, we compare the channel-level attributions from our proposed models with the existing methods (Refer Table 1 for notations). After the estimation step, we calculate channel attributions using Eq. (4). We do not include the left-over attribution presented in Eq. (5) in our results since we aim to compare the attribution results with existing marginal attribution models. In the case of Dataset-1, we calculate both order and revenue channel attributions. We present few of the scaled attribution results in Figs. 2 and 3. The results of all the three proposed models are consistent and cross-validate each other. The credit assignment to the marketing channel *travelagents* and *other-owned* channels in the order attributions of Dataset-1 are noticeable. The channels receive a higher credit than few other channels that occurred more frequently in the dataset satisfying a notion of *fairness*. In Dataset-2, we portray the extensibility of the proposed framework. Hence, we calculate channel-attributions at a more granular level, specific to the product-category using the proposed framework. We note that the channels *direct* and *display* have a higher order attribution specific to Product-1, while they have a lower order attribution in case of the online media. Such insights specific to products give the marketer an added advantage while planning his overall spend across different marketing channels

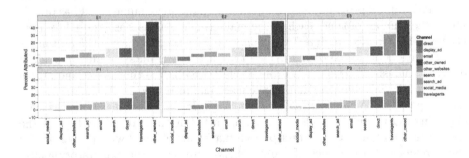

Fig. 2. Dataset-1, Channel-level order attributions compared across various attribution models defined in Table 1

Comparison. We noted that the present state of the art data-driven attribution models were proposed by Shao et al. [7] and Delassandro [5]. In both the papers, the authors compare the proposed models with traditional attribution models like Last Touch Attribution (LTA) to prove the accuracy of their models. The exploratory data analysis we performed combined with the insights from the results made in the above section, it could be deduced that our models are capturing the true attribution of channels without being guided by the frequency of the marketing channels in the datasets. Hence we do not provide a comparison of the proposed approach with LTA. Instead, since our approach leverages Shapley value, we compare the proposed models (denoted by P1, P2, P3 in the Fig. 3) with the existing parametric models defined in Table 1 (denoted by E1, E2, E3 in the Fig. 3). We observe that the results from our models are fairly consistent

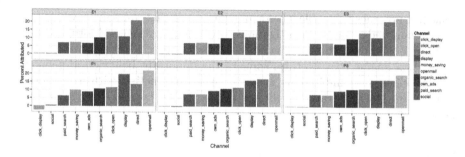

Fig. 3. Dataset-2, Product-1, Channel-level order attributions compared across various attribution models defined in Table 1

with the parametric model in case of Dataset 1 but are slightly different in case of Dataset-2 Product-1. We investigated this by observing the interaction terms β_{jl} for the marketing channels in Dataset-2 for model E1. The interaction terms were observed to be high. In the simulation study, we have established that the existing models do not perform well when there is high levels of synergy between the marketing interactions. Hence, we claim that the proposed models P1, P2, P3 would be closer to the true attributions in the case of Dataset-2 than the existing models.

7   Conclusions and Future Work

The goal of this paper it to fill important gaps in the literature on attribution modelling. Past work on attribution has used parametric approaches to estimate the likelihood of conversion due to exposure to a set of marketing channels. We show that these approaches do not work well when there are high level of synergies between the marketing channels. This is shown using a simulation study. We propose a non-parametric approach to estimate the exposure effect. The approach performs well in presence of high level of synergies between the marketing channels. We use Shapley Value to assign the surplus that is derived from the exposure effect model. The Shapley Value approach is similar to [5]. We argue that it is important to distribute the left-over return to the marketing channels. Past work has not provided any solution to this problem. We propose an approach that uses the marginal attribution solution as the input to assign left-over return to the marketing channels. We apply the proposed attribution model on two real world datasets and present the results.

In future, we can extend this work in multiple directions to account for more realistic scenarios. The effect of advertisements is believed to decay over time, in this work, we have not attempted to capture this effect in the simulation model, nor do the different existing approaches capture these effects. This is an area which may be explored in future. The computation of Shapley value increases exponentially as the number of players in the game increases. While this may be feasible when we model marketing channels (which may number between ten and twenty), it becomes infeasible when modelling marketing campaigns, which may run into hundreds. Hence, investigations to come up with closed form approximations for the proposed model could be another research direction. We provide a simulation engine that captures some of the important behavioural phenomenon known to be exhibited by customer when interacting with a marketing organisation, however, this may be further expanded by modelling more such characteristics.

References

1. Abhishek, V., Fader, P.S., Hosanagar, K.: Media exposure through the funnel: a model of multi-stage attribution. Technical report, Working paper. Carnegie Mellon University, Pittsburgh (2013)
2. Armstrong, J.S., Collopy, F.: Error measures for generalizing about forecasting methods: empirical comparisons. Int. J. Forecast. 8(1), 69–80 (1992)
3. Binmore, K., Rubinstein, A., Wolinsky, A.: The nash bargaining solution in economic modelling. RAND J. Econ. 17, 176–188 (1986)
4. Breiman, L.: Random forests. Mach. Learn. 45(1), 5–32 (2001)
5. Dalessandro, B., Perlich, C., Stitelman, O., Provost, F.: Causally motivated attribution for online advertising. In: Proceedings of the Sixth International Workshop on Data Mining for Online Advertising and Internet Economy, p. 7. ACM (2012)
6. Hastie, T., Tibshirani, R., Friedman, J., Hastie, T., Friedman, J., Tibshirani, R.: The Elements of Statistical Learning. Springer, New York (2009)
7. Shao, X., Li, L.: Data-driven multi-touch attribution models. In: Proceedings of the 17th ACM SIGKDD International Conference on Knowledge Discovery and Data Mining, pp. 258–264. ACM (2011)
8. Shapley, L.S.: A value for n-person games. Technical report, DTIC Document (1952)
9. Sinha, R., Saini, S., Anadhavelu, N.: Estimating the incremental effects of interactions for marketing attribution. In: 2014 International Conference on Behavior, Economic and Social Computing (BESC), pp. 1–6. IEEE (2014)
10. Skidmore, A.K.: A comparison of techniques for calculating gradient and aspect from a gridded digital elevation model. Int. J. Geog. Infom. Syst. 3(4), 323–334 (1989)
11. Xu, L., Duan, J.A., Whinston, A.: Path to purchase: a mutually exciting point process model for online advertising and conversion. Manage. Sci. 60(6), 1392–1412 (2014)

A Dynamic-Static Approach of Model Fusion for Document Similarity Computation

Jiyi Li[✉], Yasuhito Asano, Toshiyuki Shimizu, and Masatoshi Yoshikawa

Graduate School of Informatics, Kyoto University,
Yoshida-Honmachi, Sakyo-ku, Kyoto 606-8501, Japan
{jyli,asano,tshimizu,yoshikawa}@i.kyoto-u.ac.jp

Abstract. The semantic similarity of text document pairs can be used for valuable applications. There are various existing basic models proposed for representing document content and computing document similarity. Each basic model performs difference in different scenarios. Existing model selection or fusion approaches generate improved models based on these basic models on the granularity of document collection. These improved models are static for all document pairs and may be only proper for some of the document pairs. We propose a dynamic idea of model fusion, and an approach based on a Dynamic-Static Fusion Model (DSFM) on the granularity of document pairs, which is dynamic for each document pair. The dynamic module in DSFM learns to rank the basic models to predict the best basic model for a given document pair. We propose a model categorization method to construct ideal model labels of document pairs for learning in this dynamic module. The static module in DSFM is based on linear regression. We also propose a model selection method to select appropriate candidate basic models for fusion and improve the performance. The experiments on public document collections which contain paragraph pairs and sentence pairs with human-rated similarity illustrate the effectiveness of our approach.

Keywords: Document similarity · Dynamic model fusion · Learning to rank

1 Introduction

Computing the semantic similarity between two textual documents is a significant research topic, which can be used for various applications, such as document retrieval, clustering, recommendation, visualization, and so on [1]. Many efforts have been made on this fundamental issue in recent years. Various basic models of document representation have been proposed in existing work.

Each representation model leverages and concentrates on a certain kind of information. Vector space model [5] focuses on explicit word content; latent semantic analysis [6] and latent Dirichlet allocation [8] evaluate latent document semantic topics in word content; knowledge based graph [9], which is a state of art work, leverages entity relation knowledge in knowledge bases. However,

© Springer International Publishing Switzerland 2015
J. Wang et al. (Eds.): WISE 2015, Part I, LNCS 9418, pp. 353–368, 2015.
DOI: 10.1007/978-3-319-26190-4_24

because of the variety of textual documents in the collections, each basic model has its advantages and disadvantages, and it is difficult for a single basic model to perform well in all scenarios. For example, knowledge based graph model only uses entities extracted from documents and entity relations in knowledge bases. It does not make detailed analysis in the word content. It is able to find semantic similar documents which have many entity overlaps but few word overlaps. However, it cannot perform well if the documents have few entities or semantic dissimilar documents share same entities.

There are some approaches for utilizing multiple basic models to reach improved results. The idea of existing model selection approaches learns to select the best basic model. The idea of existing model fusion approaches such as linear regression seeks the best weights to combine these basic models. Both of these two kinds of approaches concentrate on the granularity of document collection. They are static from the perspective of document pair. Because one static model may be only proper for some document pairs, they may fail to leverage different basic models to complement each other for various scenarios in the collection.

We thus propose a dynamic idea. Our idea is that on the granularity of document pair, if we can dynamically select and fuse the models which have better performance than others, we can improve the overall performance for all document pairs in the collection. In the approach we propose, the dynamic module in our fusion model predicts the performance of basic models for a given document pair and ranks these basic models by referring the technology of learning to rank. We propose a model categorization method based on human ratings to construct the ideal model labels of document pairs for learning.

Model performance prediction for a document pair is a difficult task which results in that only a dynamic module may be not good enough. To improve the performance, we integrate a static module, which is based on linear regression, with the dynamic module into our fusion model. We also propose a model selection approach to seek the candidate basic models which are possible to improve the performance and filter the basic models with low performance. We construct experiments using human-rated document collections to verify our approach.

The contributions of our work are as follows:

- We propose a novel solution which implements a dynamic idea and generates improved results on document semantic similarity computation by dynamically fusing different document representation models on the granularity of document pair. It is able to reach better performance than the static approaches on the granularity of document collection.
- We propose a dynamic fusion module by predicting the performance of basic models to rank and select the best basic model for a document pair. The ideal model labels for learning in this dynamic module is constructed by our model categorization method which can generate theoretical optimal results of our dynamic idea when human ratings of all document pairs are available.
- We propose a Dynamic-Static Fusion Model (DSFM) which integrates both dynamic and static modules. It is able to reach better performance than the approaches with dynamic or static module only. Our approach provides a

solution of integrating learning to rank method and linear regression method. To improve the performance, we also propose a model selection method to seek appropriate candidate basic models.

The remainder of this paper is organized as follows. In Sect. 2 we review the related work. In Sect. 3, we introduce some preliminary information and analysis. We propose our dynamic module in Sect. 4 and our dynamic-static fusion model in Sect. 5. In Sect. 6 we report and discuss the experimental results. We give a conclusion in Sect. 7.

2   Related Work

First, for the existing work on document representation models, Salton et al. first proposed a vector space model in 1975 [5]. It represents text documents as vectors of index terms. Document similarity can be evaluated by cosine similarity and the term-specific weights are term frequency-inverse document frequency (tf-idf). Around 1990, the paper related to latent semantic analysis technology (or named as latent semantic indexing (LSI)) was proposed by Deerwester et al. [6]. This model explains text documents by mixing latent topics and analyze the relationships between documents and terms. Probabilistic latent semantic analysis (PLSA) [7] proposed in 1999 is based on mixture decomposition derived from a latent class analysis. PLSA is not a generative model of new documents. In 2003, latent Dirichlet allocation (LDA) was proposed to solve this shortcoming and assumed the topic distribution to have a Dirichlet prior. Recently, Schuhmacher et al. proposed a graph representation with entities extracted from a document and entity relations in a knowledge base [9] (KBG). Graph edit distance is used to compute the semantic distance of two documents. These basic models focus on one kind of information and perform different in different scenarios. Our work dynamically selects and fuses these basic models for different scenarios.

Second, there are existing work on automatically selecting machine learning methods for a data collection, e.g., a tools named AutoWeka developed by Thornton et al. [14]. Suck kinds of work are on the granularity of collection and select one model for the whole dataset. Our work selects different models for different instances in the dataset and have different granularity from these work.

Third, the machine learning technologies such as linear regression can be used to fuse basic models. However, their fusion models are static for all document pairs. Our fusion approach provides dynamic fusion for each document pair.

3   Preliminary

Before we propose the solution of document similarity computation by dynamic fusion, we describe some preliminary information. We first formulate our topic. Then we introduce the document collections that we use for experiments. We make a case study to explain the relationships between model performance and different scenarios, and the rationality of our dynamic idea in the ideal.

3.1 Notations

We denote d_i as a document in the collection $\mathcal{D}$, o_{ij} as the document pair of d_i and d_j, $\mathcal{O}$ as all document pairs in the collection, r_{ij} as the average human ratings of o_{ij}, and s_{ij}^k as the document similarity of o_{ij} by document representation model m^k from the model set $\mathcal{M}$. We also use o_x, r_x and s_x^k when we only represent a document pair in our approach without taking care of the document ID. In this paper, we name m^k as *basic model*, to differ from the *fusion model* in our approach and *ranking model* in the learning technologies.

Note that both s_x^k and s_{ij}^k in this paper are normalized scores. Because the original similarity scores of different basic models have different ranges and are not comparable, for each basic model, we normalize the original similarity scores into the range of human ratings of the collection based on the range of original similarity scores of this basic model on all document pairs in the collection.

3.2 Document Collections

We use the following two human-rated document collections for our experiments.

LP50: This public document collection contains 50 news articles selected from the Australian Broadcasting Corporation's news mail service [1]. The number of words of the documents is from 51 to 126. Totally, there are 1225 document pairs and 83 annotators. Each document pair is annotated by around 10 persons (for some document pairs, 9 or 11 persons) with ratings from 1 (highly unrelated) to 5 (highly related). The document type in this collection is a paragraph.

STS2015: This public collection is provided by the international workshop on semantic evaluation (SemEval-2015), in the task 2 on semantic textual similarity for English [2]. It utilizes the collections provided in 2012, 2013 and 2014 as training set and provides new data as the testing set in 2015. There are 12,443 document pairs in total. The range of ratings is from 0 (no relation) to 5 (semantic equivalence). In contrast to LP50, the document type in this collection is a sentence. This collection only provides the rating of document pair without the information of detailed ratings of each annotator.

3.3 Case Study: Model Performance in Different Scenarios

We use the LP50 dataset for this case study because it contains the ratings of different annotators for a document pair. The term of "scenario" can have various meanings in our topic while in this case we define it as the rating differences of human ratings for a document pair. We check these rating differences to evaluate the difficulty of semantic similarity judgment for human beings. We denote $\mathcal{R}_{ij}$ as the rating set of a document pair d_i and d_j and r_{ij} as the average rating. The measure of rating difference is defined as the difference between maximum rating and minimum rating of a document pair, $dR_{ij} = max\{\mathcal{R}_{ij}\} - min\{\mathcal{R}_{ij}\}$. When dR_{ij} is lower, it means that the ratings of the annotators are more consistent, the semantic similarity of o_{ij} is easier to be judged by human beings.

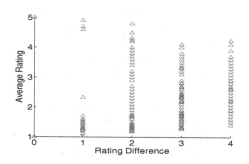

Fig. 1. Rating difference and average rating

Table 1. Model performance on entire LP50 and subset of each rating difference

Rating diff	LSA	KBG
0	**0.859**	0.714
1	**0.755**	0.622
2	**0.597**	0.567
3	0.466	**0.483**
4	0.512	**0.562**
Whole	0.575	0.570
Whole (fusion)	**0.600**	

Figure 1 shows the distribution of average ratings on rating differences. It shows that the document pairs that are easy to be judged ($d\mathcal{R}_{ij} = 0$) are the document pairs that are very similar ($r_{ij} = 5$) or very dissimilar ($r_{ij} = 1$). When r_{ij} is around 3, it is possible that the semantic similarity of o_{ij} is not easy to be judged by human beings ($d\mathcal{R}_{ij} \geq 2$). In such scenario, it may be also difficult for the document representation models to compute the similarity and generate consistent results with the human ratings which are used as ground truth.

We use two basic models, i.e., LSA [6] and KBG [9], as the examples to analyze the relationships between model performance and different scenarios. We omit the description of our implementation of these basic models, which we will describe in the experimental section. Following existing work that use the same dataset, such as [9], we use the metric of Pearson correlation coefficient to evaluate the performance of basic models. The correlation between the result set $\mathbf{s}^k = \{s_1^k, s_2^k, ..., s_{\mathcal{N}}^k\}$ of model m^k and human rating set $\mathbf{r} = \{r_1, r_2, ..., r_{\mathcal{N}}\}$ on a document pair set is defined as

$$Corr(\mathbf{s}^k, \mathbf{r}) = \frac{\sum_{x=1}^{\mathcal{N}} (s_x^k - \mu_{\mathbf{s}^k})(r_x - \mu_{\mathbf{r}})}{\sqrt{\sum_{y=1}^{\mathcal{N}} (s_y^k - \mu_{\mathbf{s}^k})^2} \sqrt{\sum_{y=1}^{\mathcal{N}} (r_y - \mu_{\mathbf{r}})^2}}$$

$\mathcal{N}$ is the number of document pairs. $\mu_{\mathbf{s}^k}$ and $\mu_{\mathbf{r}}$ are the mean scores of the results and human ratings. The range of this metric is $[-1, 1]$.

Table 1 lists the performance of each basic model. Besides the performance on the whole collection (row "whole"), we also list the performance on each rating difference. It illustrates that each basic model performs differently in different scenarios. When the rating difference increases and the difficulty of document similarity judgment by human beings increases, the performance of both two models decrease. When the rating difference is low and judging document similarity by human beings is easy, the performance of LSA is better than KBG. When the rating difference is high and judging document similarity by human beings is difficult, the performance of KBG is better than LSA. It means that LSA and KBG are possible to complement each other in some scenarios.

Base on these observations, for example, we can construct a new similarity result set named "fusion" on the whole collection with the following rules. For each document pair, if the rating differences are not higher than 2, we use the similarity of LSA; otherwise, we use the similarity of KBG. The performance of "fusion" shown in Table 1 is better than either basic models. This example shows that our dynamic idea is rational to improve the performance. It is an analysis in ideal because the human ratings of all document pairs are available. How to learn a fusion model for the document pairs without human ratings is the problem in our topic. Note that the approach we use here to construct the "fusion" set is just an intuitive example of the dynamic idea and is not the approach we propose in the following sections.

4   Dynamic Fusion by Model Performance Prediction

In this section, we propose the dynamic module in our approach. We will introduce our approach integrating both dynamic and static modules in next section.

4.1   Model Performance Prediction

Following the case study in Sect. 3.3, the concretization of the dynamic idea in our approach is that for a given document pair, we seek a basic model which may have better performance than other basic models. To reach this purpose, we predict the performance of each basic model for a document pair. We rank these basic models and use the basic model which is ranked first.

We refer to the technology of Learning To Rank (LTR) [4], which is proposed to train ranking models in ranking tasks. For example, in web search, for a given query and a document in the collection, there are many measures that can represent the relevance, such as TF-IDF, BM25, Pagerank and so on. LTR uses these measures as the feature vector of a document and the label is the rank of the document. There are several kinds of LTR methods, including pointwise, pairwise and listwise methods, which treat the ranks of documents in different ways. For the pairwise methods which we refer in this paper, the ranks of documents are transformed into partial order preference of two documents like $d_i \succ d_j$ which means d_i has higher rank than d_j.

In our work, in contrast to the web search example, the queries are document pairs from $\mathcal{O}$ and the objects that need to be ranked are the basic models from $\mathcal{M}$. We define $\mathcal{Y} = \{1, 2, ..., l\}$ as the rank labels. We use the training set as the example to describe the modelization. The validation set and testing set can be formulated as similar forms. The query set is defined as $\mathcal{O}_T = \{o_1, o_2, ..., o_{\mathcal{N}_T}\} \subseteq \mathcal{O}$. For a document pair o_x, y_x^k is the label of m^k, and $\mathbf{y}_x = \{y_x^1, y_x^2, ..., y_x^{\mathcal{N}_M}\}$ is the label set of $\mathcal{M}$. Then the training set can be denoted as $\mathcal{S}_T = \{((o_x, \mathcal{M}), \mathbf{y}_x)\}_{x=1}^{\mathcal{N}_T}$. The validation set is defined as $\mathcal{S}_V$ and The testing set is $\mathcal{S}_E$.

To carry out the LTR methods, we compute the feature vector v_x^k of each "document pair"-model pair (o_x, m^k). The features for a document pair is $\mathbf{v}_x = \{v_x^1, v_x^2, ..., v_x^{\mathcal{N}_M}\}$. We then can use $\{(\mathbf{v}_x, \mathbf{y}_x)\}_{x=1}^{\mathcal{N}_T}$ to carry out LTR methods.

v_x^k is constructed from the information of o_x and m^k. The feature used in our approach is the similarity of o_x by m^k, i.e., s_x^k. We utilize the RankSVM method which is a pairwise LTR method and implemented in the library named svm^{rank}[3]. We use RBF kernel function which can handle non-linear cases.

The rationality of this feature selection can also be explained by the case study in Sect. 3.3. In that case, LSA tends to be better when rating difference is low ($d\mathcal{R}_{ij} \leq 2$); when rating difference is very low ($d\mathcal{R}_{ij} = 0$), the average rating tends to be very low ($r_{ij} = 1$) or very high ($r_{ij} = 5$). In other words, it is possible that some basic models are proper to represent some intervals in the range of similarity (average rating). Therefore, actually the ranking model learned in our model performance prediction method evaluates the distributions of the intervals that the basic models are proper for. When ranking the basic models, it utilizes the similarities of these basic models to estimate the possible intervals and select the proper basic model. Note that the hyperplane of the intervals of two basic models may be not a single threshold in the range of similarity and thus the similarities of basic models may be not linearly separable, this is the reason that we use kernel function in our method.

In this paper, we only use the similarities of basic models as the feature for the following reasons. On one hand, we have ever tried some other features of document pairs such as document length difference, cosine similarity of word vectors and so on. Until now, we still have not found that they can significantly improve the performance. A possible reason is that these features may be implied in the similarities of basic models. On the other hand, more features will increase the time cost a lot in the training stage because of the usage of kernel functions.

An important issue is that some document pairs are exactly dissimilar. Their similarities of all basic models are zero. It means that we can select any basic model for these document pairs. We define such kind of document pairs as *zero document pairs*, and the document pairs on the contrary as *non-zero document pairs*. We do not construct feature vectors for zero document pairs. The training, validation and testing are only on non-zero document pairs. When we evaluate the performance of similarity computation, we assign the zero document pairs with the basic model which has best performance on the training set. The performance evaluation is on the set with both zero and non-zero document pairs.

4.2   Model Categorization

We propose a model categorization method on basic models for constructing ideal labels for the model performance prediction method. We estimate the model performance and rank the models for a document pair based on human ratings. Actually, after we label all document pairs in the collection, we can divide them into several categories based on the basic models. These categories can be regarded as a basic model partition on a given set of document pair which can be training, validation or testing set. Each basic model m^k corresponds to a subset $\mathbf{s}^k = \{s_1^k, s_2^k, ..., s_{\mathcal{N}^k}^k\}$ in which s_x^k is a similarity score. We define $\mathbf{r}^k = \{r_1^k, r_2^k, ..., r_{\mathcal{N}^k}^k\}$ as the set of human ratings associated with $\mathbf{s}^k$.

Therefore, the problem of model categorization can be converted to a task that given a set of document pairs, we label the document pairs with basic models to generate a basic model partition which can reach best global performance on a given metric, e.g., the Pearson correlation coefficient. We propose a method to optimize an object function which represents the global performance.

An important issue of the object function is that its values of different partitions need to be comparable. For example, an object function like $f' = \sum_k \mathcal{N}^k * Corr(\mathbf{s}^k, \mathbf{r}^k)/\mathcal{N}$ is not proper and comparable because different partitions with different number of document pairs have different division factors in the computation of Pearson correlation coefficient. A special pseudo test case is assuming to compute the similarity scores of all document pairs by same model, the values of f' are different if randomly partitioning the whole set into some subsets in different ways, though the values of f' should be same in this case.

An optional object function which satisfies this comparability requirement is as follows. It merges the similarity scores $\mathbf{s}^k$ in different subsets into one set $\mathbf{s}$. The values of f are same for different partitions in the above-mentioned pseudo test case of all similarity scores computed by same model.

$$f = Corr(\mathbf{s}, \mathbf{r}), \mathbf{s} = \cup \mathbf{s}^k, \mathbf{r} = \cup \mathbf{r}^k.$$

Actually, we can also define the object functions for other performance metrics in the same way. Our method can be adapted to generate global optimized model categorization results on a specific performance metric for a data collection of which the ground truth has been given.

The detailed optimization algorithm is as follows. It has a stochastic process because it updates the partition immediately after checking a document pair.

- *Initialization:* Construct a category a^k for each basic model m^k; randomly assign each document pair o_x with a model m^k and put it into the category.
- *Iteration:* For each document pair o_x, its category is a^{k_1}, move it to another category a^{k_2}, $a^{k_2} \neq a^{k_1}$, if and only if this movement can increase f.
- *Convergence:* In a traversal on all document pairs, there is no more movement or the increase of f is lower than a threshold.
- *Labeling:* For each document pair o_x of which the current category is a^{k_1}, for all $a^{k_2} \neq a^{k_1}$, compute the performance change δ^{k_2} if moving o_x to a^{k_2}. $\delta^{k_1} = 0$. Rank the basic models based on all δ^k in ascending order and then label a "document pair"-model pair (o_x, m^k) with rank y_x^k.

After we generate the optimized partition, we can get the tuples $((o_x, m^k), y_x^k)$ which are used as the ideal model labels in model performance prediction method. Furthermore, this model categorization method can also be regarded as an optimization method on the concretization of our dynamic idea in the case of the human ratings of all document pairs are available, while our model performance prediction method concentrates on the case that only a part of document pairs are rated by human beings.

5   Dynamic-Static Fusion Approach

In this section, we propose our solution for improving the performance of document semantic similarity computation by dynamically fusing basic models. The idea is that the parameters of our fusion model are dynamic for different document pairs, so that our approach has different behaviors in different scenarios.

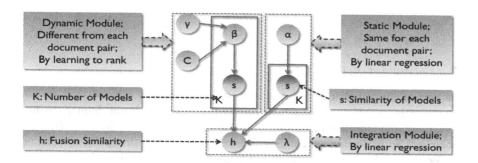

Fig. 2. Dynamic-static fusion model

Only using the dynamic module proposed in Sect. 4 may be not able to reach good performance because predicting basic model performance for a document pair is a difficult task, we thus propose a dynamic-static fusion model which integrate a static module with the dynamic module, so that we can generate better results than the results of dynamic module only or static module only. We also propose a model selection method for improving the performance.

5.1   Dynamic-Static Fusion Model

Figure 2 shows the Dynamic-Static Fusion Model (DSFM) in our approach. $\mathbf{s}_x = \{s_x^1, s_x^2, ..., s_x^{\mathcal{N}_M}\}$ denotes the similarity set of a document pair by basic models, and h_x is the result of fusion model. The computation of this fusion model can be divided into three parts.

- The left part is the dynamic module which utilizes our model performance prediction method. The parameter $\beta_x = \{\beta_x^1, \beta_x^2, ..., \beta_x^{\mathcal{N}_M}\}$ on the granularity of document pair is different for different document pairs. It is computed based on the results of the model performance prediction method which utilizes RankSVM. If m^k is labeled as the best basic model for o_x, $\beta_x^k = 1$; otherwise, $\beta_x^k = 0$. The parameters used in RankSVM are soft margin parameter C and kernel parameter γ.
- The right part is the static module which utilizes linear regression. The parameter $\alpha = \{\alpha^1, \alpha^2, ..., \alpha^{\mathcal{N}_M}\}$ on the granularity of collection is static for all document pairs.

- The bottom part is the integration module which integrates the results of dynamic and static modules with linear regression. $\lambda = \{\lambda_1, \lambda_2\}$ is on the granularity of collection. The fusion results h_x can be computed as follows. A constant term is not necessary because all results will be added same constant value and it does not influence the value of the performance metric.

$$h_x = \lambda_1 \phi_{1x} + \lambda_2 \phi_{2x}, \quad \phi_{1x} = \sum_k \beta_x^k s_x^k = \beta_x \mathbf{s}_x, \quad \phi_{2x} = \sum_k \alpha^k s_x^k = \alpha \mathbf{s}_x,$$

$$\alpha^k \in [0,1], \sum_k \alpha^k = 1; \beta_x^k \in \{0,1\}, \sum_k \beta_x^k = 1; \lambda_t \in [0,1], \sum_t \lambda_t = 1;$$

5.2 Parameter Estimation

Our performance metric is Pearson correlation coefficient defined in Sect. 3.3. Our target is to optimize this metric on the fusion results $\mathbf{h}$ and human ratings $\mathbf{r}$, i.e., $Corr(\mathbf{h}, \mathbf{r})$, on the document pair set which can be the training set $\mathcal{S}_T$, validation set $\mathcal{S}_V$ or testing set $\mathcal{S}_E$. An approximate loss function $\mathcal{G}$ for optimizing this metric can be defined as the mean squared difference of $\mathbf{h}$ and $\mathbf{r}$.

$$\mathcal{G} = \frac{1}{2\mathcal{N}} \sum_x (h_x - r_x)^2 = \frac{1}{2\mathcal{N}} \sum_x (\lambda_1 \sum_k \beta_x^k s_x^k + \lambda_2 \sum_k \alpha^k s_x^k - r_x)^2$$

Because β_x^k in the dynamic module is based on the learning and prediction results of model performance prediction method and its value is 0 or 1, it results in that $\mathcal{G}$ is not differentiable. Instead of inferencing the parameters with methods like gradient descent, we use grid search on the training set and validation set to estimate the parameters. We sample some candidate parameters, compute the performance based on them, and then select the parameters which can reach better results on both training set and validation set, i.e., the average performance on training set $\mathcal{S}_T$ and validation set $\mathcal{S}_V$.

To reduce the dimensions of parameter space, instead of optimizing $\mathcal{G}$ directly, we use an upper bound $\hat{\mathcal{G}}'$ which optimizes static module and other two modules respectively. The detailed process of parameter estimation is list in Algorithm 1.

$$\hat{\mathcal{G}} = \frac{1}{2\mathcal{N}} \sum_x (\hat{h}_x - r_x)^2 \leq \hat{\mathcal{G}}' = \frac{1}{2\mathcal{N}} \sum_x ((\lambda_1 \sum_k \beta_x^k s_x^k - \frac{1}{2} r_x) + (\lambda_2 \sum_k \alpha^k s_x^k - \frac{1}{2} r_x))^2$$

5.3 Model Selection

There are various basic models which we use in this paper. However, it is not necessary to use all these basic models for fusion. On one hand, some models are not proper for a collection and have poor performance. They will become noises in the fusion process and fusing with them may harm the performance of results, especially for the model performance prediction method. On the other hand, some models have similar results with other models and thus do not provide more

Algorithm 1. Parameter Estimation

1: Sampling candidate values of α, $\mathcal{C}$, γ and λ for grid search
2: Estimate $\hat{\alpha} = \arg\max_{\alpha} Corr(\alpha\mathbf{s}, \mathbf{r})$
3: **for** each group of $(\mathcal{C}, \gamma)$ **do**
4:     Run model performance prediction method with $(\mathcal{C}, \gamma)$, compute β_x for all o_x
5:     Estimate $\hat{\lambda}|(\mathcal{C}, \gamma) = \arg\max_{\lambda} Corr(\mathbf{h}|(\mathcal{C}, \gamma, \lambda, \hat{\alpha}), \mathbf{r})$
6: **end for**
7: Estimate $(\hat{\mathcal{C}}, \hat{\gamma}, \hat{\lambda}) = (\hat{\mathcal{C}}, \hat{\gamma}, \hat{\lambda}|(\hat{\mathcal{C}}, \hat{\gamma})) = \arg\max_{\mathcal{C}, \gamma, \hat{\lambda}|(\mathcal{C}, \gamma)} Corr(\mathbf{h}|(\mathcal{C}, \gamma, \hat{\lambda}|(\mathcal{C}, \gamma), \hat{\alpha}), \mathbf{r})$

information. Fusing with them will not significantly improve the performance and will increase the time cost on learning.

Therefore, before we carry out our DSFM, we use a model selection method to generate candidate basic models. It estimates the parameter $\theta = \{\theta^1, \theta^2, ..., \theta^{\mathcal{N}'_M}\}$ with $Corr(\theta\mathbf{s}', \mathbf{r})$ on the training set $\mathcal{S}_T$. The basic models m^k with $\theta^k > 0.1$ are selected. The selected models can be regarded as a principal component in the basic model set. We use θ instead of α used in Sect. 5.2, to discriminate that they are in different processes and the number of basic models are different. Note that all $\mathcal{M}$ used in the above sections is the candidate basic model set generated by this model selection method.

6 Experiment

6.1 Implementation of Document Representation Models

We implement four document representation models as the basic models. Without loss of generality, all this basic models are general ones and are not specially optimized for specific tasks. For example, in the task of computing the semantic similarity of sentence pairs in STS2015, some optimized basic models are proposed by utilizing additional information such as linguistics and maybe have better performance than the following basic models. How to propose an optimized basic model for a specific task is out of range of this paper. This paper focuses on a fusion model based on existing basic models.

Vector Space Model (VSM): VSM [5] represents document with vectors. We use the Natural Language Toolkit (NLTK) [10] for carrying out various natural language processing on the raw documents. The documents are tokenized into raw words. These raw words are converted based on morphology and part-of-speech tagging. They are also filtered by stop words provided by NLTK. After that, we compute the tf-idf value of each word in the documents and construct the vectors. The document similarity is computed by the cosine similarity of vectors using tf-idf as weights.

Latent Semantic Analysis (LSA): LSA [6] represents documents with vectors of latent topics, using singular value decomposition on a word-document matrix. We use the implementation in a package named gensim [11]. We use the tf-idf-based vectors computed in VSM to construct the matrix from all document in the collection. For the number of topics, we sample some numbers for each collection to select a proper number with best performance on the collection. In our experiment, the topic number for the LP50 collection is set to 100. The topic number for the STS2015 collection is set to 1000. s_{ij}^{lsa} is also computed based on the cosine similarity.

Latent Dirichlet Allocation (LDA): LDA [8] represents document using a mixture of topics with multinomial distribution and Dirichlet prior with a graphical model. We use its implementation in the gensim package. The tf-idf-based vectors and similarity s_{ij}^{lda} computation are same with that we use for LSA. The topic number for the LP50 collection is set to 100. The topic number for the STS2015 collection is set to 200. Other parameters are the defaults of the gensim function. LDA does not work well for these two collections because it needs large collections for learning and is not proper for short text.

Knowledge Based Graph (KBG): KBG [9] represents document with entity graph which is a sub-graph extracted from a knowledge base. We implement an unweighted edge version of this method, with our settings of some parameters because [9] does not describe them in detailed. We use DBpedia Spotlight [13] to annotate words and phrases in the documents with the entities in DBPedia [12]. The "confidence" parameter for using DBpedia Spotlight is set to 0.2 and the "support" parameter is set to 20. After we extract the original entity list of a document, we extend this entity list by adding the entities which directly connect to at least two original entities in DBpedia. We construct a sub-graph to represent the document with this extended entity list as nodes and their relations in DBpedia as edges. To evaluate the document similarity, we implement an approximate graph edit distance computation algorithm based on bipartite graph matching, which is used in [9].

6.2 Experimental Settings

We utilize the two collections in Sect. 3.2 for our experiments. These two collections contain two different kinds of document pairs, i.e., paragraph pairs and sentences pairs. We divide each collection into training, validation and testing subsets. For LP50, these three subsets have the percentages of 60 %, 20 % and 20 %. The numbers of paragraph pairs are 735, 245 and 245 respectively. The numbers of non-zero pairs are 617, 186 and 169. For STS2015, we construct the training set with all data used in 2012 and 2013, the validation set with the new data in 2014, and the testing set with the new data in 2015. The numbers of sentence pairs are 5693, 3750 and 3000 respectively. Almost all sentence pairs are not zero pairs and we omit this distinction. Furthermore, we use the Pearson correlation coefficient for evaluating the performance.

We carry out our model selection method in Sect. 5.3 to select candidate models for each collection. The sequence of the basic models is [VSM, LSA, LDA, KBG]. As a result, the parameter θ for LP50 is [0.625, 0.0, 0.0625, 0.3125]; the parameter θ for STS2015 is [0.692, 0.231, 0.0, 0.077]. Therefore, we select VSM and KBG for the LP50 collection, and VSM and LSA for the STS2015 collection.

The values of parameters in grid search for leaning the DSFM are set as follows, $\alpha^k = \ddot{\alpha}^k / \sum_u \ddot{\alpha}^u, \ddot{\alpha}^k \in \{0, 0.1, ..., 0.9, 1\}$; Because parameter $\mathcal{C}$ in RankSVM is always set based on the size of training set, for LP50, $\mathcal{C} \in \{2^{-6}, 2^{-5}, ..., 2^5, 2^6\} *$ 617, and for STS2015, $\mathcal{C} \in \{2^{-2}, 2^{-1}, ..., 2^1, 2^2\} * 5693$; $\gamma \in \{2^{-7}, 2^{-6}, ..., 2^6, 2^7\}$; $\lambda_1 \in \{0, 0.01, ..., 0.99, 1\}, \lambda_2 = 1 - \lambda_1$.

6.3   Experimental Results

We first analyze the performance of model categorization approach which computes the results of our dynamic idea in ideal with all human ratings of document pairs are available and then discuss the performance of our fusion approach.

Model Categorization: Table 2 lists the results on each subset. We compare our model categorization method with the following approaches. (1). Each single model, which is related to existing work that select a single model. (2). A static module only method using linear regression which is related to existing work using the static idea. (3). Random categorization which is a naive implementation of our dynamic idea. Different runs of random categorizations have different performance. The average performance of random categorization with 10,000 simulations is worse than that of our approach. Our model categorization method using random categorization as initializations can reach same categorizations with different initializations. The results show that selecting different models for different document pairs is possible to improve the overall performance of document similarity computation.

Table 2. Results and comparisons of our model categorization method

Collection	Subset	Basic models				Fusion models		
		VSM	LSA	LDA	KBG	Static	Random	Categorization
LP50	Training	0.5905	0.6031	0.0770	0.6315	0.6634	0.5967	**0.8147**
	Validation	0.5513	0.5314	0	0.5009	0.5748	0.4722	**0.8085**
	Testing	0.4683	0.4636	−0.0262	0.4076	0.4791	0.4039	**0.6339**
STS2015	Training	0.6991	0.6492	0.6222	0.3093	0.7131	0.6638	**0.8299**
	Validation	0.6430	0.5414	0.5617	0.2741	0.6499	0.5780	**0.7874**
	Testing	0.7003	0.6468	0.6679	0.2508	0.7049	0.6665	**0.7989**

Fusion Approach: Table 3 shows the experimental results of fusion approaches on the testing set. We compare our fusion approach using the dynamic-static fusion model with the following approaches. (1). Best single model approach, which is related to existing work that select a single model. (2). A static module only approach using linear regression, which is related to existing work using the static idea and same with the static module in our DSFM. (3). A dynamic module only approach which is based on our model performance prediction method without integrating a static module. Figure 3 shows the performance of candidate parameter groups. Our parameter estimation process selects the parameter groups which have best performance on both training and validation set. The selected parameter groups $(\alpha, \mathcal{C}, \gamma, \lambda)$ is ([0.625, 0.375], 9872, 128, [0.29,0.71]) for the LP50 collection and is ([0.75,0.25], 1423.25, 128, [0.17,0.83]) for the STS2015 collection.

The results illustrate that our fusion approach with dynamic idea on the granularity of document pair is able to reach better performance than the approaches on the granularity of collection, which are the best single model approach and static module only approach in these experiments. It also shows that although the dynamic module only approach may be not always better than static module only approach, when integrating it with a static module, we can generate better results than either of them. Specially, for STS2015, the performance has only improved slightly. It is because that in STS2015 if only using word list or entity list without additional information, VSM has been very good and fusing other basic models in this paper with it cannot improve the performance a lot.

In addition, Tables 2 and 3 show that there is still a gap between the results of our current fusion approach and the ideal upper limitation of our dynamic idea. The dynamic module still has a possibility of improvement.

Furthermore, Fig. 4 shows that there is a high correlation between results of dynamic module and fusion approach. This correlation is 0.7540 for the LP50 collection and 0.7840 for the STS2015 collection. When the result of the dynamic module is better, the result of fusion model is better.

In the end, on one hand, if we can improve the performance of the dynamic module in the future work, then we can improve the results of the fusion approach. On the other hand, the training statge of a dynamic module costs more time than that of a static module, in the future work, we can try to use other paramter optimization methods such as Bayesian optimization to instead grid search to tune better parameters in fewer computation.

Table 3. Results and comparisons of our fusion approach on the testing set

Data collection	Best single	Static	Dynamic	Our fusion
LP50	0.4683	0.4791	0.4783	**0.4917**
STS2015	0.7003	0.7049	0.6744	**0.7050**

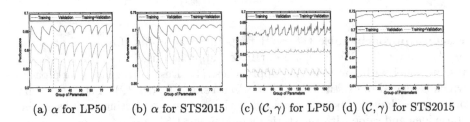

(a) α for LP50   (b) α for STS2015   (c) $(\mathcal{C}, \gamma)$ for LP50   (d) $(\mathcal{C}, \gamma)$ for STS2015

Fig. 3. Parameter estimation with training and validation set

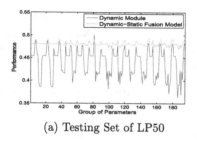
(a) Testing Set of LP50

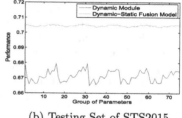
(b) Testing Set of STS2015

Fig. 4. Performance relation between our dynamic module and fusion model

7 Conclusion

In this paper, we propose an approach with a dynamic-static fusion model which generates improved results on document similarity computation by dynamically fusing different basic document representation models on the granularity of document pair. We propose a dynamic module by predicting the performance of basic models and ranking them. We also propose a model categorization method to generate ideal model labels and a model selection method to improve the performance. For the future work, we will improve the dynamic module.

References

1. Lee, M.D., Welsh, M.: An empirical evaluation of models of text document similarity. In: Proceedings of CogSci 2005, pp. 1254–1259 (2005)
2. STS2015, Semantic Textual Similarity for English in SemEval-2015. http://alt.qcri.org/semeval2015/task2/index.php?id=semantic-textual-similarity-for-english
3. Joachims, T.: Training linear SVMs in linear time. In: Proceedings of KDD 2006, pp. 217–226 (2006)
4. Li, H.: A short introduction to learning to rank. IEICE Trans. Inf. Syst. **E94–D**(10), 1854–1862 (2011)
5. Salton, G., Wong, A., Yang, C.S.: A vector space model for automatic indexing. Commun. ACM **18**(11), 613–620 (1975)
6. Deerwester, S., Dumais, S.T., Furnas, G.W., Landauer, T.K., Harshman, R.: Indexing by latent semantic analysis. J. Am. Soc. Inf. Sci. **41**(6), 391–407 (1990)
7. Hofmann, T.: Probabilistic latent semantic indexing. In: Proceedings of SIGIR 1999, pp. 50–57 (1999)
8. Blei, D.M., Ng, A.Y., Jordan, M.I.: Latent dirichlet allocation. J. Mach. Learn. Res. **3**, 993–1022 (2003)
9. Schuhmacher, M., Ponzetto, S.P.: Knowledge-based graph document modeling. In: Proceedings of WSDM 2014, pp. 543–552 (2014)
10. Bird, S., Klein, E., Loper, E.: Natural Language Processing with Python. OReilly Media Inc., Sebastopol (2009)
11. Řehůřek, R., Sojka, P.: Software framework for topic modeling with large corpora. In: Proceedings of LREC 2010 Workshop New Challenges for NLP Frameworks, pp. 46–50 (2010)

12. Auer, S., Bizer, C., Kobilarov, G., Lehmann, J., Cyganiak, R., Ives, Z.G.: DBpedia: a nucleus for a Web of open data. In: Aberer, K., Choi, K.-S., Noy, N., Allemang, D., Lee, K.-I., Nixon, L.J.B., Golbeck, J., Mika, P., Maynard, D., Mizoguchi, R., Schreiber, G., Cudré-Mauroux, P. (eds.) ASWC 2007 and ISWC 2007. LNCS, vol. 4825, pp. 722–735. Springer, Heidelberg (2007)
13. Mendes, P.N., Jakob, M., García-Silva, A., Bizer, C.: DBpedia spotlight: shedding light on the web of documents. In: Proceedings of the 7th International Conference on Semantic Systems (I-Semantics 2011), pp. 1–8 (2011)
14. Thornton, C., Hutter, F., Hoos, H., Leyton-Brown, K.: Auto-WEKA: combined selection and hyperparameter optimization of classifiaction algorithms. In: Proceedings of KDD 2013 (2013)

Genetic-Based Approach for ATS
and SLA-aware Web Services Composition

Imed Abbassi[1]($\boxtimes$), Mohamed Graiet[2], Walid Gaaloul[3],
and Nejib Ben Hadj-Alouane[1]

[1] ENIT, University of Tunis El Manar, Monastir, Tunisia
abbassi_imed@ymail.com, nejib_bha@yahoo.com
[2] High School of Computer Science and Mathematics, Monastir, Tunisia
mohamed.graiet@imag.fr
[3] Computer Science Department Telecom SudParis, Paris, France
walid.gaaloul@it-sudparis.eu

Abstract. Composite applications leveraging several functionalities offered by Web services are today the underpinnings of enterprise computing. There are two kinds of requirements of web service composition: QoS requirements and transactional requirements. The QoS requirements are recorded in service-level agreements (SLAs) contract. The transactional requirements are expressed by designers as a set of accepted termination states (ATS).

In this paper, we propose a novel selection and composition approach called ATS and SLA-aware Web services composition (WSC). Our approach aims at finding composite service compliant with ATS and better satisfying the SLA contract. As the ATS and SLA-aware WSC is NP-hard problem, an efficient algorithm is required. This work adopts genetic algorithm to deal with combinatorial problem.

1 Introduction

Web services [1] are the most famous implementation of service oriented architectures, allowing the construction and the sharing of independent and autonomous softwares. One interesting feature is the possibility to dynamically create a new added value service by composing existing web services, eventually provided by several companies and offering diverse functional (*e.g.,* ticket purchase, hotel booking), quality of service (QoS) and transactional (compensatable or not) properties. In this work, we consider the following QoS properties: response time, cost, availability and throughput.

There are two kinds of requirements of web service composition: QoS requirements and transactional requirements. The QoS requirements [2] are defined based on QoS properties. They are recorded in service-level agreements (SLAs), which is a contract specified between service providers and customers. Given a booking service, an example of constraint is that the service has to respond to the user within 5 ms and it has to be usually available. Another requirement can be specified is that the cost of a flight ticket should not exceed €1000. The transactional requirements

© Springer International Publishing Switzerland 2015
J. Wang et al. (Eds.): WISE 2015, Part I, LNCS 9418, pp. 369–383, 2015.
DOI: 10.1007/978-3-319-26190-4_25

are defined as a set of Acceptable Termination States (ATS for short) [3–6]. They define a customized transactional behavior of composite services. They express which faults are acceptable, retriable, or recoverable. For example, a service to retrieve customer details can be retried safely, but services to commit payment or place orders may require recovery or replacement upon failure.

While the service composition problem is largely studied in the last years, the proposed approaches [7–13] did not integrate ATS and SLA contract in the composition building process. There have a few works [3,5,6], which only consider ATS, while neglecting SLA contract. Integrating ATS and SLA contract in the composition building process is very important. It allows to create composite services that respond to the designers and customers' needs.

For the best of our knowledge, defining a composite service, with particular properties like ATS, and ensuring that every execution will preserve these properties remains a difficult and open problem [3,14]. Extended with SLA contract, the problem becomes complex. The problem consists in how to select the set of Web services that satisfies ATS and SLA contract.

To solve the aforementioned problem, we propose a new approach for Web service composition, called ATS and SLA-aware web service composition (WSC for short). Both ATS and SLA contract are integrated in the composition building process. Our approach allows to generate composite services that meet ATS and better satisfying the SLA contract. We adopt a Genetic Algorithm (GA) to address the ATS and SLA-aware WSC problem. In fact, the GA is a powerful tool to deal with combinatorial problem and successfully applied in many others research domains. This success motivates our choice of GA to solve ATS and SLA-aware WSC problem, which has similar properties in terms of large scale and problem complexity.

In the following, we present a review of the literature and a summary of related works of the field of service composition. Section 3 formulates the problem of ATS and SLA-aware WSC, while Sect. 4 presents an approach based on a genetic algorithm for efficiently selecting and then composing a set of Web services. Empirical studies are shown in Sect. 5. In Sect. 6, we draw some conclusions and provide some perspectives related to our future research.

2   Related Work and Motivating Example

In this section, we propose a summary of related work in the field of service composition problem. Then, describe a motivating example, that we will use throughout the paper.

2.1   Related Work

Several existing mechanisms are proposed trying to relax the atomicity property of service transactions. [4] uses the accepted termination states (ATS) properties as a criterion of correct execution. [15] proposes a method that formally verifies the transactional property of service composition by using event calculus.

At design time, the verification is performed to check the consistency of recovery. After execution, the verification is performed to notify deviations, and fix some design errors. Hence, it guarantees the reliability of service execution. In [3], authors propose a transactional approach to ensure the failure atomicity of a composite service, required by partners. They use ATS as a mean to express the required failure atomicity. Partners specify their composite service, mainly its control flow, and the required ATS. Then, they use a set of transactional rules to assist designers to compose a valid composite service with regards to the specified ATS.

Though many works have been presented on service composition [3,7–11,13]. The problem is generally tackled from the QoS aspect or from the transaction aspect separately. [10] addresses the issue of selecting and composing Web services not only according to their functional requirements, but also to their transactional and QoS properties. It proposes a selection algorithm that satisfies the user's preferences, expressed as weights over QoS criteria and as risk levels defining semantically the transactional requirements. [9] proposes a genetic approach to the selection problem, which is implemented and compared with heuristic algorithms introduced in [8]. [11] address services selection and composition in the context of a QoS-aware middleware for dynamic service environments. It proposes an efficient QoS-based selection algorithm. [12] proposes an improvement of the genetic approach by the usage of hybridization. The neighborhood of each individual of the genetic algorithm is explored iteratively to replace the actual individual with the good or the almost best neighbor.

In [16], authors propose a hybrid approach that combines global optimization with local selection in order to efficiently find a close to optimal selection. The main idea is to decompose end to-end QoS constraints to local constraints on the component service level, which can then be used to perform efficient local selection for each component independently. The decomposition of end-to-end constraints is achieved by mapping each of them for a set of precomputed local QoS levels. However, these approaches deal only with QoS dimensions and do not take into account the transactional properties of Web services that are crucial parameters to ensure service reliability. Moreover, they do not offer for designers to specify their requirements in term of control structure and fault recovery. [17] proposes an ant colony based approach, which combines transaction-aware service composition and QoS-aware service composition together. It investigates firstly transactional properties for services and present a comprehensive set of derivation rules to deduce the transactional properties of the composite services based on component service's properties and Workflow constructs. Then, it states the problem of dynamic transactional and QoS-aware service composition. [13] presents QoS-Gasp, a metaheuristic algorithm for performing QoS-aware web service composition at runtime. QoS-Gasp is a hybrid method combining GRASP with Path Relinking. Despite being interesting approaches, [13,17] are rather simple and did not consider ATS. They did not go further to satisfy the SLA contract.

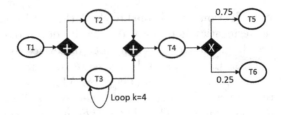

Fig. 1. The workflow process of the OTA service

2.2 Motivating Example

First, let us present a motivating example: We consider an Online Tourist-circuit Arrangement (OTA) service. The workflow process model of this service is shown in Fig. 1.

The OTA service is carried out by a composite service that is generated at runtime. It involves multiple tasks, namely flight booking (T1), car booking (T2), and hotel booking (T3) and finally online payment (T4). First, the customer performs the flight booking tasks. Each task is performed by different processing entities (Web services) but differ in the quality of service (QoS) attributes. For instance, both Flight Booking (FB) and Flight Reservation (FR) Web services can provide the flight-booking task. After executing T1, the tasks T2 and T3 are executed simultaneously. T3 is under a loop (the a maximal iteration number is 4). Only the Car Booking (CB) Web service provides the car reservation tasks. For the hotel reservation, either Hotel Reservation (HR), or Hotel Booking (HB) Services can be selected. Whenever, the hotel and the flight booking tasks are performed successfully, the customer is requested to pay. One of the following Web services: Credit Card (CC), Check (CH) and cash (SH) carries out the payment procedure. Finally, travel documents are sent to the customer by using post (T5) with probability equal to 0.75 or by email (T6) otherwise.

When the designers specify the OTA service, they expect that the composite service, which is created at runtime to perform a customer request, to be reliable. In particular, they pay attention to the failure handling. For example, they may want to be sure that the online payment task will succeed, and that it is possible to undo the effects of T3 and T1 (for instance when T2 fails). These properties define the transactional behavior of the service. This behavior is specified at runtime using transactional requirements. The transactional behavior will vary from one context to another. For instance, OTR designers may accept the failure

Table 1. ATS defined for the OTA service

Item	T1	T2	T3	T4	T5	T6
1	completed	completed	completed	completed	completed	initial
2	completed	completed	completed	completed	failed	completed
3	compensated	failed	compensated	aborted	aborted	initial
4	compensated	failed	canceled	aborted	aborted	initial

of T2 in a context, while in another one they may not tolerate such a failure at such an advanced step. So, the meaning of a reliable execution is tightly related to the transactional requirements and it may vary according to designers.

3   ATS and SLA-aware WSC Problem

In this section, we describe an ATS and SLA-aware WSC problem. We start by describing transactional services and transactional composite services. Then, we present the transactional requirements (ATS property) and QoS requirements (SLA contract). Finally, we present the formalization of the ATS and SLA-aware WSC problem.

3.1   Transactional Service

Web services [18] are self-contained, self-describing, modular applications that can be published, located, and invoked across the Web. They are eventually provided by several companies and offering diverse functional (*e.g.,* ticket purchase, hotel booking), QoS and transactional properties. In our approach, we define the Web services interface as the following function: portTypes $\in$ SERVICE $\rightarrow \mathbb{P}(\text{OPERATION})$, where **SERVICE** and **OPERATION** denote respectively sets of Web services and operations.

Transactional Properties: Transactional service is a service that emphasizes transactional properties for its characterization and usage [3]. Therefore, it can be:

- Compensatable (c) if it offers compensation policies to semantically undo its effects;
- Retriable (r) if it is sure to complete after a finite number of activations;
- Pivot (p) if once it is successfully completed, its effects remain and cannot be semantically undone.

Formally, we use the function, TP $\in$ SERVICE $\rightarrow \{c,p,r,pr,cr\}$, to define the transactional behavior of Web service.

The Web services can combine several transactional properties, which leads to a new behavioral property. For instance, a service can combine the pivot and retriable properties (TP(ts)=pr). Similarly, a service can be compensatable and retriable which leads to a new behavioral (TP(ts)=cr). However, compensatable and pivot can not be a transactional property because the effects of a pivot service can not be undone.

Termination States: The transactional behavior of Web services is described as state transition diagram (see Fig. 2) [4]. This diagram specifies the states in which a web services can terminate. In this work, we define a termination state function, TS(TS $\in$ SERVICE $\rightarrow \mathbb{P}$ (STATE)), which define the set of the

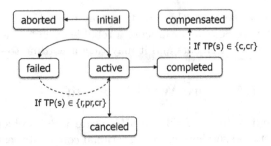

Fig. 2. The state transition diagram of web services

possible termination states of the Web services. A Web service, s, has a minimal set of termination states namely initial, aborted, canceled and completed. The failed state is a possible termination state of s if it is not retriable (failed $\in$ TS(s)), whereas compensated is a termination state of s if it is compensatable (compensated $\in$ TS(s)). This can be illustrated by the following rules:

$$TP(s) \in \{r, pr, cr\} \Leftrightarrow failed \notin TS(s) \tag{1}$$

$$TP(s) \in \{c, cr\} \Leftrightarrow compensated \in TS(s) \tag{2}$$

QoS Properties: The QoS properties of Web services describe the non-functional properties in many aspects. They include domain-independent ones such as, cost (ct), response time (rt) and availability (at) as well as domain-dependent ones like resolution or color depth for image processing services. In this work, we focus on domain-independent QoS metrics. We denote by $Q = \{q_1, ..., q_m\}$ the set of QoS properties, where q_i denotes the i-th quality parameter and m represents the number of concerned QoS properties.

The QoS attributes can be classified into two categories, namely positive denoted as Q_p and negative denoted as Q_n. For positive criteria, larger values indicate higher performance (e.g. availability), while for negative ones, smaller values indicate higher performance (e.g. cost and response time).

3.2   Transactional Composite Services

A transactional composite Web service (TCS for short)[4] is a conglomeration of existing transactional Web services working in tandem to offer a new value-added service. It takes advantage of services transactional properties to specify mechanisms for failure handling and recovery [3].

Composition Patterns: The TCSs structure can be represented as a workflow by using a set of composition patterns. For more details about the TCS definition and formalization, the reader is strictly redirected to [19]. In our work, we consider a set of patterns, which cover most of the structures specified by composition languages such as BPEL [20,21]:

- Sequence: sequential execution of web services.
- AND: parallel execution of web services.
- XOR: conditional execution of web services.
- Loop: iterative execution of web services.

Computing QoS of a TCS: The QoS of TCSs depends on the QoS values of the services components. The QoS value for a TCS is computed by the aggregation functions shown in Table 2 (taken from [11]). The aggregation formulas for the XOR pattern take into account the execution probabilities of the different branches in order to calculate the expected value for the quality dimensions. These probability data can be initialized by designers and updated based on information acquired by monitoring executions of the TCS. We assume that the expected iteration number (k) of the loop pattern is also specified. For instance, if the service under a loop construct has a cost c1, then the expected overall cost is $k * c1$.

Table 2. Aggregation functions per composition pattern and QoS attribute

QoS attributes	Sequence	XOR	AND	Loop
Response time (Rt)	$\sum_{i=1}^{n} rt(s_i)$	$\sum_{i=1}^{n} p_i * rt(s_i)$	$\sum_{i=1}^{n} rt(s_i)$	$k * t(s)$
Cost (Ct)	$\sum_{i=1}^{n} ct(s_i)$	$\sum_{i=1}^{n} p_i * ct(s_i)$	$\sum_{i=1}^{n} ct(s_i)$	$k * ct(s)$
Availability (At)	$\prod_{i=1}^{n} at(s_i)$	$\sum_{i=1}^{n} p_i * at(s_i)$	$\prod_{i=1}^{n} at(s_i)$	$at(s)^k$

3.3   Transactional Requirements

The service consistency is a crucial aspect of composite services execution. In order to meet the consistency requirements at early stages of the service composition process, we need to consider the transactional requirements as concrete parameters determining the choice of the appropriate component Web services. For this purpose, we use the Acceptable Termination States (ATS) concept as a mean to define these parameters. Formally, we define ATS as the following function: ATS $\in \mathbb{P}$ (OPERATION $\nrightarrow$ STATE). STATE is an enumerated set whose possible value is {initial, active, aborted, canceled, failed, completed, compensated}. An accepted termination state, ats (ats $\in$ ATS), is a state for which designers tolerate the termination state of a composite service to be created at runtime performing a set of functionalities.

It is therefore necessary to define a comparison criterion for ranking the Web services compliant with the consistency requirements expressed by business application designers. To do so, we introduce a new metric called **Transactional Fitness** (TF for short). TF expresses the consistency level of the transactional behavior of a given service with ATS. Formally, we define the TF of a given Web service s as follow:

$$TF(s) = \frac{\sum\limits_{ats \in ATS} sat(s, ats)}{card(ATS)} \tag{3}$$

where:

$$sat(s, ats) = \begin{cases} 1, & \text{if } s \text{ satisfies ats;} \\ 0, & \text{otherwise.} \end{cases} \tag{4}$$

A Web service, s that providing an operation op, satisfies an ats (ats $\in$ ATS) if ats(op) belongs to TS(s). All TF values belong to the interval [0,1], where 1 means a perfect satisfiability, while 0 expresses that none ats is satisfied.

For a given TCS, the overall TF can be computed based on the used composition patterns and the TF of each components service. For a XOR pattern, each Case statement is annotated with the probability to be chosen. For example, for a Workflow containing only a XOR pattern composed of n services, $s_1,..., s_n$, with probabilities $p_1,..., p_n$, the overall TF is computed as follow:

$$\sum_{i=1}^{n} p_i * TF(s_i) \tag{5}$$

The TF value of Loop pattern is computed without taking into account the factor k (the number of iteration of the loop). For example, if the Workflow contains a Loop composed of a service, s, then the estimated TF will be $TF(s)$.

The aggregation TF for the AND pattern, is essentially the same as those for the Sequence construct. For a Workflow composed of n services $s_i,..,s_n$ the overall TF is computed as follow:

$$\prod_{i=1}^{n} TF(s_i) \tag{6}$$

3.4 SLA Contract

The SLA contract specified between the TCS providers and their customers is a set of constraints, which are defined based on QoS properties. In this work, we consider the following constraints:

$$Rt(TCS) \leq rmax \tag{7}$$
$$Ct(TCS) \leq cmax \tag{8}$$
$$At(TCS) \geq amin \tag{9}$$

These constraints should be usually fulfilled in order to provide a satisfactory TCS for customers. The first and second are upper bound constraints on the negative QoS properties (Q_n). They express that the response time and cost of a TCS shall be smaller than maximal thresholds that we note by $rmax$ and $cmax$. The last one expresses a lower bound constraint on positive QoS properties (Q_p). It states that availability value of a TCS must be greater than a minimal threshold $amin$.

It is therefore necessary to define a comparison criterion for ranking the Web services compliant with the SLA contract. To do so, we introduce a new metric called **SLA Fitness** (SF for short). SF expresses the consistency level of the QoS properties of a given TCS with SLA. Let cs be a TCS, we define SF(cs) as follow:

$$SF(cs) = \alpha_1 * (cmax - Ct(cs)) + \alpha_2 * (rmax - Rt(cs)$$
$$- rmax) + \alpha_3 * (At(cs) - amin) \tag{10}$$

where α_1, α_2 and α_3 are the QoS penalty weighting factors that belonging to the interval [0,1]. They express the customers preferences and must respect the following property: $\alpha_1 + \alpha_2 + \alpha_3 = 1$. They determine which constraint of SLA contract that must be satisfied for guaranteeing customer satisfaction. For a booking service, a customer requires that the service has to respond to their request within 5 ms. Another customer requests that the service has to be usually available. This implies that first and third constraints must be satisfied. The others ones are not important for customer (can be violated).

Theorem 1. *The TCS having a positive SF necessarily meet at least one constraint of the SLA contract.*

Proof. We assume that there is a TCS whose SF is positive and does not respect the SLA contract. This implies that the quantities α_1*(cmax-Ct(TCS)), α_2*(rmax-Rt(CS)) and α_3*(At(TCS)-amin-) are strictly negative and their sum is also negative. This is not possible since the quantity α_1*(Ct(TCS)-cmax)+ α_2*(Rt(TCS)-rmax)+ α_3*(amin-At(TCS)) is positive. The TCS having the maximal positive SF more satisfies the customers requirements.

3.5   Problem Statement and Formalization

The ATS and SLA-aware WSC is a constraint optimization problem. It aims at selecting a set of web service to be composed at runtime for creating new value added services compliant with ATS and that better satisfy the SLA contract. It can be formulated as a quadruplet (WF, ATS, SLA, W), where:

- WF is a Workflow composed of n operations, OPERATION = $\{T_1,..., T_n\}$, that are connected by composition patterns.
- ATS = $\{ats_1,..., ats_d\}$, where $ats_i \in OPERATION \rightarrow STATE$.
- SLA = $\{c_1,..., c_d\}$ is a set of QoS constraints.
- W = $\{\alpha_1,..., \alpha_n\}$ denotes the QoS penalty weighting factors defined by customers, where, $\alpha_i \in [0,1]$ is the penalty weight of q_i, with $\sum_{i=1}^{m} \alpha_i = 1$.

The selection process of web services is done in two separate steps: ATS selection followed by an SLA one. The first step consists in finding all combinations compliant with ATS (*e.g.*, the TF of each combination is equal to 1), or the one that has the lower value of TF otherwise. In the second step, the combination, which has the maximal positive SF is selected. If there is no solution satisfying customers preferences, the combination maximizing the SF function will be chosen.

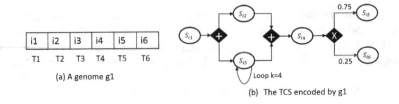

Fig. 3. A TCS is encoded by a genome

4   Solving ATS and SLA-aware WSC Problem Using GA-based Approach

As we described above, the work, presented in this paper, aims to propose an approach, based on GA, to quickly determine a TCS compliant with ATS and better satisfying the SLA contract. Such a TCS needs to:

- meet ATS, which are expressed by designers;
- optimizing the SF function.

In the sequel we shall consider a Workflow WF composed of n operations, T_1,..., T_n, whose structure is defined through some Workflow description language like BPEL. Each operation T_i can be executed by one of the m_i available Web services s_{i1},..., s_{m_i}, which are functionally equivalent.

4.1   Genome Encoding

We encode the ATS and SLA-aware WSC problem with an appropriate genome. As illustrated in Fig. 3, the genome is represented by an array of integers. It is an individual in the population, which represents a TCS (see Fig. 3b). The size of the array is equal to the number of operations belonging to WF. Each element of the array is an index of a web service. We assume the uniqueness of the index of web services providing the same operations (two web services s1 and s2 implementing the same operation must have different indexes). This allows to avoid such conflicts among web services during the generation of the solution. The indexes of all web services providing the same operation are stored in a table.

We adopt the standard two-points crossover operator used in [9]. The mutation operator randomly selects an operation T_i (i.e., a position in the genome) and randomly replaces the corresponding Web service index with another one of those available. The operations that are provided only by one available Web service are taken out from the GA evolution.

4.2   Fitness Function

The problem can now be modeled by means of a fitness function. The fitness function needs to maximize the SF function based on the customers needs.

In addition, it must penalize individual whose TF is strictly smaller than 1 (do not meet ATS), and drive the evolution towards ATS satisfaction. The fitness function for a genome g is defined as follows:

$$F(g) = SF(g) + \beta * \frac{cgen}{maxgen} * TF(g) \tag{11}$$

where β is the transactional penalty factor (real and positive weight), cgen ($cgen \in \mathbb{N}$) is the current generation and maxgen (cgen $\in$[0,maxgen]) is the maximum number of the generation.

It is necessary to define a stop condition for GA. In our work, we consider the following condition:

$$F(g) \geq \delta \tag{12}$$

where δ is the weighting quality factor (positive real value). The value of δ is determined by experimentations. A solution, which is represented by a genome g, is said to be feasible if it complies with ATS(TF(g)=1) and respects the stop condition (F(g)$\geq \delta$). In order to obtain all feasible solutions, we have adopted the following approach:

Step 1: we iterate with a maximum number of generations ($maxgen$) until TF(g) = 1. If it does not happen within $maxgen$ generations, then no solution has been found;

Step 2: once TF(g) = 1, we iterate over a further, extended number of generations $maxgen'$, which may be a percentage of $maxgen$.

Step 3: if there is no solution satisfying the stop condition (12), we take the individual with maximal SF.

5   Empirical Studies

In this section, we present the results of the experiments performed for solving the ATS and SLA-aware WSC problem using GA-based approach. By experiments, we show the effectiveness and the performance of our approach. We start by studying the impact of the value of penalty factors (δ and β) on the quality of the solution. Then, we present the evolution of the fitness parameters (SF and TF functions). Finally, we present a comparison a study of GA with Linear Integer Programming (LIP) and Ant Colony Optimization (ACO).

5.1   Experimental Setting

The experiments and simulation were performed on a computer with Pentium 4, 2.8 GHz, 4 GB of RAM, Windows 8 professional edition. Our GA was implemented in Java using a freely available library[1], while the commercial solver CPLEX 12.5 and MATLAB are respectively used for LIP and ACO. We have implemented a function, called **atsGenerator(WF)** in order to facilitate the definition of the

---

[1] http://sourceforge.net/projects/java-galib/.

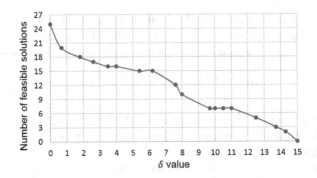

Fig. 4. The impact of the variation of δ on the quality of the solution

ATS during the experimentations. This function takes a Workflow as input parameters and returns an ats. It is especially used for iteratively regenerating a new ATS when the Workflow is modified during the experimentation.

For sake of simplicity, the weight of the SF function, α_1, α_2 and α_3, are respectively set to following value 0.5, 0.25 and 0.25. In this setting, all constraints are required and have to be satisfied, but the first one (7) is more important for customers. We set the value of β to 2. The values of δ is determined by experimentations.

We use QWS dataset [22] as a QoS dataset of candidate services. Specifically, we use response time, cost and availability as QoS parameters. In addition, we assign, for each candidate service, a transactional property (r, p, c, cr or pr).

5.2   Impact of the Quality Factor

We first vary the quality of the solution δ from 0 to 15. Then, we compute, at each case, the size of the set of the found feasible solutions.

Figure 4 presents the impact of variation of the parameter δ on the size of the set of feasible solutions. As shown, the number of feasible solutions decreases when δ increases. This is because with δ becoming stricter, the existence probability for a feasible solution declines. Some of solutions are rejected since they don't satisfy the GA stop condition (12). From the value 12, all solutions meet the constraints of the SLA contract that are required by customers. There is a risk to reject good solutions when δ exceeds 15.

As a conclusion, the quality of the generated solution is influenced by the value of δ. The fitness value of the good solutions belong to the interval [12, 14.5].

5.3   Evolution of the Fitness Parameters

In this section, we fix the value of δ to 12 for studying the evolution of SF and TF functions over generation. We consider the Workflow of the OTA service shown in Fig. 1. The designers' requirements are presented in Table 1. We assign, for each operation, a variable number of candidates (5, 10, 15, 20, 25, etc.).

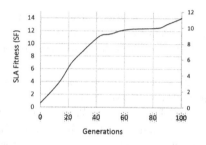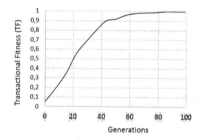

Fig. 5. Evolution of the fitness parameters

Figure 5 reports the evolution of different parameters (TF and SF) for the fitness function. The evolution shows how the GA is able to find a solution satisfying the transactional requirements ($\text{TF}(g) = 1$) and, at the same time, optimizes the QoS requirements (i.e., maximizing the SF function). According to our experimentation, the individual having the maximal positive SF more satisfy the user requirements (satisfying more constraints of the SLA contract). For instance, the individual whose SF is larger than 12 satisfies all constraints that is required by customers. The experiments are also replicated on Workflow processes of different sizes and complexity, basically confirming the results reported above.

5.4 Comparing GA with LIP and ACO

As a case study, we extend the Workflow used in the previous section by adding 4 others operations. The operation T2 is replaced by two consecutive operations T2a and T2b. Three additional operations are added to the switch constructs (T7, T8 and T9). We used the atsGenerator module that we have implemented to generate a new ATS to the updated Workflow.

Figure 6 shows the results of our comparison. When the number of Web services is small (0, 15), LIP outperforms GA and ACO, which have almost the same performance. This substantially confirms the choice made by other

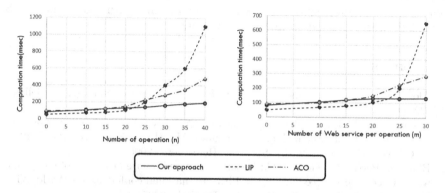

Fig. 6. Comparing GA with Ant colony optimization and Integer programing

works that adopted this kind of approach [7]. LIP is still better than ACO when the number of Web services is between 15 and 26. For about 23 Web services, GA is able to keep its timing performance almost constant. This is not the case for LIP ones for which we see an exponential grow due to the corresponding increment of the number of Web services. Then, we investigated whether the performance variation was due to the increase of the search space and the used variables. To this aim, we increased the number of Web services for a few operation only. Results obtained confirms what was described above. In addition, they were confirmed by other experiments performed with Workflow process having different size.

According to our experimentations, the ACO algorithms can efficiently approximate the optimal solution only when the problem is not complex. Nevertheless, it is poorly performed and converges slower than GA if the size of the population is larger than 100. Indeed, the size of directed acyclic graph (DAG) grows with a number of operations (in ACO, the problem must be modeled as a DAG). For example, if the Workflow contains 50 operations (the associated ATS are randomly generated using atsGenerator module in a polynomial time) and each operation can be executed by one from 20 available Web service, then the size of the resulting DAG is about 20^{50}, which can destroys the performance of ACO program and increases the computation time (time to converge). This is confirmed when the number of operations exceeds 35 (see Fig. 6). Therefore, GA should be preferred than ACO and LIP when we have a large number of available Web for each operation. This is the case of the large-scale service-oriented systems.

6   Conclusion

In this paper, we proposed an ATS and SLA-aware service composition method. ATS and SLA awareness means that the selected service must meet ATS, while satisfying SLA contract. Our approach offers more flexibility not only to customer, but also to designers to specifier their preferences in term of control structure and fault recovery.

I currently working on extending our approach by considering more QoS properties and constraints. We will work on developing an algorithm that implements our novel composition method. This algorithm return the optimal TCS (compliant with ATS and optimizing SLA contract).

References

1. Alonso, G., Casati, F., Kuno, H., Machiraju, V.: Web Services. Springer, Heidelberg (2004)
2. Chen, Y., Li, Z., Jin, Q., Wang, C.: Study on QoS driven web services composition. In: Zhou, X., Li, J., Shen, H.T., Kitsuregawa, M., Zhang, Y. (eds.) APWeb 2006. LNCS, vol. 3841, pp. 702–707. Springer, Heidelberg (2006)
3. Bhiri, S., Perrin, O., Godart, C.: Ensuring required failure atomicity of composite web services. In: Proceedings of the 14th international conference on World Wide Web. WWW 2005, pp. 138–147. ACM, NewYork (2005)

4. Bhiri, S., Gaaloul, W., Godart, C., Perrin, O., Zaremba, M., Derguech, W.: Ensuring customised transactional reliability of composite services. J. Database Manag. **22**(2), 64–92 (2011)
5. Montagut, F., Molva, R.: Augmenting web services composition with transactional requirements. In: International Conference on Web Services, ICWS 2006, pp. 91–98. IEEE (2006)
6. Montagut, F., Molva, R., Golega, S.T.: Automating the composition of transactional web services. Int. J. Web Service Res. **5**(1), 24–41 (2008)
7. Zeng, L., Benatallah, B., Ngu, A.H.H., Dumas, M., Kalagnanam, J., Chang, H.: Qos-aware middleware for web services composition. IEEE Trans. Software Eng. **30**(5), 311–327 (2004)
8. Comes, D., Baraki, H., Reichle, R., Zapf, M., Geihs, K.: Heuristic approaches for QoS-based service selection. In: Maglio, P.P., Weske, M., Yang, J., Fantinato, M. (eds.) ICSOC 2010. LNCS, vol. 6470, pp. 441–455. Springer, Heidelberg (2010)
9. Canfora, G., Penta, M.D., Esposito, R., Villani, M.L.: An approach for QoS-aware service composition based on genetic algorithms. In: GECCO, pp. 1069–1075 (2005)
10. El Hadad, J., Manouvrier, M., Rukoz, M.: TQoS: Transactional and QoS-aware selection algorithm for automatic web service composition. IEEE Trans. Serv. Comput. **3**(1), 73–85 (2010)
11. Ben Mabrouk, N., Beauche, S., Kuznetsova, E., Georgantas, N., Issarny, V.: QoS-aware service composition in dynamic service oriented environments. In: Bacon, J.M., Cooper, B.F. (eds.) Middleware 2009. LNCS, vol. 5896, pp. 123–142. Springer, Heidelberg (2009)
12. Parejo, J.A., Fernandez, P., Ruiz-Cortés, A.: QoS-aware services composition using tabu search and hybrid genetic algorithms. Actas de los Talleres de las Jornadas de Ingeniera del Software y Bases de Datos **2**(1), 51–66 (2008)
13. Parejo, J.A., Segura, S., Fernandez, P., Ruiz-Cortés, A.: QoS-aware web services composition using GRASP with path relinking. Expert Syst. Appl. **41**(9), 4211–4223 (2014)
14. Rusinkiewicz, M., Sheth, A.P.: Specification and execution of transactional workflows. Modern Database Syst. **1995**, 592–620 (1995)
15. Gaaloul, W., Bhiri, S., Rouached, M.: Event-based design and runtime verification of composite service transactional behavior. IEEE Trans. Serv. Comput. **3**(1), 32–45 (2010)
16. Alrifai, M., Risse, T., Nejdl, W.: A hybrid approach for efficient web service composition with end-to-end QoS constraints. ACM Trans. Web **6**(2), 7 (2012)
17. Wu, Q., Zhu, Q.: Transactional and QoS-aware dynamic service composition based on ant colony optimization. Future Gener. Comp. Syst. **29**(5), 1112–1119 (2013)
18. Alonso, G., Casati, F., Kuno, H., Machiraju, V.: Web Services: Concepts, Architectures and Applications. Springer, Berlin (2003)
19. Bhiri, S., Godart, C., Perrin, O.: Transactional patterns for reliable web services compositions. In: Proceedings of the 6th international conference on Web engineering. ICWE 2006, pp. 137–144. ACM, New York (2006)
20. Moscato, F., Mazzocca, N., Vittorini, V., Di Lorenzo, G., Mosca, P., Magaldi, M.: Workflow pattern analysis in web services orchestration: the BPEL4WS example. In: Yang, L.T., Rana, O.F., Di Martino, B., Dongarra, J. (eds.) HPCC 2005. LNCS, vol. 3726, pp. 395–400. Springer, Heidelberg (2005)
21. Wohed, P., Aalst, W., Dumas, M., Hofstede, A.: Pattern-based analysis of BPEL4WS (2002)
22. Al-Masri, E., Mahmoud, Q.H.: Investigating web services on the world wide web. In: WWW, pp. 795–804 (2008)

Crawling Ranked Deep Web Data Sources

Yan Wang[1]([✉]), Yaxin Li[1], Nannan Pi[1], and Jianguo Lu[2]

[1] School of Information, Central University of Finance and Economics, Beijing, China
dayanking@gmail.com, {liyaxin,pinan}@email.cufe.edu.cn
[2] School of Computer Science, University of Windsor, Windsor, Canada
jlu@uwindsor.ca

Abstract. In the era of big data, the vast majority of the data are not from the surface web, the web that is interconnected by hyperlinks and indexed by most general purpose search engines. Instead, the trove of valuable data often reside in the deep web, the web that is hidden behind query interfaces. Since the data in the deep web are often of high value, there is a line of research on crawling deep web data sources in the recent decade. However, most existing crawling methods assume that all the matched documents are returned. In practice, many data sources rank the matched documents, and return only the top k matches. When conventional methods are applied on such ranked data sources, popular queries that matches more than k documents will cause large redundancy. This paper proposes the document frequency (df) based algorithm that exploits the queries whose document frequencies are within the specified range. The algorithm is extensively tested on a variety of datasets and compared with existing two algorithms. We demonstrate that our method outperforms the two algorithms 58 % and 90 % on average respectively.

Keywords: Deep web crawling · Query selection · Estimation · Document frequency · Return limit

1 Introduction

The searchable web forms and programmable web APIs permeate the daily lives of ordinary web users as well as professional web programmers. The trove of the data hidden behind these query interfaces constitutes the deep web [1–4]. In contrast to the surface web that is connected by hyperlinks, the deep web cannot be crawled by following the hyperlinks embedded in web pages. Instead, documents in the deep web can be retrieved using queries only. For this reason, it is also called the hidden web [5–7].

The deep web is considered full of rich content that is much bigger than the surface web [1]. Nowadays, almost every web site comes with a search box. Many of them, such as twitter.com, provide in addition a programmable web API. First of all, it would be nice if those deep web documents were search engine visible. Not surprisingly general search engines, such as Google [4,8,9] and Bing [10] try to index some of these un-crawled territory. In addition, numerous applications

© Springer International Publishing Switzerland 2015
J. Wang et al. (Eds.): WISE 2015, Part I, LNCS 9418, pp. 384–398, 2015.
DOI: 10.1007/978-3-319-26190-4_26

want to tap into the rich deposit of data to build distributed search engines [11], data integration applications [12], vertical portals [13] etc. While the deep web data providers are happy to serve the data to ordinary users and even application programs, they may not want to be overloaded with automated crawlers whose target is to index or even worse to download the entire database to set up their own operation. Thus, recently more deep web providers put some constraints on their services and then ordinary data sources turn out to be *ranked* data sources.

A deep web data source is *ranked* if it sorts the matched documents and only returns top-k ones. The ranking criteria can be dynamic (e.g., tf-idf ranking) or static (e.g., date of creation). The return limit k could be any positive integer, e.g., 1,000 for Google, 10,000 for dmoz.org [14]. For ranked deep web data sources, it is difficult to excavate most documents by using most existing crawling methods.

Our task is to retrieve all documents inside a data source with the minimal cost by iteratively issuing promising queries. The crawling cost is the network transmission that can be measured by the total number of returned documents by queries [15–17]. Hence, most existing crawling methods for query selection are to maximize the coverage per unit cost, i.e., each query needs to return most *new* documents with as less redundant documents (the documents returned by previous queries) as possible. Meanwhile, the assumption of these methods is that all matched documents can be returned like ordinary data sources. However, for ranked data sources, only the k documents with the highest ranking numbers can be retrieved. This will cause that the expected new documents could not be returned by each query. It means that the promising queries from these methods for crawling ranked data sources will not reach a good performance as it works on ordinary data sources. In fact, all methods in [4, 15–20] prefer to select high or middle *document frequency* (df for short, the number of documents containing a query) queries to retrieve data. From the empirical study shown in Sect. 3, higher df queries

- lead to lower crawling coverage, i.e., only documents ranked high can be returned.
- cause more redundant retrieval. i.e., documents ranked high are much repeatedly returned by high df queries.

To address this problem, we propose a df-based crawling method for ranked data sources in this paper. The key idea of our method is to utilize middle-low df terms[1] as queries to avoid the effect of the ranking plus return limit, and then the problem of crawling ranked data sources is degenerated to crawling ordinary ones. More specifically, we firstly obtained a uniformly distributed small sample from the target data sources, then all terms of the sample are retrieved to generate a query pool. With the query pool, a df estimator is used to estimate their document frequencies and only the terms whose dfs are less than the return

---

[1] In this paper, we use the two words 'term' and 'query' interchangeably and the minor difference is that a query is an issued term.

limit k can be the query candidates. Finally, the queries are randomly selected from the candidates to retrieve the documents.

To evaluate the performance of our method, it is tested on Reuters, Wikipedia, Gov2 and Newsgroup standard data and compared with the baseline random query algorithm and the representative high-frequency query algorithm. We demonstrate that our method outperforms the two algorithms 58 % and 90 % on average respectively.

2   Related Work

In the past decade, there is a line of research for crawling deep web data sources. According to the underlying methods, they are roughly categorized into three different kinds: (1) the methods based on heuristic rules [4,9,18]; (2) the methods based on approximation algorithms for minimum set-covering problem [15–17,19]; (3) the methods based on machine learning [20–23]. And the strategies of these methods for query selection are to maximize the coverage or the coverage per unit cost.

In [18], the authors first proposed a greedy-based method to siphon deep web data sources by selecting queries with highest frequencies from a sample, and the method is composed of two phases. Phase 1 randomly selects a set of queries from the HTML search form and issues them to the target data source. By downloading and extracting all terms from the returned documents, their algorithm creates a candidate term list for crawling and all terms are ordered by their dfs in the downloaded collection. Then the process iteratively retrieves new documents and terms and update the frequency of each term by issuing the terms in the list until the number of submissions reaches the threshold. In phase 2, the method uses the highest frequency terms in the list to reach the highest coverage. Google [4] provided another heuristic rules to crawling deep web data sources. The policy of query selection is based on well-known TF-IDF evaluation system. It incrementally adds the top 25 terms of returned documents sorted by their TF-IDF values into a query pool. After eliminating popular and rare terms from the pool, the remaining terms are issued to the target data source and a new set of returned documents are downloaded. The authors of [4] attempt to use high-middle frequencies to reach a good coverage with less redundancies. Another heuristic method [9] is to exploit the query logs of Google to retrieve all entity documents in deep web data sources. Like [18], the method prefers popular queries from users to do crawling.

In fact, the crawling problem can be modelled as a set-covering problem. The universe is the set of all the documents, and each query, or the documents that contain the query, is a subset. The constraint is that all the documents need to be covered. In [19], the authors first modeled the crawling process as the set-covering problem. The document-query matrix is constructed from all downloaded documents and thus it is incremental iteratively. And the cost is network transmission. the strategy of [19] for query selection is to maximize the coverage per unit cost by using greedy algorithm. In [15–17], the crawling problem is also considered as the set-covering problem. The difference from [19] is that the algorithm in [15] is based on a fixed sample not incremental downloaded documents and the greedy algorithms used in [16,17] are further improved by adding document weights into the query selection function.

The authors in [20–23] attempt to introduce machine learning algorithms into crawling method. More specifically, the methods used machine-learning-based estimators to estimate the document frequency of each candidate term in order to select the query who can return most new documents. In [20], the authors used the reinforcement learning method to generate queries, i.e., a crawler and a target data source are considered as an agent and the environment respectively. Then its query selection strategy will be dynamically adjusted by learning previous querying results and takes account of two-step long reward.

Overall, all these methods select terms with relatively high document frequencies, without the consideration of the top-k constraint. In ranked data sources where only the top k matched documents are returned, popular terms will results in high duplicates as we shall explain in the next two sections.

3   Problem Description

Given a data source $DB = \{d_1, \cdots, d_m\}$, each d_i $(1 \leq i \leq m)$ represent a document and the subscript i is the ranking number, i.e., if $i < j$, $d_i > d_j$. Given a set of terms $Q = \{q_1, ..., q_n\}$, each document d_i contains a set of the terms. In turn, each term q_j $(1 \leq j \leq n)$ covers a set of documents in DB, and the total number of the documents is the document frequency of the term q_j denoted by $F(q_j)$. Given a return limit k, the *crawlable* relationship between the documents and the terms can be represented by the document-term matrix $A = (a_{ij})$ where

$$a_{ij} = \begin{cases} 1, & \text{if } q_j \text{ covers } d_i \text{ and } i \leq k; \\ 0, & otherwise. \end{cases} \quad (1)$$

In the matrix, each row i and each column j represent a document d_i and a term q_j respectively. Each entry $a_{ij} = 1$ if the term q_j covers the document d_i and the ranking number i should be less than (or equal to) the return limit k. The matrix demonstrates how crawlable the corresponding ranked data source by using the given set of terms.

With the matrix, our problem is to select a subset of terms Q' $(Q' \subseteq Q)$ which can cover as many documents in DB as possible with the minimal cost. Since the crawling cost is the network transmission and it is proportional to the total number of documents returned, the cost for each term is defined as following:

$$c(q_j) = \begin{cases} F(q_j), & \text{if } F(q_j) \leq k; \\ k, & \text{if } F(q_j) > k. \end{cases} \quad (2)$$

Hence, the cost of a set of terms Q' is the sum of the costs of the terms in Q':

$$c(Q') = \sum_{q_j \in Q'} c(q_j); \qquad (3)$$

In order to compare the performance of the same set of queries Q' in different data sources, we normalize the coverage and the cost by using the hit rate (HR for short) and the overlapping rate (OR for short):

$$HR = \frac{|S_{Q'}|}{|DB|},$$

$$OR = \frac{c(Q')}{|S_{Q'}|}.$$

where $|DB|$ and $|S_{Q'}|$ are the size of the data source DB and the number of unique documents covered by Q' respectively.

4 The df-Based Algorithm

4.1 Motivation

Most existing deep web crawling methods work are designed for crawling ordinary data sources in which all the matched documents are returned. However, for ranked data sources, these approaches no longer work fine because highly ranked documents tend to be retrieved quite often, while lower ranked documents have less probability of being returned even if they are matched by queries.

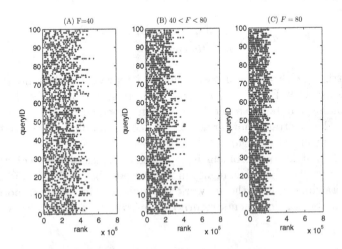

Fig. 1. Scatter plots for query results from different types of queries. For each experiment, 100 queries are sent. X axis represents the document ranking number. Sub figure (A) queries with $F = 40$; (B) queries with $40 < F < 80$; (C) queries with $F = 80$. The data source is Reuters.

To explain the phenomenon, we have conducted three experiments which are illustrated in Fig. 1. The experimental data source is Reuters corpus in local, which contains 806,791 documents. Each document is assigned a randomly generated static no duplicated ranking number. In each experiment, 100 queries are sent. The return limit k is 20, but the document frequencies F of the queries are 40, 40-80 and 80 respectively for each experiment.

From Fig. 1, we find that, if queries with large document frequencies, (1) documents ranked low may not be retrieved and thus, high coverage rates cannot be reached; (2) many documents ranked high will be repeatedly retrieved, causing high redundancy. Here, document frequencies F are considered *large* compared to the return limit k. For example, if $k = 20$, $F = 40$ is considered large.

Figure 1(B) shows the df of each query and the ranks of the retrieved documents when $k = 20$ and 100 queries with $40 < F < 80$ are sent. As we can see from this figure, documents with ranks between 430,000 and 800,000 are not retrieved at all. In addition, according to the data of this experiment, documents with ranking numbers between 0 and 410,000 are retrieved 1991 times.

More precisely, when the document frequencies of all queries in a query pool Q are greater than k, the range of the document ranks of those that can be retrieved by queries in Q can be determined by the range of the document frequencies of Q. This can be approximately expressed by the following formula:

$$max\{M_q|q \in Q\} \approx \frac{k}{min\{F(q)|q \in Q\}} \times |DB|, \tag{4}$$

where M_q is the maximum ranking number of the documents returned by issuing q.

In the first experiment (Fig. 1(A)),

$k = 20,$

$min\{F(q)|q \in Q\} = 40,$

$N = 806791.$

So we have $max\{M_q|q \in Q\} \approx (20/40) * 806791$. Thus, the biggest rank of the returned documents should be around 403,396. Most of documents whose ranking numbers bigger than this are not retrieved by these terms.

Similarly, in the third experiment (Fig. 1(C)),

$k = 20,$

$min\{F(q)|q \in Q\} = 80,$

$N = 806,791.$

So we have $max\{M_q|q \in Q\} \approx (20/80) * 806,791$. Thus, the biggest ranking number of the returned documents should be approximately 201,698. Most of documents whose ranks bigger than this are hardly retrieved by these terms.

The above formula defines a limit on the size of the documents that can be returned by Q. It was given in [24] as a special case of this formula when the document frequencies of all queries are the same and the documents are ranked according to their document degrees.

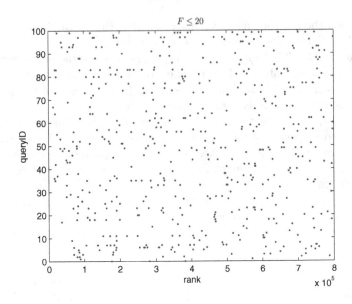

Fig. 2. The scatter plot for the query result. 100 queries whose $F \leq 20$ are sent and X axis represents document rank.

Contrast to the three experments shown in Fig. 1, we have conducted another experiment with the same conditions shown in Fig. 2. In this experiment, 100 queries whose document frequencies are less than the return limit $k = 20$ are sent. It can be seen that, compared to Fig. 1, the ranks of the retrieved documents can reach the whole range of the rank numbers.

Thus, we say, for a query with small document frequency, the documents matched by it are not affected by the return limit and ranking criteria. Here, document frequencies are considered *small* compared to the return limit k. Any numbers smaller than k, for example, can be considered small. Any numbers not much bigger than k can also be considered small. As a consequence, if there are enough queries with small dfs, most of all documents can be returned.

Similar experiments were conducted previously in [24] where the documents are ranked by their document degrees and there is no consideration of return limit k.

According to the above observations, we can draw the following conclusions as guidelines for the query selection for ranked data sources.

– *In order to reach high coverage rate, the set of selected queries should not contain only those with large document frequencies.*
  This is because, according to the observation of Fig. 1, some documents can rarely be returned by queries with large document frequencies, simultaneously, the queries with large document frequencies very often bring back same documents with small rank numbers.
– *We can reach a much better coverage rate by using queries with small document frequencies.*
  This is straightforward from the observation of Fig. 2.

The proposed crawling method is based on the above guidelines. It is introduced in the next section.

4.2 Our Algorithm

Our crawling method is shown and illustrated in Algorithm 1 and Fig. 3. It consists of four steps.

Algorithm 1. Query selection for ranked data sources.

Input: the original data source DB, the sample size m and the return limit k.
Output: a collection of URLs S.
Process:
1        $D = retrieveDocs(DB, m)$;
2        $Q = retrieveTerms(D)$;
3   foreach q in Q
               if $\widehat{F}(q) \leq k$
                      add q to Q_k;
4   while(!requirment){
               randomly select $q_u \in Q_k$ as query;
               $S = S + retrieveURLs(q_u)$;
       }

Our method is sample-based: some sample documents are randomly extracted from the original ranked data source DB (Step 1). These sample documents form the sample data source D. The size of the sample data source should meet the requirement which ensures that there are enough queries to cover DB. Our algorithm runs on the samples to generate the selected queries which are then mapped into the original data source.

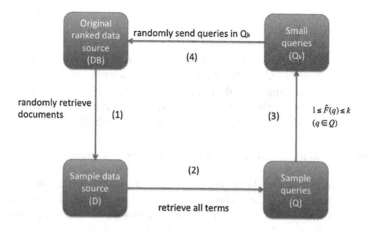

Fig. 3. Our crawling method for ranked deep web data sources

From the sample data source, we retrieve all terms to obtain the set of terms Q for selection (Step 2).

According to the discussions in the previous section, we can choose queries only among Q with small document frequencies and can reach a high coverage rate. Except the return limit k, the limit on the document frequencies to be considered as *small* depends on the given coverage rate as well. It is hard and beyond the scope of the present work to find such a limit. Here we assume that by using all queries with document frequencies less than k they can cover most of the original data source, and we select among these queries to reach a given coverage rate.

Let Q_k denote the subset of Q which contains all the queries in Q whose estimated document frequencies $\hat{F}(q)$ are no greater than k.

In Step 4, we randomly select queries from Q_k sequentially until the given coverage rate is reached.

Note that (1) in Step 1, it is not a trivial work to obtain random sample since data sources only can be accessed by queries, however, we can use the sampling method provided in [25] to generate one; (2) most of the web-access data sources do not provide the document frequency for each term inside it, and it is time-consuming to obtain the document frequencies of all queries in Q by sending them. Here we use three different df estimators to estimate the document frequencies of all terms in Q: Maximum Likelihood Estimator (MLE) [26], Simple Good-Turing Estimator (SGT) [27] and Zipf-law-based estimator(Zif for short) [11].

5   Experiments

5.1   The Data

To show the performance of our method, we run our experiments on four corpus datasets. The four corpora are Reuters, Gov, Wikipedia, and Newsgroup. They contain around 1 millions documents. Their characteristics are summarized in Table 1. These are standard test data that are used by many researchers in information retrieval.

- *Reuters* is a TREC data set that contains 806,790 news stores in English.
- *Gov* is a subset of Gov2 that contains 1 million documents. Gov2 is a TREC test data collected from .gov domain during 2004, which contains 25 million documents. We used only a subset of the data for efficiency consideration.
- *Wikipedia* is the corpus provided by wikipedia.org which contains 1.4 million documents.
- *Newsgroups* includes 1,372,911 posts in various newsgroups.

In the experiment, we built our own search engine using Lucene [28], in order to have the details of a data source such as its size. In real deep web data sources, usually the total number of documents is unknown, hence it is impossible to calculate the HR and evaluate the methods.

Table 1. Summary of test corpora

Name	Number of docs	Size in MB	Avg file size (KB)
Reuters	806,791	666	0.83
Wikipedia	1,369,701	1950	1.42
Gov	1,000,000	5420	5.42
Newsgroup	1,372,911	1080	0.73

5.2 Experimental Results

To evaluate the performance of our query selection method, we compare it with the following two existing ones: baseline *random query method*, representative *high frequency query method* that is similar with the methods provided in [18, 19].

1. *random query method*: we use a middle-sized Webster English dictionary with 51,541 words and randomly select them as queries from it;
2. *high frequency query method*: given the document frequencies of all terms in D, we directly collect all high frequency terms whose document frequencies are higher than $2k$. However we do not use stop words. Then we randomly pick these queries one by one to send to DB;
3. *our df-based methods*: each term in D whose $\widehat{F}(q) \leq k$ is sent to DB randomly until the given coverage rate is reached or all terms are used up. Here $\widehat{F}(q)$ is calculated by the three estimation methods respectively.

We have run the three methods on the four different ranked data sources. For each experiment, the return limit k is set to 1000. Each document in a corpus has a unique randomly generated ranking number. Then a document with smaller ranking number is ranked higher than a document with a bigger ranking number.

Table 2 shows the number of candidate queries of the three df estimators for our method with a 3000-document random sample derived from each corpus data source. From this table, we can see that our method with different estimation methods could have the complete same candidate queries in Q_k, such as, for Gov and Newsgroup sample data sources, the three estimators select totally same terms into Q_k (i.e., all three estimators give very close estimations on each term in the sample such that the number of queries whose $F(q) \leq k$ estimated by the three methods are identical). Thus, for our methods with the same candidate terms in Q_k, only one experiment for the three estimators is carried out.

From Fig. 4, we can see that, (1) in the presence of the return limit, the proposed method with any of the three estimation methods is much better than the random query method, which is better than the high-frequency query method; (2) the results of our df-based methods with different estimators are close to each other.

Note that all result curves in Fig. 4 look much more smooth. The reason is that, due to the return limit, the improvement of HR at each iteration could be very low (maximum 1000 new documents returned) and thus the 80 % coverages

Table 2. The number of candidate terms of the three methods in each experiment.

corpus	random query	high-frequency query	our method		
			MLE	SGT	Zipf
Reuters	51541	5441	22330	23744	23744
Wiki	51541	16030	78405	78405	83575
Gov	51541	15476	56371	56371	56371
Newsgroup	51541	15478	67860	67860	67860

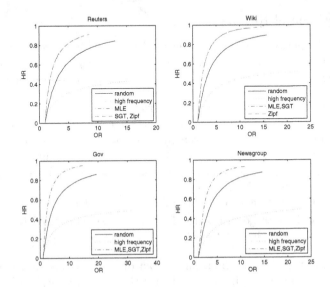

Fig. 4. The results of the three methods on different corpora. The return limit $k = 1000$. Documents are statically ranked. MLE, SGT, Zipf: our methods with MLE, SGT and Zipf-law-based estimation methods respectively.

of all results are based on a large number of sent queries issued. Figure 5 shows a zoomed-in area of each subgraph in Fig. 4 and the fluctuation of each result is demonstrated.

Since the results of our df-based methods with different estimates are close to each other, Table 3 shows the comparison of the results of the random query method, the high frequency query method and our method with MLE. From Table 3, we can see the difference between the performance of our method compared to the other two:

- The random method can also reach high coverage rate as ours, but with 80 % coverage rate, our method gives around 46 % savings on the overlapping rate.
- The high frequency query method cannot reach beyond 50 % coverage rate and our method gives around 90 % savings at its highest coverages on the four corpora.

Table 3. Comparison of the three methods ($Imp = \frac{OR_{random}(OR_{high}) - OR_{ours}}{OR_{random}(OR_{high})}$).

corpus	$HR(\%)$	Our method	Random query method		high-frequency query method	
		OR	OR	Imp(%)	OR	Imp(%)
Reuters	10	1.11	1.20	8.3	1.47	24.4
	20	1.26	1.48	14.8	2.56	50.7
	30	1.47	1.87	21.3	4.78	69.2
	40	1.73	2.40	27.9	12.55	86.2
	42	1.79	2.49	28.1	16.08	88.8
	50	2.06	2.96	30.4	-	-
	60	2.54	3.87	34.3	-	-
	70	3.33	5.33	37.5	-	-
	80	4.65	8.21	43.3	-	-
	84	5.53	12.87	57.0	-	-
Wiki	10	1.15	1.32	12.8	1.67	31.1
	20	1.33	1.65	19.3	2.55	47.8
	30	1.55	2.13	27.2	4.52	65.7
	40	1.80	2.68	32.8	8.77	79.4
	50	2.12	3.40	37.6	24.48	91.3
	60	2.54	4.41	42.4	-	-
	70	3.17	6.04	47.5	-	-
	80	4.19	9.04	53.6	-	-
	89	6.05	15.09	59.9	-	-
Gov2	10	1.22	1.36	10.2	1.81	32.5
	20	1.45	1.84	21.1	3.13	53.6
	30	1.73	2.41	28.2	6.18	72.0
	40	2.08	3.10	32.9	15.27	86.3
	45	2.28	3.52	35.2	34.56	93.4
	50	2.53	3.99	36.5	-	-
	60	3.13	5.33	41.2	-	-
	70	3.98	7.44	46.5	-	-
	80	5.53	11.75	52.9	-	-
	85	6.70	18.15	63.1	-	-
Newsgroup	10	1.19	1.29	7.75	1.62	26.5
	20	1.41	1.57	10.1	2.74	48.5
	30	1.65	1.97	16.2	5.16	68.0
	40	1.91	2.44	21.7	13.90	86.2
	44	2.01	2.68	25.0	25.66	92.1
	50	2.21	3.02	26.8	-	-
	60	2.61	3.84	32.0	-	-
	70	3.27	5.24	37.5	-	-
	80	4.39	7.75	43.3	-	-
	87	6.09	14.05	56.6	-	-

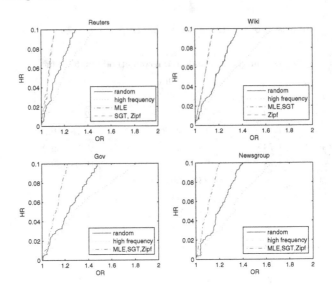

Fig. 5. The zoomed-in area of each subgraph in Fig. 4. The range of HR is up to 10 %.

As we have explained before, in the presence of a return limit and with our static ranking criterion, choosing many high frequency queries will result in low coverage rate and high overlapping rate, because (i) some documents with high ranking numbers may not be retrieved at all and as a consequence, the desired coverage rate cannot be reached; (ii) some documents with low ranking numbers may be repeatedly retrieved, causing retrieval of many redundant documents.

On the other hand, choosing many low frequency queries will result in high coverage rate and low overlapping rate, because (i) when we use queries with small document frequencies, any document can be returned by sending some of the queries; (ii) the coverage of the returned documents is more even which means less overlapping rate.

Thus, choosing many low frequency terms as we do in the present work is much better than choosing many high frequency terms as proposed in the high frequency query method both in terms of coverage rate and in terms of overlapping rate.

With the random query method, we have a mix of high and low frequency terms selected. Thus, its result both in terms of coverage rate and in terms of overlapping rate is between the high frequency query method and our method.

6   Conclusion and Future Work

In this paper, we presented a novel method to crawl ranked data source. Compared to traditional methods, our method works well in the presence of return limit and static ranking criterion. We are interested in extending this work for other ranking criteria in the original data source.

We have considered original data sources with fixed sizes and a fixed number as the return limit. For future work, we are interested in knowing the performance of this present method with the change of the size of the original data source and the change of the return limit k. Note that in particular, k is infinite corresponds to the case that there is no return limit, which is the setting used in many previous work. On the other hand, when k is pretty small and the size of ranked data source is large, it may be hard to find enough queries whose document frequencies are smaller than k in order to reach a high hit rate. In this paper, we have assumed that by using queries with document frequencies less than k we can reach a high coverage, and we could select among these queries to reach a given coverage rate. It remains an interesting problem to find the limit of the number of queries we need in order to reach a given coverage rate. In the case we have to work with a set of queries with document frequencies below this limit, we will try to use multiply key words.

Acknowledgements. This work is supported by NSFC (No.61440020 and No.6130 9029), NSERC, Programs for Innovation Research and 121 Project in Central University of Finance and Economics.

References

1. Bergman, M.K.: The deep web: Surfacing hidden value. J. Electron. Publishing **7**(1), 1–17 (2001)
2. Shestakov, D., Bhowmick, S.S., Lim, E.P.: Deque: querying the deep web. J. Data Knowl. Eng. **52**(3), 273–311 (2005)
3. He, B., Patel, M., Zhang, Z., Chang, K.C.: Accessing the deep web: a survey. Commun. ACM **50**(5), 94–101 (2007)
4. Madhavan, J., Ko, D., Kot, L., Ganapathy, V., Rasmussen, A., Halevy, A.: Google's deep-web crawl. In: Proceeding of VLDB, pp. 1241–1252 (2008)
5. Ipeirotis, P., Gravano, L., Sahami, M.: Probe, count, and classify: categorizing hidden web databases. In: proceeding of SIGMOD, pp. 67–68 (2001)
6. Raghavan, S., Molina, H.G.: Crawling the hidden web. In: Proceeding of the 27th international Conference on Very Large Data Bases (VLDB), pp. 129–138 (2001)
7. Liddle, S.W., Embley, D.W., Scott, D.T., Yau, S.H.: Extracting data behind web forms. In: Olivé, À., Yoshikawa, M., Yu, E.S.K. (eds.) ER 2003. LNCS, vol. 2784, pp. 402–413. Springer, Heidelberg (2003)
8. Madhavan, J., Afanasiev, L., Antova, L., Halevy, A.: Harnessing the deep web: present and future. In: Proceeding of CIDR (2009)
9. He, Y., Xin, D., V, G., Rajaraman, S., Shah, N.: Crawling deep web entity pages. In: Proceeding of WSDM 2013, pp. 355–364 (2013)
10. Wu, P., Wen, J.R., Liu, H., Ma, W.Y.: Query selection techniques for efficient crawling of structured web sources. In: Proceeding of ICDE, pp. 47–56 (2006)
11. Ipeirotis, P., Gravano, L.: Distributed search over the hidden web: Hierarchical database sampling and selection. In: VLDB (2002)
12. Dong, X., Srivastava, D.: Big data integration. In: ICDE, pp. 1245–1248 (2013)
13. Yang, M., Wang, H., Lim, L., Wang, M.: Optimizing content freshness of relations extracted from the web using keyword search. In: Proceeding of SIGMOND, pp. 819–830 (2010)

14. http://www.dmoz.org
15. Lu, J., Wang, Y., liang, J., Chen, J., Liu, J.: An approach to deep web crawling by sampling. In: Proceeding of Web Intelligence, pp. 718–724 (2008)
16. Wang, Y., Lu, J., Chen, J.: Crawling deep web using a new set covering algorithm. In: Huang, R., Yang, Q., Pei, J., Gama, J., Meng, X., Li, X. (eds.) ADMA 2009. LNCS, vol. 5678, pp. 326–337. Springer, Heidelberg (2009)
17. Wang, Y., Lu, J., Chen, J.: TS-IDS algorithm for query selection in the deep web crawling. In: Chen, L., Jia, Y., Sellis, T., Liu, G. (eds.) APWeb 2014. LNCS, vol. 8709, pp. 189–200. Springer, Heidelberg (2014)
18. Barbosa, L., Freire, J.: Siphoning hidden-web data through keyword-based interfaces. In: Proceeding of SBBD (2004)
19. Ntoulas, A., Zerfos, P., Cho, J.: Downloading textual hidden web content through keyword queries. In: Proceeding of the Joint Conference on Digital Libraries (JCDL), pp. 100–109 (2005)
20. Zheng, Q., Wu, Z., Cheng, X., Jiang, L., Liu, J.: Learning to crawl deep web. Inf. Syst. 38(6), 801–819 (2013)
21. Jiang, L., Wu, Z., Zheng, Q., Liu, J.: Learning deep web crawling with diverse featueres. In: WI-IAT, pp. 572–575 (2009)
22. Dong, Y., Li, Q.: A deep web crawling approach based on query harvest model. J. Comput. Inf. Syst. 8(3), 973–981 (2012)
23. Jiang, L., Wu, Z., Feng, Q., Liu, J., Zheng, Q.: Efficient deep web crawling using reinforcement learning. In: Zaki, M.J., Yu, J.X., Ravindran, B., Pudi, V. (eds.) PAKDD 2010, Part I. LNCS, vol. 6118, pp. 428–439. Springer, Heidelberg (2010)
24. Lu, J.: Ranking bias in deep web size estimation using capture recapture method. Journal of Data and Knowledge Engineering 69(8), 866–879 (2010)
25. Bar-Yossef, Z., Gurevich, M.: Random sampling from a search engine's index. In: WWW, pp. 367–376 (2006)
26. Myung, I.J.: Tutorial on maximum likelihood estimation. J. Math. Psychol. 47, 90–100 (2003)
27. Gale, W.A., Sampson, G.: Good-turing frequency estimation without tears*. J. Quant. Linguist. 2(3), 217–237 (1995)
28. Hatcher, E., Gospodnetic, O.: Lucene in Action. Manning Publications (2004)

Influence Maximization in Signed Social Networks

Chengguang Shen[1], Ryo Nishide[2], Ian Piumarta[2], Hideyuki Takada[2], and Wenxin Liang[1](✉)

[1] Dalian University of Technology, Dalian, China
shenchengguang@gmail.com, wxliang@dlut.edu.cn
[2] Ritsumeikan University, Kyoto, Japan
r_nishide@cm.is.ritsumei.ac.jp, {piumarta,htakada}@cs.ritsumei.ac.jp

Abstract. Influence Maximization is the problem of choosing a small set of seed users within a larger social network in order to maximize the spread of influence under certain diffusion models. The problem has been widely studied and several solutions have been proposed. Previous work has concentrated on positive relationships between users, with little attention given to the effect of negative relationships of users and the corresponding spread of negative opinion. In this paper we study influence maximization in signed social networks and propose a new diffusion model called LT-S, which is an extension to the classical linear threshold model incorporating both positive and negative opinions. To the best of our knowledge, we are the first to study the influence maximization problem in signed social networks with opinion formation. We prove that the influence spread function under the LT-S model is neither monotone nor submodular and propose an improved R-Greedy algorithm called RLP. Extensive experiments conducted on real signed social network datasets demonstrate that our algorithm outperforms the baseline algorithms in terms of efficiency and effectiveness.

Keywords: Influence maximization · Signed social network · Diffusion model

1 Introduction

Social networks such as Facebook, Twitter and Epinions have become important platforms for the spread of information, ideas, opinions and influence, which opens up great opportunities for large-scale viral marketing campaigns. Influence maximization is one of the most interesting problems in viral marketing and has received much research interest in recent years. Influence maximization (IM) is the problem of choosing a small set of influential users, initially targeted as adopted users, in a social network so as to maximize their aggregated influence. Kempe et al. [18] formulated IM as a directed optimization problem and proposed two basic influence diffusion models, independent cascade (IC) and linear threshold (LT). In both of these models a social network is considered

© Springer International Publishing Switzerland 2015
J. Wang et al. (Eds.): WISE 2015, Part I, LNCS 9418, pp. 399–414, 2015.
DOI: 10.1007/978-3-319-26190-4_27

as a directed graph, in which users are represented as vertices and edges reflect the social relationships between users. In both models a user is either *active* (an adopter of the product) or *inactive*. In the IC model, when a user becomes active at step t he will independently activate his inactive neighbors. In the LT model, each user has a threshold and will become active when the sum of incoming influence from his active neighbors exceeds that threshold.

Much effort [6,7,17,21] has been devoted to solving the IM problem and additional influence diffusion models [1,4,14,25] have been proposed. However, previous work only considers positive relationships, although negative relationships also exist in real social networks and have effects on influence diffusion. For example, Epinions[1] (an online review website) allows users to express trust or distrust of others, and participants on Slashdot[2] (a news discussion website) can denote others to be either 'friends' or 'foes'. Social networks having a sign, positive or negative, associated with edges are called *signed social networks* [3,16]. In signed networks the sign of each edge characterizes whether the corresponding individuals are 'friends' (positive link) or 'enemies' (negative link). Positive relationships are supposed to carry the same opinions between users in previous studies [6,7,18,25], as people more likely trust their friends and embrace the same opinions. Negative relationships carry the opposite opinions between users; if your 'enemies' choose one opinion, you are more likely to choose the opposite [3,12,16,22]. These follow the general social principles that "the friend of my friend is my friend" and "the enemy of my enemy is my friend".

Most models used for IM problems assume that users involved in propagation will always form and express positive opinions. However, in reality, users who adopted the product or information may share both positive and negative opinions with their friends. W. Chen et al. [4] extend the IC model with negative opinions and propose the IC-N model. In IC-N, a parameter q (the same across all users) is incorporated into the spread of negative opinions, which is too simplistic and not always realistic compared to real-world behaviors. Much work has also been reported in the sociology and economics literature on modeling opinion dynamics in social networks. Some models [10,11,15] incorporate both innate opinions and expressed opinions. Innate opinions are fixed for a user and can be formed according to the user's preference and history. Expressed opinions are the result of social processes. From the point view of computer science, however, no related work has been done, in particular work that combines opinion formation and its propagation through a signed network.

In this paper we propose a new model by extending the classic **Linear Threshold** model to **Signed** networks, which we call the LT-S diffusion model. In LT-S each user has a status that is either *active* or *inactive*. A user is active if the influence from his neighbors, both 'friends' and 'enemies', is greater than his threshold. When a user is in active status, he will spread his opinion to his neighbors. We denote this opinion as *expressed opinion* and distinguish it from *innate opinion* which is formed from users' preference and history. The expressed

---

[1] http://www.epinions.com/.

[2] http://slashdot.org/.

opinion of an active user is the result of the convex combination of his innate opinion and the expressed opinions from his neighbors. Each user is associated with a conformity value α_u reflecting the importance of a user's innate opinion compared to the expressed opinions coming from active neighbors. A user with a high conformity value has an expressed opinion that is largely governed by the expressed opinions of his neighbors, while a low conformity value indicates that the user's expressed opinion is governed more by his innate opinion. LT-S differs from the models proposed in [22,27] in the following ways. Li et al. [22] study influence diffusion in signed networks and extend the classic voter model to signed networks, whereas LT-S is based on the linear threshold model. Zhang et al. [27] propose an opinion-based cascading model considering individual opinions, but ignore the negative relationships between users.

Armed with the LT-S model, we formulate the influence maximization problem in signed social networks and prove it is NP-hard. In contrast to the traditional influence maximization problem, which focuses on maximizing the number of activated users, the influence spread function under our model maximizes the overall positive opinion of activated users, as positive opinion is more important from the point of view of virtual marketing. We prove that the influence spread function under the LT-S model is neither monotone nor submodular, which makes the classic greedy algorithm [18] inapplicable. In order to solve this problem, we propose an improved R-Greedy [14] algorithm, **R**-Greedy with **Live**-edge and **P**ropagation-path (RLP), which combines the R-Greedy algorithm with two efficient techniques for reducing execution time. Extensive experiments conducted on real signed social network datasets demonstrate the efficiency and effectiveness of the RLP algorithm.

The main contributions of this paper can be summarized as follows.

- We formulate the influence maximization problem in signed social networks and extend the classic linear threshold (LT) model to create a new model called LT-S. To the best of our knowledge, we are the first to study the influence maximization problem in signed social networks with opinion formation.
- We prove the influence maximization problem in signed social networks under the LT-S model is NP-hard and the influence function under LT-S is non-monotone and non-submodular, which makes classic greedy algorithms [18] inapplicable.
- We adopt the R-Greedy algorithm to solve the IM problem in signed social networks. However, using Monte-Carlo simulation makes the R-Greedy algorithm very time-consuming. To speed up the algorithm we present two efficient techniques, *live edges* and *propagation paths*, and combine them with the R-Greedy algorithm. This brings about a new algorithm called RLP (**R**-Greedy with **Live**-edge and **P**ropagation-path).
- We evaluate the performance of RLP within publicly available, real world signed social networks, and present experimental results demonstrating its efficiency and effectiveness.

The remainder of this paper is organized as follows. Related work is reviewed in the next section. In Sect. 3 we present the LT-S model and the formal problem definition. Section 4 introduces the R-Greedy algorithm and our optimized form, RLP. In Sect. 5 we describe our experiments and analyze the experimental results. Finally, Sect. 6 offers concluding remarks.

2   Related Work

Influence maximization is formulated as a discrete optimization problem by Kempe et al. [18]. They prove that the problem is NP-hard and that the influence spread function is monotone and submodular under both the IC and LT models. A set function f from sets to reals $f : 2^V \rightarrow R$ is monotone if $f(S) \leq f(T)$ for all $S \subseteq T$. The function f is submodular if $f(S \cup \{w\}) - f(S) \geq f(T \cup \{w\}) - f(T)$ for all $S \subseteq T$, $w \in V$ and $w \notin T$. Given these properties, a greedy algorithm is proposed to solve the IM problem As the greedy algorithm is very time-consuming, a number of studies are devoted to improving its performance. The CELF algorithm proposed by Leskovec et al. [21] exploits the submodularity property to achieve a 700× increase in performance compared with the original greedy algorithm. Cheng et al. [7] propose a static greedy algorithm, which reuses the generated sub-graphs to guarantee the submodularity property of the influence spread function. Several heuristic algorithms are also proposed to avoid using Monte-Carlo simulations. Chen et al. [6] restrict the computations to the local influence regions of nodes and adopt maximum influence paths to estimate influence spread. The IRIE algorithm proposed by Jung et al. [17] integrates influence ranking with influence estimation to overcome the disadvantages of pure influence ranking methods. Recently, several models have been proposed that extend IC and LT. In [25], the IM problem has been extended into continuous-time diffusion networks, Feng et al. [14] consider the problem of influence maximization with novelty decay, [1] extends the problem to competitive settings, and [5,24] study the time-constrained influence maximization problem.

Signed network analysis dates from the 1940s with the work of Heider [16] and its formalization by Harary and Carwright [3]. Signed networks have recently gained attention from computer scientists. In [19,20] researchers focus on predicting the sign of the relationship between two given entities in a signed social network. Kunegis et al. [19] study the problem using varied similarity functions, and Leskovec et al. [20] propose a solution based on machine learning. Many studies have been presented for community detection in signed networks. Yang et al. [26] propose an agent-based approach by performing a random walk on positive links, Chiang et al. [8] propose an effective low-rank modeling approach. Li et al. [22] study influence diffusion in signed networks and extend the classic voter model to incorporate negative relationships.

Several models for opinion formation have been presented in the sociology and statistics literature. The notable model proposed by Degroot [10] studies how consensus is formed when individuals' opinions are updated using the average of their neighborhood. Friedkin and Johnsen [15] firstly extend the Degroot model with the consideration of both disagreement and consensus. The other famous

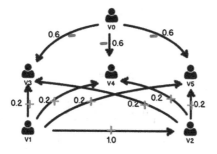

Fig. 1. An example signed social network with influence weight. Edges are labeled with their influence weight, '+' (green) denotes positive relationships, and '-' (orange) denotes negative relationships (Color figure online)

model is the voter model, proposed by Clifford et al. [9]. In the voter model, at each step a randomly-selected node chooses one of its neighbors uniformly at random and adopts that neighbor's opinion as its own.

3   Proposed Framework

3.1   Preliminaries

We describe a signed social network as a weighted, directed graph, $G = (V, E, W, S)$, where V is the set of nodes representing users and E is the set of directed edges representing relationships between users. W is the influence weight captured by the function $W : E \leftarrow [0,1]$, the weight $w_{u,v}$ associated with an edge $(u,v) \in E$ represents the influence weight of node u on v. For any $(u,v) \notin E$, $w_{u,v} = 0$, and $\sum_{u \in V} w_{u,v} \leq 1$. S is the matrix specifying the signed relationships. If $s_{u,v} = +1$, then the relationship between node u and v is positive; if $s_{u,v} = -1$, the relationship between node u and v is negative; if $s_{u,v} = 0$, there is no relationship between node u and v. Figure 1 shows an example of a simple signed social network.

In the LT model each user can be either active or inactive. Once a user becomes active they remain active forever. Each user u is uniformly assigned a threshold θ_u at random in the range $[0,1]$, and u switches from inactive to active status according to the weight of influence from their neighbors. The diffusion process proceeds in discrete time steps. At step 0, users in the seed set $S \subseteq V$ are activated while all other users are inactive. At some later step t, the user u will be activated if and only if the total weight of their active neighbors exceeds his threshold θ_u.

3.2   LT-S Model

We now extend the LT model to signed networks (LT-S) considering the opinion formation process. Due to the social interactions between users, the opinion stated by an individual is often influenced by the opinions of his neighbors.

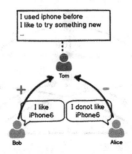

Fig. 2. An example of opinion formation. The active users (Bob and Alice) are colored red, and the inactive user (Tom) is colored blue (Color figure online)

Thus in LT-S we distinguish between the *innate opinion* and *expressed opinion* of users. Each user, whether in active status or not, has an *innate opinion* (reflecting their preference or history) that is fixed and not amenable to external influences during the diffusion process. Users who are active will form an *expressed opinion*, as a result of the social influence process, which is diffused to their neighbors. For example, in Fig. 2, Bob and Alice are in active status while Tom is inactive. Bob and Alice express their opinions about the product to their neighbor Tom. Although Tom has not been activated, he has an innate opinion about the product based on his own preference and history.

In LT-S the *innate opinion* and *expressed opinion* of user u are denoted as z_u and y_u, respectively. We encode opinion as a real quantity, $z_u \in [-1, 1]$ and $y_u \in [-1, 1]$, where $z_u > 0$ ($z_u < 0$) and $y_u > 0$ ($y_u < 0$) indicate that user u holds a positive (negative) innate opinion and a positive (negative) expressed opinion, respectively, about the product or information.

The diffusion process in LT-S works as follows. The process unfolds in discrete steps, $t = 0, 1, 2, ...$, where A_t denotes the set of users activated at step t. At step $t = 0$, a seed set $S \subseteq V$ is activated ($A_0 = S$) and each user in S has a positive expressed opinion of 1. At any later step $t > 0$, a user u is activated if and only if the total weight of his active neighbors exceeds his threshold θ_u, that is $\sum_{v \in \cup_{0 \leq i \leq (t-1)} A_i} w_{v,u} \geq \theta_u$. Although relationships can be positive or negative in signed networks, we consider that the influence from both friends and enemies has an effect.

After a user becomes active at step t, he will form his expressed opinion. The expressed opinion of a user depends on his innate opinion as well as the expressed opinions of active neighbors and their influence. The expressed opinion of a user u is calculated by Eq. 1.

$$y_u = (1 - \alpha_u) \cdot z_u + \alpha_u \cdot \sum_{v \in \cup_{0 \leq i \leq (t-1)} A_i} y_v \cdot w_{v,u} \cdot s_{v,u} \qquad (1)$$

The expressed opinion from the active neighbor v is not directly adopted by user u as it is affected by the influence weight $w_{v,u}$ and the type of relationships $s_{v,u}$ along the edge (v, u). We follow the general social principles that "the friend of my friend is my friend" and "the enemy of my enemy is my friend".

Thus positive relationships carry the opinions in a positive manner between users [6,7,18,25], as people more likely trust their friends. Conversely, negative relationships carry the opposite opinions between users. If your 'enemies' choose one opinion, you are more likely choose the opposite [3,12,16,22]. This is reflected in the $s_{v,u}$ of Eq. 1.

Each user in LT-S is associated with a conformity value α_u, in the range $[0,1]$, which modifies the importance of a user's innate opinion compared to the expressed opinions coming from active neighbors. A high conformity α_u means that the user u will be easily influenced by neighbors, and thus expressed opinions of his neighbors have a large effect on the user's final expressed opinion. A low α_u means that the user is highly self-opinionated and his expressed opinion is governed by his innate opinion. The values of *innate opinion* and *conformity* of users can be estimated from their social actions. For example, we can use a fraction of a user's tweet and retweet frequency as a proxy for conformity value and obtain the innate opinion through opinion mining [23] of recent tweets on Twitter or posts on Facebook.

When an active user forms an expressed opinion, his expressed opinion will be propagated to inactive neighbors (possibly causing their activation). The influence propagation process terminates at step t when $A_t = \emptyset$. Let $\sigma(S)$ denote the sum of positive expressed opinions of all users activated by S, $\sigma(S) = \sum_{v \in \cup_{i \geq 0} A_i} y_v$ (if $y_v > 0$). We call $\sigma(S)$ the influence spread of seed set S, and $\sigma(.)$ the influence spread function.

3.3   Problem Definition and Properties

We propose the influence maximization problem in a signed network, incorporating opinion formation and both positive and negative relationships. Our formal problem definition is based on the LT-S model.

Problem 1 (Influence Maximization in Signed Social Networks). Given a signed social network $G(V, E, W, S)$ and a parameter K ($K < |V|$), the influence maximization problem in a signed network is to find a seed set of users $S \subseteq V$ ($|S| = K$), such that by activating these users with positive opinions, that $\sigma(S)$ is maximized under the LT-S model, i.e., $S = \text{argmax}_{S \subseteq V, |S| = K} \sigma(S)$.

Theorem 1. *The influence maximization problem in signed networks under LT-S model is NP-hard.*

Proof. Consider the restricted class of instances of the problem where each node v has innate opinion $z_v = 0$ and conformity $\alpha_v = 1$. The expressed opinion of all users in active status is 1 and all edges are signed as positive. Then the problem maximizes the positive expressed opinions over this class of instance is equivalent to the classical influence maximization problem, which is proved to be NP-hard [18].

Theorem 2. *The influence spread function $\sigma(.)$ is non-monotone and non-submodular under the LT-S model.*

Unlike the classical IM problem, the influence spread function under LT-S is non-monotone and non-submodular. For the purpose of proof, we consider Fig. 1 and identify relevant cases. The innate opinion of all users is fixed at 0.2, the influence threshold is set to 0.3, and conformity is set to 0.9.

Proof. Non-monotonicity. Suppose that $S_1 = \{V_1\}$, $S_2 = \{V_1, V_0\}$, $S_3 = \{V_1, V_0, V_3\}$, then $\sigma(S_1) = 3.01$, $\sigma(S_2) = 2.92$, $\sigma(S_3) = 3.92$. As $\sigma(S_3) > \sigma(S_1) > \sigma(S_2)$, so $\sigma(S)$ is non-monotone.

Non-submodularity. Suppose that $S_1 = \{V_1\}$, $S_2 = \{V_1, V_3\}$, then $\sigma(S_1 \cup \{V_0\}) - \sigma(S_1) = -0.09$, and $\sigma(S_2 \cup \{V_0\}) - \sigma(S_2) = 0.27$. Because $S_1 \subset S_2$ and $\sigma(S_1 \cup \{V_0\}) - \sigma(S_1) < \sigma(S_2 \cup \{V_0\}) - \sigma(S_2)$, $\sigma(S)$ is non-submodular.

4 Proposed Algorithm

4.1 R-Greedy Algorithm

As the influence spread function $\sigma(.)$ under LT-S is non-monotone and non-submodular, the traditional greedy algorithm proposed by Kempe et al. [18] is inapplicable. Some investigation has been made for functions that are submodular but non-monotone [2,13], but comparatively little work has been presented toward maximizing a function which is both non-monotone and non-submodular. Feng et al. [14] proposed a restricted greedy (R-Greedy) algorithm to solve the problem.

The main idea of R-Greedy is to choose the first K nodes having maximal marginal influence, and then pick the set of seed nodes with the largest influence spread. R-Greedy is illustrated in Algorithm 1, in which S_k denotes the set of selected seeds, s_k is a single selected seed node at round k, and Inf_k^u denotes the influence after adding u to the selected seed set S_{k-1}. R-Greedy uses the dynamic pruning optimization to remove nodes whose influence is smaller than *maxMargin* (Line 6). For a node checked in round $(k-1)$, the upper bound of its marginal influence is $(Inf_{k-1}^u + \sigma(\{s_{k-1}\}) - \sigma(S_{k-1}))$. If the upper bound is not larger than *maxMargin*, the node is also ignored (Line 7). To obtain the influence spread, R-Greedy uses Monte-Carlo simulation which makes it extremely inefficient.

4.2 Techniques for Improving R-Greedy

We present two techniques to speed up the R-Greedy algorithm. The propagation-path based technique removes users with small influential ability and the live-edge based technique reduces the number of Monte-Carlo simulations.

Propagation-Path Based Technique. Instead of considering all users in the signed network, we consider only those with higher influence when choosing seed users. The influence from user u is represented by the paths originating from u, so we propose the propagation path based technique to calculate influential ability of users and choose those with higher influence ability as seed users.

Algorithm 1. R-Greedy Algorithm

Input: G, K
Output: S
1: **for** $v = 1$ to $|V|$ **do**
2:    calculate $\sigma(v)$ and insert $(v, \sigma(v))$ into Q_0
3: **end for**
4: **for** $k = 1$ to K **do**
5:    $maxMargin \leftarrow -\infty$
6:    **for** $u \in V \backslash S_{k-1}$, $\sigma(u) \geq maxMargin$ **do**
7:       **if** $u \in Q_{k-1}$ and $(Inf_{k-1}^u + \sigma(\{s_{k-1}\})) - \sigma(S_{k-1}) < maxMargin)$ **then**
8:          Continue
9:       **else**
10:          Calculate Inf_k^u and insert (u, Inf_k^u) into Q_k
11:          **if** $Inf_k^u - \sigma(S_{k-1}) > maxMargin$ **then**
12:             $maxMargin \leftarrow Inf_k^u - \sigma(S_{k-1})$
13:             $s_k \leftarrow u$
14:          **end if**
15:       **end if**
16:    **end for**
17:    $S_k \leftarrow S_{k-1} \cup \{s_k\}$
18:    $\sigma(S_k) \leftarrow \sigma(S_{k-1}) + maxMargin$
19: **end for**
20: return S

Definition 1 (Propagation Path). *A propagation path from node u to node v ($v \neq u$) is defined as $P = \langle v_1 = u, v_2, ..., v_m = v \rangle$, where $m > 1$. The influence probability of path P is given by $Pr(P) = \prod_{i=1}^{m-1} w_{v_i, v_{i+1}}$.*

A propagation path P from node u to node v provides a possible way for u to influence v with probability $Pr(P)$, thus the influence probability of the node u can be represented as all propagation paths starting from u. We note that the influence probability of the propagation path will diminish rapidly if the path length increases, as the influence weight w among edges is $w \in [0, 1]$. We therefore choose a threshold θ and prune any path whose $Pr(P)$ is smaller than θ. The calculation of the influential ability of a user is described in Algorithm 2.

Algorithm 2 recursively calculates the influence probability of user u until the path terminates or the influence probability of the path is smaller than θ. Algorithm 3 describes the process of obtaining seed users through propagation path-based technique. At Line 1 we initialize a max heap H to store the influence probability of users. Lines 2–5 calculate the influence probability of all users in the graph G and inserts them into H. At Line 3, when we invoke Algorithm 2, we initialize the influence probability to 1.0 as node u influences itself with probability 1.0. At Line 6, we choose the top $1/2^t \cdot |V|$ users from H as the candidate set and then apply the R-Greedy algorithm with the candidate set as input users to obtain the final seed set S (Line 7).

Algorithm 2. Calculate the influence probability of users

Input: u, u', p, θ
Output: inf_u, influence probability of u
1: $visit_{u'} \leftarrow$ true
2: $inf_u \leftarrow inf_u + p$
3: **for** each $v \in Adj(u')$ **do**
4:     $p' \leftarrow w_{u',v} \times p$
5:     **if** $visit_v$ is false and $p' \geq \theta$ **then**
6:         $callnf(u, v, p', \theta)$
7:         $visit_v \leftarrow$ false
8:     **end if**
9: **end for**
10: return inf_u

Algorithm 3. Propagation-path algorithm

Input: G, K, t, θ
Output: S
1: Initialize a max-heap H
2: **for** each u in V **do**
3:     $inf_u \leftarrow callnf(u, u, 1.0, \theta)$
4:     insert inf_u into H
5:     $visit_u \leftarrow$ false
6: **end for**
7: $CandidateSet \leftarrow$ choose the top $1/2^t \cdot |V|$ users from H
8: $S \leftarrow$ use R-Greedy with $CandidateSet$ as input users
9: return S

Live-Edge Based Technique. We use the live-edge based technique to generate live-edge graphs and calculate the influence spread in these graphs. Kempe et al. [18] show that the LT model is equivalent to reachability in 'live-edge' graphs, where each node $v \in V$ selects at most one of its incoming edges at random such that edge (u, v) is selected with probability $w_{u,v}$ and no edge is selected with probability $1 - \sum_u w_{u,v}$. The selected edges are called *live* and all other edges are called *blocked*. Cheng et al. [7] also point out that if the computation of influence spread is limited to the smaller number of live-edge graphs, the computational expense can be reduced without loss of accuracy. The first part of the LT-S model is similar to LT, so we can begin by generating smaller live-edge graphs instead of performing a larger number of simulations.

In LT-S a user u will form an expressed opinion when he becomes active, according to his relationships with active neighbors in Eq. 1. When traversing a live-edge graph we therefore need to record the order of active users. We use a breadth-first search (BFS) strategy to traverse the live-edge graph G and create an ordered record of active users. The details are described in Algorithm 4. In lines 2–4 we initialize the expressed opinion and active status of users in S. When visiting the ith layer of G_r we use a queue Q_i to record the reachable active users in order. In the computation of the expressed opinion of active user u we therefore know which users are the active neighbors of u (Lines 9–18).

Algorithm 4. Live-edge algorithm

Input: G, S, R (# live-edge graphs)
Output: $\sigma(S)$
 1: **for** $r = 1$ to R **do**
 2: **for** u in S **do**
 3:    $y_u \leftarrow 1$, $active_u \leftarrow$ true, enqueue u into Q_0
 4:    $\sigma(S) \leftarrow \sigma(S) + y_u$
 5: **end for**
 6: **while** $Q_i \neq \emptyset$ **do**
 7:    **while** Q_i is not empty **do**
 8:      $u \leftarrow$ dequeue from Q_i
 9:      **for** v in users reachable from u in G_r **do**
10:        $y \leftarrow 0$
11:        **for** each $t \in Adj(v)$ **do**
12:          **if** $active_t$ **then**
13:            $y \leftarrow y + y_t \times w_{t,v} \times s_{t,v}$
14:          **end if**
15:        **end for**
16:        $y_v \leftarrow \alpha_v \times z_v + (1 - \alpha_v) \times y$
17:        $active_v \leftarrow$ true, enqueue v into Q_{i+1}
18:        $\sigma(S) \leftarrow \sigma(S) + y_v$
19:      **end for**
20:    **end while**
21:    $i \leftarrow i + 1$
22: **end while**
23: **end for**
24: return $\sigma(S)/R$

4.3 Analysis of Algorithms

Let n (m) be the number of nodes (edges) in G. The R-Greedy algorithm uses Monte-Carlo simulation to estimate the influence spread. Its time complexity is therefore $O(KnR(n + m))$, where R is the number of simulations, generally set to $10,000$.

The R-Greedy algorithm using the propagation-path technique has two parts. The first part calculates the influential ability of users and chooses the most influential; the second part applies R-Greedy to the most influential users. The time complexity is therefore $O(n\bar{n} + KnR(n' + m))$, where $\bar{n}$ is the average number of nodes in the local region within threshold θ, and n' is the number of 'most influential' users.

The time complexity of the R-Greedy algorithm using the live-edge technique also has two parts. Firstly, the complexity of generating R' live-edge graphs is $O(R'm)$; secondly, it takes $O(KnR'm')$ time to select seed nodes using R-Greedy in R' live-edge graphs, where m' is the average number of live-edges. Thus, the total time complexity is $O(R'm + KnR'm')$, where R' (as suggested in [7]) is 100.

5 Experiments

In this section we describe experiments conducted on several real-world, signed social networks[3] to assess the performance of different algorithms. All algorithms are implemented in C++ and measured on a 2.66GHz Intel Xeon server with 24GB of main memory.

5.1 Experiment Setup

Datasets

- **Epinions** is a general consumer product review site in which users can either trust or distrust other's reviews. The dataset contains 131,000 nodes and 841,000 edges.
- **Slashdot** is a news discussion web site in which users can tag each other as 'friend' or 'foe'. The dataset contains 77,000 users and 526,000 edges.

Algorithms

- **CELFGreedy [CELF]**. Original greedy algorithm with CELF optimization [21], denoted as **CELF**. Following the literature we set $R =10,000$ which means that, for each seed set S, 10,000 Monte-Carlo simulations are conducted to obtain an accurate result.
- **R-Greedy [RG]**. The restricted greedy algorithm proposed in [14], designed for an influence spread function which is non-monotone and non-submodular. R is set to be the same value as in **CELF**.
- **R-Greedy with Propagation-Path [RP]**. R-Greedy using the propagation-path technique. We set t to 5 and θ to 0.003.
- **R-Greedy with Live-Edge [RL]**. R-Greedy using the live-edge technique for influence spread estimation. R is set to 100, as suggested in [7].
- **R-Greedy with Live-Edge and Propagation-Path [RLP]**. R-Greedy using both live-edge and propagation-path techniques. R is the same as in RL; t and θ are the same as in RP.
- **Max Weight Degree [Degree]**. Select the K nodes with the largest degrees.

Influence Models

- **Weighted Model** sets the weight of every incoming edge of v to be $1/d_v$, where d_v is the indegree of v.
- **Trivalency Model** sets the weight of edges randomly from $\{0.1, 0.01, 0.001\}$ then normalizes the weights of all incoming edges to each node so that they sum to 1.

In both models, the innate opinion of each user is generated uniformly at random from $[-1, 1]$ and the conformity from $[0, 1]$.

---

[3] http://snap.stanford.edu/data/index.html.

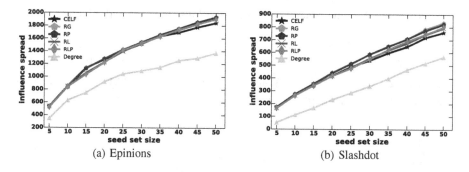

Fig. 3. Influence spread achieved by various algorithms under the weighted model

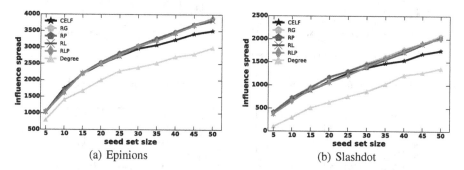

Fig. 4. Influence spread achieved by various algorithms under the trivalency model

5.2   Experimental Results and Discussion

The evaluation metrics include influence spread and running time. To obtain influence spread, we run Monte-Carlo simulation $10,000$ times for each seed set. The seed size K is in the range 1 to 50 and we compare running time using the case where $K = 50$.

Influence Spread. Figures. 3 and 4 show the influence spread under the weighted model and trivalency model, respectively. From the results we see that **RG** and the improved algorithms, **RP**, **RL** and **RLP**, achieve the most stable and accurate results. Our algorithms, **RP**, **RL** and **RLP**, have similar influence spread to **RG**. Conversely, **CELF** is very unstable compared with **RG**, **RP**, **RL** and **RLP**. The results of **Degree** tell us that simply choosing high-degree nodes is not effective.

Running Time. Table 1 shows the running times of different algorithms on the Epinions and Slashdot datasets. **RL** and **RLP** are several orders of magnitude more efficient than **RG**. Compared to **RG**, **RLP** obtains similar results but runs approximately 900 times faster.

Table 1. Running time of different algorithms (in hours and minutes)

	Weighted model					Trivalency model				
	CELF	RG	RP	RL	RLP	CELF	RG	RP	RL	RLP
Epinions	121.7h	111.6h	29.3h	8.5m	6.8m	224.2h	216.1h	46.1h	16.2m	14.8m
Slashdot	48.h	45.1h	19.1h	4.6m	3.1m	201.4h	195.2h	38.2h	15.4m	10.5m

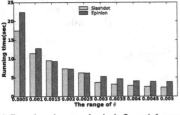
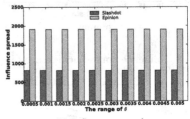

(a) Running time to obtain influential users   (b) Influence spread

Fig. 5. The effect of θ on running time and influence spread under the weighted model

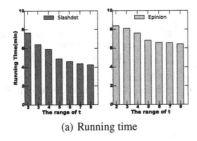
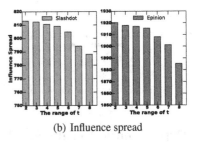

(a) Running time   (b) Influence spread

Fig. 6. The effect of t on running time and influence spread under the weighted model

Parameter Settings

(I) Effect of different values of θ on **RLP**.

Figure 5 shows the effect of different values of θ on the **RLP** algorithm. In Algorithm 2 the parameter θ controls the length of propagation paths, and in Fig. 5(a) we see the time taken to obtainin influential users using **RLP**. We notice that varying θ has little effect on influence spread, but reducing θ decreases the running time. We apply $\theta = 0.003$ across all the other experiments to balance the influence spread and running time.

(II) Effect of different values of t on **RLP**.

Figure 6 shows the effect of different values of t on the **RLP** algorithm. In Algorithm 3 the parameter t controls the size of the candidate users set. In Fig. 6(a) we see that as t increases the candidate set becomes smaller and the running time decreases. However, in Fig. 6(b) we see that influence spread also decreases as t increases, so we choose $t = 5$ as a compromise between influence spread and running time.

6   Conclusion

In this paper we considered the influence maximization problem in signed social networks. We extended the classical linear threshold (LT) model to create a new model LT-S that incorporates both opinion formation and signed relationships. Based on the LT-S model we formulated the influence maximization problem in signed social networks. We showed that the influence spread function under LT-S is neither monotone nor submodular and proposed an improved R-Greedy algorithm, R-Greedy with Live-edge and Propagation-path (RLP), which extends R-Greedy with two efficient techniques. We conducted extensive experiments on datasets taken from large, real-world, signed social networks, and presented results that demonstrate the effectiveness and efficiency of the RLP algorithm.

References

1. Bharathi, S., Kempe, D., Salek, M.: Competitive influence maximization in social networks. In: Deng, X., Graham, F.C. (eds.) WINE 2007. LNCS, vol. 4858, pp. 306–311. Springer, Heidelberg (2007)
2. Buchbinder, N., Feldman, M., Naor, J., Schwartz, R.: Submodular maximization with cardinality constraints. In: SODA, pp. 1433–1452 (2014)
3. Cartwright, D., Harary, F.: Structural balance: a generalization of Heider's theory. Psychol. Rev. **63**(5), 277–293 (1956)
4. Chen, W., Collins, A., Cummings, R., Ke, T., Liu, Z., Rincon, D., Sun, X., Wang, Y., Wei, W., Yuan, Y.: Influence maximization in social networks when negative opinions may emerge and propagate. In: SDM, pp. 379–390 (2011)
5. Chen, W., Lu, W., Zhang, N.: Time-critical influence maximization in social networks with time-delayed diffusion process. In: AAAI (2012)
6. Chen, W., Wang, C., Wang, Y.: Scalable influence maximization for prevalent viral marketing in large-scale social networks. In: KDD, pp. 1029–1038 (2010)
7. Cheng, S., Shen, H., Huang, J., Zhang, G., Cheng, X.: Staticgreedy: solving the scalability-accuracy dilemma in influence maximization. In: CIKM, pp. 509–518 (2013)
8. Chiang, K.Y., Hsieh, C.J., Natarajan, N., Dhillon, I.S., Tewari, A.: Prediction and clustering in signed networks: a local to global perspective. J. Mach. Learn. Res. **15**(1), 1177–1213 (2014)
9. Clifford, P., Sudbury, A.: A model for spatial conflict. Biometrika **60**(3), 581–588 (1973)
10. DeGroot, M.H.: Reaching a consensus. J. Am. Stat. Assoc. **69**(345), 118–121 (1974)
11. DeMarzo, P.M., Vayanos, D., Zwiebel, J.: Persuasion bias, social influence, and unidimensional opinions. Q. J. Econ. **118**(3), 909–968 (2003)
12. Fan, P., Wang, H., Li, P., Li, W., Jiang, Z.: Analysis of opinion spreading in homogeneous networks with signed relationships. J. Stat. Mech. Theory Exper. **2012**(08), P08003 (2012)
13. Feige, U., Mirrokni, V.S., Vondrak, J.: Maximizing non-monotone submodular functions. SIAM J. Comput. **40**(4), 1133–1153 (2011)
14. Feng, S., Chen, X., Cong, G., Zeng, Y., Chee, Y.M., Xiang, Y.: Influence maximization with novelty decay in social networks. In: AAAI, pp. 37–43 (2014)

15. Friedkin, N.E., Johnsen, E.C.: Social influence and opinions. J. Math. Sociol. **15**(3–4), 193–206 (1990)
16. Heider, F.: Attitudes and cognitive organization. J. Psychol. **21**(1), 107–112 (1946)
17. Jung, K., Heo, W., Chen, W.: Irie: scalable and robust influence maximization in social networks. In: ICDM, pp. 918–923 (2012)
18. Kempe, D., Kleinberg, J., Tardos, É.: Maximizing the spread of influence through a social network. In: KDD, pp. 137–146 (2003)
19. Kunegis, J., Lommatzsch, A., Bauckhage, C.: The slashdot zoo: mining a social network with negative edges. In: WWW, pp. 741–750 (2009)
20. Leskovec, J., Huttenlocher, D., Kleinberg, J.: Predicting positive and negative links in online social networks. In: WWW, pp. 641–650 (2010)
21. Leskovec, J., Krause, A., Guestrin, C., Faloutsos, C., VanBriesen, J., Glance, N.: Cost-effective outbreak detection in networks. In: KDD, pp. 420–429 (2007)
22. Li, Y., Chen, W., Wang, Y., Zhang, Z.L.: Influence diffusion dynamics and influence maximization in social networks with friend and foe relationships. In: WSDM, pp. 657–666. ACM (2013)
23. Liu, B.: Sentiment analysis and opinion mining. Synth. Lect. Hum. Lang. Technol. **5**(1), 1–167 (2012)
24. Liu, B., Cong, G., Xu, D., Zeng, Y.: Time constrained influence maximization in social networks. In: ICDM, pp. 439–448 (2012)
25. Rodriguez, M.G., Schölkopf, B.: Influence maximization in continuous time diffusion networks. In: ICML, pp. 313–320 (2012)
26. Yang, B., Cheung, W.K., Liu, J.: Community mining from signed social networks. IEEE Trans. Knowl. Data Eng. **19**(10), 1333–1348 (2007)
27. Zhang, H., Dinh, T.N., Thai, M.T.: Maximizing the spread of positive influence in online social networks. In: ICDCS, pp. 317–326. IEEE (2013)

Analyzing File-to-File Relation Network in Malware Detection

Lingwei Chen[1], William Hardy[1], Yanfang Ye[1（✉）], and Tao Li[2]

[1] Department of Computer Science and Electrical Engineering,
West Virginia University, Morgantown, WV 26506, USA
{lgchen,whardy2}@mix.wvu.edu, yanfang.ye@mail.wvu.edu
[2] School of Computer Science, Florida International University,
Miami, FL 33199, USA
taoli@cs.fiu.edu

Abstract. Due to its major threats to Internet security, malware detection is of great interest to both the anti-malware industry and researchers. Currently, features beyond file content are starting to be leveraged for malware detection (e.g., file-to-file relations), which provide invaluable insight about the properties of file samples. However, we still have much to understand about the relationships of malware and benign files. In this paper, based on the file-to-file relation network, we design several new and robust graph-based features for malware detection and reveal its relationship characteristics. Based on the designed features and two findings, we first apply Malicious Score Inference Algorithm (MSIA) to select the representative samples from the large unknown file collection for labeling, and then use Belief Propagation (BP) algorithm to detect malware. To the best of our knowledge, this is the first investigation of the relationship characteristics for the file-to-file relation network in malware detection using social network analysis. A comprehensive experimental study on a large collection of file sample relations obtained from the clients of anti-malware software of Comodo Security Solutions Incorporation is performed to compare various malware detection approaches. Promising experimental results demonstrate that the accuracy and efficiency of our proposed methods outperform other alternate data mining based detection techniques.

Keywords: File-to-file relation network · Malware detection · Social network analysis

1 Introduction

Malware (short for *mal*icious soft*ware*), is software disseminated by an attacker in the hopes of causing harm (e.g., viruses, worms, backdoors, spyware, trojans) [7]. This causes many Internet users emotional and financial troubles, especially when private information is divulged. To put this into perspective, according to a recent CSI survey, the average loss caused by malware attacks is about $345,000

© Springer International Publishing Switzerland 2015
J. Wang et al. (Eds.): WISE 2015, Part I, LNCS 9418, pp. 415–430, 2015.
DOI: 10.1007/978-3-319-26190-4_28

dollars per incident [5]. Thus, the detection of malware is of great interest to both the anti-malware industry and researchers in order to curb the major threats to Internet security it poses.

Currently, most malware detection is done by anti-malware software products (e.g., Symantec, Kaspersky, Comodo), which typically use the signature-based method [8]. A signature is a short sequence of bytes unique to each known malware, which allows newly encountered files to be correctly identified with a small error rate [11]. However, driven by economic benefits, today's malware are created at a rate of thousands per day [28]. Meanwhile, malware disseminators easily evade this method through techniques such as obfuscation, polymorphism, and encryption [19]. In order to remain effective, new, intelligent malware detection techniques need to be investigated. As a result, many research efforts have been conducted on applying data mining techniques for intelligent malware detection [1,13,26,27]. Most existing researchers utilize local features of malware samples, either static or dynamic representations (e.g. binary n-grams [1], system calls [17], or behavior based features [8]), and apply specific classification/clustering methods. Currently, features beyond file content are starting to be leveraged for malware detection [2,10,22,28], such as machine-to-file relations [2] and file-to-file relations [22,28], which provide invaluable insight about the properties of file samples. In our previous work [3,28], we proposed a semi-parametric classification model for combining file content and file relations together for malware detection [28], and then we further adopted Belief Propagation algorithm to detect malware based on the constructed file relation graphs [3].

However, much needs to be done to understand about the relationships of malware and benign files. Based on the file-to-file relationships, current studies [22,28] merely employ the superficial graph properties without delving deeper into the features of the files' mutual relationships as a network. As is investigated, helpful information can be gleaned from the relationships between malware and benign files. Based on the constructed file-to-file relation graphs, it is of interest to know:

- *What graph-based features can be employed for malware detection?*
- *Is the legitimacy of a file affected by indirectly connected files?*
- *Is each malware of equal importance? If not, what are the differences between the important malware and non-important ones?*

In this paper, resting on the constructed file-to-file relation graph, we design several new and robust graph-based features for malware detection and investigate its relationship characteristics. Based on the designed features and two findings, we propose inference learning algorithms composed of Malicious Score Inference Algorithm (MSIA) and Belief Propagation (BP) algorithm to detect malware from unknown file collection. To the best of our knowledge, this is the first work of investigating the relationship characteristics of the file-to-file relation network in malware detection using social network analysis. The major contributions of our work can be summarized as follows:

- *Graph-based feature design for malware detection:* We formalize our malware detection as a problem of analysis and identification for malware's social network and design five graph-based features to represent each file.

- *Provide insight into the relationship characteristics of the file-to-file relation network:* Based on the constructed file-to-file relation graph, we further analyze its relationship characteristics and have two findings: ***Finding 1:*** A file can greatly inherit the indirect influences from other files. ***Finding 2:*** (1) The importance of each file is different; (2) The neighbors of the important malware are associated through it, while the neighbors of the non-important malicious file are inclined to be a clique. We also use graph metrics to quantitatively validate these findings (See Sect. 3.3 for details).
- *Build effective graph inference algorithms for malware detection:* Based on the designed features and two findings, we first apply Malicious Score Inference Algorithm (MSIA) to select the representative samples from the large unknown file collection for labeling, and then use Belief Propagation (BP) algorithm to detect malware.
- *Comprehensive experimental study on a real dataset from an anti-malware industry company:* Resting on 7,093 clients, we obtain the file relations between 9,893 malware samples, 19,402 benign files, and 31,429 unknown files from Comodo Cloud Security Center. We then build a practical solution for malware detection based on our proposed algorithms and provide a comprehensive experimental study.

The rest of the paper is organized as follows. Section 2 discusses the related work. Section 3 describes the designing of graph-based features for malware detection and reveals relationship characteristics of the file-to-file relation network. Section 4 introduces graph inference algorithms for malware detection. In Sect. 5, we systematically evaluate the effectiveness and efficiency of our proposed method in comparison with other alternate methods for malware detection. Finally, Sect. 6 concludes.

2   Related Work

In the area of malware detection, limited researches have been done using features beyond file content [2,3,22,28], such as machine-to-file relations [2] and file-to-file relations [22,28], which provide invaluable insight about the properties of file samples. However, these researches still lack insights into the characteristics of file-to-file relation network.

An Online Social Network (OSN) is a convenient communication platform for Internet users which creates an indispensable environment. However, some users exploit this platform for nefarious goals, e.g. propagating advertisements, reposting duplicate microblogs, or following a large number of users to gain more themselves [12,14,23]; these users are considered spammers and parallel malicious files in our domain. Many research efforts have been devoted to spammer detection [9,12,15,20,25] for OSNs, which successfully set examples in designing robust graph-based features for spammer detection [24], and understanding the characteristics of criminal accounts' social relationships [25], etc.

Yang et al. [24] proposed several robust graph-based features, such as local clustering coefficient, betweenness centrality, and bi-directional link ratio, to detect Twitter spammers. The reasoning for these features was that it was difficult for spammers to change their positions in the graph even when they change their behaviors. Ting et al. [21] represented several frequently used technologies in social networks analysis where degree centrality and closeness centrality were used to specify each node's importance. Chen et al. [4] proposed coefficients of reposting and commenting to detect the user influence when a new post was published. Steep rise in reposting or commenting implied the account was legitimate, otherwise spammer.

Using these features, a number of spammer detection methods were designed. In [9,12,24,30], classical data mining based methods (e.g., Support Vector Machine, Decision Tree, and Naïve Bayes) were applied for spammer detection. Moh et al. [15] used trust propagation to infer whether a user is a spammer or a legitimate user. Tang et al. [20] proposed trust index utilizing the idea of page rank for criminal account detection. Yang et al. [25] proposed a malicious relevance score propagation algorithm to extract criminal supporters. Hu et al. [9] presented a unified model to effectively integrate both social network information and content information for spammer detection.

Different from previous work on malware detection [22,28], based on the file-to-file relation network, we would like to know more about its relationship characteristics. The researches on spammer detection for OSNs have inspired us to obtain the solutions and we attempt to transfer some of the researches done on spammer detection for malware detection.

3   Investigation of File-to-File Relation Network

In this section, we empirically analyze the malware's characteristics based on the constructed file-to-file relation network using the data collected from Comodo Cloud Security Center by: (1) visualizing its relationship graph, (2) designing its graph-based features, and (3) revealing its relationship characteristics.

3.1   Visualization of File Relation Graph

To achieve our research goals, we analyze the dataset obtained from Comodo Cloud Security Center, which contains the relationships between 60,724 files (9,893 malware, 19,402 benign files and 31,429 unknown files) on 7,093 clients. Based on the collected data, we construct a file-to-file relation graph to describe the relations among file samples. Generally, two files are related if they are shared by many clients (or equivalently, file lists). The social graph is defined as $G = (V, E)$, where V is the set of file samples and E denotes the relations between file samples. Given two file samples v_i and v_j, let C_i be the set of clients containing v_i and C_j be the set of clients containing v_j. $|.|$ represents the size of a set. The connectivity between v_i and v_j is computed as

$$con(v_i, v_j) = \frac{|C_i \bigcap C_j|}{|C_i \bigcup C_j|}. \tag{1}$$

If the connectivity between a pair of file samples is greater than the specified threshold (e.g., 0), then there is an edge between them. Each file is in a state $S \in \{s_m, s_b, s_u\}$ (s_m: malicious, s_b: benign, s_u: unknown). Assume that v_i is in state s_i and v_j is with state s_j, the weight of the edge between v_i and v_j which infers the probability that node i and node j can be connected together is defined as

$$w(v_i, v_j) = \frac{|E_{s_i,s_j}|}{|E|}, \qquad (2)$$

where $|E_{s_i,s_j}|$ is the number of the edges between all the files with states s_i and s_j, and $|E|$ is the number of all the edges. The weight of node v_i which denotes its popularity can be defined as

$$w(v_i) = \frac{|C_i|}{|C|}, \qquad (3)$$

where C is the set of all the clients.

For the file relations collected from 7,093 clients, we construct the graph consisting of 60,724 nodes and 3,471,288 edges. Figure 1(a) shows a part of the constructed graph, while Fig. 1(b) and (c) give examples of a malware relation graph and a benign file relation graph with one-hop information respectively.

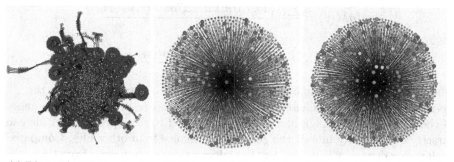

(a) File-to-File Relation Graph   (b) Malware Relation Graph   (c) Benign File Relation Graph

Fig. 1. Visualization of file-to-file relation graphs. (Red nodes denote malware, green nodes represent benign file, and yellow nodes are unknown file.) (Color figure online)

3.2 Designing Graph-Based Features for Malware Detection

To counter malware's evasion tactics, after the construction of the file-to-file relation graph, we further investigate several robust graph-based features for malware detection. Ideal features are either difficult or costly to evade, even when malware is obfuscated. In this section, on the basis of special characteristics of the file relationships between malware and benign files, we design five robust and representative graph-based features for malware detection, which are described in details in the followings.

Vertex Degree. The degree of a vertex in a graph is the number of edges incident to the vertex, which can specifically represent the association between the vertex and its neighbors [6]. In the file relation graph, we use the degree of malware (DoM) and degree of benign files (DoB) to capture the association between the file and its neighbors. These two metrics can be calculated as

$$DoM(v) = |\delta_m^v|, \ DoB(v) = |\delta_b^v|, \tag{4}$$

where $|\delta_m^v|$ is the total number of vertex v's malicious neighbors, and $|\delta_b^v|$ is the total number of vertex v's benign neighbors. As the moral says that "man is known by the company he keeps", it's easy to understand that malware is more likely to have a larger DoM than DoB, and vice versa. To further support this point, we calculate the degree for each file in the collected dataset described above: 53.75 % of malware have larger DoM than DoB; while only 3.10 % of benign files have larger DoM than DoB.

Influence Coefficient. For spammer detection, in [4], the authors used reposting and commenting coefficients to indicate the ability that a user affects others to repost or comment. In malware detection, we define the influence coefficient of malware and benign files by Eqs. 5 and 6.

$$IoM(v) = \frac{\sum_{i=1}^{N} \log(Malware\_Count(v_i) + 1)}{N}, \tag{5}$$

$$IoB(v) = \frac{\sum_{i=1}^{N} \log(Benign\_Count(v_i) + 1)}{N}, \tag{6}$$

where N denotes the number of vertex v's neighbors and v_i denotes the ith neighbor of v. $Malware\_Count(v_i)$ and $Benign\_Count(v_i)$ represent the number of the malware and benign files directly connected to v_i respectively. A file can directly or indirectly inherit the goodness or malice from other files. Compared with vertex degree, which considers the information one-hop away from the node, the feature of influence coefficient takes the indirect influence from other files into consideration.

Local Clustering Coefficient. The local clustering coefficient of a vertex in a graph specifies how close vertices in its neighborhood are to being a clique [24]. For each vertex in the constructed file relation graph, its local clustering coefficient can be calculated as [24]

$$LCC(v) = \frac{2|e^v|}{k_v(k_v - 1)}, \tag{7}$$

where $|e^v|$ is the total number of edges built by all v's neighbors, and k_v is the degree of the vertex v. For benign files, different users may install different sets of applications according to their occupations, ages, etc. And these applications are unnecessary to have associations with each other. However, for malware, just specified groups of users would be infected by malware. When infected, not only one malicious software would appear in the client, but also its related files would be released or downloaded. For example, variants of trojans will always come together with trojan-downloader and co-exist in the clients. Therefore, malware will have a larger local clustering coefficient than benign files. To quantitatively validate this, we calculate the local clustering coefficient for each file in the collected dataset described above: the average LLC for malware is 0.9387, while the average LLC for benign files is 0.7573.

Degree Centrality. Degree centrality of a vertex is determined by the number of vertices adjacent to it. The larger the degree, the more important the vertex is [18]. In malware detection, degree centrality can be used to quantify the importance of a file, which can be computed as [18]

$$DC(v) = \frac{\delta(v)}{n-1}, \tag{8}$$

where $\delta(v)$ is the degree of the vertex v, and n is the number of vertices in the graph.

Closeness Centrality. Closeness centrality measures the importance of vertices by quantifying their centrality. Central vertices tend to reach the whole graph more quickly than non-central vertices [18]. Closeness centrality factors in how close a vertex is to other vertices, which is computed as [18]

$$CC(v) = \frac{1}{n-1} \sum_{\substack{u \neq v}}^{n} g(u, v), \tag{9}$$

where $g(u, v)$ is the distance between the vertex u and vertex v, and n is the number of vertices in the graph. Malware attackers always use a shotgun approach to find victims and allure them to download variants of malicious files (e.g., trojans, adware). These files in the victim clients are always connected through the downloaders. Thus, the closeness centrality of those downloaders will be high.

3.3   Revealing Relationship Characteristics

After visualizing the constructed file-to-file relation network and designing the graph-based features, we further analyze its relationship characteristics, and give the following observations.

Finding 1: A file can greatly inherit the indirect influences from other files in the file-to-file relation network. Again, as the moral says "man is known by the company he keeps", in malware detection, a file's goodness or malice can be judged by the other files that always co-exist with it in the clients.

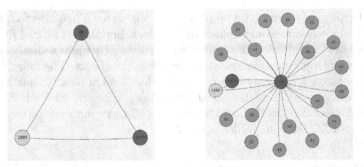

(a) The direct influences from its neighbors (b) The indirect influences from other files

Fig. 2. The indirect influences superior than direct influences for file 1880 (yellow node) (Color figure online)

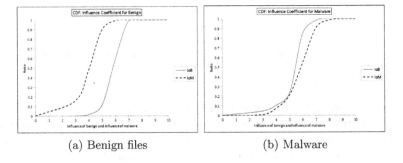

(a) Benign files   (b) Malware

Fig. 3. The comparison of benign files and malware in IoB and IoM measures

However, sometimes, a file can not only be directly influenced by its neighbors, but also greatly inherit the influences from other files (e.g., its neighbors' neighbors). Figure 2 shows an example that the indirect influences is superior than the direct influences for file 1880 (marked in yellow node). To quantitatively validate this finding, we use the features of influence of benign files (IoB) and influence of malware (IoM) designed in the above section for measure. For file 1880, its IoB is 1.6290, while its IoM is just 0.6931, which means this file is more likely to be influenced by benign files, even though all the files it directly connects with are malware.

To further illustrate, based on the collected dataset described in Sect. 3.1, we measure the indirect influences from other files for each node. Figure 3 displays the Cumulative Distribution Function (CDF) of IoB and IoM for both malware and benign files, which shows that both benign files and malware can greatly inherit the goodness and malice indirectly from other files.

Possible Factor: To disseminate the malicious files, it is not uncommon for malware to be packaged into a software product (especially when it is free and open source) by the attackers. This would cause such kind of benign software to be closely related to malicious files, however, their neighbors of neighbors would not necessarily be. On the other hand, variants of online game trojans may have indirect associations through the same kind of online game applications, since they target on stealing specific kind of online game accounts' information, but they are unnecessary to co-exist in the same clients.

From the observation above, we can see that a file's goodness or malice not only depends on its neighbors, but also greatly inherit the indirect influences from other files (e.g., its neighbors' neighbors). Furthermore, we are also interested to know: *(1) Is each malware of equal importance? (2) If not, what are the differences between the important malware and non-important ones?*

Finding 2: In the file-to-file relation network, (1) the importance of each file is different; (2) the neighbors of the important malware are associated through it, while the neighbors of the non-important malicious file are inclined to be a clique. To initially evaluate the importance of each node, we use degree centrality for measure. Based on the collected dataset described in Sect. 3.1, we calculate the degree centrality of each file: about 2 % of the malware have the degree centrality over 0.01, which are 10 – 1000 times larger than the remaining 98 % ones. From this analysis, we can see that the importance of each malware is different: the larger the degree, the more important the vertex is [18]. Note that there is another interesting observation that those malware with larger degrees also have higher node weight values in the graph, which means the "important" malware are always with higher popularity. We mark those 2 % malicious files with higher degree centrality as "important" malware, compared with the remaining 98 % ones. Figure 4(a) displays the CDF of degree centrality for the important malware and non-important ones.

To further analyze the different characteristics of the important and non-important malware, we take insights to their graph structures. Figure 5(a) illustrates an example of the relationship between an important malware A and its neighbors, while Fig. 5(b) shows the relations between its neighbors. From Fig. 5(a) and (c), we can see that both important and non-important malicious nodes with one-hop information have the star-structures, but the degree centralities of them are different. From Fig. 5(b) and (d), we can see that, the neighbors of the important malware are associated through it (the closeness centrality of the important malware A is 0.25), while the neighbors of the non-important

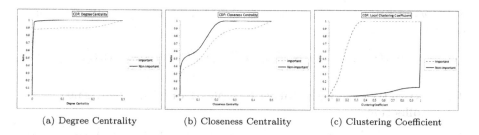

| (a) Degree Centrality | (b) Closeness Centrality | (c) Clustering Coefficient |

Fig. 4. The comparisons of "important" malware and "non-important" ones.

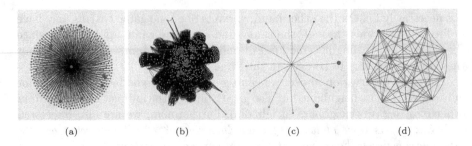

(a)   (b)   (c)   (d)

Fig. 5. Graph structure comparisons of "important" and "non-important" malware. (a) An important malware A and its neighbors; (b) File relations between A's neighbors; (c) A non-important malware B and its neighbors; (d) File relations between B's neighbors.

malicious file are inclined to be a clique (the local clustering coefficient LLC of it is equal to 1). Figure 4(b) and (c) display the CDF of local clustering coefficient and closeness centrality for the important malware and non-important ones respectively, which also validate the *Finding 2*.

Possible Factor: The importance of each malware is different, since the impacts different malicious files play are different. For example, a popular trojan or adware downloader will infect more clients, compared with the specific kind of trojan or adware variants. The files co-exist with the popular downloader in different clients are unnecessary to have a close relationship among them, but are associated through the downloader; while the files co-exist with the variants of same trojan or adware are prone to be a clique, since they tend to be the same or similar kind of applications those trojans or adware target on.

4   Inference Learning Algorithms for Malware Detection

Via empirical analysis for the file-to-file relation network, each node v_i (i.e., a file sample) in the constructed graph can be represented by its relations with other nodes and its graph-based features designed in Sect. 3.2, denoted as $Fv_i = \langle Rv_i, Gv_i \rangle$. Rv_i can be defined as

$$Rv_i = \langle v_{1i}, v_{2i}, ..., v_{ni} \rangle, \tag{10}$$

where $v_{ji} = \{0, 1\}$ (i.e., if $(v_j, v_i) \in E$, $v_{ji} = 1$; otherwise, $v_{ji} = 0$). Gv_i can be defined as

$$Gv_i = \langle DoM(v_i), DoB(v_i), IoM(v_i), IoB(v_i), LCC(v_i), DC(v_i), CC(v_i) \rangle. \tag{11}$$

4.1   Extracting Representative Samples from Unknown File Collection

Based on the *Finding 2* that the importance of each file is different and the neighbors of the important malware are associated through them, therefore, selecting those representative malware from the large unknown file collection for labeling is very important to further improve the detection accuracy. In spammer detection, Yang et al. [25] proposed a Malicious Relevance Score Propagation Algorithm (Mr. SPA) to extract criminal supporters. In this section, we propose a Malicious Score Inference Algorithm (MSIA), which adapts and improves Mr. SPA [25] to assign a malicious score for each file to quantify its importance.

Given a constructed file relation graph $G = (V, E)$, let n be the number of nodes (files) in the graph, and $I(v_i, v_j)$ be the indicator to denote whether $(v_i, v_j) \in E$ (i.e., if $(v_i, v_j) \in E$, $I(v_i, v_j) = 1$; otherwise, $I(v_i, v_j) = 0$). At each step, for each node v_i, its malicious score $M(v_i)$ can be calculated as [25]

$$M(v_i) = \alpha \cdot \sum_{j=1}^{n} I(v_i, v_j) W(v_i, v_j) M(v_j), \tag{12}$$

where α is an adjustable factor, and $W(v_i, v_j)$ is the weight between v_i and v_j which reflects the coordination between each pair of nodes. For each node v_i, we calculate the similarity between itself and each of its neighbors v_j based on their presented features described above, denoted as $sim(Fv_i, Fv_j)$. Then, the weight $W(v_i, v_j)$ between node v_i and v_j is computed as

$$W(v_i, v_j) = \frac{sim(Fv_i, Fv_j)}{\sum_{e_{v_k v_j} \in E} sim(Fv_k, Fv_j)}. \tag{13}$$

In our application, we initialize $M^0(v_i) = \{0, 1\}$ (i.e., if v_i is malicious, $M^0(v_i) = 1$; otherwise, $M^0(v_i) = 0$).

With the consideration of the historical score record for each node, at each step $t(t \geq 1)$, an initial score bias $(1 - \alpha) \cdot M_i^0$ is added to its malicious score. Thus the malicious score vector $\overrightarrow{M^t}$ for all nodes at step $t(t \geq 1)$ can be computed as [25]

$$\overrightarrow{M^t} = \alpha \cdot \overrightarrow{I} \cdot \overrightarrow{M^{t-1}} + (1 - \alpha) \cdot \overrightarrow{M^0}. \tag{14}$$

The algorithm MSIA stops when the updates of malicious score vector converge or a maximum number of the iterations has finished. The higher the malicious score it has, the more important the file is.

4.2   A Belief Propagation Algorithm for Malware Detection

Based on *Finding 1*, which states a file's goodness or malice not only depends on its neighbors, but also indirectly on other files (e.g., its neighbors' neighbors), we apply Belief Propagation (BP) algorithm for malware detection, since BP algorithm can propagate the indirect influences from other files for each node. BP algorithm is a promising method for solving inference problems over graphs and it has also been successfully used in many domains (e.g., computer vision, coding theory) [29]. Nodes of the graph perform as a local summation operation by iterations using the prior knowledge from their neighbors and then pass the information to all the neighbors in the form of messages [16]. In our previous work [3], we proposed a variant of Belief Propagation (BP) algorithm for malware detection. The message update equation is defined as [3]

$$m_{i->j}(v_j) = \frac{1}{\beta} \sum_{v_i \in S} f_{i->j}(v_i, v_j) g_i(v_i) \frac{\sum_{k \in N(i)/j} m_{k->i}(v_i)}{p}, \qquad (15)$$

where $m_{i->j}(v_j)$ is the probability node i believes that the neighbor node j being a benign file; $f_{i->j}(v_i, v_j)$ is the probability that node i and node j can be connected together, which is equal to the weight of edge described in Eq. 2; $g_i(v_i)$ is the prior probability node i being a benign file [3]; p equals to the number of the neighbors of node i (excluded node j) and β is a normalizing constant. The belief read-out equation is designed as follow [3]

$$b_i(v_i) = \frac{1}{\gamma} g_i(v_i) \frac{\sum_{k \in N(i)} m_{k->i}(v_i)}{p}, \qquad (16)$$

where γ is an adjustable constant.

5   Experimental Results and Analysis

In this section, we conduct three sets of experiments based on the collected dataset obtained from Comodo Cloud Security Center described in Sect. 3.1: (1) In the first set of experiments, we use MSIA to evaluate the effectiveness of the designed features; (2) In the second set of experiments, we further evaluate our proposed inference learning algorithms in malware detection; (3) In the last set of experiments, we compare our proposed algorithms with other classification methods (i.e., SVM, Decision Tree, and Naïve Bayes).

We evaluate the malware detection performance of different methods using the measures of True Positive (TP), True Negative (TN), False Positive (FP), False Negative (FN), and Accuracy (ACY) [3]. All the experiments are conducted under the environment of 64 Bit Windows 7 operating system with 4th Generation Intel Core i7 Processor (Quad Core, 8 MB Cache, up to 4.0 GHz w/Turbo Boost) plus 16 G of RAM using Apache Pig, MySQL and C++.

5.1   Evaluation of the Designed Features

For each sample in the constructed file-to-file relation network, we extract its relations with other samples described in Sect. 3.1 and its graph-based features designed in Sect. 3.2 for representation. In this section, we conduct the experiments using MSIA to evaluate the effectiveness of the designed features. The collected dataset from Comodo Cloud Security Center contains 60,724 files: 9,893 are malware, 19,402 are benign files, and 31,429 are unknown (with the analysis by anti-malware experts of Comodo Security Lab, 470 of them are labeled as malware and 1,273 of them are benign files). Those 9,893 malware and 19,402

Table 1. Evaluation of the designed graph-based features

Training	TP	FP	TN	FN	ACY
MSIA(RF)	7,392	1,301	18,101	2,501	0.8702
MSIA(GF)	6,960	2,410	16,992	2,933	0.8176
MSIA($RF + GF$)	7,890	1,371	18,031	2,003	0.8848
Testing	TP	FP	TN	FN	ACY
MSIA(RF)	293	78	1,195	177	0.8537
MSIA(GF)	179	122	1,151	291	0.7631
MSIA($RF + GF$)	315	78	1,195	155	0.8663

Remark 1. *RF* denotes the file relation features of the samples, while *GF* denotes the graph-based features of the samples.

benign files are used for training, while 470 malware and 1,273 benign files from the unknown file collection which are labeled by anti-malware experts are used for testing. The results in Table 1 demonstrate the effectiveness of the designed features in malware detection.

5.2 Evaluation of the Proposed Inference Learning Algorithms in Malware Detection

In this section, we further evaluate our proposed inference learning algorithms in malware detection: (1) Based on the training and testing sets described in Sect. 5.1, we compare the performance of MSIA and BP in malware detection; (2) To further improve the detection accuracy, we first apply MSIA for representative samples selection (193 samples are selected from the unknown file collection for labeling), and then use BP for detection. The results in Table 2 show that our proposed algorithms composed of MSIA and BP (MSIA+BP) can greatly improve the accuracy in malware detection, compared with using MSIA and BP respectively, or BP after randomly selecting 193 samples from the unknown file collection for labeling (Random+BP).

5.3 Comparisons of the Proposed Algorithms with Other Classification Methods

In this section, we further compare our proposed algorithms with other classification methods (Support Vector Machine (SVM), Decision Tree (DT), and Naïve Bayes (NB)) based on the same testing dataset described in Sect. 5.1. The results in Table 3 and the ROC curves based on the testing set in Fig. 6(a) demonstrate that our proposed algorithm composed of MSIA and BP (MSIA+BP) is superior to SVM, DT, and NB in malware detection. Figure 6(b) shows that the detection efficiency of our proposed algorithms outperform other classification methods.

Table 2. Evaluation of the proposed inference learning algorithms in malware detection

Training	TP	FP	TN	FN	ACY
MSIA	7,890	1,371	18,031	2,003	0.8848
BP	9,881	866	18,536	12	0.9700
Random+BP	10,059	870	18,545	14	0.9701
MSIA+BP	10,060	851	18,564	13	0.9707
Testing	TP	FP	TN	FN	ACY
MSIA	315	78	1,195	155	0.8663
BP	411	119	1,154	59	0.8979
Random+BP	462	204	1,069	8	0.8784
MSIA+BP	437	100	1,173	33	0.9236

Table 3. Comparisons of different detection methods

Training	TP	FP	TN	FN	ACY
SVM	8,661	797	18,618	1,412	0.9251
DT	8,308	1,761	17,654	1,765	0.8804
NB	5,288	502	18,913	4,785	0.8207
MSIA+BP	10,060	851	18,564	13	0.9707
Testing	TP	FP	TN	FN	ACY
SVM	452	172	1,101	18	0.8910
DT	412	241	1,032	58	0.8285
NB	159	31	1,242	311	0.8037
MSIA+BP	437	100	1,173	33	0.9236

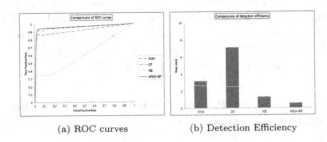

(a) ROC curves   (b) Detection Efficiency

Fig. 6. Comparisons of ROC curves and detection efficiency of different methods

6 Conclusion and Future Work

In this paper, resting on the constructed file-to-file relation graphs, we design several new and robust graph-based features for malware detection and reveal the characteristics of file relation network. Based on the designed features and two findings, we propose effective inference learning algorithms for malware detection. To the best of our knowledge, this is the first investigation of the relationship characteristics for the file-to-file relation network in malware detection using social network analysis. Due to the difficulty in thoroughly obtaining the social interactions and motivations of malware, we recognize that the validations on some proposed explanations are not entirely rigorous. However, we believe that our novel analysis of those phenomena still yields great value and unveils a new avenue for better understanding malware's file relation ecosystem. In our future work, we will further investigate the distinctions between malware and benign files in the relationship graph, as well as design a full detection solution by combining file relations with other detection features to further improve the detection accuracy.

Acknowledgments. The authors would also like to thank the anti-malware experts of Comodo Security Lab for the data collection, as well as the helpful discussions and supports.

References

1. Bailey, M., Oberheide, J., Andersen, J., Mao, Z.M., Jahanian, F., Nazario, J.: Automated classification and analysis of internet malware. In: Kruegel, C., Lippmann, R., Clark, A. (eds.) RAID 2007. LNCS, vol. 4637, pp. 178–197. Springer, Heidelberg (2007)
2. Chau, D., Nachenberg, C., Wilhelm, J., Wright, A., Faloutsos, C.: Polonium: terascale graph mining for malware detection. In: SIAM International Conference on Data Mining (SDM), pp. 131–142 (2011)
3. Chen, L., Li, T., Abdulhayoglu, M., Ye, Y.: Intelligent malware detection based on file relation graphs. In: 9th IEEE International Conference on Sematic Computing, pp. 85–92 (2015)
4. Chen, K., Zhu, P., Xiong, Y.: Mining spam accounts with user influence. In: International Conference on ISCC-C, pp. 167–173 (2013)
5. Computer Security Institute: 12th annual edition of the CSI computer crime and security survey. Technical report, Computer Security Institute (2007)
6. Diestel, R.: Graph Theory, vol. 173, 4th edn. Springer, Heidelberg (2010)
7. Egele, M., Scholte, T., Kirda, E., Kruegel, C.: A survey on automated dynamic malware analysis techniques and tools. ACM CSUR **44**(2), 6:1–6:42 (2008)
8. Filiol, E., Jacob, G., Liard, M.L.: Evaluation methodology and theoretical model for antiviral behavioural detection strategies. J. Comput. Virol **3**(1), 27–37 (2007)
9. Hu, X., Tang, J., Zhang, Y., Liu, H.: Social spammer detection in microblogging. In: Proceedings of the 23rd IJCAI, pp. 2633–2639 (2013)
10. Karampatziakis, N., Stokes, J.W., Thomas, A., Marinescu, M.: Using file relationships in malware classification. In: Flegel, U., Markatos, E., Robertson, W. (eds.) DIMVA 2012. LNCS, vol. 7591, pp. 1–20. Springer, Heidelberg (2013)
11. Kephart, J., Arnold, W.: Automatic extraction of computer virus signatures. In: Proceedings of 4th Virus Bulletin International Conference, pp. 178–184 (1994)
12. Lin, C., Zhou, Y., Chen, K., He, J., Yang, X., Song, L.: Analysis and identification of spamming behaviors in Sina Weibo microblog. In: SNAKDD 2013 (2013)

13. Masud, M.M., Al-Khateeb, T.M., Hamlen, K.W., Gao, J., Khan, L., Han, J., Thuraisingham, B.: Cloud-based malware detection for evolving data streams. ACM TMIS **2**(3), 16:1–16:27 (2008)
14. Mislove, A., Marcon, M., Gummadi, K.P., Druschel, P., Bhattacharjee, B.: Measurement and analysis of online social networks. In: Proceedings of the 7th ACM SIGCOMM, pp. 29–42 (2007)
15. Moh, T.-S., Murmann, A.J.: Can you judge a man by his friends? Enhancing spammer detection on the twitter microblogging platform using friends and followers. In: Prasad, S.K., Vin, H.M., Sahni, S., Jaiswal, M.P., Thipakorn, B. (eds.) ICISTM 2010. CCIS, vol. 54, pp. 210–220. Springer, Heidelberg (2010)
16. Noorshams, N., Wainwright, M.J.: Belief propagation for continuous state spaces: stochastic message-passing with quantitative guarantees. J. Mach. Learn. Res. **14**(1), 2799–2835 (2013)
17. Park, Y., Zhang, Q., Reeves, D., Mulukutla, V.: AntiBot: clustering common semantic patterns for bot detection. In: IEEE 34th Annual Computer Software and Applications Conference, pp. 262–272 (2010)
18. Scott, J.: Social Networks Analysis: A Hand Book, 2nd edn. SAGE Publications Ltd, Thousand Oaks (2000)
19. Sung, A., Xu, J., Chavez, P., Mukkamala, S.: Static analyzer of vicious executables (save). In: Proceedings of the 20th ACSAC, pp. 326–334 (2004)
20. Tang, R., Lu, L., Zhuang, Y., Fong, S.: Not every friend on a social network can be trusted: an online trust indexing algorithm. In: IEEE/WIC/ACM International Conferences on Web Intelligence and Intelligent Agent Technology (WI-IAT), pp. 280–285 (2012)
21. Ting, I.H., Wang, S.L.: Content matters: a study of hate groups detection based on social networks analysis and web mining. In: IEEE/ACM ASONAM, pp. 1196–1201 (2013)
22. Tamersoy, A., Roundy, K.A., Chau, D.: Guilt by association: large scale malware detection by mining file-relation graphs. In: ACM SIGKDD (2014)
23. Weng, J., Lim, E.P., Jiang, J., He, Q.: Twitterrank: finding topic-sensitive influential twitterers. In: Proceedings of the Third ACM WSDM, pp. 261–270 (2010)
24. Yang, C., Harkreader, R.C., Gu, G.: Die free or live hard? Empirical evaluation and new design for fighting evolving twitter spammers. In: Proceedings of the 14th International Conference on Recent Advances in Intrusion Detection, pp. 318–337 (2011)
25. Yang, C., Harkreader, R., Zhang, J., Shin, S., Gu, G.: Analyzing spammer's social networks for fun and profit: a case study of cyber criminal ecosystem on twitter. In: Proceedings of the 21st International Conference on World Wide Web (WWW 2012), pp. 71–80 (2012)
26. Ye, Y., Wang, D., Li, T., Ye, D., Jiang, Q.: An intelligent PE-malware detection system based on association mining. J. Comput. Virol. **4**, 323–334 (2008)
27. Ye, Y., Wang, D., Li, T., Ye, D.: IMDS: Intelligent malware detection system. In: Proceedings of the 13th ACM SIGKDD, pp. 1043–1047 (2007)
28. Ye, Y., Li, T., Zhu, S., Zhuang, W., Tas, E., Gupta, U., Abdulhayoglu, M.: Combining file content and file relations for cloud based malware detection. In: Proceedings of the 17th ACM SIGKDD, pp. 222–230 (2011)
29. Yedidia, J. S., Freeman, W.T., Weiss, Y.: Understanding belief propagation and its generalizations. Mltsubishl Electric Research Laboratories (2001)
30. Zhang, C., Niu, K., He, Z.: Dynamic detection of spammers in Weibo. In: 4th IEEE IC-NIDC, pp. 112–116 (2014)

Similarity-Based Context-Aware Recommendation

Yong Zheng[✉], Bamshad Mobasher,
and Robin Burke

Center for Web Intelligence, School of Computing, DePaul University,
243 South Wabash Ave, Chicago, IL 60604, USA
{yzheng8,mobasher,rburke}@cs.depaul.edu

Abstract. Context-aware recommender systems (CARS) take context into consideration when modeling user preferences. There are two general ways to integrate context with recommendation: contextual filtering and contextual modeling. Currently, the most effective context-aware recommendation algorithms are based on a contextual modeling approach that estimate deviations in ratings across different contexts. In this paper, we propose context similarity as an alternative contextual modeling approach and examine different ways to represent context similarity and incorporate it into recommendation. More specifically, we show how context similarity can be integrated into the sparse linear method and matrix factorization algorithms. Our experimental results demonstrate that learning context similarity is a more effective approach to context-aware recommendation than modeling contextual rating deviations.

Keywords: Recommender system · Context · Context-aware · Matrix factorization

1 Introduction

Introducing contexts to recommender systems (RS) contributes to building new types of applications, such as context-aware recommendation [2] and context suggestions [21]. Context-aware recommender systems (CARS) have been a topic of considerable recent research interest in RS. Considering context is shown to be useful in a variety of recommendation domains including tourism [4,17], music [3], and restaurants [13]. In many application domains, including these, it is reasonable to assume that users' preferences may change from context to context, even for the same item. For example, a user may choose a different movie if he or she is going to see the movie with kids, rather than on a romantic date. In this case, the *companion* is the influential contextual variable.

The standard formulation of the recommendation problem begins with a two dimensional (2D) matrix of ratings, organized by user and item: *Users* × *Items* → *Ratings*. The key insight of context-aware recommender systems is that users' preferences for items may be a function of the context in which those items are

© Springer International Publishing Switzerland 2015
J. Wang et al. (Eds.): WISE 2015, Part I, LNCS 9418, pp. 431–447, 2015.
DOI: 10.1007/978-3-319-26190-4_29

encountered. Incorporating contexts requires that we estimate user preferences using a multidimensional rating function – R: *Users* × *Items* × *Contexts* → *Ratings* [2].

Contextual modeling is a popular general architecture for developing context-aware recommender systems. The most effective contextual modeling approaches, such as context-aware matrix factorization [5], try to adapt to contextual preferences by modeling contextual rating deviations: how do ratings for one context differ generally from those in another context. However, researchers ignore the reason for such deviations, namely that contexts may be inherently similar, and contextual similarity may be a better framework for understanding and representing contextual effects.

In this paper, we propose the notion of similarity-based contextual recommendation. The basic assumption behind this paradigm is that recommendation lists should be similar if their contextual situations are similar. Furthermore, we provide specific ways to model the context similarity and incorporate it into two recommendation models respectively: sparse linear method (SLIM) [12] and matrix factorization (MF) [10]. Our experimental results over multiple context-aware data sets demonstrate that similarity-based context-aware recommendation algorithms are able to outperform the ones based on modeling contextual rating deviations, offering a promising and novel way to further develop context-aware recommendation algorithms.

2   Related Work

Context is usually defined as any information that can be used to characterize the situation of an entity [1]. In CARS, the contextual variables are usually from the attributes of the activity itself [16], e.g., *time* and *companion* are two variables which can be considered as the attribute of the activity "watching a movie". Other attributes from users or items, such as user gender and movie genre, are usually deemed as contents in RS.

There are generally two ways to incorporate contexts into recommender systems – contexts can be used either as filters in the algorithms (i.e., *contextual filtering*), such as differential context modeling [17,18] and context-aware splitting [6,20], or as one part of the predictive functions in the recommendation process (i.e., *context modeling*).

There are usually two categories in context modeling: independent models and dependent models. For example, tensor factorization (TF) [9] directly considers contexts as additional dimensions in the multidimensional rating space, which is an independent model assuming contextual effects are not dependent with users or items. On the other hand, context-aware matrix factorization (CAMF) [5] and contextual sparse linear method (CSLIM) [22,23] are two dependent models which try to adapt to contextual preferences by modeling contextual rating deviations, where those deviations are usually dependent with either user or item dimension in the rating data set. Previous work [5,23] has demonstrated that dependent models usually outperform the independent models in most cases. Thus, much of the research in contextual modeling explores various ways to model context dependencies.

Current dependent models rely on the dependencies between contexts and user (or item) dimensions by modeling rating deviations in different contexts. We call these algorithms as *Deviation-Based* context-aware recommendation algorithms; However, such algorithms ignore the relationships between context themselves. Context similarity was explored in some previous work – it is calculated based on either semantics [8,11] or co-ratings in the same contexts [7]. Semantic context similarity is derived from the applications in natural language processing and information retrieval. It is usually useful in hierarchical semantic structure and not general enough to any cases, which limits its application, e.g., it is difficult to measure the similarity between *at home* and *at cinema* from semantics. Calculations based on co-ratings are usually unreliable, since the number of co-ratings in the same context is limited.

2.1 Matrix Factorization and Deviation-Based CAMF

Matrix factorization (MF) [10] is one of the most effective recommendation algorithms in the traditional, non-contextual, recommender systems. Simply, both users and items are represented by vectors, e.g., $\vec{p_u}$ is used to denote a user vector, and $\vec{q_i}$ as an item vector. As a result, the rating prediction can be described by Eq. 1.

$$\hat{r}_{ui} = \vec{p_u} \cdot \vec{q_i} \tag{1}$$

$$\hat{r}_{ui} = \mu + b_u + b_i + \vec{p_u} \cdot \vec{q_i} \tag{2}$$

More specifically, the values in the user or item vectors indicate the weights on K (e.g., $K = 5$) latent factors. The weights in $\vec{p_u}$ can be viewed as the degree to which those latent factors represent user's interests, and the weights in $\vec{q_i}$ represent how the specific item is associated with the latent factors. Therefore, the dot product of those two vectors can be used to indicate how much the user likes this item, where users' preferences on items are captured by the latent factors. In addition, user and item rating biases can be added to the prediction function, as shown in Eq. 2, where μ denotes the global average rating in the data, b_u and b_i represent the user and item bias respectively (Table 1).

Table 1. Contextual ratings on movies

User	Item	Rating	Time	Location	Companion
U1	T1	3	Weekend	Home	Alone
U1	T1	5	Weekend	Cinema	Girlfriend
U1	T1	?	Weekday	Home	Family

Assume there are one user $U1$, one item $T1$, and three contextual dimensions – Time (weekend or weekday), Location (at home or cinema) and Companion (alone, girlfriend, family) as shown in the table above. In the following discussion, we use *contextual dimension* to denote the contextual variable, e.g. "Location". The term *contextual condition* refers to a specific value in a dimension, e.g. "home" and "cinema" are two contextual conditions for "Location". A *context* or *contextual situation* is, therefore, a set of contextual conditions, e.g. {*weekend, home, family*}. More specifically, we use c_k and c_m to denote two different contextual situations. In other words, c_k is composed by a set of contextual conditions. Let $c_{k,l}$ denote the lth contextual condition in the context c_k. For example, if $c_k = \{weekend, home, alone\}$, then $c_{k,2}$ is "home".

The rating prediction function in CAMF [5] can be shown in Eq. 3.

$$\hat{r}_{uic_{k,1}c_{k,2}...c_{k,L}} = \mu + b_u + \sum_{j=1}^{L} B_{ijc_{k,j}} + \overrightarrow{p_u} \cdot \overrightarrow{q_i} \tag{3}$$

Assume there are L contextual dimensions in total, $c_k = \{c_{k,1}c_{k,2}...c_{k,L}\}$ is used to describe the contextual situation, where $c_{k,j}$ denotes the contextual condition in the jth context dimension. Therefore, $B_{ijc_{k,j}}$ indicates the contextual rating deviation associated with item i and the contextual condition in the jth dimension.

A comparison between Eqs. 2 and 3 reveals that CAMF simply replaces the item bias b_i by a contextual rating deviation term $\sum_{j=1}^{L} B_{ijc_{k,j}}$, and it assumes that the contextual rating deviation is dependent on items. Therefore, this approach is denoted as CAMF_CI. In a similar manner, the deviation can also be viewed as being dependent on users, which replaces b_u by the contextual rating deviation term and helps formulate the CAMF_CU. Finally, the CAMF_C variant of the algorithm assumes that the contextual rating deviation is independent of users and items.

As the typical optimization in matrix factorization, the parameters, such as the user and item vectors, user biases and rating deviations, can be learned by the stochastic gradient descent (SGD) method to minimize the squared rating prediction errors. In early work [5], CAMF was demonstrated to outperform the independent contextual modeling approaches, such as the tensor factorization [9].

2.2   Sparse Linear Method and Deviation-Based CSLIM

Sparse linear method (SLIM) shown in Fig. 1 is an approach designed for top-N recommendations in traditional RS. It improves upon the traditional item-based K-nearest neighbor collaborative filtering by learning a sparse matrix of aggregation coefficients between items that are analogous to the traditional item-item similarities [12].

The ranking score in SLIM for user u_i on item t_j is represented by $\widehat{S}_{i,j}$, which can be estimated by Eq. 4 – where N is the total number of items, W is a matrix estimating the coefficients between items, and t_h corresponds to the items rated by u_i rather than item t_j. $R_{i,h}$ therefore refers to the known rating given by u_i on other items. In other words, $\widehat{S}_{i,j}$ is calculated by a sparse aggregation

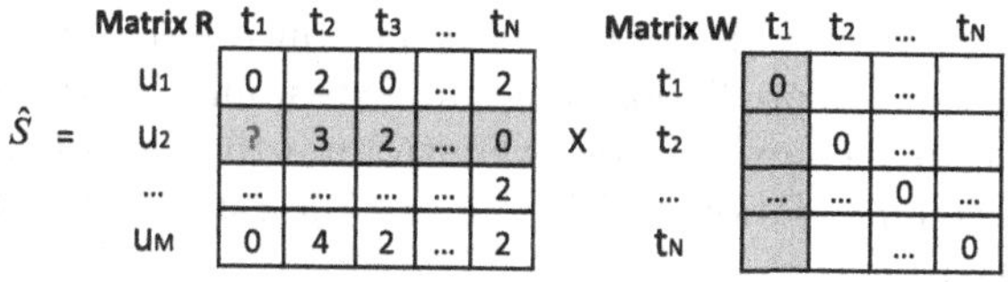

Matrix R	t₁	t₂	t₃	...	t_N

Fig. 1. Example of SLIM

(i.e., aggregated by the coefficient $W_{h,j}$) of the ratings on the other items (i.e., $R_{i,h}$) that have been rated by u_i.

$$\widehat{S}_{i,j} = R_{i,:}.W_{:,j} = \sum_{\substack{h=1 \\ h \neq j}}^{N} R_{i,h} W_{h,j} \quad (4)$$

From another perspective, this function is analogous to the rating prediction function in ItemKNN, where the normalization term is removed. In addition, the coefficients between items are learnt based on the loss function shown by Eq. 5, instead of calculations in ItemKNN (e.g., using Pearson correlations or cosine similarity). β_1 and β_2 are the regularization parameters. Both ℓ_F terms (e.g. $\|W\|_F^2$) and ℓ_1 terms (e.g. $\|W\|_1$) are included, where the ℓ_1 regularization term is usually applied for sparse models.

$$\underset{W}{Minimize} \ \frac{1}{2} \left\| R_{i,j} - \widehat{S}_{i,j} \right\|_F^2 + \frac{\beta_2}{2} \|W\|_F^2 + \beta_1 \|W\|_1 \quad (5)$$

Previously, SLIM model has been extended to incorporate contextual information. These extensions include a deviation-based general contextual SLIM (GCSLIM) [23] approach which learns and estimates the contextual rating deviations from a context to another. More specifically, in GCSLIM, the goal is to create a prediction function shown in Eq. 6 to estimate the ranking score, $\widehat{S}_{i,j,c_k}$, for u_i on item t_j in contexts c_k. Again, t_h belongs to the set of items rated by u_i ($h \neq i$), and c_m is the contextual situation where u_i placed rating on t_h (note: it is allowed that $c_m = c_k$). Assume there are L contextual dimensions in total. Then $c_{m,l}$ denotes the contextual condition in the lth dimension in context c_m. The function Dev measures the contextual rating deviation between two contextual conditions – it is zero if $c_{m,l} = c_{k,l}$. The matrix W continues to measure the coefficient between two items. In contrast to the SLIM approach, GCSLIM additionally learns the deviations between two contextual conditions in the algorithm, where the optimization goal is to minimize the ranking score prediction error and the loss function can be generated accordingly by adding the deviation terms to Eq. 5.

$$\widehat{S}_{i,j,c_k} = \sum_{\substack{h=1 \\ h \neq i}}^{N} \left(R_{i,h,c_m} + \sum_{l=1}^{L} Dev(c_{m,l}, c_{k,l}) \right) W_{h,j} \quad (6)$$

In a summary, both CAMF and CSLIM introduce a contextual rating deviation term as part of the prediction function, extending the original recommendation algorithm. In the next section, we introduce the idea of similarity-based context-aware recommendation which replaces the rating deviation term by a context similarity term.

3   Similarity-Based Contextual Modeling

Given a particular user and multiple contexts in which recommendations can be provided, we assume that the greater the similarity between two contexts, the greater the similarity between recommendation lists that should be delivered. Therefore, contextual similarity becomes a constraint of the delivery of recommendations.

3.1   Similarity-Based CSLIM Approach

Inspired by the prediction function in Eq. 6, we can derive a similarity-based CSLIM approach which was first proposed in [14] and initially explored in [24]. The corresponding ranking score prediction can be described as follows.

$$\widehat{S}_{i,j,c_k} = \sum_{\substack{h=1 \\ h \neq j}}^{N} R_{i,h,c_m} \times W_{h,j} \times Sim(c_k, c_m) \tag{7}$$

Specifically, we aggregate the ranking score by the contextual rating score R_{i,h,c_m} with the coefficients between item t_j and t_h multiplying by the similarity between c_k and c_m. We set $h \neq j$ to avoid bias by using u_i's other contextual ratings on t_j. This strategy will ensure that we learn the coefficients between as many different items as possible. Using this approach, it is possible to learn the coefficients between items and also the similarity between different contexts by SGD accordingly.

3.2   Similarity-Based CAMF Approach

The idea above can also be applied to matrix factorization:

$$\hat{r}_{uic_k} = \vec{p_u} \cdot \vec{q_i} \cdot Sim(c_k, c_E) \tag{8}$$

Due to that $\vec{p_u} \cdot \vec{q_i}$ represents a non-contextual rating, the context similarity turns to measure the similarity between a contextual situation and a non-contextual situation. We use c_E to denote the empty context (i.e. non-contextual) situation – the value in each contextual dimension is empty or "N/A"; that is, $c_{E,1} = c_{E,2} = ... = c_{E,L} = N/A$. Therefore, the function $Sim(c_k, c_E)$ actually estimates the correlation between the c_E and the contextual situation c_k where at least one contextual condition is not empty or "N/A". Note that in Eq. 3, the contextual rating deviation can be viewed as the deviation from the empty contextual situation to a non-empty contextual situation.

$$Minimize_{p,q,Sim} \frac{1}{2} \|(r_{uic_k} - \hat{r}_{uic_k})\|_F^2 + \frac{\alpha}{2}(\|\overrightarrow{p_u}\|^2 + \|\overrightarrow{q_i}\|^2 + Sim^2) \qquad (9)$$

Accordingly, the user and item vectors, as well as the contextual similarity can be learned by SGD, where the loss function is shown in Eq. 9 and α denotes the regularization parameter. Note that this is the general form for the loss function, where the term "Sim^2" should be specified and adjusted accordingly when it is modeled in different ways. Here, we will introduce three context similarity models as follows.

3.3 Modeling Context Similarity

Apparently, the form of the context similarity will vary if it is represented in different ways. The remaining challenge is how to represent or model the similarity of contexts in the prediction functions. This decision will directly influence the performance of the learning algorithm, thus it is necessary to discuss and explore different options. In this paper, we introduce three ways to represent similarity of contexts as follows. Note that we introduce those notions in general – the context similarity or correlation is assumed to be measured between any two contextual situations c_k and c_m. However, in order to prevent having to compute an quadratic number of between-context similaritiers, we measure similarity between c_k and c_E, where c_E is the empty context.

Independent Context Similarity (ICS). An example of a similarity matrix can be seen in Table 2. With Independent Context Similarity, we only measure the similarity between two contextual conditions when those they lie on the same contextual dimension, e.g., we never measure the similarity between "Time = Weekend" and "Location = Home", since they are from two different dimensions. Each pair of contextual dimensions are assumed to be independent. In this case, the similarity between two contexts can be represented by the product of the similarities among different dimensions. For example, assume c_k is {Time = Weekend, Location = Home}, and c_m is {Time = Weekday, Location = Cinema}, the similarity between c_k and c_m can be represented by the

Table 2. Example of a similarity matrix

	Time = Weekend	Time = Weekday	Location = Home	Location = Cinema
Time = Weekend	1	0.54	N/A	N/A
Time = Weekday	0.54	1	N/A	N/A
Location = Home	N/A	N/A	1	0.82
Location = Cinema	N/A	N/A	0.82	1

similarity of <Time = Weekend, Time = Weekday> multiplied by the similarity of <Location = Home, Location = Cinema>, since the two dimensions, "Time" and "Location" are assumed as independent.

Assuming there are L contextual dimensions in total, the similarities can be depicted by Eq. 10, where $c_{k,l}$ is used to denote the value of contextual condition in the lth dimension in context c_k, and the "similarity" function is used to represent the similarity between two contextual conditions, which is also what to be learnt in the optimization. In other words, the similarity between two contexts is represented by the multiplication of the individual similarities between contextual conditions on each dimension.

$$\text{Sim}(c_k, c_m) = \prod_{l=1}^{L} similarity(c_{k,l}, c_{m,l}) \tag{10}$$

These similarity values (i.e., $similarity(c_{k,l}, c_{m,l})$) can be learned by the optimization process via the base algorithm (either SLIM or MF). The risk of this representation is that some information may be lost if similarities are not in fact independent in different dimensions. For example, if users usually go to cinema to see romantic movies with their partners, the "Location" (e.g. at cinema) and "Companion" (e.g. partners) may have significant correlations as a result.

Latent Context Similarity (LCS). As noted earlier, contextual rating data is often sparse, since it is somewhat unusual to have users rate items repeatedly within multiple contextual situations. This poses a difficulty when new contexts are encountered. For example, the similarity between a *new pair* of contexts <"Time = Weekend", "Time = Holiday"> may be required in the testing set, but it may not have been learned from the training data due to the sparsity problem. But, the similarity for two *existing pairs*, <"Time = Weekend", "Time = Weekday"> and <"Time = Weekday", "Time = Holiday">, may have been learned. In this case, this representation suffers from the contextual rating sparsity problem. Treating each dimension independently prevents the algorithm from taking advantage of comparisons that might be made across dimensions.

To alleviate this situation, we represent each contextual condition by a vector of weights over a set of latent factors (we use 5 latent factors in our experiments), where the weights are initialized at the beginning and learnt by the optimization process. The dot product between two vectors can be used to denote the similarity between each pair of contextual conditions. As a result, even if the newly observed pair does not exist in the training data, the weights in the vectors representing the two conditions (i.e., "Time = Weekend" and "Time = Holiday") will be learned and updated by the learning process over *existing pairs*, and the similarity for the *new pair* can be easily computed using the dot product. The similarity is given by

$$similarity(c_{k,l}, c_{m,l}) = V_{c_{k,l}} \bullet V_{c_{m,l}} \tag{11}$$

where $V_{c_{k,l}}$ and $V_{c_{m,l}}$ denote the vector representation for the contextual condition $c_{k,l}$ and $c_{m,l}$, respectively, over the space of latent factors. We then use the same similarity product calculation as in Eq. 10. We call this approach the *Latent Context Similarity* (LCS) model. This approach was able to improve the performance of deviation-based CSLIM algorithms [15]. In contrast to the independent context similarity approach, LCS provides more flexibility, but it also has the added computational costs associated with learning the latent factors. In LCS, what to be learnt in the optimization process are the vectors of weights representing each contextual condition.

3.4 Multidimensional Context Similarity (MCS)

In the multidimensional context similarity model, we assume that contextual dimensions form a multidimensional coordinate system. An example is depicted in Fig. 2.

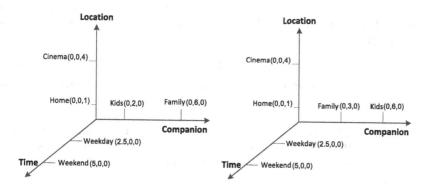

Fig. 2. Example of multidimensional coordinate system

Let us assume that there are three contextual dimensions: time, location and companion. We assign a real value to each contextual condition in those dimensions, so that each condition can locate a position in the corresponding axis. In this case, a context (as a set of contextual conditions) can be viewed as a point in the multidimensional space. Accordingly, the distance between two such points can be used as the basis for a similarity measure. In this approach, the real values for each contextual condition are the parameters to be learned in the optimization process. For example, the values for "family" and "kids" are updated in the right-hand side of the figure. Thus, the position of the data points associated to those two contextual conditions will be changed as well as the distance between the corresponding two contexts.

To avoid unconstrained distance measures, the values assigned to the contextual conditions can be normalized to be within the range [0, 1]. As a result, the data points can be represented as a cube where the length of each side equals 1. The similarity can be measured as the inverse of the distance between two data points. In our experiments, we use Euclidean distance to measure the distances, though other distance measures can also be used. The computational cost is directly associated with the number of contextual dimensions and conditions, which may make this approach the highest-cost model. Again, the number of contextual conditions can be reduced by context selection.

3.5 An Example: Constructing Algorithms

We can incorporate those three representations for context similarity (i.e., ICS, LCS and MCS) to the prediction functions in similarity-based CSLIM and CAMF respectively. Here, we give examples of how to build similarity-based CAMF. Other algorithms, using SLIM as the base, can be built in a similar manner.

As mentioned before, the prediction function in similarity-based CAMF can be shown in Eq. 8, and the loss function is described by Eq. 9. Assume, here c_k equations to $<$"Time = Weekday", "Location = Home"$>$, and c_E is the empty contexts $<$"Time = N/A", "Location = N/A"$>$. In ICS, contextual dimensions are assumed as independent, so the context similarity between c_k and c_E can be represented by follows.

$$sim_1 = similarity(\text{``}Time = Weekday\text{''}, \text{``}Time = N/A\text{''}) \tag{12}$$

$$sim_2 = similarity(\text{``}Location = Home\text{''}, \text{``}Location = N/A\text{''}) \tag{13}$$

$$Sim(c_k, c_E) = sim_1 \times sim_2 \tag{14}$$

Given the loss function in Eq. 9, what to be learnt are user and item vectors, sim1 and sim2. Based on stochastic gradient descent, those parameters can be updated as follows, where α is the regularization parameter and β is the learning rate.

$$err = r_{uic_k} - \hat{r}_{uic_k} \tag{15}$$

$$\vec{p_u} = \vec{p_u} + \beta \cdot (err \cdot \vec{q_i} \cdot Sim(c_k, c_E) - \alpha \cdot \vec{p_u}) \tag{16}$$

$$\vec{q_i} = \vec{q_i} + \beta \cdot (err \cdot \vec{p_u} \cdot Sim(c_k, c_E) - \alpha \cdot \vec{q_i}) \tag{17}$$

$$sim_1 = sim_1 + \beta(err \cdot (\vec{p_u} \cdot \vec{q_i}) \cdot sim_2 - \alpha \cdot sim_1) \tag{18}$$

$$sim_2 = sim_2 + \beta(err \cdot (\vec{p_u} \cdot \vec{q_i}) \cdot sim_1 - \alpha \cdot sim_2) \tag{19}$$

In LCS, each context condition is represented by a vector. Assume the vectors for "Time = Weekday", "Time = N/A", "Location = Home", "Location = N/A" can be denoted by $\vec{V_{Tw}}$, $\vec{V_{Tna}}$, $\vec{V_{Lh}}$ and $\vec{V_{Lna}}$ respectively. The context similarity is still calculated by Eq. 27, but the individual similarity in each context dimension is switched to be:

$$sim_1 = \vec{V_{Tw}} \cdot \vec{V_{Tna}} \tag{20}$$

$$sim_2 = \vec{V_{Lh}} \cdot \vec{V_{Lna}} \tag{21}$$

What is being learned in LCS are the user and item vectors, as well as those four vectors in this example. Accordingly, the updating functions for user and item vectors are the same as shown in Eqs. 22–24. The four vectors can be updated as follows:

$$\overrightarrow{V_{Tw}} = \overrightarrow{V_{Tw}} + \beta(err \cdot (\overrightarrow{p_u} \cdot \overrightarrow{q_i}) \cdot sim_2 \cdot \overrightarrow{V_{Tna}} - \alpha \cdot \overrightarrow{V_{Tw}}) \tag{22}$$

$$\overrightarrow{V_{Tna}} = \overrightarrow{V_{Tna}} + \beta(err \cdot (\overrightarrow{p_u} \cdot \overrightarrow{q_i}) \cdot sim_2 \cdot \overrightarrow{V_{Tw}} - \alpha \cdot \overrightarrow{V_{Tna}}) \tag{23}$$

$$\overrightarrow{V_{Lh}} = \overrightarrow{V_{Lh}} + \beta(err \cdot (\overrightarrow{p_u} \cdot \overrightarrow{q_i}) \cdot sim_1 \cdot \overrightarrow{V_{Lna}} - \alpha \cdot \overrightarrow{V_{Lh}}) \tag{24}$$

$$\overrightarrow{V_{Lna}} = \overrightarrow{V_{Lna}} + \beta(err \cdot (\overrightarrow{p_u} \cdot \overrightarrow{q_i}) \cdot sim_1 \cdot \overrightarrow{V_{Lh}} - \alpha \cdot \overrightarrow{V_{Lna}}) \tag{25}$$

In MCS, we use a real value to represent the position of each context condition, e.g., T_1 positions "Time = Weekday", T_0 denotes "Time = N/A", L_1 positions "Location = Home" and L_0 indicates "Location = N/A". The context similarity is described as:

$$Dist = \sqrt{(T_1 - T_0)^2 + (L_1 - L_0)^2} \tag{26}$$

$$Sim(c_k, c_E) = 1 - Dist \tag{27}$$

To make sure the similarity values are in the range $[0, 1]$, the position of each context condition should be limited to $[0, \frac{1}{\sqrt{D}}]$, where D is the number of context dimensions. Or, a better solution is to use bounded (or projected) gradient method. Therefore, what to be learnt in MCS are user and item vectors and those real values representing the positions of each context condition in the multidimensional context space. The updating function for those positions can be described as follows:

$$T_1 = T_1 + \beta(err \cdot (\overrightarrow{p_u} \cdot \overrightarrow{q_i}) \cdot \frac{T_1 - T_0}{Dist} - \alpha \cdot T_1) \tag{28}$$

$$T_0 = T_0 - \beta(err \cdot (\overrightarrow{p_u} \cdot \overrightarrow{q_i}) \cdot \frac{T_1 - T_0}{Dist} + \alpha \cdot T_0) \tag{29}$$

$$L_1 = L_1 + \beta(err \cdot (\overrightarrow{p_u} \cdot \overrightarrow{q_i}) \cdot \frac{L_1 - L_0}{Dist} - \alpha \cdot L_1) \tag{30}$$

$$L_0 = L_0 - \beta(err \cdot (\overrightarrow{p_u} \cdot \overrightarrow{q_i}) \cdot \frac{L_1 - L_0}{Dist} + \alpha \cdot L_0) \tag{31}$$

4 Experimental Evaluation

The number of context-aware data sets is quite limited because ratings in multiple contexts are difficult to collect and user privacy is often a concern. In CARS research, most publicly available data sets are collected from surveys, resulting in small and sparse data sets. In this paper, we select three context-aware data sets with different numbers of contextual dimensions and conditions. *Restaurant* data [13] is comprised of users' ratings on restaurants in city of Tijuana, Mexico. *Music* data [3] captures users' ratings on music tracks in different driving and traffic conditions. The *Tourism* data [4] collects users' places of interest (POIs) from mobile applications. The characteristics of these data sets are summarized in Table 3. For more specific information about the contextual dimensions and conditions, please refer to the original papers using those data sets.

Table 3. Context-aware data sets

	Restaurant	Music	Tourism
Rating profiles	50 users, 40 items 2314 ratings	40 users, 139 items 3940 ratings	25 users, 20 items 1678 ratings
# of contextual dimensions	2	8	14
# of contextual conditions	7	34	67
Rating scale	1–5	1–5	1–5

4.1   Evaluation Protocols

We use five-fold cross validation on our data sets, performing top 10 recommendation task and using precision and mean average precision (MAP) as the evaluation metrics. Precision is defined as the ratio of relevant items selected to number of items recommended in a specific context. MAP is another popular ranking metric which additional takes the ranks of the recommended items into consideration. It is calculated by Eq. 32, where M denotes the number of the users, and N is the size of the recommendation list, where $P(k)$ means the precision at cut-off k in the item recommendation list, i.e., the ratio of number of users followed up to the position k over the number k, where m in $ap@N$ denotes the number of relevant items.

$$MAP@N = \sum_{i=1}^{M} ap@N/M, \text{ where } ap@N = \frac{\sum_{k=1}^{N} P(k)}{\min(m, N)} \tag{32}$$

For comparison purposes, tensor factorization (TF) [9] is chosen as a baseline since it is an independent contextual modeling method. The proposed similarity-based CSLIM and CAMF were built upon the SLIM and MF approaches, therefore we would like to compare them with the deviation-based algorithms. More specifically, CAMF [5] is selected as the baseline. We tried all three variants: CAMF_CI, CAMF_CU and CAMF_C, and only present the best performing one, denoted as "CAMF-Dev", in the following sections. Similarly, we choose the best performing GCSLIM [23], denoted as "CSLIM-Dev". For the similarity-based context-aware recommendation algorithms, we simply use "Algorithm-SimilarityModel" to present the algorithm, e.g., "CAMF-ICS" denotes the similarity-based CAMF using independent context similarity modeling. "CSLIM-MCS" indicates the similarity-based CSLIM using multidimensional context similarity.

4.2   Analysis and Findings

The experimental results by similarity-based CSLIM approaches can be described by the Fig. 3, and the results based on similarity-based CAMF can be shown in Fig. 4.

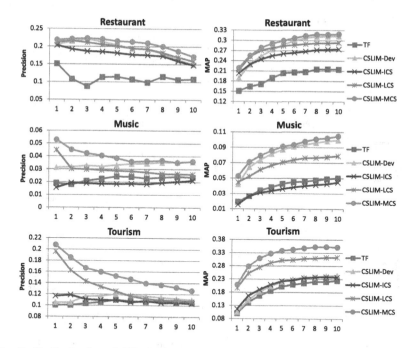

Fig. 3. Similarity-based CSLIM (x-axis denotes the number of recommendations.)

Overall Comparisons. The best performing algorithm is "CSLIM-MCS" which denotes the similarity-based CSLIM using multidimensional context similarity model. In Fig. 3, we can see that CSLIM-Dev always outperforms TF. Not all the similarity-based CSLIM approaches are able to beat TF, e.g., in the music data, CSLIM using the independent context similarity works worse than TF. However, we can always find one similarity-based CSLIM that outperforms both the TF and deviation-based CSLIM, suggesting that the context similarity works well if appropriately designed. The similar pattern can be found in the similarity-based CAMF approaches shown in Fig. 4.

In general, recommendation models using latent context similarity always outperform the ones using independent context similarity, which reveals that the LCS is able to successfully alleviate the sparsity problem mentioned previously. In addition, multidimensional context similarity is the best representation for context similarity at the price of increased computational effort, although this can be alleviated through careful context selection or merging.

Comparisons Between Deviation-Based and Similarity-Based Models. Similarity-based context-aware recommendation models are able to outperform the best performing deviation-based algorithms when the context similarity is appropriately represented. Generally, CSLIM-MCS is able to beat CSLIM-Dev, and CAMF-MCS always outperforms CAMF-Dev in our three data sets. This confirms that learning context similarity is a better way than modeling the con-

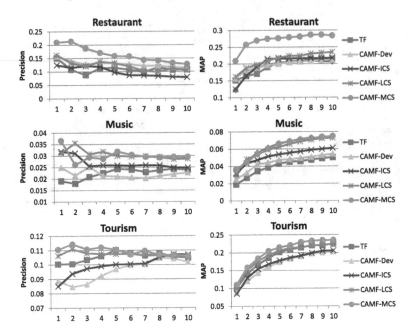

Fig. 4. Similarity-based CAMF (x-axis denotes the number of recommendations.)

textual rating deviations in developing dependent context modeling approaches. We expect similar benefits if the approach were applied to other recommendation algorithms, e.g. the slope-one recommendation model.

Table 4. Comparison between CAMF and CSLIM

		Restaurant				Music				Tourism			
		CAMF-Dev	CAMF-MCS	CSLIM-Dev	CSLIM-MCS	CAMF-Dev	CAMF-MCS	CSLIM-Dev	CSLIM-MCS	CAMF-Dev	CAMF-MCS	CSLIM-Dev	CSLIM-MCS
Precision	@5	0.1309	0.1586	**0.2044**	**0.2151**	0.0210	0.0385	**0.0332**	**0.0385**	0.0968	0.1105	**0.1200**	**0.1522**
	@10	0.1130	0.1276	**0.1496**	**0.1723**	0.0226	0.0361	**0.0359**	**0.0361**	0.1018	0.1061	**0.1101**	**0.1265**
MAP	@5	0.2001	0.2765	**0.2889**	**0.2993**	0.0474	0.0661	**0.0871**	**0.0913**	0.1740	0.2148	**0.2199**	**0.3379**
	@10	0.2122	0.2840	**0.3128**	**0.3187**	0.0542	0.0747	**0.0995**	**0.1047**	0.2026	0.2350	**0.2442**	**0.3518**

Comparisons Between CSLIM and CAMF. To compare CSLIM and CAMF in both their deviation-based and similarity-based formulations, we extracted the top-5 and top-10 precision and MAP values shown in Table 4. (Only the best performing models are shown.) From the table, we can see that CSLIM outperforms CAMF significantly when the same strategy (either deviation or correlation modeling) is applied. CSLIM-MCS works the best in general. It is not surprising, since earlier work [12] has demonstrated that SLIM is able to outperform matrix factorization in many settings. The result that CSLIM outperforms CAMF further confirms this pattern.

Summary. Similarity-based context-aware recommendation algorithms are able to outperform the deviation-based algorithms and the independent contextual modeling approach (i.e., TF). CSLIM always outperform the CAMF when the same strategy (either deviation-based or similarity-based modeling) is applied. Choosing the appropriate representation for context similarity is important, since it directly influences the performance of the similarity-based context-aware recommendation. In general, the multidimensional context similarity is the best way to model the similarity of contexts at the price of computational costs, but these costs can be alleviated by context selection.

5   Conclusions and Future Work

In this paper, we proposed the idea of similarity-based contextual recommendation where context similarity is incorporated into the recommendation algorithm. Specifically, we chose sparse linear method and matrix factorization as two base algorithms. We then developed similarity-based contextual recommendation algorithms by modeling the similarity of contexts in three different ways. Our experimental results demonstrate that a multidimensional approach is the best representation for context similarity, and that CSLIM-MCS works the best among all compared recommendation algorithms.

One direction that we will explore in future work is to explore pre-selection or other pre-processing approaches to reduce noise and improve the efficiency of the multidimensional similarity model. Also, we believe the context similarity can be incorporated into more other recommendation algorithms, such as slope one recommender, which we will explore in the future. Furthermore, the correlations or similarities between contexts could be used for interpretations, such as interpreting emotional effects [19] in RS.

References

1. Abowd, G.D., Dey, A.K.: Towards a better understanding of context and context-awareness. In: Gellersen, H.-W. (ed.) HUC 1999. LNCS, vol. 1707, pp. 304–307. Springer, Heidelberg (1999)
2. Adomavicius, G., Mobasher, B., Ricci, F., Tuzhilin, A.: Context-aware recommender systems. AI Mag. **32**(3), 67–80 (2011)
3. Baltrunas, L., Kaminskas, M., Ludwig, B., Moling, O., Ricci, F., Aydin, A., Lüke, K.-H., Schwaiger, R.: InCarMusic: context-aware music recommendations in a car. In: Huemer, C., Setzer, T. (eds.) EC-Web 2011. LNBIP, vol. 85, pp. 89–100. Springer, Heidelberg (2011)
4. Baltrunas, L., Ludwig, B., Peer, S., Ricci, F.: Context-aware places of interest recommendations for mobile users. In: Marcus, A. (ed.) HCII 2011 and DUXU 2011, Part I. LNCS, vol. 6769, pp. 531–540. Springer, Heidelberg (2011)
5. Baltrunas, L., Ludwig, B., Ricci, F.: Matrix factorization techniques for context aware recommendation. In: Proceedings of the Fifth ACM Conference on Recommender Systems, pp. 301–304. ACM (2011)

6. Baltrunas, L., Ricci, F.: Experimental evaluation of context-dependent collaborative filtering using item splitting. User Model. User-Adap. Inter. **24**(1–2), 7–34 (2014)

7. Chen, A.: Context-aware collaborative filtering system: predicting the user's preference in the ubiquitous computing environment. In: Strang, T., Linnhoff-Popien, C. (eds.) LoCA 2005. LNCS, vol. 3479, pp. 244–253. Springer, Heidelberg (2005)

8. Codina, V., Ricci, F., Ceccaroni, L.: Local context modeling with semantic prefiltering. In: Proceedings of the 7th ACM Conference on Recommender Systems, pp. 363–366. ACM (2013)

9. Karatzoglou, A., Amatriain, X., Baltrunas, L., Oliver, N.: Multiverse recommendation: n-dimensional tensor factorization for context-aware collaborative filtering. In: Proceedings of the Fourth ACM Conference on Recommender Systems, pp. 79–86. ACM (2010)

10. Koren, Y., Bell, R., Volinsky, C.: Matrix factorization techniques for recommender systems. IEEE Comput. **42**(8), 30–37 (2009)

11. Liu, L., Lecue, F., Mehandjiev, N., Xu, L.: Using context similarity for service recommendation. In: 2010 IEEE Fourth International Conference on Semantic Computing (ICSC), pp. 277–284. IEEE (2010)

12. Ning, X., Karypis, G.: SLIM: sparse linear methods for top-n recommender systems. In: 2011 IEEE 11th International Conference on Data Mining, pp. 497–506. IEEE (2011)

13. Ramirez-Garcia, X., Garca-Valdez, M.: Post-filtering for a restaurant context-aware recommender system. In: Castillo, O., Melin, P., Pedrycz, W., Kacprzyk, J. (eds.) Recent Advances on Hybrid Approaches for Designing Intelligent Systems, vol. 547, pp. 695–707. Springer, Heidelberg (2014)

14. Zheng, Y.: Deviation-based and similarity-based contextual SLIM recommendation algorithms. In: Proceedings of the 8th ACM Conference on Recommender Systems, pp. 437–440. ACM (2014)

15. Zheng, Y.: Improve general contextual SLIM recommendation algorithms by factorizing contexts. In: Proceedings of the 30th Annual ACM Symposium on Applied Computing, pp. 929–930. ACM (2015)

16. Zheng, Y.: A revisit to the identification of contexts in recommender systems. In: Proceedings of the 20th ACM Conference on Intelligent User Interfaces Companion, pp. 133–136. ACM (2015)

17. Zheng, Y., Burke, R., Mobasher, B.: Differential context relaxation for context-aware travel recommendation. In: Huemer, C., Lops, P. (eds.) EC-Web 2012. LNBIP, vol. 123, pp. 88–99. Springer, Heidelberg (2012)

18. Zheng, Y., Burke, R., Mobasher, B.: Recommendation with differential context weighting. In: Carberry, S., Weibelzahl, S., Micarelli, A., Semeraro, G. (eds.) UMAP 2013. LNCS, vol. 7899, pp. 152–164. Springer, Heidelberg (2013)

19. Zheng, Y., Burke, R., Mobasher, B.: The role of emotions in context-aware recommendation. In: ACM RecSys 2013, Proceedings of the 3rd International Workshop on Human Decision Making in Recommender Systems, pp. 21–28. ACM (2013)

20. Zheng, Y., Burke, R., Mobasher, B.: Splitting approaches for context-aware recommendation: an empirical study. In: Proceedings of the 29th Annual ACM Symposium on Applied Computing, pp. 274–279. ACM (2014)

21. Zheng, Y., Mobasher, B., Burke, R.: Context recommendation using multi-label classification. In: Proceedings of the 13th IEEE/WIC/ACM International Conference on Web Intelligence, pp. 288–295. IEEE/WIC/ACM (2014)

22. Zheng, Y., Mobasher, B., Burke, R.: CSLIM: contextual SLIM recommendation algorithms. In: Proceedings of the 8th ACM Conference on Recommender Systems, pp. 301–304. ACM (2014)
23. Zheng, Y., Mobasher, B., Burke, R.: Deviation-based contextual SLIM recommenders. In: Proceedings of the 23rd ACM Conference on Information and Knowledge Management, pp. 271–280. ACM (2014)
24. Zheng, Y., Mobasher, B., Burke, R.: Integrating context similarity with sparse linear recommendation model. In: Ricci, F., Bontcheva, K., Conlan, O., Lawless, S. (eds.) UMAP 2015. LNCS, vol. 9146, pp. 370–376. Springer, Heidelberg (2015)

ECS: A Framework for Diversified and Relevant Search in the Internet of Things

Ali Shemshadi[✉], Lina Yao, Yongrui Qin, Quan Z. Sheng, and Yihong Zhang

School of Computer Science, The University of Adelaide, Adelaide, SA 5005, Australia
{ali.shemshadi,lina.yao,yongrui.qin,
michael.sheng,yihong.zhang}@adelaide.edu.au

Abstract. Things search engines play a key role in increasing the visibility of the emerging Internet of Things (IoT) paradigm. Developing an innovative search approach is a fundamental step to lay the foundations of future IoT search engines. Currently, the most adopted approach for searching things is based on keyword search. Unfortunately, keyword search does not provide enough functionality for an IoT search engine. Correlating things based on their attributes is an emerging approach which can potentially improve the IoT search process. Since in reality there might exist a number of different correlations between a pair of everyday objects, integrating and applying them in IoT search is challenging. In this paper, we propose the ECS (*Extract, Cluster, Select*) framework. Our framework contains a novel approach to extract and integrate different types of correlation graphs with a spectral clustering method and a selection method to improve the coherence and the diversity of top-k results for a given search query. We evaluate our framework through extensive experiments using real-world datasets from different domains of IoT applications. The results show that the quality of search results improves greatly after we diversify the results of IoT data queries.

Keywords: Internet of Things · Clustering · Search engine · Correlation graph

1 Introduction

Internet of Things (IoT) is a booming research and application area which aims to connect everyday physical things to a global scale network via wireless and identification technologies such as sensors and RFID. It revolutionizes a large number of current applications in a wide variety of real-world domains. IoT based solutions potentially provide competitive advantages in various key sectors of society [6] such as smart cities, smart homes/smart building management, health-care and security and surveillance. The IoT paradigm is not a hypothetical concept any more as nowadays a large number of services provide tools for publishing sensor data on the Internet. However, a major milestone in improving the visibility of the IoT is to create future search engines for it.

© Springer International Publishing Switzerland 2015
J. Wang et al. (Eds.): WISE 2015, Part I, LNCS 9418, pp. 448–462, 2015.
DOI: 10.1007/978-3-319-26190-4_30

Some of the notable works in this area are as follows [20]: (1) Snoogle and Microsearch [12,15]; (2) Dyser [8]; and (3) SPITFIRE [10]. Current search engines perform the search using keyword based filtering. This kind of approaches have limited applications in the context of the IoT due to the dynamic nature of IoT. Some other recent approaches such as Context Aware Search [9] and Sensor Similarity Search [13] provide more functionality to search in the IoT. However, a glimpse into IoT applications reveals that supporting the correlations between things yet remains undiscovered in the IoT search.

The formation of Social IoT [1] and the Semantic IoT [11] trigger the idea of establishing correlation graphs for things inspired by social networks. Technically, it bears a good potential to support more complex queries and design more effective search approaches by adopting the correlation graphs between things.

Although this idea may seem similar to making connections between people through social networks, correlations in IoT are not established in the same way. In IoT, correlations are extracted based on the attributes of things while in social networks users contribute in establishing their relationships. Different correlations between things are expressed and understood differently and thus, can fall into different categories based on their definition and scope. For example, a correlation that is defined based on location attributes is different than a correlation which is defined based on ownership attributes. We consider *Things Correlation Graphs* (TCGs) to represent things correlation networks. Each vertex in a TCG represents a thing in IoT and each pair of vertices may be connected through an edge of a specific type, which represents a correlation of that type. Establishing correlations of different types can improve the depth of our model from the real world. It becomes possible to apply more filters to get the related things from the huge pool of things. Each type of edge is formed based on a specific type of correlation. A non-exhaustive list of correlation types include [1]: (1) Co-Location Object Relationship (CLOR): when a pair of things are in the same location or very close to each other; (2) Parental Object Relationship (POR): when a pair of things belong to the same production batch or the same kind; (3) Social Object Relationship (SOR): when a pair of things belong to a correlated pair of people; (4) Co-Work Object Relationship (C-WOR): when a pair of things can generate/consume feed for/of each other; and (5) Ownership Object Relationship (OOR): when a pair of things belong to the same person.

From the correlations mentioned above, researching SOR requires accessing social network data and the ownership relations between users and objects. Due to the security and privacy concerns, IoT and ownership data would not be found easily. The same reason, affects our ability to research OOR as well. Furthermore, researching C-WOR in which objects that can feed (I/O) each other are targeted, requires a generic framework to match sensors and actuators. Therefore, as many requirements are missing for other types of correlations, we limit the scope of this paper to CLOR and POR only. However, extracting and maintaining different types of correlations and searching things is still challenging due to the specific nature of smart things in ubiquitous computing environment. An important problem is how we can effectively manipulate correlations in the IoT to obtain better search results when it is not clear that what kind of correlations are more

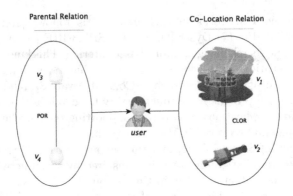

Fig. 1. Example of various relationships between things and users in IoT

important for users search? If we pick only one type of correlation, would it harm the *diversity* of the search results? The following example explains this issue.

Assume a user simply enters a query such as `<"air","USA">`. As shown in Fig. 1, keyword filtering can yield four different types of things such as an oil rig (v_1) and a Buoy (v_2) close to each other in the gulf of the Mexico as well as two air quality eggs (v_3 and v_4) in two different places. We can see that v_1 and v_2 are almost in the same location and thus they are connected through a co-location edge. This indicates a CLOR relationship between them. Meanwhile, v_3 and v_4 belong to same type of sensors, and thus they are connected through a parental relationship edge. This indicates a POR relationship between them. Now, if we want to limit the number of things in the result to two items only, how can we select them to best balance coherence and diversity? A selection of $\{v_3, v_4\}$ or $\{v_1, v_2\}$ will improve the *coherence*, but the search result is limited to just one type of correlation, leading to less *diversity*. In contrast, the selection of $\{v_1, v_3\}$ or other similar combinations will increase the *diversity* of the final result but with less *coherence*. Thus, there will be a tradeoff between *coherence* and *diversity* in the selection process.

In response to the above challenge, our main contributions are as follows:

- We formally define the characteristics of search results (coherence and diversity) and use them to increase the quality of search in IoT. To the best of our knowledge, our work is the first piece of research that investigates and formulates different types of relationships.
- We propose a new framework, the ECS (Extract, Cluster, Select) framework, to manage the coherence and diversity of search results. Our framework contains a novel approach to extract and integrate different types of correlation graphs with a spectral clustering method and a selection method to improve the coherence and the diversity of top-k results for a given search query.
- We conduct extensive experiments to validate the effectiveness of our approach using real-world datasets.

The rest of this paper is organized as follows. In Sect. 2, we use a graph based data structure to formally define the problem and provides an overview of our methodology. Section 3 presents the technical details of our ECS approach. We describe the results of experimental evaluation in Sect. 4. In Sect. 5, we review the related work. Finally, we conclude the paper in Sect. 6.

2 Problem Statement

2.1 Problem Formulation

In this section we formally define various types of correlations and then the problem. In this paper we focus mainly on two types of correlations: *Co-Location Object Relationship (CLOR)* and *Parental Object Relationship (POR)*.

Definition 1 (Thing). *Suppose that V is the universal set of things and a thing $v \in V$ is an object connected to IoT where (1) $(v.lat, v.lon) = v.l$ denotes the location of the object composed of latitude $v.lat$ and longitude $v.lon$; (2) $v.dt = \{dt_i\}_{i=1}^{m}$ and $dt_i \in T_d$ denotes a set of descriptive tags about v, where T_d is the universal set of keywords (descriptive tags); (3) $v.lt = \{lt_i\}_{i=1}^{n}$ and $lt_i \in T_l$ where T_l is an alphabet consisting of location tags such as road names.*

Definition 2 (Parental Object Relationship – POR). *If $\sigma^P(v, v') \in [0, 1]$ and $\tau \in [0, 1]$ denote the strength of POR and the similarity threshold, between each given pair $(v, v') \in V \times V$, a POR exists if the following equation is satisfied:*

$$\sigma^P(v, v') = \frac{|v.dt \cap v'.dt|}{|v.dt \cup v'.dt|} \geq \tau \tag{1}$$

Definition 3 (Co-Location Object Relationship – CLOR). *Between each given pair $(o, o') \in V \times V$, a CLOR exists if the following equation is satisfied:*

$$\sigma^L(v, v') = \frac{|v.lt \cap v'.lt|}{|v.lt \cup v'.lt|} \geq \tau \tag{2}$$

where $\sigma^L(v, v') \in [0, 1]$ is the strength of CLOR and τ is the similarity threshold.

Definition 4 (Things Correlation Graph (TCG)). *A TCG is an undirected graph denoted as $G = (V, E, w)$ where (1) V is the universal set of things; (2) $E \subseteq V \times V$ is the set of edges and each edge $(v_1, v_2) \in E$ indicates a correlation (either POR or CLOR) between v_1 and v_2; and (3) $w : E \to [0, 1]$ is the weight function and $w(v_1, v_2)$ represents the weight of relationship of edge (v_1, v_2).*

Our goal is that given the set of *TCGs* for two types of relationships (*POR* and *CLOR*), retrieve the most coherent and diverse results for a search query issued by a user. The problem is formally defined as:

For a given query $Q = $ `<keyword,location>`, we select the set of *Things* ($\mathcal{V} \subseteq V$) with highest *coherence* and *diversity* which are ordered non decreasingly by their *matching scores*. We compute the matching score from the following:

$$\mu(Q, v) = w_1 \frac{|Q.dt \cap v.dt|}{|Q.dt \cup v.dt|} + w_2 \frac{|Q.lt \cap v.lt|}{|Q.lt \cup v.lt|} \tag{3}$$

where $Q.dt$ and $Q.lt$ refer to the descriptive (keyword) and location tags, respectively. We obtain the coherence of the result set from the following equation:

$$\phi(\mathcal{V}) = \frac{1}{|\mathcal{V}|^2} \sum_{v,v' \in \mathcal{V}} w_1 \sigma^L(v, v') + w_2 \sigma^P(v, v') \tag{4}$$

We define search results diversity based on the selection size as follows:

$$\delta(\mathcal{V}) = \frac{k \times |C^*|}{n \times |C|} \tag{5}$$

where C^* denotes the clusters with selected nodes, C denotes the set of all clusters, k is the number of nodes in the top-k result and n denotes the number of selected nodes.

For $\alpha, \beta \in [0,1]$ s.t. $\beta = 1 - \alpha$, the objective function $f : 2^V \to [0,1]$ is defined as follows:

$$f = \alpha \times \phi(\mathcal{V}) + \beta \times \delta(\mathcal{V}) \tag{6}$$

2.2   Methodology

Figure 2 shows the general structure of our framework and how it is evolved from the traditional approach.

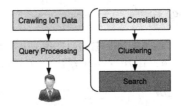

Fig. 2. Components of the ECS framework

As mentioned, current IoT search engines deploy a simple keyword based search strategy based on the crawled data from the IoT. The initial steps are *TCG Construction* and then *Clustering*. The next step is *Selection* which is performed on clusters. In our framework, we integrate different types of *TCGs* and cluster the vertices into a set of different clusters. Then, in the next step, we prepare the top-k results for a given search query.

3   ECS Approach

3.1   TCG Construction

Given a set of things represented by $\mathcal{V} = \{v_{i=1}^n\} \in \mathbb{R}^m$, their correlation graph TCG (Definition 5) is an undirected weighted graph $TCG = < V, E, \mathbf{W} >$, where V denotes all the things, $E \subseteq V \times V$ is the edges of TCG, the weighted matrix

---

Algorithm 1. ℓ_1-based Thing Correlation Graph Construction

Input: A collection of feature vectors of things $\mathcal{V} = \{v_1, .., v_n\}, v_i \in \mathbb{R}^m$
Output: Similarity matrix $\mathbf{W} \in \mathbb{R}^{n \times n}$

1: Each v_i is normalized to be $v_i = \dfrac{v_i}{||v||_2^2}$;

2: **for** $i = 1 \to n$ **do**
3:     $\mathcal{V}_i = [\mathcal{V} \backslash v_i | \mathbf{I}] \in \mathbb{R}^{m \times (m+(n-1))}$;
4:     $\min_{\hat{\mathbf{w}}_i} ||\mathbf{w}||_1$ s.t., $||v_i - \mathbf{B}\mathbf{w}_i||_2 = 0, \mathbf{w} \geq 0$;
5:     **for** $j = 1 \to i - 1$ **do**
6:         **if** $(1 \leq j \leq i - 1)$ **then**
7:             $\mathbf{W}(i, j) = \mathbf{w}_i^j$;
8:         **if** $(i + 1 \leq j \leq n)$ **then**
9:             $\mathbf{W}(i, j) = \mathbf{w}_i^{j-1}$;
10: **return** $\mathbf{W}$

$\mathbf{W}$ can be obtained using ℓ_1-based one-to-all sparse graph based reconstruction, in which each thing can be considered as a linear span of other things in the dataset. Given a set of things $\mathcal{V} = \{v_{i=1}^n\} \in \mathbb{R}^m$, for each thing v_i, its similarity/affinity with other things can be obtained by $\mathbf{v}_i = \mathcal{V}_i \mathbf{w}$, where $\mathbf{v}_i \in \mathbb{R}^m$ is the sample to be reconstructed, $\mathbf{w} \in \mathbb{R}^n$ is the reconstruction coefficients, $\mathcal{V}_i = \mathcal{V} \backslash v_i = [v_1, ..., v_{i-1}, v_{i+1}, ..., v_n] \in \mathbb{R}^{m \times (n-1)}$, which is formed by other objects in the dataset except for object v_i. The reconstruction coefficients of v_i can be computed using the objective function:

$$\min_{\hat{\mathbf{w}}} ||\hat{\mathbf{w}}||_1, \qquad s.t. \quad v_i = \mathcal{V}_i \mathbf{w} \tag{7}$$

where $||\cdot||_1$ is the ℓ_1 norm, tending to minimize the ℓ_1 norm of reconstruction error. Since the relations of nodes on a graph is supposed to be non-negative, we impose an extra constraints on Eq. 7, which is formulated as:

$$\min_{\hat{\mathbf{w}}} ||\hat{\mathbf{w}}||_1, \qquad s.t. \quad v_i = \mathcal{V}_i \mathbf{w}, \quad \mathbf{w}_i^j \geq 0 \tag{8}$$

The construction process is summarized in Algorithm 1. Algorithm 2 summarizes the clustering process using the weighted matrix generated by Algorithm 1. In this way, the algorithm can adaptively select the neighbors for each data point and at the same time the similarity matrix indicating pointwise similarity is automatically derived from the calculation of these sparse representations. It automatically leads to a sparse solutions, which means the thing correlation graph would be a sparse graph.

3.2   Clustering

After the previous step, we obtain a sparse representation for each data points whose nonzero elements ideally correspond to points from the same subspace. To infer the cluster of each data points on TCG into different subspaces using the sparse coefficients $\hat{\mathbf{w}} \in \mathbf{W}$, we first employ spectral clustering techniques to extract more informative structures by computing the first K eigenvectors of

---

Algorithm 2. Subspace Clustering on Weighted Matrix $\mathbf{W}$ of TCG

Input: Weighted matrix $\mathbf{W}$ of TCG
Output: Clustering membership for each data point v_i

1 Normalizing columns of $\mathbf{W}$ as $\mathbf{W}(:, i) \leftarrow \dfrac{\mathbf{W}(:, i)}{||\mathbf{W}(:, i)||_\infty}$;

2 Symmetrizing $\mathbf{W} = \dfrac{1}{2}(\mathbf{W} + \mathbf{W}^T)$;

3 Computing graph Laplacian matrix $\mathbf{L} = \mathbf{I} - \mathbf{D}^{-1/2}\mathbf{W}\mathbf{D}^{-1/2}$, where $\mathbf{D} = [\mathbf{d}_{ij}]$ is a diagonal matrix with $\mathbf{d}_{ii} = \sum_j w_{ij}$;

4 Computing k eigenvectors of $\mathbf{L}$ corresponding to k largest eigenvalues, and form the matrix $\mathbf{C} = [\mathbf{c}_1, ..., \mathbf{c}_k]$;

5 Performing K-means using $\mathbf{C}$;

6 Assign each data point v_i to cluster j if the i-th row of the matrix $\mathbf{C}$ is assigned to the cluster j;

7 **return** *the set of clusters;*

---

graph Laplacian of $\mathbf{W}$, the graph Laplacian can be computed as

$$\mathbf{L} = \mathbf{I} - \mathbf{D}^{-1/2}\mathbf{W}\mathbf{D}^{-1/2} \tag{9}$$

where $\mathbf{D} = [\mathbf{d}_{ij}]$ is a diagonal matrix with $\mathbf{d}_{ii} = \sum_j w_{ij}$. $\mathbf{D}^{-1/2}$ indicates the inverse square root of $\mathbf{D}$. The Laplacian is symmetric positive semidefinite, and its first K eigenvectors corresponding to K smallest eigenvalues can be computed. Thus, we can perform $k$means to cluster all the data points.

Time Complexity Analysis. Sparse similarity $\mathbf{W}$ construction and k-means based spectral clustering would be two main stages of consuming computational resource. As we know, efficiency is crucial, especially we will deal with a large amount of ubiquitous things. We briefly draw some analysis of time complexity of our method and how to adapt our proposed approach in a scalable way to match the needs of IoT.

We adopt the ℓ_1 sparse representation to decode the similarity of queries and the training dataset, which is the most expensive part of our proposed method. Given a collection of data points (i.e., each object can be considered as a virtual data point), ℓ_1-graph finds a sparse representation for the object using all other objects in the dataset and builds a similarity graph using the representation coefficients. The sparse similarity graph needs an iterative optimization process for respective objective function, and has no closed form like ℓ_2 norm. In our implementation, we adopt Homotopy algorithm based fast solver, where each step of the algorithm involves the rank-one update of a linear system. If the whole procedure stops in K steps, yielding a solution with K nonzeros, its overall complexity is bounded by $4Kd'^2/3 + Kd'n + \mathcal{O}(Kn)$, where d' is the dimension of the reduced features and n is the number of the training samples. To further reduce the complexity, we sparsely reconstruct each data point from its k nearest neighbors in feature space instead of using all the other samples. In this way, the searching space of constructing TCG will be reduced from $n - 1$ to k ($k < n$).

On the other hand, k-means algorithm employs an iterative procedure. At each iteration, one finds each data point's nearest center and assigns it to the corresponding cluster. Cluster centers are then recalculated. The procedure stops

---

Algorithm 3. Node Selection

Input: A collection of Clusters $C = \{C_i\}_{i=0}^N$; k the size of the result set; Query Q; and α, β
Output: An ordered set of things $\mathcal{V}$ for which $|\mathcal{V}| = k$
1: Filter the nodes in clusters set C based on the features specified in the query Q;
2: **for** $i = 1 \to |C|$ **do**
3:     $m = \mathsf{random}(v_x) \in C_i$;
4:     **for** $v_z \in C_i$ **do**
5:        **if** $\sum_{v_y \in C_i} \mathbf{W}(z, y) > \sum_{v_y \in C_i} \mathbf{W}(z, y)$ **then**
6:           $m = v$;
7:     $\mathcal{V} \leftarrow \mathcal{V} + m$;
8:     $C_i \leftarrow C_i \backslash m$;
9: **if** $(|\mathcal{V}| < k)$ **then**
10:     Add select(updated $C, k - |\mathcal{V}|, Q, \alpha, \beta$) to $\mathcal{V}$;
      return $\mathcal{V}$;

---

after reaching a stable error function value. Since the algorithm evaluates the distances between any point and the current k cluster centers, the time complexity of k-means is $\mathcal{O}(nk^2)$, where n is the number of data points and k number of clusters. However, since its weight matrix $\mathbf{W}$ is a sparse similarity matrix as well, we can use sparse eigensolvers to accelerate the computation of top K eigenvectors from Laplacian matrix $\mathbf{L}$, e.g., ARPACK [4], which can reduce the time complexity of computing eigenvectors from $\mathcal{O}(m^3)$ to $\sim \mathcal{O}(nm)$.

3.3 Selection

To make this selection, we can pick a single node from each cluster and add it to the result set. We can use different strategies in the selection step. For example, one can choose elements based on their similarity values, or the medoid element of each cluster. However, the quality of the final results highly depends on the quality of the selection strategy [14]. We propose to implement a selection strategy in such a way that can satisfy the following conditions: (i) The set of things in each cluster have the most similarity to each other while the inter-cluster similarity is minimized; and (ii) The returned results maximize the objective function which aggregates the *coherence* and the *diversity* of the result set. We select the nodes with maximum similarity index to intra-cluster nodes.

Algorithm 3 summarizes the steps of our selection strategy. The algorithm takes a set of parameters including the clusters from the clustering step, the expected size of the result k, the query Q and the *tradeoff* coefficients α and β. The algorithm initializes $\mathcal{V}$ in the first turn (Lines 1–8) to cover conditions (1) and (2). First, the set of nodes are filtered based on the keywords from the query (Line 1). Then the algorithm iterates through all of the clusters with a for loop (Line 2), picks a random node from the cluster C_i as its medoid (Line 3), and iterates through the remaining nodes in the cluster (line 4) to find nodes which have less intra-cluster dissimilarity with other nodes. Once a new optimal node is found, it is replaced by the selected medoid (Lines 5–8). Later the medoid is added to $\mathcal{V}$ and removed from C_i (Lines 9, 10). The algorithm recursively calls itself with subtracted $\mathcal{V}$ from the clusters k is greater than k (Lines 12–14).

Finally, set $\mathcal{V}$ is returned. The order of execution for the Algorithm 3 for a small k is $n + \frac{K\mathcal{O}(kn)}{k}$ as it is mainly composed of keyword based filtering, a loop through all clusters and their nodes and a recursive call. This order for the case of $K = k$ is reduced to $n + \mathcal{O}(kn)$ as the need for a recursive call is removed.

4   Experimental Results

We perform extensive experiments to evaluate the effectiveness of our approach. We use real-world datasets to ensure that the proposed approach works under different conditions. We implemented our framework and experiments in R programming language with *ggmap*[1]. The experiments were performed on a computer with 2.5 GHz core-i5 processor and 8 GB of RAM. In this section, we first describe the characteristics of datasets we used, and then present the results from experimental evaluation and discuss the findings of our research.

4.1   Datasets

Nowadays, many of IoT applications such as intelligent transport systems and environmental sensing are publishing their data on the Internet. IoT search engines such as *Thingful*[2], rely on this type of data. We use the following real-world datasets to evaluate our work:

1. Vehicle Trajectory Data: It contains nearly 16,000,000 trajectories for 10,357 taxicabs in the city of Beijing in a one week period [19]. We model the area of the city with a 76×76 grid of cells.
2. Weather Sensor Data: We also used a dataset known as LinkedSensorData[3], which is an RDF dataset containing expressive descriptions of nearly 10,000 weather stations in the USA. The data originated at MesoWest[4]. It contains the observations from weather sensor measurements such as temperature, pressure, wind speed, humidity, etc. The dataset contains 1,730,284,735 triples from 159,460,500 observations during major hurricanes. We accompanied every sensor's data with location tags from GeoNames[5].
3. Query Data: we used a real-world dataset consisting of 136,746 queries between Feb, 2014 to Feb, 2015 from Thingful search engine. This dataset is not IoT data, but we used it for query generation and analysis. Figure 3 shows distribution of most popular keywords in the queries dataset.

First we performed a dataset analysis on each dataset we have used. After computing the weights matrix $\mathbf{W}$, we normalized the matrices and obtained the graph for both of the real-world datasets. Figure 4(a) shows a partition of the integrated TCG for the vehicles trajectory dataset. Comparing with Fig. 4(b) which shows a partition from weather stations dataset, the graph of the first dataset is far more scarce than the graph for the first dataset.

---

[1] http://cran.r-project.org/web/packages/ggmap/index.html.
[2] http://thingful.net.
[3] http://wiki.knoesis.org/index.php/SSW_Datasets.
[4] http://mesowest.utah.edu/index.html.
[5] http://www.geonames.org/.

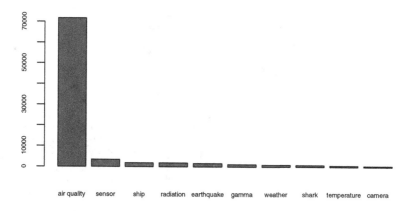

Fig. 3. Distribution of query keywords from *Thingful* dataset

Figure 4(c) depicts a map of Beijing and the trajectory data along with the status of the vehicles. Figure 4(d) shows the result of clustering using 10 nearest neighbors and can be compared to Fig. 4(e) which depicts the same dataset clustered with 30 nearest neighbors. As the figures show, we end up with a less number of same color patches in the second figure. For the weather station dataset, Fig. 4(f) depicts the location and altitude of weather station across North America. Clustering this dataset is more challenging as things have more common features and as Fig. 4(i) shows, the density of 3D location distribution for a large part of the data is concentrated around two points. Figure 4(g) shows the result of clustering using 5 nearest neighbors and can be compared to Fig. 4(h) which depicts the same dataset clustered with 200 nearest neighbors. We observed that the geographical distribution of clusters increases when we increase the number of nearest neighbors.

4.2   Results

In our experiments, our target is to investigate the answers to the following questions: (i) How do different methods react to different ratios between α and β? (ii) What would be the outcome for different amount of k in the initial clustering? (iii) Which selection approach would be more effective?

We compared our approach with other thing selection approaches including *Fixed Length Selection* (**FLS** - selecting random k nodes), *Maximum Similarity Selection* (**MSS** - selecting things with highest augmented μ), k-*Medoids Selection* (**KMS** - our approach) and *Plain Selection* (**PS**) which is the current approach used by *Thingful* and other works.

For varying α and β, in Fig. 5(a), we compare the final ranking scores for different ratios of α and β between different other selection approaches. As we increase the ratio from 1 to 10, the final score for *MSS* approach increases to meet *FLS* score. As *FLS* and *KMS* strategies both experience a slight increase in their final scores for greater amounts of α, it shows that the *MSS* is better

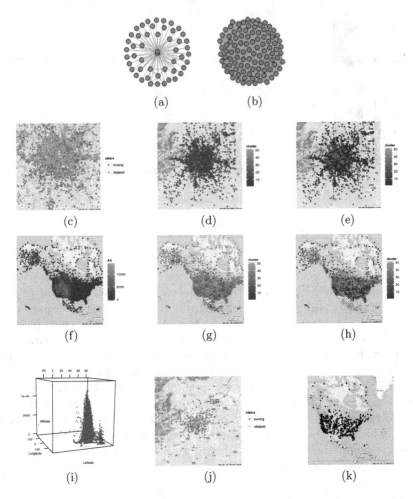

Fig. 4. (a) TCG for trajectory dataset; (b) TCG for weather stations dataset (c) Trajectory dataset mapped onto Beijing area; (d) Results of clustering with 10 nearest neighbors; (e) Results of clustering with 30 nearest neighbors; (f) Weather stations dataset mapped onto North America; (g) Results of clustering with 5 nearest neighbors; (h) Results of clustering with 200 nearest neighbors; (i) Distribution of 3D location data for weather stations; (j) Selection results from the trajectory dataset; (k) Selection result from the weather stations dataset

in providing more coherent results but the *KMS* performs stronger in overall distribution of diversity and coherence.

To answer the last question, we investigated the coherence and diversity of results for each one of the object selection strategies. Figure 5(b) compares the coherence of the outcomes for varying k. As shown, the *KMS* method outperforms other methods in terms of results coherence. While the coherence scores of PS are quite steady, *MSS* and *FLS* methods fluctuate dramatically.

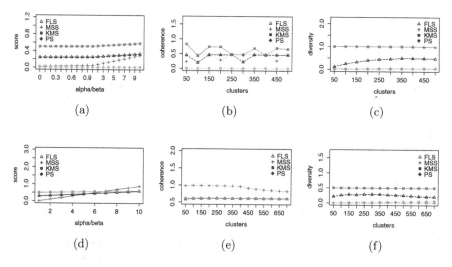

Fig. 5. (a) Trajectory dataset selection score analysis for different ratios of α and β; (b) Trajectory dataset coherence for different amounts of k; (c) Trajectory dataset diversity for different amounts of k;(d) Weather stations dataset selection score analysis for different ratios of α and β; (e) Weather stations dataset coherence for different amounts of k; (f) Weather stations dataset diversity for different amounts of k

Finally, Fig. 5(c) compares the diversity of the outcomes for different methods. The diversity index for the results of *KMS* method constantly remains as 1.00 for varying amounts of k as one vertex is selected from each cluster. For *MMS* method, the diversity of results improves to a limit as k is increased. As the figure shows, both of the *PS* and *FLS* methods perform very poorly in terms of diversity. Due to the fact that the weather sensors dataset generates a much denser graph than the trajectory dataset, it is more challenging for our approach to improve the *coherence*, the *diversity* and as a result the overall score of the selected k things. However, as Fig. 5(d) shows, the *KMS* method outperforms other methods when the coherence has a higher weight ($\geq 1/7\beta$). For higher values of α, the *MSS* method outperforms *KMS* as the graph is very dense and its nodes have a very strong correlation. Figure 5(e) compares the coherence of the results for a varying amount of k. As the Figure shows, the *MSS* approach outperforms other approaches in terms of coherence but their difference decreases as k is increased. Figure 5(e) compares the diversity of the outcomes for different methods. The diversity index for the results of *KMS* method remains 1.00 for varying amounts of k as one vertex is selected from each cluster.

To conclude, the experimental results show that our approach improves the *coherence* and *diversity* compared to the current and baseline approaches. Our findings show that our approach performs better on sparse graphs although it improves the overall results score from dense graphs as well.

We feature the nodes along with filtering criteria based on different types of correlations to simplify the final filtering for users (Fig. 4(j) and (k)). Thus, the users will not have to go through all existing things in the dataset.

5   Related Work

In this section we review works related to search based on social relationships and IoT search engines.

In [2] the details of anatomy of a social search engine, namely Aardvark, have been discussed. The search process in Aardvark is based on the village search paradigm in which people use natural language to ask questions and answers are generated in real-time by someone from the village and trust is based on intimacy. Main components of Aardvark are as follows: Crawler and indexer, Query analyzer, Ranking function, UI. To index the relationships and affiliation of users, social graph structure has been deployed. Aardvark is a social search engine designed to index people and propose the best users to answer a question given by a user and the results are ranked based on factors such as topic expertise, intimacy (connectedness) and activity (availability).

Microsoft SensorMap [7] and linked sensor middleware [3] support search for sensors based on textual metadata that describes the sensors (e.g., type and location of a sensor, measurement unit, object to which the sensor is attached). Such metadata is often manually entered by the person who deploys the sensors. Other users can then search for sensors with certain metadata by entering appropriate keywords. There are efforts to provide a standardized vocabulary to describe sensors and their properties such as SensorML[6] or the Semantic Sensor Network Ontology (SSN)[7]. Unfortunately, these ontologies and their use are rather complicated. It is problematic that end users are able to provide correct descriptions of sensors and their deployment context without the help from experts. In other words, this type of solutions are time-consuming as they require the prior knowledge, e.g., define the descriptions of things and their corresponding characteristics under a uniform format. Furthermore, these solutions do not exploit the rich information about user's historical interactions with things, containing implicit relations of different entities, e.g., if some users have the similar usage pattern on some things, which may indicate close connection of those things.

Another alternative approach for searching things is based on search-by-example. The work in [13] adopts this approach to sensors, i.e., a user provides a sensor, respectively a fraction of its past output as an example, and requests sensors that produced similar output. Ostermaier et al. [8] propose a real-time search engine for Web of Things, which allows searching real-world entities having certain properties. They associate a Web page to a real-world entity (e.g., a meeting room) containing additional structured metadata about the sensors connected to it. This method takes care of the valuable information of historical data, but misses the relations among contextual information. Maekawa et al. [5] propose a context-aware web search in ubiquitous sensor environments, and Yao et al. [16,18] construct the models that captures the pairwise relations between things via mapping the contextual information into separate graphs. However, more complex relations between heterogeneous objects can not be captured in these

---

[6] http://www.opengeospatial.org/standards/sensorml.

[7] http://www.w3.org/2005/Incubator/ssn/ssnx/ssn.

works. Hypergraph-based model is another approach to capture the high-order and complex relations among things [17]. However, the complexity of processing hypergraph problems is more than the complexity of graph problems and those models, can be considered as a future extension for correlation management in large scale systems.

6   Conclusion

Building enterprise search services is considered as one of the most significant steps in IoT research area. We proposed a novel design for framework components using techniques such as k-Nearest Neighbors, Spectral Clustering and k-Medoids Clustering. We assessed the proposed framework using a number of the real-world datasets and showed that our approach can improve the *coherence* and *diversity* compared to the current and possible substitute approaches. Our framework supports two types of correlations including *Co-Location Object Relationship* and *Parental Object Relationship*. One of the future directions for this research is to extend the framework to support other types of correlations such as *Social Object Relationship*. We also target to extend the solution over time series and uncertain IoT data to provide more accurate search results.

References

1. Atzori, L., Iera, A., Morabito, G., Nitti, M.: The social internet of things (SIoT)-when social networks meet the internet of things: concept, architecture and network characterization. Comput. Netw. **56**(16), 3594–3608 (2012)
2. Horowitz, D., Kamvar, S.: The anatomy of a large-scale social search engine. In: Proceedings of the 19th International Conference on World Wide Web, pp. 431–440. ACM, April 2010
3. Le-Phuoc, D., Quoc, H.N.M., Parreira, J.X., Hauswirth, M.: The linked sensor middleware-connecting the real world and the semantic web (2011)
4. Lehoucq, R.B., Sorensen, D.C., Yang, C.: ARPACK users' guide: solution of large-scale eigenvalue problems with implicitly restarted Arnoldi methods, vol. 6. SIAM (1998)
5. Maekawa, T., Yanagisawa, Y., Sakurai, Y., Kishino, Y., Kamei, K., Okadome, T.: Context-aware web search in ubiquitous sensor environments. ACM Trans. Int. Technol. (TOIT) **11**(3), 12 (2012)
6. Miorandi, D., Sicari, S., De Pellegrini, F., Chlamtac, I.: Internet of things: vision, applications and research challenges. Ad Hoc Netw. **10**(7), 1497–1516 (2012)
7. Nath, S., Liu, J., Zhao, F.: Sensormap for wide-area sensor webs. Computer **40**(7), 0090–93 (2007)
8. Ostermaier, B., Romer, K., Mattern, F., Fahrmair, M., Kellerer, W.: A real-time search engine for the web of things. In: Proceedings of the 2nd International Conference on the Internet of Things (IOT), pp. 1–8. IEEE, November 2010
9. Perera, C., Zaslavsky, A., Christen, P., Compton, M., Georgakopoulos, D.: Context-aware sensor search, selection and ranking model for internet of things middleware. In: Proceedings of the 14th International Conference on Mobile Data Management (MDM), vol. 1, pp. 314–322. IEEE (2013)

10. Pfisterer, D., Romer, K., Bimschas, D., Kleine, O., Mietz, R., Truong, C., Hasemann, H., Kroller, A., Pagel, M., Hauswirth, M., et al.: SPITFIRE: toward a semantic web of things. IEEE Commun. Mag. **49**(11), 40–48 (2011)

11. Song, Z., Cárdenas, A.A., Masuoka, R.: Semantic middleware for the internet of things. In: Proceedings of the 2nd International Conference on the Internet of Things (IOT), pp. 1–8. IEEE, November 2010

12. Tan, C.C., Sheng, B., Wang, H., Li, Q.: Microsearch: a search engine for embedded devices used in pervasive computing. ACM Trans. Embed. Comput. Syst. (TECS) **9**(4), 43 (2010)

13. Truong, C., Romer, K., Chen, K.: Sensor similarity search in the web of things. In: Proceedings of the 13th International Symposium on a World of Wireless, Mobile and Multimedia Networks (WoWMoM), pp. 1–6. IEEE, June 2012

14. Vieira, M.R., Razente, H.L., Barioni, M.C.N., Hadjieleftheriou, M., Srivastava, D., Traina, A., Tsotras, V.J.: On query result diversification. In: Proceedings of the 27th IEEE International Conference on Data Engineering (ICDE), pp. 1163–1174. IEEE, April 2011

15. Wang, H., Tan, C., Li, Q.: Snoogle: a search engine for pervasive environments. IEEE Trans. Parallel Distrib. Syst. **21**(8), 1188–1202 (2010)

16. Yao, L., Sheng, Q.Z.: Exploiting latent relevance for relational learning of ubiquitous things. In: Proceedings of the 21st ACM International Conference on Information and Knowledge Management (CIKM), pp. 1547–1551. ACM, October 2012

17. Yao, L., Sheng, Q.Z., Falkner, N.J., Ngu, A.H.: ThingsNavi: finding most-related things via multi-dimensional modeling of human-thing interactions. In: Proceedings of the 11th International Conference on Mobile and Ubiquitous Systems: Computing, Networking and Services (Mobiquitous), pp. 20–29. ICST, July 2014

18. Yao, L., Sheng, Q.Z., Gao, B., Ngu, A.H.H., Li, X.: A model for discovering correlations of ubiquitous things. In: Proceedings of the 2013 IEEE International Conference on Data Mining (ICDM), December 2013

19. Yuan, J., Zheng, Y., Xie, X., Sun, G.: Driving with knowledge from the physical world. In: Proceedings of the 17th International Conference on Knowledge Discovery and Data Mining (SIGKDD), pp. 316–324. ACM, August 2011

20. Zhang, D., Yang, L.T., Huang, H.: Searching in internet of things: vision and challenges. In: Proceedings of the 9th International Symposium on Parallel and Distributed Processing with Applications (ISPA), pp. 201–206. IEEE, May 2011

Sense and Focus: Towards Effective Location Inference and Event Detection on Twitter

Yihong Zhang[(✉)], Claudia Szabo, and Quan Z. Sheng

School of Computer Science, The University of Adelaide, Adelaide, SA 5005, Australia
{yihong.zhang,claudia.szabo,michael.sheng}@adelaide.edu.au

Abstract. Twitter users post observations about their immediate environment as a part of the 500 million tweets posted everyday. As such, Twitter can become the source for invaluable information about objects, locations, and events, which can be analyzed and monitored in real time, not only to understand what is happening in the world, but also an event's exact location. However, Twitter data is noisy as sensory values, and information such as the location of a tweet may not be available, e.g., only 0.9 % of tweets have GPS data. Due to the lack of accurate and fine-grained location information, existing Twitter event monitoring systems focus on city-level or coarser location identification, which cannot provide details for local events. In this paper, we propose SNAF (Sense and Focus), an event monitoring system for Twitter data that emphasizes local events. We increase the availability of the location information significantly by finding locations in tweet messages and users' past tweets. We apply data cleaning techniques in our system, and with extensive experiments, we show that our method can improve the accuracy of location inference by 5 % to 20 % across different error ranges. We also show that our prototype implementation of SNAF can identify critical local events in real time, in many cases earlier than news reports.

Keywords: Twitter · Microblog content classification · Location inference · Event detection

1 Introduction

A micro-blogging service such as Twitter allows its users to conveniently create and publish *tweets*, which are short messages of maximum 140 characters and are immediately available for anyone to read online. As Twitter has gained enormous popularity over the past five years, the number of users and published tweets has increased significantly. Nowadays, 284 million active Twitter users generate 500 million tweets every day[1]. Moreover, as a direct consequence of the increasing popularity of smartphone, 80 % of Twitter users use their mobile phones to create tweets (See footnote 1). The use of mobile platform for tweeting implies that users can conveniently report current events and objects in their physical vicinity.

---

[1] https://about.twitter.com/company.

© Springer International Publishing Switzerland 2015
J. Wang et al. (Eds.): WISE 2015, Part I, LNCS 9418, pp. 463–477, 2015.
DOI: 10.1007/978-3-319-26190-4_31

In this paper, we assume that a user is an ordinary person, as opposed to a community or a corporation. A user's tweet can be a description of the user's mood, a message to a friend, or a link to an online article the user wants to share. A user may also post observations of surrounding environments, such as *"the air is fresh"*, or *"I just saw a homeless man outside of the church"*. We call this type of tweets *immediate observations*, because they describe the state of the observed object, person, or event, at the time and location when and where the user posts the message.

The immediate observations of Twitter users can be seen as reading values of some sensory devices [16]. Take, for example, a user posting a tweet in Times Square on Tuesday morning, saying *"the air is fresh"*. This can be seen as an air quality sensor installed in Times Square that just recorded a good reading on Tuesday morning. An immediate observation posted on Twitter can be converted to sensory values, if the time and location of the tweet are known. Given the high density of Twitter users in urban areas, the perspective of Twitter users as sensors instantly provides a dense virtual sensor network, by which some events and objects of interest can be monitored. Several works have proposed using tweets as sensory values to monitor environmental disasters [14, 16]. However, due to the lack of accurate and fine-grained location information, existing event monitoring systems for Twitter data only identifies city-level or coarser locations, and cannot provide accurate details for local events [14, 22].

In this paper, we propose SNAF (Sense and Focus), an event monitoring system that captures local events reported on Twitter and accurately infers event locations. SNAF implements a location inference mechanism based on a large gazetteer and distance-based data cleaning algorithms. It also takes the poster's past tweets as sources to increase the availability of location information. Given the accurate and high-recall location information provided by the mechanism, our system is able to successfully capture local events, even before news reports. The main contributions of our work are summarized as follows:

- We propose a method that converts immediate observations of events and objects of interest reported on Twitter into sensory values. We employ a classifier-based filter that collects relevant tweets, organize a gazetteer generated from DBpedia data, and use users' past tweets as sources of location information. By identifying place names in tweet messages and users' tweet histories, we increase the availability of location information significantly.
- We select and implement two existing distance-based sensor data cleaning algorithms for improving the accuracy of location inference, and conduct comprehensive experiments to test them. Our experimental results with real Twitter data show that, comparing to the existing approaches, our method improves the accuracy of location inference by 5 % to 20 % across different error ranges.
- We show the potential of further use of converted sensory values from observation tweets, by designing and implementing an event monitoring system, which aggregates immediate observations of an event or an object of interest based on their locations. We verify the system by comparing the detected events to

relevant news articles, and find that the event detected by our system are in many cases several hours faster than news articles, while the locations inferred are accurate comparing to the locations reported in the news.

The remainder of this paper is organized as the following: in Sect. 2, we discuss the related works. In Sect. 3, we present SNAF and the details of its components. In Sect. 4, we report our experiments for immediate observation classification accuracy, location inference accuracy, event detection for real world events. Finally, Sect. 5 provides some discussions on the proposed system and Sect. 6 offers some concluding remarks.

2 Related Work

Our work follows a current research trend of converting microblog messages to sensory values. Sakaki et al. [16] proposed and investigated the idea of using tweets as sensory values for environmental event monitoring. Their results show that the use of tweets as sensory values is feasible. In a later work, they extended the system to predict typhoon movements using tweet messages [17].

Current researches on location inference on Twitter focus on different granularities. A category of research aims at inferring city-level locations [4,5,13]. Li et al. [13] inferred tweet location by first inferring the user's home location, relying on the home location entries in Twitter's user profiles, which are usually entered as cities. Graham et al. [4] also exploited home location entries in user profiles for inferring the location of the tweets and users.

Another category of research aims at inferring finer-grained locations [6,12, 14,18]. Li et al. [12] trained a classifier to associate placenames found in tweet messages with formal names, based on Foursquare data. Ikawa et al. [6] inferred tweet locations by matching the tweet with tweets with known locations using cosine similarity, and also based on Foursquare data. Foursquare is a location-based service that provides place-coordinate association for commercial places such as hotels and restaurants. Schulz et al. [18] leveraged DBpedia for identifying places in tweet messages, but resulted in large errors, with the median error distance of $1,100$ km. DBpedia is a large user-contributed database for name-entities used in the Web [1]. Using a generation tool called DBpedia Spotlight [3], we can generate a gazetteer containing more than 800,000 places with respective GPS coordinates, which covers a large number of street-level places. However, as we will discuss below, such large gazetteer will make location information very noisy.

The earthquake and typhoon detection system proposed in [17] takes all tweets containing the keyword as the tweets for a single event, assuming that earthquakes and typhoons do not occur frequently. Local events such as vehicle crashes occur in much higher frequencies with smaller impact areas. Local event detection thus requires clustering, preferably with location information [13,15,22]. Unankard et al. [22] proposed an event detection method based on clustering of words. They relate events to locations, but only at a country-level. Li et al. [13] proposed a classification-based event detection method for crime and disaster events, with a location extraction component focuses on city-level locations. The lack of accurate and fine-grained location information is the main issue that prevents existing event monitoring systems from obtaining local details.

Noisy sensor readings have been extensively studied in sensor networks. Basic techniques include temporal and spatial aggregation using mean or median [7,23,24]. Particularly, the median is effective for avoiding extreme outlier values in the dataset [23]. More advanced techniques identify and remove outliers in the dataset, based on distance measures or clustering. Since it is straightforward to interpret the distance of two location points, the distance-based outlier detection is particularly suitable for location data. Subramaniam et al. [20] and Branch et al. [2] both used k-Nearest Neighborhood (KNN) for identifying outliers. An issue for using KNN is to choose the proper distance threshold that excludes the outliers. Sheng et al. [19] proposed two KNN-based outlier detection algorithms, using a fixed distance threshold and a relative distance threshold, respectively.

3   Sense and Focus: Event Monitoring with Location Inference

SNAF consists of a Sense component and a Focus component, as illustrated in Fig. 1. To Sense, SNAF filters immediate observations of certain events or objects of interest from Twitter streams. To Focus, SNAF infers the location of each report, and aggregates reports based on their locations for detecting local events. As mentioned earlier, the main issue for event detection system for Twitter data is the lack of accurate and fine-grained location information. SNAF challenges this issue by using a large gazetteer and users' past tweets to increase the availability of location information, and applying sensor data cleaning techniques to remove the noises in the data. In this section, we discuss the Sense component's *immediate observation filtering*, and the Focus component's *placename extraction*, *location resolution using past locations*, and *aggregation for event detection*.

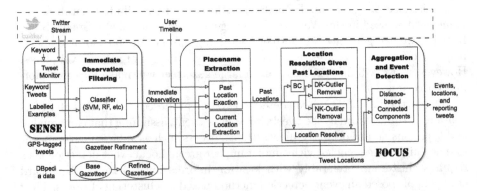

Fig. 1. The overall architecture of SNAF: sense and focus

3.1 Filtering Immediate Observation Reports

Immediate observation reports are tweets describing an event or an object of interest that users found in their immediate environment. Twitter allows a developer to monitor all tweet traffic using keywords, but not all tweets containing a chosen keyword are immediate observation reports. The purpose of our filter is to select, as accurate as possible, immediate observation reports from other tweets containing the keyword. Although observations on different events and objects of interest have similar expressions, different keywords have very different ways of being mentioned. For example, *crash* often refers to a character in a popular TV show in recent tweets, and *shooting* is often mentioned in messages about movie productions, instead of crimes we want to monitor. While it is possible to make a general filter that applies to all immediate observation tweets, to achieve higher accuracy, we prefer to produce different filters for different keywords.

The accuracy of a classifier for natural language processing strongly depends on the features selected. We observe that the two words preceding and following the keyword is often sufficient to determine the meaning of the keyword in the tweet. Immediate observation reports usually have similar words just before the keyword, such as *"saw a"*, or *"to a"*. Also the hashtags contained in a tweet message can be helpful for deciding true or false-positive. For example, the hashtag *"#NYMed"* is associated with a medical documentary series broadcasted on TV, and tweets referring to homeless people appeared in it are not what we consider as immediate observations. But a hashtag representing a city name, such as *"#Manchester"* or *"#Boston"*, is often associated with an immediate observation, as a way the user expresses the location of the observation. We choose five features for training our classifier, two words before the keyword, two words after the keyword, and the first hashtag word. If no such a word is present, *null* is placed. A similar approach is used in [17].

3.2 Location Extraction for a Single Tweet

To infer the location of a report, we consider the tweet text as well as past tweets of the user who made the report as sources of locations. For each of these tweets, we run a gazetteer-based name extraction by comparing the words in the tweet text with gazetteer entries. We generated our base gazetteer using the latest version of DBpedia geo-coordinate dataset[2] and the DBpedia Spotlight name generation tool [3]. DBpedia is a user-contributed name-entity database for Internet-of-Things data lookup, and in a recent version (See footnote 2) includes 4.58 million names or things. The geo-coordinate dataset contains a subset of these names that are associated with a geo-coordinate. DBpedia Spotlight can track the usage of these names in Web documents such as Wikipedia pages, and produce a gazetteer containing formal and informal names. Our base gazetteer contains more than 800,000 entries and covers a large number of street-level places, including their formal and informal names.

---

[2] http://wiki.dbpedia.org/Datasets.

Using a large gazetteer such as our base gazetteer has a significant computation overhead. Furthermore, it also contains a large number of names that are usually not referring to a place when people mention them on Twitter. Therefore removing less meaningful entries from the dictionary helps both the efficiency and accuracy of the system. We use a heuristic to automatically refine the gazetteer. First we collect a set of GPS-enable tweets and extract locations from the messages using the base gazetteer. The extracted locations are compared with the GPS data. If the difference between the extracted location and the GPS data is larger than an acceptable threshold λ, we count a hit for the location. If the difference is larger than a rejection threshold θ where $\theta > \lambda$, we count a miss for the location. Usually a common location word will be extracted more than once. After the process, each extracted location will have a hit number and a miss number. Now we pick the locations with at least one hit to put in the refined gazetteer, but if the number of misses is more than half of the total extraction number, we remove it from the refined gazetteer. The purpose of this heuristic is to keep all the gazetteer entries relevant to the GPS data, with respect to the training dataset. In our experiments we chose $\theta = 10\,\mathrm{km}$ and $\lambda = 300\,\mathrm{km}$. These numbers are relatively stricter because we focus on smaller granularity, but different numbers can be chosen with different requirements in error tolerance.

3.3   Location Resolution Given Past Locations

When considering the past tweets from a user, we can generally extract a handful of past locations. The past tweets of Twitter users are publicly accessible through the timeline API provided by Twitter[3], with a restricted availability of a maximum of 3,000 most recent tweets. About 5 % tweets have a placename that can be identified using DBpedia-based gazetteer [18]. We conducted an experiment to analyze 1,000 random tweets and found that, while only 0.9 % of tweets have GPS tags, 87 % of the 982 users who made these tweets have at least one extractable placename in their past tweets, and 68 % of them have more than ten. By using placenames extracted from past tweets to infer the current location, we can effectively associate location information with a significant portion of tweets.

Various techniques have been proposed to infer the current location of an object based on past locations, such as Kalman filters and Particle filters. Since it is difficult to define a dynamic model for the movement of Twitter users, the Bayesian Filter-type inference techniques are ineffective. In sensor networks, a common way to aggregate spatial data is using the average. Particularly, median aggregation is effective for avoiding extreme values in the dataset [23]. More advanced techniques identify outliers in the dataset, which are removed before the data aggregation. Since it is straightforward to interpret the distance between location points, we choose two representative distance-based outlier detection techniques from the sensor network literature, called DK-Outlier and NK-Outlier [19]. Both outlier detection techniques are based on the k-Nearest

---

[3] https://dev.twitter.com/rest/public/timelines.

Neighbor (KNN). Given a set of locations H, and a location $l \in H$, let $V(l)$ be the distances from l to all other locations in H, sorted in ascending order. Then $V_k(l)$ is the distance from l to its k-th nearest neighbor. Introduced in [9], given an acceptable distance threshold d, a location l is a DK-Outlier if $V_k(l) \geq d$.

NK-Outlier is a new type of outlier proposed by [19]. NK-Outlier detection sets a relative distance threshold based on comparison between locations in the dataset. Given n, the number of other locations in the data a location needs to be comparable in order to be considered acceptable, a location l is a NK-Outlier if there are no more than $n - 1$ other locations p, such that $V_k(l) < V_k(p)$. The choice of d, n and k depends on the desired data accuracy and expected errors in the dataset. The algorithms for removing DK-Outlier and NK-Outlier are shown below as Algorithms 1 and 2.

Algorithm 1. Removing DK-Outlier

1: $H \leftarrow$ extracted past locations of the user
2: $R \leftarrow null$
3: **for each** $l \in H$ **do**
4:     $V \leftarrow$ distances from l to all other locations in H
5:     $VS \leftarrow V$ sorted in ascending order
6:     **if** $VS_k < d$ **then**
7:         add l to R
8:     **end if**
9: **end for**
10: remove all $l \in R$ from H

Algorithm 2. Removing NK-Outlier

1: $H \leftarrow$ extracted past locations of the user
2: $R \leftarrow null$
3: **for** $i \leftarrow 1$ **to** $|H|$ **do**
4:     $V \leftarrow$ distances from H_i to all other locations in H
5:     $VS \leftarrow V$ sorted in ascending order
6:     $K_i \leftarrow VS_k$
7: **end for**
8: **for** $i \leftarrow 1$ **to** $|H|$ **do**
9:     $count \leftarrow 0$
10:     **for** $j \leftarrow 1$ **to** $|H|$ **do**
11:         **if** $K_j \geq K_i$ **then**
12:             count++
13:         **end if**
14:     **end for**
15:     **if** $count < n$ **then**
16:         add l to R
17:     **end if**
18: **end for**
19: remove all $l \in R$ from H

When examining the data, we notice that users may have a burst of interest in a certain location, and post messages about this location intensively within a short period of time. This phenomenon often distorts the location inference significantly. Therefore we propose a compression algorithm, called *burst compression*, which basically removes consecutive occurrences of a location within three days from the first occurrence, as shown in Algorithm 3. This algorithm can be added to the above outlier removal algorithms as a pre-processing step.

The final inference of the location is thus the median location of all remaining locations extracted from the user's past tweets, after outliers are removed.

Algorithm 3. Burst Compression

1: $H \leftarrow$ extracted past locations of the user
2: $P \leftarrow$ days lapsed for each $l \in H$
3: $R \leftarrow null$
4: $curLoc \leftarrow H_1$
5: $curDay \leftarrow P_1$
6: **for** $i \leftarrow 2$ **to** $|H|$ **do**
7:     **if** $H_i = curLoc$ **and** $P_i - curDay \leq 3$ **then**
8:         add H_i to R
9:     **else**
10:         $curLoc \leftarrow H_i$
11:         $curDay \leftarrow P_i$
12:     **end if**
13: **end for**
14: remove all $l \in R$ from H

3.4 Realtime Event Monitoring

Events such as car crashes and shooting incidents attract attention easily, especially if they happen in urban areas. In areas where Twitter is popular, such events usually trigger multiple immediate observation reports posted to Twitter. By aggregating these reports, we can potentially detect the event and raise early alarms.

The final module of SNAF focuses on aggregation-based event detection, which uses the location inference described above to find out the correct location of the event. The aggregator is based on the *connected component* of the graph theory. In our approach, a connected component connects reports that are geographically close to each other. Reports and their locations are added to the graph one-by-one. An alarm is raised to indicate that an event is detected, if the connected component after adding a new report now has enough connected nodes. We set 5 as the number of reports needed for raising an alarm. Figure 2 illustrates this process.

Currently the relevant information such as all reporting tweets for the event are written to a log file when an alarm is raised, but it is trivial to write it to an online platform such as a website, making detected events publicly accessible

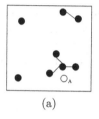 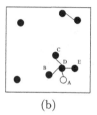

(a)   (b)

Fig. 2. Event detection based on connected components. The figure indicates observation reports and their geographical locations. (a) Adding report A. (b) A is connected to the nearest report. The connected component now has five nodes, so an alarm is raised, with location of the event calculated as the mean location of nodes A, B, C, D, and E.

in realtime. To maintain the timeliness of the monitoring system, we only store the most recent tweets, and reports older than a day are removed. Algorithm 4 shows details of our method.

Algorithm 4. Event Detection and Alarming

1: initialize an empty graph $G = \{V, E\}$
2: **for each** incoming report r **do**
3:      find $p \in V$ geographically nearest to r
4:      **if** distance from r to p is less than θ **then**
5:          add $\{r, p\}$ to E
6:      **end if**
7:      add r to V
8:      $CC \leftarrow$ the connected component r is connected to
9:      $s \leftarrow$ the number of nodes in CC
10:     **if** $s = 5$ **then**
11:         compute location as the median of locations of all nodes in CC
12:         output an alarm, record time, location, and connected nodes to the log file
13:     **end if**
14: **end for**
15: **for** once each hour **do**
16:     remove all nodes older than 24 h from G
17: **end for**

Our event detection can leverage distance-based connected components because we are confident about the accuracy of our location inference at fine granularity. Since the reports are added to the system individually, SNAF is also capable of processing realtime streams, making it a realtime event detection system.

4   Experimental Analysis

To test and validate our approach, we collected two sets of data from Twitter, following *homeless* and *shooting* reports. For each dataset, we used the filter to generate a set of immediate observation reports, then used location inference methods to find out the location of each report. The accuracy of each method based on the comparison between the inferred locations and the GPS data was recorded. The inferred location were then used in our event detection system. In this section, we will discuss our experiment setup and results for accuracy tests and event detections.

4.1   Datasets

We collected two sets of tweets by monitoring Twitter using the keywords *homeless* and *shooting*, through Twitter's public stream API[4]. The homeless dataset contains around one million tweets dated from August 5 to September 9, 2014. The shooting dataset contains around two million tweets dated from September 24 to October 17, 2014.

We produced a classifier for filtering immediate observations in each dataset with different training data. We manually labelled 400 tweets, which consist of nearly half true-positives and half false-positives, for training and testing. We tested four well-known learning models for each classifier, including *Support Vector Machine* (SVM), *Bagging*, *Maximum Entropy* (ME), and *Random Forrest* (RF) [8]. The precision and recall for each model measured using three-fold cross validation [10] is shown in Table 1. Since the RF classifier provides the most accurate classification, in subsequent experiments, we used true-positive tweets classified by RF classifier as the experimental data.

Table 1. Precision and recall for various classifiers

Dataset: Homeless				Dataset: Shooting			
	Precision	**Recall**	*F*-score		**Precision**	**Recall**	*F*-score
SVM	0.57	**0.59**	0.57	SVM	0.54	0.07	0.13
Bagging	0.61	0.37	0.46	Bagging	0.77	0.24	0.36
ME	0.45	0.53	0.49	ME	0.34	**0.53**	**0.41**
RF	**0.65**	0.55	**0.60**	RF	**0.83**	0.25	0.39

Filtering the homeless dataset generated 20,229 reports, 1,696 of which contained GPS data. Filtering the shooting dataset generated 10,216 reports, 359 of which contained GPS data. For each report, we collected the timeline of the user who made the report, using Twitter's timeline API, unless the report itself contains location information.

4.2   Measurements and Baseline Methods

We set five distance error buckets of 3, 5, 10, 30, and 100 km, and counted an inferred location in a bucket if the difference between the inferred location and the GPS data of the report was within that error range. If the error of the

---

[4] https://dev.twitter.com/streaming/public.

inferred location was more than 100 km, we counted it as failed. The accuracy of a method in an error range is calculated as the percentage of reports counted in that distance error bucket, in all reports considered. If no location information was found for a report, we still counted it as failed. Since we count all the reports, the accuracy we use here is the same as *recall* in information retrieval. Because we can extract a location from most of the reports, the precision would be very close to the recall, and is not included in the measurement.

In addition to DK-Outlier removal (DKO), NK-Outlier removal (NKO), and their combination with burst compression (DKO+BC and NKO+BC), as defined in the previous section, we also tested two baseline methods for comparison, the *median* cleaning method, and the method proposed by Ikawa et al. in [6], which we term *Ikawa* method. The median method, which uses the median of all past locations as the tweet location, is a popular data cleaning method in sensor networks. The Ikawa method associates messages with locations using messages containing location words, and then infers the location of a new message by matching the message with the trained messages, based on cosine similarity and term frequency.

4.3   Location Accuracy Results

We applied the location inference methods on 1,696 homeless reports and 359 shooting reports, and generated a location for each report. The accuracy of the proposed inference methods for each of the two datasets is shown in Table 2. The highest accuracy in each distance error bucket is highlighted in bold font.

Table 2. Accuracy of location inference methods

Error	Dataset: Homeless					Dataset: Shooting				
	≤3 km	≤5 km	≤10 km	≤30 km	≤100 km	≤3 km	≤5 km	≤10 km	≤30 km	≤100 km
Median	0.086	0.113	0.192	0.327	0.432	0.091	0.116	0.194	0.298	0.387
Ikawa	0.084	0.113	0.184	0.284	0.363	0.066	0.089	0.155	0.256	0.334
DKO	0.093	0.129	0.211	0.332	0.430	0.091	0.122	0.211	0.300	0.389
NKO	**0.100**	**0.135**	0.231	0.353	0.453	0.089	0.133	**0.217**	0.311	0.395
DKO+BC	0.084	0.120	0.208	0.333	0.441	**0.100**	0.130	0.214	0.295	0.389
NKO+BC	0.096	0.132	**0.232**	**0.356**	**0.459**	**0.100**	**0.142**	**0.217**	**0.314**	**0.406**

From the table, we can see that for the homeless dataset, NKO achieved the highest accuracy in smaller error ranges, while NKO+BC achieved higher accuracy in larger error ranges. For the shooting dataset, NKO+BC achieved the highest accuracy in all error ranges. Comparing to the median method, our methods, which essentially cleaned the data before applying median, successfully improved the accuracy by 5 % to 20 % across different error ranges. Our best achieving method was also more accurate than the Ikawa method in all error ranges.

Table 3. Examples of detected events

Detected event: Indiana State University Shooting, 27/09/14	
Time alarmed	27/09/14 17:43
Location inferred	39.469222724999995,-87.41235737
Earliest News article	http://www.wthr.com/story/26644661/2014/09/27/isu-reports-shooting-near-lincoln-quad-residence-hall
News article time	27/09/14 17:52
News article location	Indiana State University (39.471345, -87.408071)
Time before news article	**+9 minutes**
Location error	**0.43km**
Reporting Tweets	
27/9 16:28 *Shooting on campus.. #MoreToBlue*	
27/9 16:26 *An actual shooting on campus. On family day. What has happened to Indiana State.*	
27/9 16:42 *ok basically witnessed someone die earlier and now there's a shooting on campus.*	
27/9 16:58 *Niggas really out here shooting on campus in terre haute*	
27/9 17:43 *WTF is terre haute coming to there was just a shooting at isu..*	

Detected event: Fern Creek High School Shooting, 30/09/14	
Time alarmed	30/09/14 14:09
Location inferred	38.1597,-85.5877
Earliest News article	http://www.tristatehomepage.com/story/d/story/shooting-confirmed-at-louisville-high-school/27354/ izZd6FZn8kaM9yx3V2PFoQ
News article time	30/09/14 13:37
News article location	Fern Creek High School (38.1563, -85.5923)
Time before news article	**-32 minutes**
Location error	**0.5km**
Reporting Tweets	
30/09 13:28 *Damn shooting at fern creek*	
30/09 14:09 *One student suffered injuries not believed to be life-threatening, WLKY reports.*	
30/09 13:57 *Fern Creek High still on lockdown after reports of shots fired. Student text parent that there was someone shooting at another person.*	
30/09 13:47 *Damn, kids shooting at fern creek. Smh*	
30/09 13:51 *damn, shooting at fern creek*	

Detected event: Marysville Shooting, 15/10/14	
Time alarmed	16/10/14 00:38
Location inferred	48.062,-122.163
Earliest News article	http://www.marysvilleglobe.com/news/279423142.html
News article time	16/10/14 07:00
News article location	64th Ave, Marysville (48.059219, -122.144871)
Time before news article	**+6 hours 28 minutes**
Location error	**1.38km**
Reporting Tweets	
16/10 00:22 *Damn, I guess there was a guy shooting at police and and drove through marysville*	
16/10 00:09 *Gunman shot at Granite Falls and Lake Stevens police stations, shooting at pursuing officers in Marysville.*	
16/10 00:05 *Another lake Stevens shooting going on as we speak.*	
16/10 00:38 *That's so scary. I told my dad I heard like 15 gun shots go off and it was that guy shooting up the lake Stevens police department*	
16/10 00:29 *There's a police chase on state and the guy is shooting at the cops #MarysvilleProbs*	

4.4   Event Detection Results

In this experiment, we tested the event detection of the SNAF system using the *Shooting* dataset, which contains 10,216 single reports. We manually examined 100 alarms and found that 54 of them contained reports of actual shooting incidents. Table 3 shows three instances of the detected shooting incidence compared to corresponding news articles.

For each instance, the *time alarmed* is the time of the last report that triggered the alarm. The *earliest news article* is the earliest online news article about the event we found using our best effort. We found these news articles mostly on *Google News*[5], making use of their "sort by date" feature. The *news article time*

[5] http://news.google.com.

is the time of the publication of the news. The *time before news article* indicates how much faster our alarm was than the publication of the earliest news article, measured in hours. A positive number means the alarm is faster than news, while a negative one means the alarm is slower. All times are converted to the local time where the event happened. *Supported tweets* shows tweet messages of the reports in the connected component that triggered the alarm.

In the *Indiana State University Shooting* case, although the news reacted quickly and reported less than one hour and a half after the incident, our system was able to capture the incident 9 min earlier than the news. From the supporting tweets we also identified a mistake in the news, that the incident happened not at 6:30, but at 16:30. In the *Fern Creek High School Shooting* case, although we were able to detect it just one hour after the shooting, it was half an hour slower than the news. In the third instance, the Marysville Shooting on the night of October 15, 2014, which happened around midnight, was not reported by the news until next morning. However, our monitoring system was able to detect it at mid-night, only an hour after the shooting. In all three instances, the event locations the system inferred were very close to the reported location, with distance errors between a few hundred meters and 1.5 km. The inferred location for the last event was less accurate because the event involved a series of sub-events each had a different location.

5   Discussion

SNAF currently assumes that there is only one location for the monitored events. However, in many real-world cases, events change their locations over time, or occur in multiple locations at the same time. For example, a shooter may be running through an area and causing shooting in a series of locations. In a public demonstration, the police may clash with protesters, at several locations at the same time. In such cases, providing a single location would not be accurate.

Recognizing if an event has multiple locations, and inferring these locations accordingly, remain a future study topic. Unankard et al. [21] proposed to divide an event into sub-events based on geo-location. Although their focus is on national-level events such as elections, they provided interesting insights on location-based sub-events. Lee et al. [11] investigated the evolution of topic discussion on social networks, which also provides some interesting insights for evolving events that potentially involve location changes.

6   Conclusion

Twitter can be seen as a dense sensor network in urban areas, and each Twitter user can be seen as a sensor. Such a sensor network can be used for monitoring events and objects of interest, as Twitter users post observations of their immediate environments continuously and constantly. This information can be collected for detecting local events, potentially much earlier than any news reports or official announcements. In many cases, it is important to know the location of an observation, which is also the location of the event or the object of interests. Unfortunately, location information is sparse on Twitter, and current works proposed to infer locations from tweets usually focus on coarse-grained locations.

In this paper, we propose Sense and Focus (SNAF), an event monitoring system that infers fine-grained locations for immediate observations posted on Twitter, based on a comprehensive place gazetteer and users' past tweets. We increase the availability of location information by using location names in tweet messages and users' past tweets. By applying sensor data cleaning techniques, we remove the noises in the location data and improve the accuracy of the location inference over existing location inference approaches. For various error ranges between 3 km and 100 km, our method improve the location accuracy by 5 % to 20 %. Taking shooting incidents as the target event, our prototype event monitoring system captures local events and provides accurate location information based on the inferred locations, in many case with less than 1 km error. Based on the effective report classification and message location inference, the event detected by our system were actual event of interest more than half of the time. In the future, we plan to extend our approach for monitoring events with multiple, continuously changing locations.

References

1. Auer, S., Bizer, C., Kobilarov, G., Lehmann, J., Cyganiak, R., Ives, Z.: DBpedia: a nucleus for a web of open data. In: Aberer, K., et al. (eds.) The Semantic Web. LNCS, vol. 4825, pp. 722–735. Springer, Heidelberg (2007)
2. Branch, J., Szymanski, B., Giannella, C., Wolff, R., Kargupta, H.: In-network outlier detection in wireless sensor networks. In: Proceedings of the 26th IEEE International Conference on Distributed Computing Systems (2006)
3. Daiber, J., Jakob, M., Hokamp, C., Mendes, P.N.: Improving efficiency and accuracy in multilingual entity extraction. In: Proceedings of the 9th International Conference on Semantic Systems (2013)
4. Graham, M., Hale, S.A., Gaffney, D.: Where in the world are you? Geolocation and language identification in twitter. Prof. Geogr. **66**(4), 568–578 (2014)
5. Hong, L., Ahmed, A., Gurumurthy, S., Smola, A.J., Tsioutsiouliklis, K.: Discovering geographical topics in the twitter stream. In: Proceedings of the 21st International World Wide Web Conference (2012)
6. Ikawa, Y., Enoki, M., Tatsubori, M.: Location inference using microblog messages. In: Proceedings of the 21st International World Wide Web Conference Companion (2012)
7. Jeffery, S.R., Alonso, G., Franklin, M.J., Hong, W., Widom, J.: Declarative support for sensor data cleaning. In: Fishkin, K.P., Schiele, B., Nixon, P., Quigley, A. (eds.) PERVASIVE 2006. LNCS, vol. 3968, pp. 83–100. Springer, Heidelberg (2006)
8. Jurka, T.P., Collingwood, L., Boydstun, A., Grossman, E., van Atteveldt, W.: Rtexttools: Automatic text classification via supervised learning. R package version 1.3, 9 (2012)
9. Knox, E.M., Ng, R.T.: Algorithms for mining distance based outliers in large datasets. In: Proceedings of 24th International Conference on Very Large Data Bases (1998)

10. Kohavi, R.: A study of cross-validation and bootstrap for accuracy estimation and model selection. In: Proceedings of the Fourteenth International Joint Conference on Artificial Intelligence (1995)

11. Lee, p., Lakshmanan, L.V., Milios, E.E.: Incremental cluster evolution tracking from highly dynamic network data. In: Proceedings of the 30th International Conference on Data Engineering (2014)

12. Li, C., Sun, A.: Fine-grained location extraction from tweets with temporal awareness. In: Proceedings of the 37th International ACM SIGIR Conference on Research and Development in Information Retrieval (2014)

13. Li, R., Lei, K.H., Khadiwala, R., Chang, K.-C.: Tedas: a twitter-based event detection and analysis system. In: Proceedings of 28th International Conference on Data Engineering (2012)

14. Lingad, J., Karimi, S., Yin, J.: Location extraction from disaster-related microblogs. In: Proceedings of the 22nd International World Wide Web Conference Companion (2013)

15. McMinn, A.J., Moshfeghi, Y., Jose, J.M.: Building a large-scale corpus for evaluating event detection on twitter. In: Proceedings of the 22nd ACM International Conference on Information and Knowledge Management, pages 409–418. ACM (2013)

16. Sakaki, T., Okazaki, M., Matsuo, Y.: Earthquake shakes Twitter users: Real-time event detection by social sensors. In: Proceedings of the 19th International World Wide Web Conference (2010)

17. Sakaki, T., Okazaki, M., Matsuo, Y.: Tweet analysis for real-time event detection and earthquake reporting system development. IEEE Trans. Knowl. Data Eng. 25(4), 919–931 (2013)

18. Schulz, A., Hadjakos, A., Paulheim, H., Nachtwey, J., Mühlhäuser, M.: A multi-indicator approach for geolocalization of tweets. In: Proceedings of the Seventh International Conference on Weblogs and Social Media (2013)

19. Sheng, B., Li, Q., Mao, W., Jin, W.: Outlier detection in sensor networks. In: Proceedings of the 8th ACM International Symposium on Mobile Ad Hoc Networking and Computing (2007)

20. Subramaniam, S., Palpanas, T., Papadopoulos, D., Kalogeraki, V., Gunopulos, D.: Online outlier detection in sensor data using non-parametric models. In: Proceedings of the 32nd International Conference on Very Large Data Bases (2006)

21. Unankard, S., Li, X., Sharaf, M., Zhong, J., Li, X.: Predicting elections from social networks based on sub-event detection and sentiment analysis. In: Benatallah, B., Bestavros, A., Manolopoulos, Y., Vakali, A., Zhang, Y. (eds.) WISE 2014, Part II. LNCS, vol. 8787, pp. 1–16. Springer, Heidelberg (2014)

22. Unankard, S., Li, X., Sharaf, M.A.: Emerging event detection in social networks with location sensitivity. World Wide Web J. (2015, in press)

23. Wen, Y.-J., Agogino, A.M., Goebel, K.: Fuzzy validation and fusion for wireless sensor networks. In: Proceedings of the ASME International Mechanical Engineering Congress (2004)

24. Zhang, Y., Meratnia, N., Havinga, P.: Outlier detection techniques for wireless sensor networks: A survey. IEEE Commun. Surv. Tutor. 12(2), 159–170 (2010)

A Hybrid Feature Selection Method
for Predicting User Influence on Twitter

Yan Mei[1(✉)], Zizhu Zhang[2], Weiliang Zhao[1], Jian Yang[1],
and Robertus Nugroho[1]

[1] Macquarie University, Sydney, Australia
{yan.mei,robertus.nugroho}@students.mq.edu.au,
{weiliang.zhao,jian.yang}@mq.edu.au
[2] University of International Business and Economics, Beijing, China
zizhu07.zhang@gmail.com

Abstract. This paper proposes a hybrid feature selection method for predicting user influence on Twitter. A set of candidate features from Twitter is identified based on the five attributes of influencers defined in sociology. Firstly, less relevant features are filtered out with a feature-weighting algorithm. Then the Sequential Backward Floating Selection is utilized as the search strategy. A Back Propagation Neural Network is employed to evaluate the feature subset at each step of searching. Finally, an optimal feature set is obtained for predicting user influence with a high degree of accuracy. Experimental results are provided based on a real world Twitter dataset including seven million tweets associated with 200 popular users. The proposed method can provide a set of features that could be used as a solid foundation for studying complicated user influence evaluation and prediction.

Keywords: Hybrid · Feature selection · User influence

1 Introduction

Nowadays, Online Social Networks (OSNs), such as Facebook, Twitter, Google+ and LinkedIn, have gained increasing attention from all over the world, and they are among the most popular websites on the Internet. OSNs not only provide individual users a platform to share information and keep in touch with their friends, but also become an important marketing channel for companies and organizations. It is crucial for companies to establish themselves in OSNs. At the same time it is important for them as well to identify influential people as the marketing targets in OSNs. Both academic community and industry have shown great enthusiasm on the study of user influence in OSNs.

The study of influence originated from psychology and sociology. Generally, influence means "change in a person's cognition, attitude, or behavior, which has its origin in another person or group" [1]. Merriam-Webster dictionary defines influence as "the power to change or affect someone or something; the power

© Springer International Publishing Switzerland 2015
J. Wang et al. (Eds.): WISE 2015, Part I, LNCS 9418, pp. 478–492, 2015.
DOI: 10.1007/978-3-319-26190-4_32

to cause changes without directly forcing them to happen"[1]. When influence is studied in OSNs, e.g. Twitter, researchers have also given their own explanations. Cha et al. [2] focus on "an individual's potential to lead others to engage in a certain act". Leavitt et al. [3] describe influence on Twitter as "the potential of an action of a user to initiate a further action by another user". Rosenman Evan TR [4] interprets the term to mean "the ability to, through one's own behaviour on Twitter, promote activity and pass information to others".

Briefly speaking, influence is the ability to cause a change in others' thoughts or actions. Due to this nature, it is difficult to define a quantitative measure for influence in OSNs. In the context of Twitter, we are unable to detect whether there are some changes in others' mind, and despite no online actions (such as reply or retweet) observed, there might be some offline actions (such as buying a product). Therefore, to predict user influence on Twitter, we can only speculate from available features, including characteristics of a user and explicit actions from others.

Intuitively we can say that a user with a large number of followers is influential. Therefore a user's follower count is generally considered as an important predictor for influence. However, estimating influence only by followers may introduce noises. For example, it is quite possible that some followers of a particular user are from faked accounts or even spammers. The study by Cha et al. [2] claimed that the top influencers showed a stronger correlation with retweets and mentions than followers. A similar metric for measuring user influence is the ratio between follower count to friend count. This ratio probably describes the types of users in the community, but it is imprecise to measure user influence [3].

In recent years, some Influence Scoring Services (ISS) have gained attention by offering numerical scores that quantify users' social media influence. Klout[2] utilizes social media analytics to rank users according to their online social influence via Klout Score, which is a numerical value between 1 and 100. Klout Score turns out to be a good reference for user influence and gets widely used in industry. Some famous brands in the world, such as Sony, Nike, Disney, are using Klout for business to run successful marketing campaigns by targeting valuable influencers. Kred[3] also measures user influence in online communities. It has dual metrics for Influence and Outreach. Kred influence score ranges from 1 to 1,000. PeerIndex[4] is another social ranking site, which provides social media analytics based on footprints from use of major social media services, and works out PeerIndex scores to indicate how influential an individual is. Another popular online influence tool is Followerwonk[5], which also offers its own measures of user influence (called as Social Authority) on Twitter.

Many researchers have also proposed their own algorithms for predicting user influence on Twitter [2,3,5–9]. However, most studies directly utilize their

---

[1] http://www.merriam-webster.com/.
[2] https://klout.com/.
[3] http://kred.com/.
[4] http://www.peerindex.net/.
[5] https://followerwonk.com/.

own pre-defined features to build the model without a pre-evaluation process for these selected features. No literature has extensively investigated and evaluated the potential features that can be used to measure user influence in Twitter environment. We believe that identifying important features that are crucial for influence measurement is the first step towards influence model construction, and this is what we are going to investigate in this paper.

A feature selection method is proposed to obtain an optimal feature set for predicting user influence on Twitter. The main features of this work are summarized as follows:

- The method inherits the advantages of commonly used *filter* and *wrapper* approaches to achieve a high degree of efficiency and accuracy in the optimization.
- To the best of our knowledge, this work is the first one to intensively study the feature selection for evaluating/predicting the online user influence. This work can provide a solid foundation for further analysis of user influence to cover complicated situations.
- This work employs the five attributes of influencers defined in sociology as the criteria to explore the candidate features in online social networks. Experiments based on a real world Twitter dataset show the effectiveness of the proposed method.

The remainder of the paper is as follows. Section 2 summarizes the existing approaches of predicting user influence in OSNs and the general feature selection methods. Section 3 investigates the candidate features and describes our hybrid feature selection method in the context of Twitter. Section 4 provides the detailed description of our experiments and discusses our experimental results. Finally, in Sect. 5, we conclude the paper with a brief discussion on future work.

2   Related Work

In the past decade, researchers have proposed many algorithms to predict user influence on Twitter. A typical kind of algorithms is based on the follow-relationship on Twitter, which can be represented by a topological graph. Several algorithms have been proposed with the similar method in Google's *PageRank* algorithm [10] for ranking the popularity of web pages. Kwak et al. [11] ranked Twitter users by follower count and *PageRank*-like algorithm (follow-network analysis), and found that these two rankings were similar. *TwitterRank* [5], an extension of *PageRank* algorithm, measures user influence taking both the topical similarity and link structure into account. Similarly, some other algorithms have been proposed, such as *TunkRank* [6], *KHYRank* [7] and *InfluenceRank* [8].

In addition, there are some measurements relying on users' actions, since researchers believe others' actions (such as reply, mention or retweet) are explicit signs for predicting user influence. Cha et al. [2] defined three types of influence: *Indegree influence* (i.e. the number of followers), *Retweet influence* and *Mention influence*. Leavitt et al. [3] categorized users' actions from the perspectives of conversation and content, mapping to replies and retweets respectively. Then user influence was measured based on the ratio of actions caused to the tweets the user posted. *IARank* [9] is a model to continuously rank influential Twitter users in real-time, based on a concept of "information amplification". The information amplification is characterized by three activities: event activity, attention obtained and social connectivity.

Although there have been many studies on quantifying user influence on Twitter, it is lacking of an extensive investigation for an optimal feature set that is crucial in the evaluation of user influence. Before the existing algorithms were proposed, the reason why they adopted those features was not well explained. Besides, even though people consider the popular ISS as good reference for user influence, their algorithm details are almost black-boxes.

Feature selection is an important problem in the area of machine learning. The aim of feature selection is to find an optimal feature subset, which has significant impact on the target variables (i.e. user influence in this paper), from the complete feature set. According to their evaluation criteria, existing feature selection methods can be divided into two main categories: *Filter* methods [12] and *Wrapper* methods [13]. Filter methods select features based on performance evaluation functions calculated directly from the data. The advantage of filter methods is the high computational efficiency. However, the effectiveness is sometimes unsatisfactory since they separate the feature selection from model building. A wrapper method utilizes a predetermined learning algorithm and uses its estimated performance as the evaluation criterion for feature selection. Normally, a wrapper method can provide more accurate result but it has a higher degree of computational complexity in comparison with a filter method.

In this paper, we focus on the feature selection for predicting user influence. We extensively investigate the candidate features in Twitter environment. A hybrid filter-wrapper method is proposed to obtain the optimal feature subset with a high degree of efficiency and accuracy.

3   Methodology

3.1   Determining Candidate Features

As we mentioned earlier, algorithms have been developed for predicting user influence based on different features, such as followers, retweets, mentions, tweet content, etc. In this section, we firstly introduce a background theory from sociology. Then we select corresponding features from Twitter based on this theory.

A. Five Attributes of Influencers. In sociology area, there are numerous studies on identifying influencers. For example, Keller and Berry [14] define the influentials from five attributes:

– **Activists:** Influencers are active in their communities. They attend community events, serve on committees, and persuade others of their opinions.

- **Connected:** Influencers have richer social connections than the average. Their contacts are likely to bring them into connection with more people, in an ever-widening network.
- **Impact:** Influencers are people others look up to for advice. They are trustworthy and reliable because of their reputation or expertise.
- **Active minds:** Influencers have interests in many areas, and they always share their new experience or ideas with others.
- **Trendsetters:** Influencers tend to be early adopters in markets. A new fashion or trend usually originates from some influential people.

The study of the traditional influence in sociology starts much earlier and goes further than that of online influence. Thus, we consider these five attributes of influencers as our basis for feature selection. However as we observed, influence and characteristics of influencers might appear differently in different environments.

B. Candidate Features from Twitter. Firstly, we select candidate features which are possible predictors for target value (i.e. user influence) as the original feature set. We capture the relevant features from Twitter, and map them to the five social attributes mentioned before. And these features are available through public Twitter APIs.

User influence is changing over time. For instance, it is often seen that a user's influence suddenly increases due to some emerging news, or gradually drops as a result of low level of engagement. Therefore, we do not consider only the *long-term* features which reflect a user's accumulative efforts or achievements, but also the *short-term* features which reflect a user's dynamic situation in a recent period of time.

Based on the above thoughts, we select 17 candidate features (listed in Table 1) for predicting user influence on Twitter. All the features starting with "New" are short-term features. Besides, the other five features (*Topic Diversity, Average Length of Tweets, Original Tweets, Original Tweet Ratio* and *Average Retweets of Original Tweets*) are also calculated based on the tweet data in a specific period of time. Note that the complete historical tweet data is not available through Twitter APIs.

These candidate features are briefly explained as follows. The features mapped to *Activists* attribute include:

(1) **Tweet Frequency:** the average number of tweets a user posted per month, since his/her account was created. It represents the active level of the user.
(2) **New Tweets:** the number of tweets issued by a user during a recent period of time. We understand that influence is time sensitive, since user influence in a social network is changing over time. If an influential user does not post any tweet for a period of time, his/her influence probably starts declining.

The features mapped to *Connected* attribute include:

Table 1. Candidate features for predicting user influence on twitter

Social attributes	Features on twitter
Activists	Tweet frequency
	New tweets
Connected	Followers
	Friends
Impact	Verified
	Public lists
	New public lists
	New followers
	New mentions
	New retweets
	Followers to friends ratio
Active Minds	Topic diversity
	Average length of tweets
Trendsetters	Account age
	Original tweets
	Original tweet ratio
	Average retweets of original tweets

(3) **Followers:** the total number of followers a user has. It is an explicit metric of connectivity. Generally, more followers mean more potentials of a high degree of influence.

(4) **Friends:** the total number of friends a user has. It is a metric of outgoing connection.

The features mapped to *Impact* attribute include:

(5) **Verified:** whether it is a verified account. Twitter verifies accounts on an ongoing basis, focusing on popular users in interest areas, such as music, acting, politics, media, sports, business and others.

(6) **Public Lists:** the number of public lists which a user is a member of. Being included in public lists indicates a user is visible and people show interest in the user.

(7) **New Public Lists:** the number of new public lists which include the user in a recent period of time. An increase of public lists implies a user's continued impact.

(8) **New Followers:** the number of new followers during a recent period of time. Whether a user's follower count is increasing or decreasing is a good metric for user influence.

(9) **New Mentions:** the number of new mentions or replies to a user during a recent period of time, i.e., the number of tweets including "@username" (excluding "RT @username"). Mention is an explicit signal reflecting a user's impact to others.

(10) New Retweets: the number of new retweets of the tweets created by a user during a recent period of time, i.e., the number of tweets including "RT @username". Retweeting indicates the retweeter has been influenced by the original author, no matter positive or negative impact.

(11) Followers to Friends Ratio: the ratio of a user's follower count to friend count. The higher the ratio is, the more people are interested in the user's status updates without the user showing interest in return.

The features mapped to *Active Minds* attribute include:

(12) Topic Diversity: a metric measuring how many different topics a user's tweets might cover. To understand topic diversity, we train a Latent Dirichlet Allocation (LDA) model based on the corpus of tweets. We combine all tweets posted by a user during one-month period as one document and obtain the document-topic distribution by training LDA model with Stanford Topic Modeling Toolbox[6]. Then the entropy of document-topic distribution is computed to represent the topic diversity. If a user only concentrates on one or two topics, the entropy is relatively small, while larger entropy indicates more diverse topics are covered.

(13) Average Length of Tweets: The activists tend to share their ideas with more words, and the average length of tweets is an obvious indicator. Studies have shown that on Twitter with 140-character limit, too short text conveys little information and informative content with enough words is critical to gaining attention [15].

The features mapped to *Trendsetters* attribute include:

(14) Account Age: the number of months since a user's Twitter account was created. It reflects whether a user is an early adopter on Twitter.

(15) Original Tweets: the number of original tweets (excluding replies and retweets) a user created during a recent period of time. A trendsetter should have some original thoughts, rather than always joining conversations or forwarding information.

(16) Original Tweet Ratio: the ratio of original tweets to total tweets. It reflects the relative originality based on the user's total tweets.

(17) Average Retweets of Original Tweets: the average number of retweets obtained per original tweet. This metric reflects a user's performance in trendsetting.

3.2 A Hybrid Feature Selection Method

There are two main categories of feature selection methods: filter methods and wrapper methods, which are demonstrated in Fig. 1. A filter method directly evaluates the quality of features according to their data values. A wrapper method employs learning algorithms as the evaluation criteria to select optimal feature subsets. Comparing with a filter method, a wrapper method is more

---

[6] http://nlp.stanford.edu/software/tmt/tmt-0.4/.

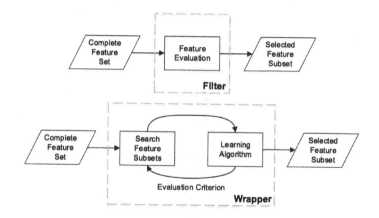

Fig. 1. The processes of filter and wrapper methods

effective, but it often brings in a higher degree of computational complexity. In this paper, we combine the advantages of these two types of methods and propose a hybrid filter-wrapper method for predicting user influence on Twitter. The filter method provides a quick way to eliminate the less relevant features and then the wrapper method is employed to achieve a high accuracy.

A. the Proposed Method. The proposed method is illustrated in the flow-chart (Fig. 2).

We explain the detailed procedure in the following seven steps.

Step 1: Determine the candidate feature set $F = \{F_1, F_2, ..., F_f\}$.

Step 2: Utilize the feature weighting algorithm (*RReliefF*) to compute the weights for each feature, and filter out the features which have little relevance to the target user influence. The remaining features form a reduced feature set $F' = \{F_1, F_2, ..., F_n\}(n <= f)$, in which features are sorted in descending order based on their weights.

Step 3: Train prediction model with a learning algorithm (BPNN) on training set and calculate Mean Square Error (MSE) on testing set, denoted as e. Meanwhile, set $i = n$.

Step 4: Remove the feature F_i which is the last feature in F', set $F' = F' - \{F_i\}$, then train prediction model based on F' and calculate new MSE, denoted as e'. If $e' < e$, then set $e = e'$ and go to step 6.

Step 5: Move the feature F_i back into F' as the new first feature, and set $F' = \{F_i\} + F'$.

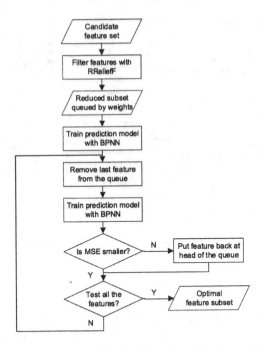

Fig. 2. Our proposed hybrid feature selection method

Step 6: Set $i = i - 1$, if $i > 0$, go back to Step 4.

Step 7: The feature subset F' is the optimal feature subset.

B. Filter - Feature Ranking. Estimating the quality of features is critical in feature selection. A robust feature weighting technique is the *Relief* algorithmic family [16]. *Relief* algorithms estimate how well the features' values distinguish between instances. The output of *Relief* algorithms is a set of numerical weights representing the percentages of the features' contribution to the variance in dependent variable. Features that are assigned weights larger than zero can cause the dependent variable to vary, while those features with zero or negative weights are believed to have no contribution to the variance of the dependent variable. *Relief* algorithms perform well even when strong dependencies exist between features and have been used successfully in a variety of contexts. In our method, we employ *RReliefF* [17], which is designed for regression problems, to rank all the candidate features.

Assume $I_1, I_2, ..., I_j$ are the instances with n features $F_1, F_2, ..., F_n$ and target values. To estimate the weights of all features (denoted by $W[F]$), *RReliefF* starts with selecting k nearest instances around a randomly selected instance I_i and then updates the weight estimation $W[F]$ for all features F based on probabilities of difference. This whole process is repeated for m times. Here k

and m are user-defined parameters, and $W[F]$ is calculated as below:

$$W[F] = \frac{P_{diffC} \mid_{diffF} P_{diffF}}{P_{diffC}} - \frac{(1 - P_{diffC} \mid_{diffF})P_{diffF}}{1 - P_{diffC}}$$

where

$$P_{diffF} = P(\text{diff. value of } F \mid \text{nearest instances})$$

$$P_{diffC} = P(\text{diff. prediction} \mid \text{nearest instances})$$

$$P_{diffC} \mid_{diffF} = P(\text{diff. prediction} \mid \text{diff. value of } F \text{ and nearest instances})$$

We implement *RReliefF* in the Weka tool[7] (a popular software tool of machine learning) to calculate the feature weights. Due to the problem of under-estimating numerical attributes shown in the work of [16] when both numerical and nominal features are in the feature set, we initially remove the nominal feature *Verified* and put the remaining sixteen features into the *RReliefF* algorithm. We set the parameter k, which is the number of nearest neighbours, to 10 as proposed in [16] and keep other parameters as the default settings in Weka.

C. Wrapper - Feature Search Strategy and Learning Algorithm. As we show in Fig. 1, feature search strategy and learning algorithm are the two main parts in wrapper methods. Various search strategies have been proposed under two main ideas: exhaustive and heuristic searches. Exhaustive search can guarantee the optimal feature subset but with a high complexity (2^N possible feature subsets, where N is the number of features), while heuristic search can achieve near optimality more efficiently. Among the heuristic search strategies, floating search methods, including Sequential Forward Floating Selection (SFFS) and Sequential Backward Floating Selection (SBFS), are proven to be qualified [18], since they can provide near-optimum or optimum results in most situations.

We go through the feature subsets with SBFS strategy, starting from the full feature set. All the candidate features are removed one by one according to the feature weights (from smallest to largest). If the training model performs better without a certain feature, then this feature will be deleted and the model's performance will be set as the current optimum. However, if the model performs worse, then the tentatively-deleted feature will be kept and put back into the feature set. The search stops when all the features have been examined once.

Neural Networks (NN) are important learning algorithms for modelling complex non-linear systems [19]. A basic NN model has three layers: input layer, hidden layer and output layer. On each layer, there are a number of nodes (or neurons). Nodes on input layer are connected to nodes on hidden layer, and nodes on hidden layer are connected to nodes on output layer. These connections between nodes represent the weights. Back Propagation Neural Network (BPNN) is one of the most popular NN algorithms. The main idea in BPNN is that, the network output is compared to the target output, and the errors propagate backwards from the output nodes to the input nodes. If results are

---

[7] http://www.cs.waikato.ac.nz/ml/weka/.

not satisfactory, the connections (weights) between layers are modified and this process is repeated again and again until some stopping criterion is satisfied.

We implement the wrapper method with Neural Network Toolbox[8] in MAT-LAB to select the optimal feature subset from the remaining features after filtering. Tan-Sigmoid is selected as the transfer function for the hidden layer. It is defined as:

$$f(x) = \frac{2}{1 + e^{-2x}} - 1$$

And we select the linear transfer function for the output layer, which is defined as:

$$f(x) = purelin(x) = x$$

All the samples are randomly divided to three sample sets: training set (70 %), validation set (15 %) and testing set (15 %). Training stops when generalization stops improving, as indicated by an increase in the mean square error of the validation samples, or the maximum iteration limit is reached. Since training multiple times generates different results, we train the BPNN model for 20 times and utilize the average Mean Square Errors (MSE) to evaluate the performance. The MSE is calculated as below:

$$E = \frac{1}{n} \sum_{k=1}^{n} e(k)^2 = \frac{1}{n} \sum_{k=1}^{n} (t(k) - a(k))^2$$

Here n is the number of samples in testing set, $t(k)$ is the target output and $a(k)$ is the network output. If the MSE becomes smaller than the current optimum, the out feature will be deleted; otherwise, the out feature will be put back into the feature set.

4 Experiment and Analysis

4.1 Experiment Setup

We select 200 most-followed Twitter users in Australia, who explicitly indicate "Australia" in their location profiles. All the tweets posted by these users and all the responses (including replies, mentions, retweets) are collected during one month period (from 12 January to 12 February in 2015). There are totally 6,770,715 tweets (around 36.8 gigabytes data) captured in the month.

Different types of features are retrieved or calculated from different Twitter APIs. For example, user profile data are captured through Twitter REST APIs[9]. All the long-term features are calculated from user profile data. Real-time tweet data are captured through Twitter Streaming APIs[10]. Most of the short-term features (e.g. *New Retweets*) and tweet content related features (e.g. *Topic Diversity*) are calculated by analyzing these tweet data.

---

[8] http://www.mathworks.com/products/neural-network/.

[9] https://dev.twitter.com/rest/public.

[10] https://dev.twitter.com/streaming/overview.

In our study, we consider the average of standardized scores of the four popular ISS (i.e. *Klout, Kred, PeerIndex, Followerwonk*) as the relative truth of user influence. We believe that, each system has its own rationality in the scoring scheme, but the average influence score should better reflect the level of user influence from more comprehensive aspects. This is also used as the desired output in the supervised learning algorithms involved (such as BPNN).

4.2 Results and Discussion

Firstly, we pre-process the candidate features with the filter algorithm *RReliefF*. The results of calculated feature weights are shown in Table 2.

Table 2. Feature weights computed by RReliefF algorithm

Rank	Features	Weights
1	Average length of tweets	0.016720
2	Account age	0.015140
3	Followers	0.012932
4	Public lists	0.012165
5	New retweets	0.011357
6	New followers	0.010440
7	New mentions	0.009830
8	Average retweets of original tweets	0.007138
9	New public lists	0.006059
10	Original tweet ratio	0.003991
11	Followers to friends ratio	0.000187
12	Friends	−0.00101
13	Tweet frequency	−0.00622
14	New tweets	−0.01013
15	Original tweets	−0.01056
16	Topic diversity	−0.01494

Based on the ideas of *RReliefF*, the last five features with negative weights are filtered out, since they have no contribution to the variance of user influence. Then the eleven remaining features are tested by feature search strategy SBFS. For each feature subset, we train with BPNN algorithm and compute MSE on testing set for twenty times. The average MSE for each subset is shown in Table 3.

As we can see from Table 3, the average MSE reaches the minimum in the eleventh iteration of the loop. Therefore, this feature subset is the optimal feature subset, which includes seven features: *Average Length of Tweets, Followers, Public Lists, New Retweets, New Mentions, Average Retweets of Original Tweets* and *New Public Lists*.

Table 3. Average MSE for each iteration of the loop

# of Loop	1	2	3	4	5	6
Avg. MSE	0.0334	0.0282	0.0277	0.0299	0.0280	0.0305
# of Loop	7	8	9	10	11	12
Avg. MSE	0.0261	0.0273	0.0338	0.0271	**0.0260**	0.0315

In view of the mapping between social attributes and candidate features (shown in Table 1), we find that most of the features belonging to the *Impact* attribute are included in the optimal feature subset, and all features belonging to the *Activists* attribute are not included. According to the feature weights calculated by *RReliefF* filter algorithm (shown in Table 2), we believe that, in the Twitter environment, the social attribute *Impact* plays the most important role, followed by *Active minds, Connected* and *Trendsetters*. The attribute *Activists* seems to contribute little to user influence.

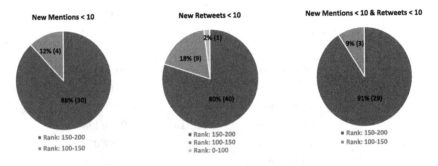

Fig. 3. Influence analysis with few mentions and retweets

The identified seven features can be used as the starting point to model the user influence. The influence models will be built up in different ways such as the linear regression or the rule-based evaluation with specific constraints on selected features. For example, we group the 200 users into three influence levels: top 100 influencers, users ranked between 100 and 150, users ranked between 150 and 200. The distributions of users with less than 10 mentions or/and retweets at different influence levels are shown in Fig. 3. Over 90 % of the users with less than 10 mentions and retweets are ranked between 150 and 200. These users have quite low influence scores as discussed in Subsect. 4.1. While identifying influencers, these users can be filtered out at an early stage.

5   Conclusion and Future Work

This paper has proposed a hybrid feature selection method for predicting user influence on Twitter. Based on the five attributes of influencers defined in sociology, we have explored the candidate features from Twitter and selected seventeen ones as the starting point. We collected the experimental dataset through public Twitter APIs including all tweets associated with the 200 most-followed users in Australia. We employed the *RReliefF* algorithm (a filter method) to evaluate the quality of features. As the result, a reduced feature subset was obtained. Following the principles of wrapper methods, we developed our method by utilizing the SBFS search strategy and assessing the feature subset at each searching step by employing the BPNN learning algorithm.

The proposed method produce an optimal feature set for predicting user influence on Twitter. These features include: *Average Length of Tweets, Followers, Public Lists, New Retweets, New Mentions, Average Retweets of Original Tweets* and *New Public Lists*. Some of these optimal features such as *Average Length of Tweets* and *Public Lists* have rarely been discussed in the existing literature. These features will be used as the basis for further study about incentive mechanisms of user influence and methods to build up user influence on social networks. We will develop user influence models and study topic popularity by considering the characteristics of social network structures.

Acknowledgments.. The work presented in this paper was partially supported by Macquarie University Research Excellence Scholarship (Allocation No.2013115), an Australian Research Council Linkage Project (LP120200231) and the China Scholarship Council. We also thank anonymous reviewers for their valuable comments.

References

1. Raven, B.H.: Social influence and power. Technical report, DTIC Document (1964)
2. Cha, M., Haddadi, H., Benevenuto, F., Gummadi, P.K.: Measuring user influence in twitter: the million follower fallacy. ICWSM **10**, 30 (2010)
3. Leavitt, A., Burchard, E., Fisher, D., Gilbert, S.: The influentials: new approaches for analyzing influence on twitter. Web Ecol. Proj. **4**, 1–18 (2009)
4. Rosenman, E.T.: Retweetsbut not just retweets: quantifying and predicting influence on twitter. Ph.D. thesis, Bachelors thesis, applied mathematics. Harvard College, Cambridge (2012)
5. Weng, J., Lim, E.P., Jiang, J., He, Q.: Twitterrank: finding topic-sensitive influential twitterers. In: Proceedings of the third ACM International Conference on Web Search and Data Mining, pp. 261–270. ACM (2010)
6. Tunkelang, D.: A twitter analog to pagerank. The Noisy Channel (2009)
7. Yu, A., Hu, C.V., Kilzer, A.: Khyrank: Using retweets and mentions to predict influential users (2011)
8. Chen, W., Cheng, S., He, X., Jiang, F.: Influencerank: an efficient social influence measurement for millions of users in microblog. In: 2012 Second International Conference on Cloud and Green Computing (CGC), pp. 563–570. IEEE (2012)
9. Cappelletti, R., Sastry, N.: Iarank: Ranking users on twitter in near real-time, based on their information amplification potential. In: 2012 International Conference on Social Informatics (SocialInformatics), pp. 70–77. IEEE (2012)
10. Page, L., Brin, S., Motwani, R., Winograd, T.: The pagerank citation ranking: Bringing order to the web (1999)

11. Kwak, H., Lee, C., Park, H., Moon, S.: What is twitter, a social network or a news media? In: Proceedings of the 19th international conference on World wide web, pp. 591–600. ACM (2010)
12. John, G.H., Kohavi, R., Pfleger, K., et al.: Irrelevant features and the subset selection problem. In: Machine Learning: Proceedings of the Eleventh International Conference, pp. 121–129 (1994)
13. Kohavi, R., John, G.H.: Wrappers for feature subset selection. Artif. intell. **97**, 273–324 (1997)
14. Keller, E., Berry, J.: The influentials: one American in ten tells the other nine how to vote, where to eat, and what to buy. Simon and Schuster, New York (2003)
15. Naaman, M., Boase, J., Lai, C.H.: Is it really about me? Message content in social awareness streams. In: Proceedings of the 2010 ACM Conference on Computer Supported Cooperative Work, pp. 189–192. ACM (2010)
16. Robnik-Šikonja, M., Kononenko, I.: Theoretical and empirical analysis of relieff and rrelieff. Mach. Learn. **53**, 23–69 (2003)
17. Robnik-Šikonja, M., Kononenko, I.: An adaptation of relief for attribute estimation in regression. In: Machine Learning: Proceedings of the Fourteenth International Conference (ICML 1997), pp. 296–304 (1997)
18. Jain, A., Zongker, D.: Feature selection: evaluation, application, and small sample performance. IEEE Trans. Pattern Anal. Mach. Intell **19**, 153–158 (1997)
19. Goh, A.: Back-propagation neural networks for modeling complex systems. Artif. Intell. Eng. **9**, 143–151 (1995)

Classifying Perspectives on Twitter: Immediate Observation, Affection, and Speculation

Yihong Zhang$^{(\boxtimes)}$, Claudia Szabo, Quan Z. Sheng,
and Xiu Susie Fang

School of Computer Science, The University of Adelaide,
Adelaide, SA 5005, Australia
{yihong.zhang,claudia.szabo,michael.sheng,xiu.fang}@adelaide.edu.au

Abstract. Popular micro-blogging services such as Twitter enable users to effortlessly publish observations and thoughts about ongoing events. Such social sensing generates a very large pool of rich and up-to-date information. However, the large volume and a fast rate of posting make it very challenging to read through the posts and find out useful information in relevant tweets. In this paper, we propose an automated tweet classification approach that distinguishes three perspectives in which a Twitter user may compose messages, namely *Immediate Observation*, *Affection*, and *Speculation*. Using tweets made about the Ukraine Crisis in 2014, our experimental results show that, with the right choice of features and classifiers, we can generally obtain very satisfying results, with the classification precisions in many cases higher than 0.8. We show that the classification results can be used in event time and location detection, public sentiment analysis, and early rumor detection.

Keywords: Twitter · Social media · Data mining · Short message classification

1 Introduction

Micro-blogging services, which have gained high popularity in recent years, offer a convenient way for individuals to create and share information online. An example is Twitter, which allows its users to create and publish short messages with a maximum of 140 characters, called *tweets*. Nowadays, Twitter generates 500 million tweets from its 288 million active users every day[1]. There is no cost associated with publishing tweets, and the messages entered by a user are immediately available for anyone to read online. Currently, 80 % of Twitter users use their mobile phones to tweet (see Footnote 1). Making tweets using mobile devices allows users to post messages about anything that is happening in front of them, at any given moment. Twitter as a news source thus has attracted a number of researches, and the users who post event-related messages on Twitter are sometimes termed *citizen reporters* [2,12].

---

[1] https://about.twitter.com/company.

© Springer International Publishing Switzerland 2015
J. Wang et al. (Eds.): WISE 2015, Part I, LNCS 9418, pp. 493–507, 2015.
DOI: 10.1007/978-3-319-26190-4_33

A recent example is the Ukraine Crisis in 2014[2], during which numerous users have made firsthand observations about ongoing events. When the protesters gathered around city hall in Kiev in early February, and when the clashes erupted between the police and protester on 18th February, a large number of firsthand observations were published on Twitter by people from the crowd, describing what was happening. Later events involving military actions around the boarder of Ukraine also led to numerous tweets made by people who were experiencing them firsthand, expressing their observations, emotional reactions, and speculations[3]. These messages provided invaluable information at a time when no credible news source had published related stories.

Analyzing an event as it is reported on Twitter by a variety of users can reveal a range of information, from the location of the event, the number of people participated and the actions they are taking, to the predicted outcome of the event [7,14,20]. However, the large variety of tweet messages can be distracting and distorting for the analysis. For example, Maddock et al. showed that, a controversial story may invoke hedges, questions, speculations, misinformation, corrections and many other unrelated messages on Twitter [11]. Therefore it is invaluable if the perspectives of the tweet messages can be first provided. In this paper, we distinguish among three perspectives that a Twitter user might employ when composing messages that are helpful for us to understand ongoing events, namely, *Immediate Observation, Affection,* and *Speculation.* Table 1 shows examples of each of the perspectives.

Immediate Observations are observations made immediately at the scene of the event. Through Immediate Observations, we can gain direct observation of events that are not easily accessible. In addition, since the messages are associated with time and location data, we can also infer the time and location of the event from the information indirectly provided by the user. Affection tweets are messages about how an event impacts users emotionally. They are important for understanding the public consensus about controversial events. Speculation tweets are users' attempt to guess or predict the nature and future of uncertain events. A key characteristic of speculations is that the information they provided has not been confirmed as true or otherwise. While messages themselves have little value for understanding the event, the speculations that contain false information may invoke public panic when widely circulated. Therefore tracking speculations may help one to identify which false stories are gaining popularity.

Given a large number of Twitter posts, it is difficult for anyone to read through the tweets manually, identify these perspectives, and further analyze the data. In our experiments, we collected tweets related to Ukraine Crisis over a period of 56 days, and the dataset contains over 950,000 tweets. An automated classification method that helps identify the perspectives of the tweets is crucial. Based on existing machine learning approaches [17], our goal with this paper is to test the feasibility of using automatic learning models for classifying tweet

---

[2] http://www.bbc.com/news/world-europe-26270866.

[3] http://www.washingtonpost.com/blogs/worldviews/wp/2014/02/18/
the-16-essential-twitter-accounts-to-follow-ukraines-unfolding-crisis/.

Table 1. Examples of tweets in a given perspective

Immediate observation	Snipers are shooting deputies standing on stage, one of them is injured,ask med.volunteers to come
	Number of individuals passing #Russia troops and equipment heading for #Ukraine. Told not to photograph. Tweeting instead.
	Moscow is currently shrouded in a mysterious gas that smells like rotten eggs.
Affection	Sadly, in #Ukraine and not only #Russia, standard practice to hide true military casualty figures.
	I pity anyone trying to justify #Odessa horror.
	I am surprise that the #EU are doing absolutely nothing to help the #Ukraine Government
Speculation	So-called decent West with its fixed high opinion of itself has been supporting a regime in #Ukraine using cluster bombs to kill civilians.
	In this environment, if Ukraine manages to pull off presidential elections, it'll be a miracle.
	BREAKING: Multiple credible reports say #Russia has sent in an armored column of 70 AFVs into #Ukraine with 400 more to follow.

perspectives. To identify the most suitable learning techniques, we select a number of lexical and textual features that can be obtained from tweet messages, and test them on various model building techniques, including SVM, Naive Bayes, and Random Forest.

Following the classification, we introduce three scenarios in which the classification results can be used. Particularly, the Immediate Observations can be used for identifying the time and location of the reported events. The Affection tweets can be aggregated as a form of public opinion polls. The Speculation tweets can be clustered to track the popular stories involving an event or object of interest.

The remainder of this paper is organized as the following: in Sect. 2, we discuss the related works. In Sect. 3, we present our feature selection and classification method. In Sect. 4, we report our experiments for testing the classification accuracy of our method. We introduce the application scenarios in Sect. 5. Finally, Sect. 6 offers some conclude remarks.

2   Related Work

Twitter as a public media and news source has been studied in several works. Wu et al. [24] investigated the demographics of influential Twitter users, whom they grouped into media, organization, celebrity, and blogger. They also found that bloggers, who are not tied to particular organizations and express mostly personal views, are the producers of the most influential tweets. Vis [23] studied Twitter as a reporting tool during 2011 UK riots, author identifying several actor types for user who involved in event reporting, including journalists, activists, mainstream media, police, and fake accounts. Metaxas et al. [12] studied tweets reporting criminal activities in Mexico in 2010 and 2011. They used peaks in tweet posting for detecting events, while users involved are grouped into different actor roles. These studies are based on the roles of reporting users, which we consider inappropriate, because an ordinary user does not need to have a consistent role when tweeting about an event. In contrast, while not proposing an automated classification method, Maddock et al. [11] studied misinformation, speculation, question, and other categories of information reported on Twitter when an important event happened, labelling the category of each tweet.

While the reports posted on Twitter by ordinary users are very often informal and incomplete, many researchers have found invaluable information from them in scenarios such as election prediction, disaster location, object tracking, and event detection. Tumasjan et al. [20] found that the sentiments users express on Twitter correspond closely to actual election results, based on Germany federal election in 2009. Sang et al. [16] conducted a content analysis and also found that sentiments expressed on Twitter correspond closely to election results, based on Dutch Senate election in 2011. Their sentiment analysis is based on LWIC categories [19]. Unankard et al. [21] deployed sentiment analysis and clustering techniques to predict Australian federal election in 2013 based on Twitter messages. Lingad et al. [10] studied using locations mentioned in tweet messages for locating natural disasters and affected areas. Sakaki et al. [14] developed a system that tracks the movement of earthquakes and typhoons based on the reports detected on Twitter. Li et al. [7] studied using tweets for detecting crime and disaster events reported on Twitter. Unankard et al. [22] developed an event detection system for local, time-sensitive events using messages posted on Twitter. Lin et al. [9] proposed an event-based point-of-interest (POI) detection system based on Twitter message. Kwon et al. [6] identifies several signatures that can be used to distinguish false rumors among stories circulating on Twitter. These works showed a range of possibilities that the content of reports on Twitter can be used.

Given that tweet messages are usually informal, unorganized texts, it is difficult to apply traditional model-based natural language processing techniques. Currently, the most used method for tweet content analysis is machine-learning-based classification. Such a method involves devising a set of features that can be extracted from tweet messages, and using the extracted feature strings as the input to certain machine learning models. For example, Sakaki et al. [15] classified tweets into positive and negative reports, based on a set of relatively simple features, including the word count and the position of the query word. Li et al. [7] used location words, URLs and reply signs in tweet messages and an automatic classifier model in their event detection system. Castillo et al. [1] studied user credibility on Twitter. They included text, topic, and propagation features in automatic classifiers for determining credibility in tweet messages. Sriram et al. [18] provided a solution to tweet classification for the purpose of understanding user intention. They classified five types of tweets: news, opinions, deals, events, and private messages, based on features such as the presence of time-event phrase, opinion words, and dollar signs. Gupta [4] developed an online system for identifying tweet credibilities, based on a SVM classifier and a number of features including character and word count, and the linguistic category of each word. However, to the best of our knowledge, there are no solution proposed for automatically classifying perspectives of the tweets reporting complex events, which can induce different types of responses.

3   Methodology

We aim to find an automated classification method that can accurately classify event-related tweets according to the perspectives they are posted in. Our method involves first extracting meaningful features from the text messages, then training a classifier using machine learning approaches. We design the features based on the lexical and textual analysis on the texts. In this section, we discuss our approach in detail.

3.1   Tokenization of Tweet Message

Each tweet message consists of up to 140 characters. Since our analysis focuses on words, we break each tweet message up by spaces and punctuations into a series of words. For example, the tweet "`Yesterday, 32 monuments of Lenin were toppled around Ukraine`" is tokenized into "Yesterday", "32", "monuments", "of", "Lenin", "were", "toppled", "around", and "Ukraine". These words are the information unites in our approach, which basically involves counting words in different categories. As an example, in this tweet message, there is one word related to time, "Yesterday", and one word related to space, "around".

3.2   Lexical Categories

We use the categories defined in *Linguistic Inquiry and Word Count* (LIWC) for lexical analysis. LIWC provides a dictionary for different categories of English words, such as functional words, common verbs, and emotional words, and is one of the most used tools in natural language processing [19]. The dictionary contains four main categories and 64 sub-categories, and the number of entries exceeds 10,000. The entries in different categories may overlap.

We use L1 to L64 as the code name of the 64 LWIC categories. Not all categories of words are helpful for distinguishing tweets of a certain perspective from other tweets. For example, emotional words are clear signatures of the affection perspective, but they are less meaningful for identifying observation and speculation tweets. Therefore we first test the presence of each of the 64 sub-categories defined in LIWC 2007 using the following method. Given a set of

Table 2. Effective lexical categories for different perspectives

Perspective	Category code
Immediate observation	L1, L2, L4, L9, L22, L23, L27, L29, L33, L51
Affection	L1, L2, L10, L11, L13, L22, L23, L27, L28, L51
Speculation	L1, L10, L11, L13, L16, L22, L28, L31, L33, L53

Table 3. Selected lexical categories and example words

Code	Lexical category	Examples
L1	Function Words	*about, back, can, despite, each*
L2	Personal Pron	*he, she, I, we, ours*
L4	1st Plural	*lets, our, ourselves, us, we*
L9	Articles	*a, an, alot, the*
L10	Common Verbs	*accept, become, call, depends, eaten*
L11	Auxiliary Verbs	*are, can, do, gonna, had*
L13	Present Tense	*ask, asks, become, becomes*
L16	Prepositions	*about, before, down, except, for*
L22	Total Pron	*anybody, he, I, ourselves, that*
L23	Social Process	*admit, baby, captain, deal, email*
L27	Affective Processes	*abuse, bad, care, danger, eager*
L28	Positive Emotion	*accept, beauty, careful, daring, easy*
L29	Negative Emotion	*abuse, bad, careless, danger, enemy*
L31	Anger	*abuse, bastard, cheat, destroy, enrage*
L33	Cognitive Processes	*absolute, ban, cause, deduce, effect*
L51	Relativity	*above, before, carry, dance, early*
L53	Space	*air, bend, capacity, deep, east*

tweets T whose perspectives we already knew, for each perspective, we calculate the effect of each lexical category e_l as $e_l = |p_{l,1} - p_{l,0}|/|T|$, where $p_{l,1}$ is the count of words in the lexical category l in all tweets in that perspective, $p_{l,0}$ is the count of words in the lexical category l in all tweets not in that perspective, and $|T|$ is the total number of tweets. Once we have the effect of each lexical category $\{e_1, e_2, e_3, ..., e_{64}\}$, we take the 10 most effective lexical categories for each perspective. Table 2 shows the categories we obtained. Table 3 shows the category names defined in LIWC 2007 and some example dictionary entries of these categories. The tables show some special characteristics of messages expressing in each perspective. For example, first person plural nouns, articles, and negative emotions have particular strong presences in immediate observations.

Table 4. Textual features

Code	Textual features	Explanation
T1	Word Count	the number of words in the tweet message
T2	Emphasis	the number of words in capital letters, e.g., GREAT
T3	Exclaim Mark	the number of exclaim marks
T4	Question Mark	the number of question marks
T5	Hashtag Presence	whether a hashtag is present, 1=present, 0=otherwise

3.3 Textual Features

In addition to lexical categories, we choose five textual features that are expressive and may reveal the perspective of a tweet. These features include *word count, emphasis count, exclaim mark count, question mark count,* and *hashtag presence*. Table 4 gives brief explanations for these features. *Hashtags* is a special textual mark used on Twitter to indicate the topic of the message, and the presence of a hashtag may reveal the user's relationship with the event. The textual features are analyzed for all three perspectives.

3.4 Feature String and Model Building

A supervised learning model is generated from a set of labelled data, called *training set*. For each perspective, we supply a set of tweets as labelled either positive or negative. Positive for a perspective means that the tweet is composed in that perspective, and negative means the tweet is not composed in that perspective. A separate classifier is built for each perspective using the respective training data. The tweet messages are transformed into features using the method mentioned in previous sections, before processed by the classifiers. The training data for each perspective consists of 15 features, the word count in ten most effective lexical categories for that perspective, and five textual features shared by all perspectives.

As an example, Table 5 shows the feature string for the tweet "Shell hits school in #Donetsk, eastern #Ukraine, killing two children amid renewed violence". Since we are interested in whether this message is an immediate observation, we select the Immediate Observation feature set shown in Table 2. The textual feature set shown in Table 4 is also used. In this example, the tweet contains two Functional words, two Social Processes words, one Affective Processes words, one Negative Emotion words, two Relativity words, and the count for words in other lexical categories is zero. The total word count in the message is 13, and there are hashtags in the message, no exclaim mark, question mark, or words in capital letters.

Various techniques have been established for building the supervised learning model, such as *Support Vector Machine* (SVM), *Naive Bayes, Random Forest,*

Table 5. Example feature string

L1	L2	L4	L9	L22	L23	L27	L29	L33	L51	T1	T2	T3	T4	T5
2	0	0	0	0	2	1	1	0	2	13	0	0	0	1

and *Linear Discriminant Analysis* (LDA), and have been frequently used in existing works for tweet classification [1,14]. A SVM classifier finds a hyperplane that separates data in different classes with a maximum margin [5]. It usually maps data into a higher-dimension space to improve efficiency, using different kernel functions, such as linear and polynomial. A Naive Bayes classifier is trained using Bayes Theorem, and is based on the assumption that features in the dataset are strongly independent [3]. A Random Forest classifier is built from multiple decision trees, which are sets of rules that separate data in different classes generated from the training data [8]. A LDA classifier finds a linear combination of features that separates data in different classes [13]. In our experiments, we trained classifiers using all four techniques and compared their classification accuracies. For the SVM classifier, we tested both the linear and polynomial kernels.

4   Experimental Results

We conducted a set of experiments for testing the classification accuracies of different combinations of features and machine learning models, using a real dataset we collected from Twitter. In this section, we will present our experiment setup, measurement methods, and experimental results.

4.1   Data Collection and Preparation

We collected two sets of tweets related to the Ukraine Crisis using Twitter's Stream API[4]. The first set contains tweets posted in February and March 2014, over a period of 40 days. The second set contains tweets posted in November and December 2014, during a period of 56 days. The two sets of data contain over a million tweets. We randomly selected a subset of them and manually labelled them according to their perspectives. For each perspective we labelled 1,000 tweets. For the immediate observation perspective, we labelled 500 positive and negative examples. For the affection perspective, we labelled 418 positive examples and 582 negative examples. For the speculation perspective, we labelled 520 positive examples and 480 negative examples. As a result, for each perspective, we have a separate dataset, although the same tweet may be present in more than one datasets.

---

[4] https://dev.twitter.com/streaming/overview.

4.2 Measurements

We used three-fold cross-validation when testing classifiers. For each test, we divided the labelled data into three parts, and used two for training the classifier, and one for testing. After the classifier produced the prediction for the testing set, the predicted labels were compared with the original labels. If the classifier predicted positive for a perspective for a tweet, and the tweet was labelled positive in the dataset, we counted the prediction as a true positive. We measured the *precision* and *recall* for the positive results. The classification precision was calculated as the percentage of true-positives in all predicted positives. The classification recall was calculated as the percentage of true-positives in all labelled positives in the dataset for that perspective. As an indicator of general accuracy, the F-value is calculated as $\frac{2 \times precision \times recall}{precision + recall}$. In this study we do not consider negative predictions. The use of negative reports for event analysis will be a future study topic.

4.3 Results and Discussion

We tested the classifiers in each of the three datasets. First we tested them using lexical features only, and the results are shown in Table 6 and Fig. 1. The accuracy of best performing algorithm in each test is highlighted in bold.

Table 6. Classification accuracy results with lexical features

	Immediate observation			Affection			Speculation		
	prec.	recall	f-value	prec.	recall	f-value	prec.	recall	f-value
SVM Linear	0.527	0.622	0.57	0.656	0.627	0.641	0.631	0.743	0.682
SVM Polynomial	**0.585**	0.38	0.46	0.651	0.57	0.603	0.603	**0.854**	**0.706**
Naive Bayes	0.54	**0.806**	**0.646**	**0.663**	0.54	0.595	0.625	0.79	0.696
Random Forest	0.529	0.58	0.535	0.624	**0.737**	**0.673**	**0.672**	0.573	0.616
LDA	0.566	0.538	0.551	0.657	0.673	0.622	0.64	0.712	0.673

Then we tested them using both lexical and textual features, and the results are shown in Table 7 and Fig. 2.

According to the results, for the Immediate Observation and Affection perspective, using lexical features only allowed a higher classification recall. However, adding textual features increased the classification precision significantly. For the Immediate Observation perspective, adding textual features increased the classification precision from 0.585 to 0.837. For the affection perspective, the precision was also increased to above 0.8. With over 0.8 precision, we can confidently trust that the classifier will provide true-positives most of the time. For the Speculation perspective, the addition of textual features was less significant, but it generally improved the classification accuracy for all classifiers, as evident in the increased F-values.

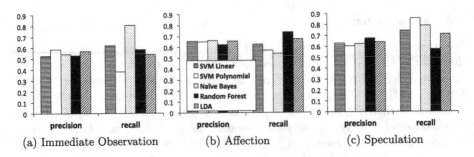

Fig. 1. Precisions and recalls with lexical features

Table 7. Classification accuracy results with lexical and textual features

	Immediate observation			Affection			Speculation		
	prec.	recall	f-value	prec.	recall	f-value	prec.	recall	f-value
SVM Linear	0.704	0.689	0.695	0.778	0.635	0.697	0.64	0.792	0.707
SVM Polynomial	0.821	0.325	0.464	0.784	0.535	0.636	0.622	**0.889**	**0.731**
Naive Bayes	**0.837**	0.339	0.473	**0.804**	0.507	0.606	0.612	0.864	0.716
Random Forest	0.694	**0.7**	**0.697**	0.742	**0.733**	**0.765**	**0.659**	0.573	0.609
LDA	0.792	0.485	0.6	0.765	0.643	0.698	0.64	0.728	0.681

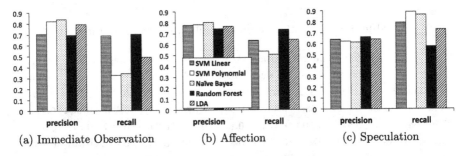

Fig. 2. Precisions and recalls with lexical and textual features

Among the classifiers, the Naive Bayes classifier generally provided the highest classification precision for the Immediate Observation and Affection perspectives, while Random Forest achieved the highest recall and F-value. Interestingly, for the Speculation perspective, the Random Forest classifier achieved the highest precision, while SVM with polynomial kernel achieved the highest recall.

We conclude that using both lexical and textual features generally improves classification accuracy than using only the lexical features, and that the choice of classifier depends on the goal of the task. If the aim is to collect immediate observations with as few false positives as possible, the Naive Bayes classifier should be used. If the aim is to find as many as possible users' emotional reactions towards an event, and wrongly classified messages are more tolerable, then the Random Forest classifier should be deployed.

5   Application Scenarios

In above experiments, we show that automated machine learning methods can classify event-related tweets into immediate observations, affections, and speculations, with relatively satisfying accuracy. The automated classification helps users who want to find out event-related information on Twitter to focus on a particular perspective. It also enables further automated data analysis to exact a particular type of information from the classification results. In this section, we introduce three scenarios in which the classification results can be further analyzed to generate interesting information. In particular, the immediate observations can be used to infer the time and location of the event. The affection tweets can be used to understand public sentiments toward an event. The speculation tweets can be used to detect false rumors in early stages.

5.1   Inferring Time and Locations of an Event

Immediate observations are observations made immediately at the scene of the event. They describe the state of the observed object, person, or event, at the time and location when and where the user posts the message. Provided that Twitter provides timestamp and GPS data with the tweets post, even if a message itself does not contain time and location information, we can implicitly obtain this information. For example, when we collect a tweet post in Times Square on Tuesday morning, saying "the air is fresh", which is associated with a timestamp and GPS data, we obtain the air quality information in Times Square on Tuesday morning, even if the message itself does not provide this information.

Table 8 shows some examples of immediate observations found in our dataset and their timestamps and GPS data as locations. The timestamps uses GMT timezone. The GPS data is in $\{latitude, longtitude\}$ format.

By inputting the GPS data in a digital map system, such as Google Map[5], we can clearly identify the location where the observations and thus the events happened. The first tweet indicates a troop invasion of a military base near Sevastopol on 19th March, 2014, even though this information was not shown explicitly in the message. The second tweet provides the time and location of another alleged invasion. The third message provides some information about a protest, and its GPS data indicates that the protest happened in the center of Kyiv.

Immediate observations conveniently allow the time and location of the tweet to be used as the time and location of the event. By classifying and aggregating immediately observations, we can not only find out time and location of an event quickly, but also track the location changes of the event over time.

---

[5] https://maps.google.com/.

Table 8. Example immediate observations

Message	Time	Location
Russian troops without insignia showed up overnight followed by irregulars who then stormed the base, broke down doors, threw smoke grenades	19/3/2014 9:15	33.51153234, 44.58660487
Ukrainian base at Perevalnoye surrounded by hundreds of Russian troops. Commander Sergei Storozhenko says the Russians are from Sevastopol	2/3/2014 11:46	34.33016684, 44.84080314
Many @lentaruofficial journos leaving to protest editor's firing. Among them: @A3AP, whose great Ukraine coverage aroused the Kremlin's ire	12/3/2014 15:00	30.51913980, 50.45470648

Table 9. Example affection tweets

Message	VN	N	P	VP
during #euromaidan I was afraid to wake up in the morning, to find out there was crackdown.	7	55	5	1
Though things are more or less calm now,some numbers are still shocking-298 protesters are missing	9	67	3	1
Difficult days are coming for my country, my city...WAR IS starting Again...	9	53	10	3
lucky I had chance to work w/ great journalists at #techcamp and gained many insights into crisis in #ukraine from journos working there	2	10	44	6

5.2   Public Sentiments Toward an Event

Sometimes it is important to know the public sentiments toward an event for government decisions and policymaking, particularly when the event is controversial and the public sentiments are diverse. Twitter users express opinion about events in many ways, and some users may express opinions about future events or events they are not directly involving in. Aggregating sentiments with these opinions may make the results noisy and inaccurate. By limiting the scope to the affections of the users who are directly involving in and impacted by the event using automated classification, we can obtain a much more accurate understanding of how the event actually affected users.

As a demonstration, Table 9 shows sentiment analysis results of the affection tweets in our dataset using StanfordNLP online tool[6]. The tool analyzes sentences and short message and provides scores for Very Negative (VN), Negative (N), Positive (P), and Very Positive (VP) sentiments.

---

[6] http://nlp.stanford.edu:8080/sentiment/rntnDemo.html.

Through automated sentiment analysis of a large number of affection tweets like the ones shown in Table 9, we quickly obtain the overall sentiments Ukraine people had for the series of events including conflicts, protests, and journalist activities. The aggregated form of these results, which is similar to an opinion poll, can be used to aid policy making. It will be particular effective when the classification is used with topic-based clustering techniques, and the affection tweets are assigned to specified topics and sub-events.

5.3 Early Rumor Detection

When an important event happens and many facts are unclear, it is guaranteed that it will invoke wide speculations. Usually the speculations will be mixed in numerous discussions, opinions, emotional statements, and facts, which makes them difficult to be identified. By separating from other information the speculations, which are based on individual judgments instead of facts, we can quickly identify possible misinformation. The classified speculations can then be further clustered to find popular stories involving particular events and objects of interest. For demonstration purpose, we run a simple clustering algorithm based on hashtags on our dataset. The algorithm groups together messages containing a same hashtag. Table 10 shows the messages containing the hashtag #MH17. The number of retweets (# ret) was updated on 29th May, 2015.

In 17th July, 2014, Malaysia Airlines Flight 17 (MH17) was reportedly being shot down at the Ukraine-Russia boarder. Because it was a shocking news with many unclear facts, it has invoked a large number of speculations, even long after the time the event happened. The messages in the table tell stories that are yet to be confirmed at the time of posting. Combining the classification and clustering helps us to identify false stories involving a particular event or object of interest. Furthermore, when monitoring in realtime, it can show us which stories are gaining popularity, and have the potential to become a widespread

Table 10. Example speculation tweets

Message	Timestamp	# ret
BREAKING: Remainder of Malaysian #MH17 Boeing wreckage in #Ukraine may start to be removed in next two weeks	10/11/2014 9:55	6
Soros claims that Putin ordered Yanukovych to fire on the maidan protesters. #Ukraine #MH17 #Russia	10/11/2014 10:10	0
Meanwhile, #Putin has deployed more troops to the #Ukraine. And if #MH17 is any metric, they're in ''combat roles'' #UkraineUnderAttack #tcot	7/11/2014 21:21	2
NRO #MH17 National Reconnaissance Office Flight 17 downed by Kiev forces #Ukraine	8/11/2014 14:23	4

rumor. This can be done, for example, by tracking the increases of the retweet numbers of the speculation tweets.

6   Conclusion

Twitter offers ordinary users a convenient means to publish their observations, reactions and opinions for ongoing events. Since messages post on Twitter are available for access immediately, many people nowadays choose Twitter as the news source when an important event happened. For a complex event, there are many perspectives in which a user may post messages, among which are the *immediate observations*, expressions of *affections*, and *speculations*. While understanding the perspectives can be invaluable for further analysis of the event based on these tweets, it is difficult for anyone to read through the tweets manually and identify their perspectives, given the large volume of posts on Twitter.

In this paper, we experimented automatic classification methods based on a set of lexical and textual features and various machine learning models. Our experimental results show that, with the right choice of features and classifiers, we can generally obtain satisfying classifying results. We showed that the automated classification method for immediate observations and affection tweets can achieve a precision higher than 0.8, while a recall higher than 0.8 can be achieved for the speculation tweets. In the future we plan to use the classified Immediate Observation and Affection reports to automatically generate detailed descriptions for ongoing events.

References

1. Castillo, C., Mendoza, M., Poblete, B.: Information credibility on Twitter. In: Proceedings of the 20th International World Wide Web Conference (2011)
2. Chung, D.S., Nah, S.: Media credibility and journalistic role conceptions: views on citizen and professional journalists among citizen contributors. J. Mass Media Ethics 28(4), 271–288 (2013)
3. Frank, E., Bouckaert, R.R.: Naive Bayes for text classification with unbalanced classes. In: Proceedings of 10th European Conference on Principles and Practice of Knowledge Discovery in Databases (2006)
4. Gupta, A., Kumaraguru, P., Castillo, C., Meier, P.: TweetCred: a real-time web-based system for assessing credibility of content on twitter (2014). arXiv preprint arXiv:1405.5490
5. Joachims, T.: Text categorization with support vector machines: learning with many relevant features. In: Nédellec, C., Rouveirol, C. (eds.) ECML 1998. LNCS, vol. 1398, pp. 137–142. Springer, Heidelberg (1998)
6. Kwon, S., Cha, M., Jung, K., Chen, W., Wang, Y.: Prominent features of rumor propagation in online social media. In: Proceedings of 13th International Conference on Data Mining (2013)
7. Li, R., Lei, K.H., Khadiwala, R., Chang, K.-C.: Tedas: a twitter-based event detection and analysis system. In: Proceedings of 28th International Conference on Data Engineering (2012)

8. Liaw, A., Wiener, M.: Classification and regression by randomforest. R. News **2**(3), 18–22 (2002)

9. Lin, C.-S., Chiang, M.-F., Peng, W.-C., Chen, C.-C.: An event-based poi service from microblogs. In: Proceedings of 13th Asia-Pacific Network Operations and Management Symposium (APNOMS) (2011)

10. Lingad, J., Karimi, S., Yin, J.: Location extraction from disaster-related microblogs. In: Proceedings of the 22nd International World Wide Web Conference Companion (2013)

11. Maddock, J., Starbird, K., Al-Hassani, H., Sandoval, D.E., Orand, M., Mason, R.M.: Characterizing online rumoring behavior using multi-dimensional signatures. In: Proceedings of the 18th ACM Conference on Computer Supported Cooperative Work & Social Computing (2015)

12. Metaxas, P., Mustafaraj, E.: The rise and the fall of a citizen reporter. In: Proceedings of the 5th Annual ACM Web Science Conference (2013)

13. Phan, X.-H., Nguyen, L.-M., Horiguchi, S.: Learning to classify short and sparse text & web with hidden topics from large-scale data collections. In: Proceedings of the 17th International World Wide Web Conference (2008)

14. Sakaki, T., Okazaki, M., Matsuo, Y.: Earthquake shakes Twitter users: real-time event detection by social sensors. In: Proceedings of the 19th International World Wide Web Conference (2010)

15. Sakaki, T., Okazaki, M., Matsuo, Y.: Tweet analysis for real-time event detection and earthquake reporting system development. IEEE Trans. Knowl. Data Eng. **25**(4), 919–931 (2013)

16. Sang, E.T.K., Bos, J.: Predicting the 2011 Dutch senate election results with Twitter. In: Proceedings of the Workshop on Semantic Analysis in Social Media, pp. 53–60. Association for Computational Linguistics (2012)

17. Sebastiani, F.: Machine learning in automated text categorization. ACM Comput. Surv. **34**(1), 1–47 (2002)

18. Sriram, B., Fuhry, D., Demir, E., Ferhatosmanoglu, H., Demirbas, M.: Short text classification in Twitter to improve information filtering. In: Proceedings of the 33rd International ACM SIGIR Conference on Research and Development in Information Retrieval (2010)

19. Tausczik, Y.R., Pennebaker, J.W.: The psychological meaning of words: LIWC and computerized text analysis methods. J. Lang. Soc. Psychol. **29**(1), 24–54 (2010)

20. Tumasjan, A., Sprenger, T.O., Sandner, P.G., Welpe, I.M.: Predicting elections with Twitter: what 140 characters reveal about political sentiment. In Proceedings of the Fourth International Conference on Weblogs and Social Media (2010)

21. Unankard, S., Li, X., Sharaf, M., Zhong, J., Li, X.: Predicting elections from social networks based on sub-event detection and sentiment analysis. In: Benatallah, B., Bestavros, A., Manolopoulos, Y., Vakali, A., Zhang, Y. (eds.) WISE 2014, Part II. LNCS, vol. 8787, pp. 1–16. Springer, Heidelberg (2014)

22. Unankard, S., Li, X., Sharaf, M.A.: Emerging event detection in social networks with location sensitivity. World Wide Web J. (2015, in press)

23. Vis, F.: Twitter as a reporting tool for breaking news: journalists tweeting the 2011 UK riots. Digit. Journal. **1**(1), 27–47 (2013)

24. Wu, S., Hoffman, J.M., Mason, W.A., Watts, D.J.: Who says what to whom on Twitter. In: Proceedings of the 20th International World Wide Web Conference (2011)

Evaluating k Nearest Neighbor Query on Road Networks with no Information Leakage

Lu Wang, Ruxia Ma, and Xiaofeng Meng[(✉)]

Information School, Renmin University of China, Beijing, China
{luwang,maruxia,xfmeng}@ruc.edu.cn

Abstract. The development of positioning technologies and pervasiveness of mobile devices make an upsurge of interest in location based services (LBS). The k nearest neighbor(kNN) query in road networks is an important query type in LBS and has many real life applications, such as map service. However, such query requires the client to disclose sensitive location information to the LBS. The only existing method for privacy-preserving kNN query adopts the cloaking-region paradigm, which blurs the location into a spatial region. However, the LBS can still deduce some information (albeit not exact) about the location. In this paper, we aim at strong privacy wherein the LBS learns nothing about the query location. To this end, we employ private information retrivial (PIR) technique, which accesses data pages anonymously from a database. Based on PIR, we propose a secure query processing framework together with flexible query plan for arbitrary kNN query. To the best of our knowledge, this is the first research that preserves strong location privacy for network kNN query. Extensive experiments under real world and synthetic datasets demonstrate the practicality of our approach.

Keywords: Location privacy · Private information retrieval · kNN query · Spatial networks

1 Introduction

With the popularity of mobile devices and development of the positioning technologies, location based service(LBS) is becoming more and more popular. To provide users with location based service, LBS system (e.g., Map Quest and Google Maps for mobile users) has been widely deployed by mobile users. The nearest neighbor queries in LBS occupy an extremely important position. For example, client traveling on the road may want to get the nearest gas station, or tourist may hope to learn the nearest restaurant from his current location. LBS

This research was partially supported by the grants from the Natural Science Foundation of China (No. 61379050, 91224008); the National 863 High-tech Program (No. 2013AA013204); Specialized Research Fund for the Doctoral Program of Higher Education(No. 20130004130001); the Fundamental Research Funds for the Central Universities, and the Research Funds of Renmin University(No. 11XNL010).

© Springer International Publishing Switzerland 2015
J. Wang et al. (Eds.): WISE 2015, Part I, LNCS 9418, pp. 508–521, 2015.
DOI: 10.1007/978-3-319-26190-4_34

gives us more convenience, however, also causes sensitive privacy problems. Once the client requests a query, he must submit his location to LBS, which leads to the leakage of location privacy, even personal information such as health status, economic conditions, shopping habits, etc. [25].

Therefore, there exist lots of approaches for privacy-aware kNN query [1–4]. However, these works only consider Eculidean distance rather than road network distance. The only existing work of privacy-aware kNN query under road network follows the location obfuscation approach [5] which blurs client's exact location into a cloaked region and computes network k nearest neighbors by network voronoi diagram [22]. However, the method reveals certain location information of client to the LBS.

To guarantee strong location privacy, a promising cryptography tool is private information retrieval (PIR) [11]. PIR allows a data item (e.g., a disk page) to be retrieved from a server without leaving any clue of the item being retrieved. PIR was considered to be resource-intensive, but thanks to the recent progress in cryptography, practical software or hardware PIR solutions have been proposed [14]. Since then it has been successfully applied to spatial queries, such as kNN, BRNN and shortest path search [7,9,13].

In this paper, our goal is to investigate privacy-preserving kNN query on road network without the LBS inferring any information about the query. To this end, we adopt practical PIR techniques that retrieve a single data page as the building block. The challenges of a PIR-based kNN solution lie in the following aspects: (1) although PIR guarantees secure access of a single page from the server, the variation of the number of page accesses from different queries may reveal information about the query point. Further, when user desires to propose queries with varied k, our processing must be safe for arbitrary k, which makes the problem more challenging. (2) as the database contains voluminous points, directly applying PIR for the kNN query on road network is inefficient, thus calling for an integration with spatial index. To address these challenges, we propose a PIR-based kNN query processing framework that guarantees strong privacy. Concretely, we design index structure and deduce query plans for arbitrary k, which means adversary cannot deduce any information from arbitrary query. To summarize, we have three main contributions as follows:

(1) To the best knowledge, this is the first research evaluating k nearest neighbor query on road network with no information leakage.
(2) We deduce the fixed query plans for arbitrary fixed k and thus guarantees the strong privacy for arbitrary k nearest neighbor query on road network.
(3) We conduct extensive experiments under real-world and synthetic datasets, which shows our proposed approach is practical.

The rest of the paper is arranged as follows. Related works are surveyed in Sect. 2. In Sect. 3, we define our security model and prove its security. We then present our solutions for the PIR-based kNN query processing in Sect. 4. The solutions are evaluated by experiments in Sect. 5. Section 6 concludes this paper.

2   Related Work

In this section, we review related works in the following two areas: (1) priavcy-aware k nearest neighbor query on road network and (2) the application of PIR based approaches on spatial query.

2.1   Privacy-Preservation for kNN on Road Network

There are several existing network nearest neighbor query processing methods in the literature, such as the network expansion based methods [19,20], solution based methods [22–24] and hierarchical road networks based methods [21]. The classic method without privacy is the network voronoi nearest neighbor based solution which is proposed by Kolahdouzan [22]. It utilizes network voronoi diagram to partition the network into cells to reduce computation cost and communication cost. The only existing work of private network NN query based on spatial network follows the location obfuscation approach [5] which blurs client's exact location into a cloaked region and computes network k nearest neighbors by network voronoi diagram. However, the method reveals certain location information of client to the LBS.

So far, PIR technology is the only tool to guarantee strong privacy which means server cannot deduce any information about the query. There has been no works on applying PIR-based method to network nearest neighbor query to provide strong privacy guarantee. As the hardware PIR based method requires different queries execute the same query plan which implies that every query incur the same processing cost, the existing methods above cannot apply directly to our PIR-based private network nearest neighbor query.

2.2   Application of PIR

PIR is a type of technology that can request a data item on a database and does not let the database know which item is requested [11]. To make oblivious data item access in malicious server, various Private Information Retrieval (PIR) technology have been widely adopted since its first proposal [6]. Then, there are three streams of relevant research: (1) information based PIR theoretic [10,11]; (2) computational PIR [6,12] and (3) secure hardware based PIR [14]. In this paper, we adopt the secure hardware as its implementation. The secure hardware relies on a temper-resistant CPU which is positioned at the server and is trusted by the clients. It is considered as an interface that supports oblivious data page access. The overhead of one PIR access involves two parts: (1) the online cost which represents the overhead of retrieving, re-encrypting and storing the data page (2) the offline cost which is taken to reorganize data pages in the data structure. All the online and offline cost grow sub-linearly to the space size. This fact explains why we mainly focus on reducing the number of PIR accesses rather than saving the storage space in later design.

For spatial NN query, to prevent location information leakage, [8] presents a novel LBS privacy preserving approach based on computational PIR for

NN query. Later, based on some computing impractical problem(e.g.,Quadratic Residuosity Assumption, QRA), nontrivial implementation for PIR is proposed. Then some works utilize Oblivious Transfer [15] or Paillier encryption scheme [16] integrating with computational PIR technology to protect location privacy of the spatial NN query [17,18]. The only existing spatial query with hardware based PIR methods is PIR-based kNN query [7] and PIR-based BRNN query [13]. To guarantee equal number of PIR access for query proposed by any location, all these methods figure out a maximal number of PIR access after pre-computation over the dataset.

3 Problem Definition

In this section we review the preliminaries of network nearest neighbor query, and then describe our system model and security model.

3.1 System Model

The kNN query on road network has received much attention in research community since its seminal work [26]. A road network is modeled as a graph $G(V, E)$, where a vertex $v \in V$ denotes a road junction or point of interest(POI) and an edge $e \in E$ denotes the path between two vertices; and the weight of the edge denotes the network distance of the two points. A k nearest neighbor query issued at q on road network returns k POIs that are the closest to q in terms of network distance. Without considering the privacy protection, the client poses network kNN query to LBS, and LBS reports the results back to the client based on G.

To guarantee strong privacy, a naive solution is transferring the whole dataset to the client when a query is processing so that the server cannot get extra information about the query except just a query occurring. However, this way is not practical due to heavy communication cost. Thanks to the private information retrieval(PIR) technology, we can design index structure and query plan to combine with it to reduce both the communication and computation cost. In this paper, we adopt the secure co-processor($SCOP$) [9] which is installed at LBS to execute PIR functionality. It offers a PIR interface that can allow clients to retrieve data pages from the database of LBS. The interface can be trusted by the clients as it support complete tamper detection. Figure 1 shows our system model. There exists two parts, the client and the LBS which deploy $SCOP$. Both plaintext of road network G and the encrypted PIR-based index are hosted by LBS. And the indexing information is encrypted by $SCOP$ after being organized in equal-sized data pages. When clients issues a query, he need to follow the query plan to retrieve multi-rounds data pages from the encrypted PIR-based index by $SCOP$.

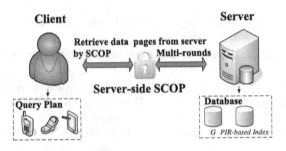

Fig. 1. System Model

3.2   Security Model

Without loss of generality, we assume that LBS is the adversary as it may know the client's identity (e.g.,via user log-in) or may infer it. Also, we assume the adversary is curious, but not malicious, that is, it executes page access routines correctly with no falsified data and wishes to gain extra information about the client's query. The adversary is also aware of the processing protocol in use and its computational power is polynomially bounded.

Our objective is to create PIR protocol for processing network nearest neighbor queries at the LBS without the latter deducing any information about the queries. We assert that every network nearest neighbor query follows the same query plan which is necessary to achieve our privacy goal. Specifically, the query plan needs to ensure each query (i) executes in the same number of rounds, (ii) in each round it accesses the same index in the same order, and (iii) from each index accessed in a specific round, it retrieves the same number of pages. In our paper, we name PIR-based index as *database*. And commonly, we need to design more than one *database* to improve the query performance. For example, if the protocol confirms that 3 pages are fetched from *database* DB_1 and 10 from DB_2(in this order), each query must fetch 3 pages from DB_1 and 10 pages form DB_2. If some query may need fewer than the determined number of pages, the protocol will pad its requests with dummy page accesses in order to conform to the query plan. The following theorem proves that our methodology achieves the security objective.

Theorem 1. *The network nearest neighbor query processing methodology that combines PIR technology with common query plan can achieve strong privacy. Equivalently, it leaks no information to the adversary about query location.*

Proof. In our methodology, each data page requested from database via PIR protocol. Therefore, the adversary is oblivious of which page of the database is being read. What is only visible to the LBS is the number of data pages being accessed in the database. Since all queries follow the same query plan, the number of pages retrievals in the database is identical for all queries. Consequently, adversary cannot tell any two of them apart.

4   PIR-based kNN Processing Framework

In this section, we describe PIR-based kNN Processing Framework to provide strong privacy. Recall our security model, any query processing must follow the same query plan. And, a group of moderate indices stored at LBS are needed to accelerate the query process. In our paper, we split the whole dataset into i databases DB_1, DB_2, ..., DB_i which can help reduce the update cost and the communication cost. So the query plan $[cnt_1, cnt_2, ..., cnt_i]$ represents the maximal number of PIR based page accesses for each *database*.

4.1   Preliminaries

The Network Voronoi Diagram (NVD) has shown to be successful to solve spatial queries such as kNN on road networks. As Fig. 2 illustrates, v_1–v_{10} are vertices of the road networks, wherein v_1–v_4 are POIs. Each cell of the Voronoi Diagram is centered by one POI and contains the locations that are closest to this POI than any other POIs. In road networks, neighboring voronoi cells are separated by border points, such as b_1 to b_6. For example, voronoi cells centerd at v_1 and v_3 are separated by border points b_1, b_3, and b_5. For each voronoi cell, its border points construct a region. The distance between the voronoi cell center and any query point of the cell can be computed by given all edges in such region.

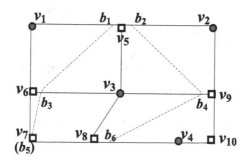

Fig. 2. Example for voronoi diagram in road network

Given properties of the NVD that are described in [22], we can easily compute the kNN query in road network:

(1) The 1NN of query point q is the center of the voronoi cell that q locates in. For example, we assume the query q located at v_8, and v_8 is in the voronoi cell of v_3, then q's nearest neighbor object is v_3.
(2) The k_{th} NN lies in the neighbor of previously found voronoi cells of the (k-1) NN results. That means, if q's nearest neighbor object is v_3, then q's second nearest neighbor object must be the center of neighboring voronoi cells v_1, v_2 and v_4. We further develop efficient method to determine which voronoi cells to fetch without actually obtaining their network information.

4.2   Three Databases

To enhance PIR performance, we first construct a spatial index to partition the whole road networks so that the candidate voronoi cells of q can be located efficiently. To achieve high space utilization, we use the widely adopted KD-tree to partition the whole map. The KD-tree leaf node splits when the voronoi cell overlapping with it occupies more space than a data page. Note that if only one voronoi cell takes more than one page, the leaf node will not split and this corner case is handled by augmenting the leaf node with linked overflow data pages. As Fig. 3(a) illustrates, the four dotted rectangle N_1, N_2, N_3 and N_4 represent the leaf nodes of KD-tree. Each node contains 3 to 5 edges. Correspondingly, in Fig. 3(b), DB_1 occupies 4 data pages referring to the four leaf nodes in KD Tree. Each page records the ID of the voronoi cell residing in the leaf node. We assume the client issues a query at q(the black star). According to the location of q, client can access the leaf node N_3's record A_3, and then client can get the candidate voronoi cell q located in: V_1, V_3 and V_4. Once we obtain the distance between the query point q with these candidate voronoi cell centers, we can know q's 1NN. As such, we design the second structure DB_2.

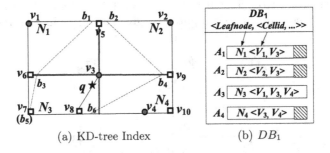

(a) KD-tree Index   (b) DB_1

Fig. 3. Example for spatial network partitioning

As illustrated in Fig. 4(a), DB_2 stores the network information of each network voronoi cell including the vertices, edges and border points. In the example above, the client can access the records B_1, B_3 and B_4 from DB_2. With these network information of V_1, V_3 and V_4, the client can compute which voronoi cell q located in by employing distance computation algorithms under road networks, such as Dijkstra algorithm [19].

According to the property (2) of NVD, the next nearest neighbor of q resides in a number of candidate voronoi cells. Note that we have obtained their border points because they are neighboring to our obtained voronoi cells. Since we have obtained the distance between q to all border points, if we know the distance between each border point to the voronoi cell center on the other side of the edge (voronoi cell on this side of the edge has been obtained), we can determine which voronoi cell center is the next nearest neighbor without fetching all their network information.

As Fig. 4(b) illustrates, we assume each border point b belongs to m_b voronoi cells, so DB_3 stores m_b distance lists for each border point b. Each distance list contains two parts: one is the distance between the border point b_i and the center of voronoi cell V_j it belongs to; The other is the distance between each border point $b_i \in B_{j-}$ falling on the same voronoi cell (B_{j-} represents all the border points fall on the same voronoi cell V_j). Therefore DB_3 can help to compute the minimum distance between q and candidate voronoi cell centers. In the example above, the 2NN of q is the voronoi cell center v_4 adjacent to the q's 1NN via border point b_6 with minimal overall distance $dist(q,b_6) + dist(b_6,v_4)$.

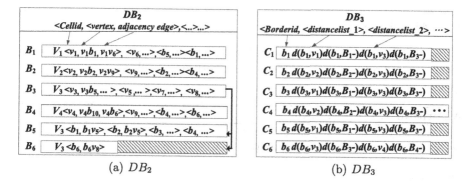

Fig. 4. Example for DB_2 and DB_3

4.3 Query Plan

To achieve the security goal, we determine the query plan $[cnt_1, cnt_2, cnt_3]$ which represents the maximal number of PIR based data page accesses for each database. For ease of description, we use n_i to represent the maximum number of data pages for a single record in DB_i. Take Figs. 3 and 4 as an example, $n_1 = n_3 = 1$, $n_2 = 3$. According to the rational of PIR, each query must follow query plan to retrieval data pages from DB_1, DB_2 and DB_3 respectively. The implementation of the algorithm is as follows:

(1) For each query q, we use DB_1 and DB_2 to compute which network voronoi cell q locates in, and the center of this network voronoi cell is the 1NN of q. Then $cnt_1 = n_1$. For cnt_2, we assume that the maximal number of voronoi cells in each record in DB_1 is c_2, so $cnt_2 = n_2 \times c_2$.

(2) according to the DB_3, the client computes q's next nearest neighbor as its $2NN$ which takes the minimal distance between q and $1NN$'s neighbor centers. Recursively, we can repeat the step (2) to compute the $3 - k$ NN as the query result. In this process, the client needs to maintain the distance of q to every border point of the voronoi cells obtained before.

For cnt_3, we assume that each record in DB_2 refers to maximal c_3 border points. Then, each query result needs $cnt_3 = n_3 \times c_3$ data page accesses. To get the complete result set, there needs $k-1$ iterations.

Overall, the deduced retrieval plan for kNN query on road network requires n_1 PIR based data page accesses for DB_1, $n_2 \times c_2$ page accesses for DB_2 to obtain the 1NN and (k-1) rounds to obtain the rest nearest neighbors. And each round takes $n_3 \times c_3$ page accesses to obtain all distance related to these border points and the new voronoi centers for the next round. In this way the trivial query plan requires $n_1 + n_2 \times c_2 + (k-1) \times n_3 \times c_3$ PIR accesses for arbitrary kNN query on road network.

4.4 Algorithm

In this following, we present our PIR-kNN algorithm. According to the three databases and query plan, we design our algorithm as follows:

Algorithm 1. PIR-kNN algorithm

Input: Query point q, query parameter k
Output: network k nearest neighboring object points, R
1: $R = \emptyset$
2: $C = \emptyset$
3: Fetch entries corresponding to voronoi cells from DB_1, denoted by E^{DB_1}, by locating the leaf node in KD Tree that contains q via cnt_1 PIR page accesses
4: Fetch detailed contents of such voronoi cells E^{DB_1}, denoted by E^{DB_2} from DB_2 by cnt_2 PIR page accesses
5: **for** each record $e \in E^{DB_2}$ **do**
6:     $distance_e = dist(e, q)$
7:     $C.push(e, distance_e)$
8: **end for**
9: $cc = \{C.top()\}$
10: **if** $k == 1$ **then**
11:     $R = cc$
12:     return R
13: **end if**
14: **for** $i = 2$ to k **do**
15:     **for** each border point b of cc **do**
16:         Fetch all pre-computed distance of border b via cnt_3 PIR page accesses
17:         **for** each b's relevant voronoi center vc **do**
18:             $distance_e = mindist(q, vc)$
19:             $C.push(vc, distance_e)$
20:         **end for**
21:         $cc = C.top()$
22:         $R = R \cup \{cc\}$
23:     **end for**
24: **end for**
25: return R

As Algorithm 1 illustrates, the first step is to get the 1NN of q (Line 3–13). There needs n_1 PIR data page accesses for DB_1 and $n_2 \times c_2$ page accesses for DB_2. Then, we fetch all the border points and their associated distance to neighboring voronoi cells from DB_3 via $(k-1) \times n_3 \times c_3$ PIR accesses (Line 14–22). Note that in this step, we only need to obtain the detailed distance information of the new border points obtained in the last iteration. Obtaining all the distance, we can determine the next nearest neighbor (Line 18–19). Until we obtain k nearest neighbor, the algorithm terminates (Line 25).

5 Experimental Evaluation

In this section, we conduct experiments under real world and synthetic datasets to demonstrate the effectiveness of our PIR-based kNN approach. We also compare the performance with a weaker location privacy preservation approach — the cloaking region-based kNN method on road networks(CR-kNN) [5] and show our algorithm is of great practical value.

5.1 Experiment Settings

Datasets. We conduct our experiments on two public real-world networks, namely California map (CA) and New York map (NY). Both datasets are collected from Open Street Map[1]. Both datasets have relatively uniform distribution, while the junctions and roads are more denser in NY than in CA. We summarise the statistics of our datasets in Table 1.

Table 1. Statistics of our datasets.

Dataset	# Edges	# Junctions	# Point of Interests
CA	$47,185$	$20,997$	$84,328$
NY	$56,263$	$14,890$	$60,327$

As for the synthetic dataset, we scatter 10^6 point of interests on aforementioned CA map to simulate different data distribution. To emulate a skewed distribution, a portion $f \in (0,1]$ of these points are distributed on edges in a skewed way, while the rest $1 - f$ portion of points are uniformly generated on edges.

All algorithms are implemented in C# and run on a machine with an Intel Core2 Quad CPU 2.53 Ghz and 4 GByte of RAM. As with previous hardware-based PIR methods, we assume the IBM 4764 PCI-X Cryptographc Coprocessor as the SCOP and strictly simulate its performance. The client communicates with the LBS using a link with round trip time of 700ms and bandwidth 384 Kbit/s, which emulates a moving client connected via a 3G network.

---

[1] www.openstreetmap.org.

5.2   Performance Comparison

In this section, we compare the performance of our PIR-kNN method with the CR-kNN method under both real world datasets. The latter method fetches all POIs that overlap with the client-issued cloaking region and the candidate voronoi cells, and then returns all these POIs to the client. Note that the performance of CR-kNN is plotted only for reference, as it still discloses a cloaking region to the LBS.

Figure 5 illustrates that when $k = 1$, PIR-kNN approach takes more time to return the nearest point of interests from the road network than cloaking region based approach. This is because PIR-kNN first requires to locate the voronoi cell in road network, and when multiple voronoi cell overlaps in a rectangle in the map, all map contents in these cells must be fetched. Thus, this routine consumes much running time. Interestingly, when $k > 1$, PIR-kNN approach gradually outperforms CR-kNN approach and the performance gap enlarges when k gets larger. This is because after the voronoi cell in the road network is located for the query point, each increment of k only incurs one extra fetch for pre-computed distance information via PIR interface. While, for CR-kNN, as k increases, POIs locating in larger map area must be fetched.

We can also see that in NY dataset, where the junctions and roads are denser, it takes more time to return the query result. This is because each POI has more neighbor POIs. Note that the CR-kNN approach also takes more time in NY dataset. This is because the cloaking region with the same size now contains more point of interests.

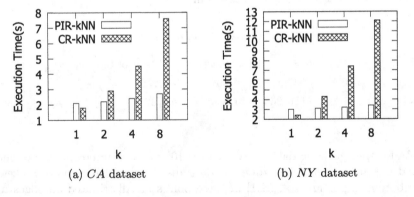

(a) *CA* dataset   (b) *NY* dataset

Fig. 5. Performance comparison under real world datasets.

We validate this argument by the more detailed measurement in Fig. 6. In Fig. 6, we can clearly see that the network overhead for our PIR-kNN approach is much less than that of CR-kNN approach. This demonstrates that there are significant unnecessary POIs are transferred from LBS to the client. In contrast, our PIR-kNN approach seldom conveys unnecessary data and the major overhead comes from the online and offline processing routine in SCOP to implement oblivious fetch.

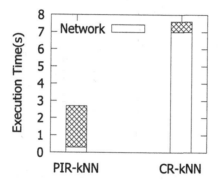

Fig. 6. Execution time proportion over network communication on the CA dataset.

5.3   Detailed Analysis of Our Method

In the following, we give a detailed analysis of our method. In specific, the performance of our PIR-kNN approach under different data distribution and its scalability are evaluated.

Effect of Data Distribution. First, we compare the performance under CA and NY datasets. It should not be surprising that in Fig. 7(a), when k is less than 8, the execution time for our PIR-kNN approach is significantly shorter under CA dataset than under NY dataset. As we have mentioned, this is because the initial locating for the query point tends to fetch more neighboring voronoi cells in NY dataset. So when k is larger, the latter needs to fetch more extra distance information than the former to determine the kNN query processing.

Under synthetic dataset, we can see from Fig. 7(b) that when f increases, the maximal number of PIR accesses for one voronoi cell increases, the overall execution time increases as well. This is intuitive because our query plan should cover the worse case in terms of the PIR accesses.

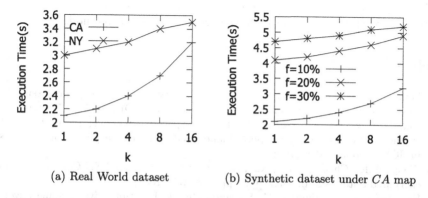

(a) Real World dataset   (b) Synthetic dataset under CA map

Fig. 7. Effect of data distribution.

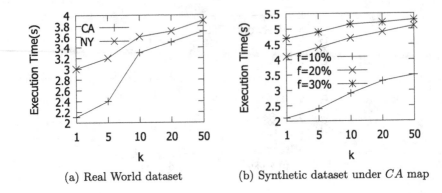

(a) Real World dataset   (b) Synthetic dataset under CA map

Fig. 8. Effect of scalability.

Evaluation of Scalability. Finally, we evaluate the scalability of our approach. Regardless of whether real world data or synthetic dataset, we can see from Fig. 8 that the execution time increases linearly to the query parameter k. Further, we can see that the increasing rate of the execution time gets slower as k gets larger than 10. This is because the voronoi cells that contain much more POIs or edges than normal voronoi cells have been considered by a smaller threshold k and when k goes beyond this threshold, the number of PIR accesses required for increased k is much less.

6   Conclusion

In this paper we introduce the novel problem of PIR-based kNN query on road networks with strong privacy guarantee, where an adversary cannot distinguish a kNN query from any other query in the network space. This is the first work that applies PIR to network kNN query. Further, we design the data structure to fetch only necessary data and deduce a query plan for arbitrary query parameter k. Finally, we evaluate our method on real world dataset and synthetic dataset. Extensive experiments demonstrate the practicality of our method.

References

1. Mokbel, M.F., Chow, C.Y., Aref, W.G.: The new casper: query processing for location services without compromising privacy. In: VLDB, pp. 763–774 (2006)
2. Yiu, M.L., Jensen, C., Huang, X., Lu, H.: Spacetwist: managing the trade-offs among location privacy, query performance, and query accuracy in mobile systems. In: ICDE, pp. 366–375 (2008)
3. Wong, W.K., Cheung, D.Q., Kao, B., Manoulis, N.: Secure kNN Computation on encrypted databases. In: SIGMOD, pp. 139–152 (2009)
4. Khoshgozaran, A., Shahabi, C., Shirani-Mehr, H.: Location privacy: going beyond k-anonymity, cloaking and anonymizers. KAIS **26**(1), 435–465 (2011)

5. Jung-Ho, U., Yong-Ki, K., Hyun-Jo, L., Miyoung, J., Jae-Woo, C.: K nearest neighbor query processing algorithm for cloaking regions towards user privacy protection in location-based services. J. Syst. Archit. EUROMICRO J. **58**(9), 354–371 (2012)
6. Kushilevitz, E., Ostrovsky, R.: Replication is NOT needed: SINGLE database, computationally-private information retrieval. In: FOCS, pp. 364–373 (1997)
7. Papadopoulos, S., Bakiras, S., Papadias, D.: Nearest neighbor search with strong location privacy. Proc. VLDB **3**, 619–629 (2010)
8. Ghinita, G., Kalnis, P., Khoshgozaran, A., Shahabi, C., Tan, K.-L.: Private queries in location based services: anonymizers are not necessary. In: SIGMOD (2008)
9. Mouratidis, K., Yiu, M.L.: Shortest path computation with no information leakage. PVLDB **5**(1), 692–703 (2012)
10. Beimel, A., Ishai, Y., Kushilevitz, E., Raymond, J.-F.: Breaking the O(n1/(2k-1)) barrier for information-theoretic private information retrieval. In: Proceedings of FOCS, pp. 261–270 (2002)
11. Chor, B., Kushilevitz, E., Goldreich, O., Sudan, M.: Private information retrieval. JACM **45**(6), 965–981 (1998)
12. Gentry, C., Ramzan, Z.: Single-database private information retrieval with constant communication rate. In: Caires, L., Italiano, G.F., Monteiro, L., Palamidessi, C., Yung, M. (eds.) ICALP 2005. LNCS, vol. 3580, pp. 803–815. Springer, Heidelberg (2005)
13. Wang, L., Meng, X., Hu, H., Xu, J.: Bichromatic reverse nearest neighbor query without information leakage. In: Renz, M., Shahabi, C., Zhou, X., Cheema, M.A. (eds.) DASFAA 2015. LNCS, vol. 9049, pp. 609–624. Springer, Heidelberg (2015)
14. Williams, P., Sion, R.: Usable PIR. In: NDSS (2008)
15. Naor, M., Pinkas, B.: Oblivious transfer with adaptive queries. In: Wiener, M. (ed.) CRYPTO 1999. LNCS, vol. 1666, pp. 573–590. Springer, Heidelberg (1999)
16. Paillier, P.: Public-key cryptosystems based on composite degree residuosity classes. In: Stern, J. (ed.) EUROCRYPT 1999. LNCS, vol. 1592, pp. 223–238. Springer, Heidelberg (1999)
17. Paulet, R., Kaosar, Md.G., Yi, X., Bertino, E.: Privacy-preserving and content-protecting location based queries. In: ICDE, pp. 44–53 (2012)
18. Yi, X., Paulet, R., Bertino, E., Varadharajan, V.: Practical k nearest neighbor queries with location privacy. In: ICDE, pp. 640–651 (2014)
19. Dijkstra, E.W.: A note on two problems in connexion with graphs. Numer. Math. **1**(1), 269–271 (1959)
20. Goldberg, A.V., Harrelson, C.: Computing the shortest path: A* search meets graph theory. In: SODA, pp. 156–165 (2005)
21. Lee, K.C.K., Lee, W.-C., Zheng, B.: Fast object search on road networks. In: EDBT, pp. 1018–1029 (2009)
22. Kolahdouzan, M., Shahabi, C.: Voronoi-based K nearest neighbor search for spatial network databases. In: VLDB, pp. 840–851 (2004)
23. Hu, H., Lee, D.-L., Xu, J.: Fast nearest neighbor search on road networks. In: Ioannidis, Y., Scholl, M.H., Schmidt, J.W., Matthes, F., Hatzopoulos, M., Böhm, K., Kemper, A., Grust, T., Böhm, C. (eds.) EDBT 2006. LNCS, vol. 3896, pp. 186–203. Springer, Heidelberg (2006)
24. Safar, M.: K nearest neighbor search in navigation systems. Mob. Inf. Syst. **1**(3), 207–224 (2005)
25. Narayanan, A., Shmatikov, V.: Robust De-anonymization of large sparse datasets. In: IEEE Symposium on Security and Privacy, pp 111–125 (2008)
26. Jensen, C., Kolarvr, J., Pedersen, T.B., Timko, I.: Nearest neighbor queries in road networks. In: GIS, pp. 1–8 (2003)

Similarity Search over Personal Process Description Graph

Jing Ouyang Hsu[✉], Hye-young Paik,
and Liming Zhan

The University of New South Wales, Sydney, NSW, Australia
{jxux494,hpaik,zhanl}@cse.unsw.edu.au

Abstract. People are involved in various processes in their daily lives, such as cooking a dish, applying for a job or opening a bank account. With the advent of easy-to-use Web-based sharing platforms, many of these processes are shared as step-by-step instructions (e.g., "how-to guides" in eHow and wikiHow) on-line in natural language form. We refer to them as *personal process descriptions.* In our early work, we proposed a graph-based model named Personal Process Description Graph (PPDG) to concretely represent and query the personal process descriptions. However, in practice, it is difficult to find identical personal processes or fragments for a given query due to the free-text nature of personal process descriptions. Therefore, in this paper, we propose an idea of similarity search over the "how-to guides" based on PPDG. We introduce the concept of "similar personal processes" which defines the similarity between two PPDGs by utilizing the features of both PPDG nodes and structure. Efficient and effective algorithms to process similarity search over PPDGs are developed with novel pruning techniques following a filtering-refinement framework. We present a comprehensive experimental study over both real and synthetic datasets to demonstrate the efficiency and scalability of our techniques.

Keywords: How-to guides · Personal process description graphs · Similarity search

1 Introduction

People are engaged in all kinds of *processes* all the time, such as cooking a dish, applying for a bank account or filing a tax return. Although the expertise in the area of workflow management and business process management (BPM) [4] has produced solutions for modelling, automating and managing much of the business and organizational workflows, still significant portion of the processes that people experience daily exist outside the realm of these technologies.

However, with the advent of easy-to-use Web-based sharing platforms, people often share their experience/knowledge about a process on the Web, in the form of how-to guides or step-by-step instructions. Although these are primarily

© Springer International Publishing Switzerland 2015
J. Wang et al. (Eds.): WISE 2015, Part I, LNCS 9418, pp. 522–538, 2015.
DOI: 10.1007/978-3-319-26190-4_35

describing a process, without the modelling expertise, they are normally written in natural language as a sequence of sentences forming the process steps. In order to distinguish these texts from the conventional business workflow models, we refer to them as *personal process descriptions*. A plethora of examples on personal process descriptions can be found in cooking recipes, how-to guides or Q&A forums.

The texts in natural language format are not precise enough to be useful in utilizing the process information presented in them. For example, the state-of-the-art for search technologies over the existing personal process descriptions are still a keyword/phrase-based search and users would have to manually investigate the results. A *process-aware* technique should be able to (i) produce the overall activity structure, (ii) show the dependencies between data and actions, and (iii) compare and contrast different paths to accomplish the goal.

In our previous work [13], we proposed a simple structured query language designed to perform exact-match search over the personal process descriptions. The language is supported by a graph-based, light-weight process model called PPDG (Personal Process Description Graph) which concretely represents the personal process description texts.

In this paper, we extend our query technique to return *similar* process descriptions to a query input. This technique is particularly important because the PPDGs are obtained from texts, and being able to cope with semantic similarity in words as well as the similarity in the structure of the activities in a process can return more relevant results. Our contributions are summarized as follows:

- We formally define the *similarity over PPDG*. PPDG is a graph-based process description language that presents both control-flow and data-flow in a personal process description.
- We propose effective and efficient algorithms to perform similarity search over PPDGs. The similarity algorithms deal with personal processes from different perspectives in that, unlike existing approaches, both control-flow and data-flow are considered.
- We further improve the performance of the algorithms by utilizing three novel pruning techniques.

The paper is organized as follows: Sects. 2 and 3 define the preliminary concepts and the problems. Section 4 describes the efficient algorithms to process the PPDG similarity search. Then we present the experiment results in Sect. 5. The related work is discussed in Sect. 6 followed by a conclusion in Sect. 7.

2   Preliminaries

In this section, we briefly introduce the personal process description graph (PPDG) and querying PPDGs. The full descriptions of these concepts are presented in [13].

2.1 Personal Process Description Graph

A PPDG represents a personal process description as a labeled directed graph. It describes the whole process of performing a personal process placing *equal emphasis* on both actions and input/output data relating to each action. Figure 1 depicts a PPDG of a PhD admission process experienced by an international applicant. Actions can be a one-off action, repeated action, or duration action. Data elements are represented by hexagonal nodes. A data element can be either basic or composite (i.e., composition of basic data). In order to make the visualization of the graph simple, all types of actions are represented by using the same notation. The details are stored in the schema associated with each PPDG. The same principle applies to the data.

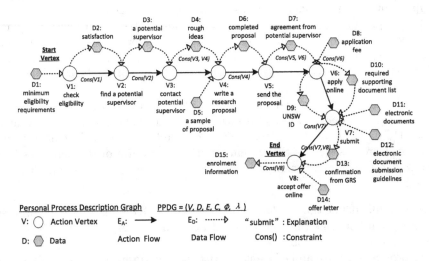

Fig. 1. How to apply for PhD admission at UNSW by an international applicant

The data elements and actions are connected to form 'action flow' and 'data flow'. Action flows, represented by solid lines, describe temporal sequence of the actions. For example, in Fig. 1, 'V_1: *check eligibility*' takes place before 'V_2: *find a potential supervisor*'. Data flows, represented by dotted lines, keep track of data sources and denote the relationships between the data and actions. For example, 'V_4: *write a research proposal*' takes two data inputs 'D_4: *rough idea*' and 'D_5: *a sample of proposal*' and produces one data output 'D_6: *completed proposal*'.

A PPDG also stores constraints/conditions relating to an action, data or the flows. For example, a condition may specify a location or the time an action takes place. We define PPDG more formally as follows.

Definition 1. *A personal process description graph PPDG is a tuple* $PPDG := (A, D, E_A, E_D, C, \phi, \lambda)$ *where:*

- *A is a finite set of nodes a_0, a_1, a_2,... depicting the starting action (a_0) and actions (a_1, a_2, ...).*

- D is a finite set of nodes d_0, d_1, d_2,... depicting the data input/output of an action.
- E_A is a finite set of directed action-flow edges ea_1, ea_2,..., where $ea_i = (a_j, a_k)$ leading from a_j to a_k ($a_j \neq a_k$) is an action-flow dependency. It reads a_j takes place before a_k. Each node can only be the source/target of at most one action-flow edge: $ea = (a_i, a_j) \in E_A : ea' = (a_k, a_l) \in E_A \setminus ea : a_i \neq a_k$ and $a_j \neq a_l$.
- E_D is a finite set of directed data-flow edges ed_1, ed_2,..., where $ed_i = (a_j, d_k)$ leading from a_j to d_k is a data-flow dependency. It reads a_j produces d_k. $ed_l = (d_m, a_n)$ leading from d_m to a_n is a data-flow dependency. It reads a_n takes d_m.
- C is a finite set of conditions $c_1, c_2,...$ with $c_i = (< name, descr >, x_j)$ being associated to $x_j \in \{A, D, E_A, E_D\}$ and having name and description of the condition.
- ϕ: a function that maps Action Label to action nodes.
- λ: a function that maps Data Label to data nodes.

In this work, we have not yet considered constraints in PPDG. For simplicity, we remove C (conditions) from PPDG here on.

Definition 2 (PPDG Query Graph). A PPDG query graph is a tuple $PPDG\text{-}Q = (QA, QD, QE_A, QE_D, Q_\phi, Q_\lambda, QP, \Delta)$ where:

- QA is a finite set of action nodes in a query.
- QD is a finite set of data nodes in a query.
- $QE_A \subseteq QA \times QA$ is the action flow relation between action nodes in a query.
- $QE_D \subseteq QA \times QD$ is the data flow relation between action nodes and data nodes in a query.
- Q_ϕ: a function that maps Action Label to action nodes.
- Q_λ: a function that maps Data Label to data nodes.
- QP is the path relation between action nodes which includes data nodes and data edges corresponding to each action node in query.
- Δ: $QP \rightarrow \{true, false\}$

Figure 2 shows an example of PPDG query input and output. A PPDG query graph consisting of "collect dress" action, immediately followed by "take photos" action with input data "dress", followed by a path query edge leading to "attend ceremony" action. The symbol "‖" is used to represent a path query between action nodes. In the above example, the path will match any action as well as connected data nodes from "take photos" to "attend ceremony". The subgraph inside dotted box in Fig. 2 is the result of the query.

However, in practice, it is difficult to find out the identical personal processes by a given query graph due to natural property of personal process. The similarity search technique proposed in this paper will see the query results expanded to include subgraphs that are similar to the query graph.

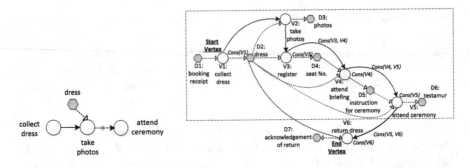

Fig. 2. An example of PPDG query input and output (an exact match)

2.2 Constructing PPDG and PPDG Repository

The PPDGs are constructed via *PPDG Builder*, a prototype system is implemented as part of a companion project which aims to establish a Web-based repository of PPDGs. The system relies on the Stanford's NLP parser and POS (Part-Of-Speech) tagger[1] to extract potential pairs of action and data (e.g., {book, academic dress}, {pay, cash}). We then build a smart *error correction and management* layer by using the well-known knowledge acquisition technique, RDR (Ripple Down Rules). The rules are incrementally built and managed to correct the extraction errors from the NLP parser and tagger outputs. Based on this, we are currently developing a graphical tool, *PPDG Editor*. The editor provides the graphical notations for the syntax elements in PPDG. For each personal process description, the users will see suggested action/data listing from the PPDG Builder. Although we do not assume that the PPDG construction process can be totally automated, our aim is to automate the mapping of the suggested action/data to PPDG nodes and labels as much as possible over time.

3 Problem Definition

The query techniques proposed in our earlier work implemented an exact match between a query graph and PPDGs. In this paper, we investigate the problem of similarity search over PPDGs. This is an important extension to PPDG queries because the construction process of PPDGs from the text-based sources generates nodes and labels that are semantically 'similar', but appear different. In similarity search, for a given PPDG query Q, our system returns similar PPDGs ranked by similarity scores from the PPDG repository. Without loss of generality, we assume the size of Q is not larger than that of any PPDG $P \in \mathcal{P}$. We define the similarity between Q and P, denoted as $Sim(Q, P)$, in this section.

To get the similarity score $Sim(Q, P)$, we separately consider the nodes (both action and data), and the directed edges linking the nodes[2]. Hence, we compute

---

[1] http://nlp.stanford.edu/software/, Stanford NLP Group.

[2] Throughout the paper, we sometimes refer to the directed edges to/from nodes as simply *graph structure*.

two separate similarity scores: one based on the labels of the nodes, the other on the edges from/to the nodes. We then combine the two to obtain $Sim(P, Q)$.

We note that, different from a PPDG, a PPDG query Q may include a *path* between two action nodes. If a *path* exists in Q, that is $\Delta = true$, we consider the *path* as a special action edge and Q as a special PPDG. Q with *path* is special case and we will discuss it in Sect. 4. For the sake of simplicity, from here on, let us treat Q as a PPDG. Now we formally define the similarity between two PPDGs P and P' as follows.

Graph Nodes Similarity. First, we consider the similarity between the nodes in P and P'. Since the label of each node n in a PPDG is composed by a set W of words, we can calculate the label similarity between two nodes by comparing the corresponding words sets.

Definition 3 (Label Similarity). *Given two nodes n_1, n_2, let ω be the function to separate the label of a node into a set of words[3], we get $W_1 = \omega(n_1)$ and $W_2 = \omega(n_2)$. Let $\mathcal{M}$ be a function that returns the number of exact matching words, $\mathcal{S}$ be a function that returns the number of synonymous words, we define the label similarity between the two nodes as follows:*

$$lSim(n_1, n_2) = \frac{2(\mathcal{M}(W_1, W_2) + \mathcal{S}(W_1, W_2))}{|W_1| + |W_2|} \tag{1}$$

Example 1. Considering the two nodes n_1, n_2, their labels are "seek a potential supervisor" and "search for a supervisor", we get the two word sets $W_1=\{$seek, potential, supervisor$\}$ and $W_2=\{$search, supervisor$\}$. Suppose we know "seek" and "search" are synonymous, then according to Eq. 1, the $lSim(n_1, n_2) = \frac{2(1+1)}{3+2} = 0.8$.

According to Definition 3, we can obtain all label similarity scores between the nodes of P and nodes of P'. For a node $n_P \in P$, we choose the best matching node $n'_P \in P'$, where the label similarity $lSim(n_P, n'_P)$ is the highest among all similarity scores between n_P and all nodes in P'. Then we are able to define the graph nodes similarity $nSim(P, P')$ as follows.

Definition 4 (Graph Nodes Similarity). *Given two PPDGs $P = (A, D, E_A, E_D, \phi, \lambda)$ and $P' = (A', D', E'_A, E'_D, \phi', \lambda')$ with $|A| \leq |A'|$, let a_i, d_j, a'_k, and d'_l represent the nodes of A, D, A', and D', respectively. Then the graph nodes similarity between P and P' is:*

$$nSim(P, P') = \frac{u \cdot \sum_{i=1}^{|A|} \max_{k=1}^{|A'|}(lSim(a_i, a'_k)) + v \cdot \sum_{j=1}^{|D|} \max_{l=1}^{|D'|}(lSim(d_j, d'_l))}{u \cdot |A| + v \cdot |D|} \tag{2}$$

where u and v are weights for action nodes and data nodes, and $u, v \in [0, 1]$.

---

[3] Note that the common auxiliary words, such as "a", "for" and "of", are not included.

The weight u and v are defined by users to indicate which type of node is more important for them (i.e., action or data).

Graph Structure Similarity. From the processing of graph nodes similarity, we obtain the similar nodes mapping of P and P'. Then we consider the similarity of the directed edges between the similar nodes of P and P'. Particularly, we compute the similarity of the edges from/to the action nodes separately from that of the data nodes. Hence, the similarity score is computed as follows:

Definition 5 (Graph Structure Similarity). *Given two PPDGs $P = (A, D, E_A, E_D, \phi, \lambda)$ and $P' = (A', D', E'_A, E'_D, \phi', \lambda')$ with $|A| \leq |A'|$, the similar nodes mapping of P and P' : $A \leftrightarrow A'$ and $D \leftrightarrow D'$, we get the edges matching $M_{E_A} : E_A \leftrightarrow E'_A$ and $M_{E_D} : E_D \leftrightarrow E'_D$. Then the structure similarity between P and P' is:*

$$sSim(P, P') = \frac{u \cdot |M_{E_A}| + v \cdot |M_{E_D}|}{u \cdot |E_A| + v \cdot |E_D|} \tag{3}$$

where u and v are weights of the two types of matchings.

Graph Similarity. Based on the two types of similarities between P and P', we utilize the *Harmonic Mean* to compute the graph similarity between P and P'.

Definition 6 (Graph Similarity). *Given two PPDGs P and P', we obtain their nodes similarity $nSim(P, P')$ and structure similarity $sSim(P, P')$. Then the graph similarity between P and P' is computed by the Harmonic Mean as:*

$$Sim(P, P') = \frac{2 \cdot nSim(P, P') \cdot sSim(P, P')}{nSim(P, P') + sSim(P, P')} \tag{4}$$

Problem Statement. Given a query graph Q, a set $\mathcal{P}$ of PPDGs, and a threshold η, a PPDG similarity search returns all PPDGs from the set $\mathcal{P}$, such that the similarity between Q and $P \in \mathcal{P}$ is no less than η, i.e. $Sim(Q, P) \geq \eta, P \in \mathcal{P}$.

4   Similarity Search

In this section, we present the efficient algorithms to process the PPDG similarity search. Given a PPDG query Q and a set $\mathcal{P}$ of PPDGs, the straightforward approach to find the similarity graphs to Q is to compute the graph similarity between Q and each $P \in \mathcal{P}$. Firstly, we join the node set of Q and P to get a set of node pairs - each pair $\{n_Q, n_P\}$ contains one node from Q and the other from P, and calculate the label similarity between the nodes in each pair by Eq. 1. Then for each n_Q, we find the *most similar node pair* with highest label similarity among all pairs containing n_Q. Next, after obtaining all the *most similar node pairs* between Q and P, we compute the nodes similarity $nSim(Q, P)$ by Eq. 2. Thirdly, based on the *most similar node pairs*, we find the matched edges of Q and P and compute the structure similarity $sSim(Q, P)$ by Eq. 3. Finally, we get the graph similarity $Sim(Q, P)$ by Eq. 4. The naive approach is costly because we need to iterate all the nodes and edges of one PPDG. Therefore, we propose efficient and effective techniques to process similarity search over PPDGs following a filtering-refinement framework. We first perform the nodes similarity to get a set C of candidate PPDGs, and then refine C by structure similarity to get the search results. Section 4.1 presents the filtering technique with efficient and effective pruning rules to get the candidate set. Section 4.2 refines the candidates by computing graph structure similarity.

4.1   Filtering by Graph Nodes

For the query Q and a PPDG P, the computing of graph nodes similarity $nSim(Q, P)$ is based on the label similarities between their nodes. In this subsection, we first propose the technique to compute label similarities between the nodes of Q and P. Then the graph nodes similarity $nSim(Q, P)$ is obtained according to the label similarities.

To compute the label similarities between the nodes of Q and P, we first decompose the label of each node into a set of words, and store the words set in corresponding node, i.e. n_Q. W for one node $n_Q \in Q$ and n_P. W for one node $n_P \in P$. Then we create a set of nodes pairs, each consisting of two nodes - one from Q and the other from P, denoted as $\{n_Q, n_P\}$. In each pair, n_Q and n_P are the same type of nodes. Recall that there are two types of node - action node and data node - existing in PPDG. For each pair, we can compute the label similarity score between the two nodes.

To identify whether two words are synonymous, we use the electronic lexical database - WordNet [5] to match words. We also build a local dictionary to store the synonymous words to save the processing time of our algorithm. In this paper, we assume the dictionary is already built locally and it contains all the synonymous words in all graphs.

Theorem 1. *Given two nodes n_1 and n_2, let W_1 and W_2 to be their words sets respectively, then the similarity between the two nodes is:*

$$lSim(n_1, n_2) \leq \frac{2 \cdot \min(|W_1|, |W_2|)}{|W_1| + |W_2|} \tag{5}$$

which is the upper bound of $lSim(n_1, n_2)$, denoted as $lSim(n_1, n_2)^{up}$.

Proof. According to Definition 3, if all words are matched, the similarity score is $\frac{2 \cdot \min(|W_1|, |W_2|)}{|W_1| + |W_2|}$. $\square$

Theorem 1 gives an upper bound of the label similarity between two nodes, and we can use the bound to prune some node pairs without calculating their label similarities.

Pruning rule 1. For a node $n_Q \in Q$, we want to find a *matching* node $n_P \in P$ with the highest label similarity score. Once we get the label similarity $lSim(n_Q, n_{Pi})$ for one node pair $\{n_Q, n_{Pi}\}$, then any node pair $\{n_Q, n_{Pj}\}$ with $lSim(n_Q, n_{Pj})^{up} < lSim(n_Q, n_{Pi})$ can be pruned safely. Therefore, we perform the nodes similarity between Q and P as following steps. (1) we first extract words from the label of each node in Q and P. (2) For each node $n_Q \in Q$, we join it with all nodes in P that are of the same type as n_Q, to create a set of node pairs. Recall that there are two types of node - action node and data node - in one PPDG. (3) All pairs are sorted by the upper bound of their label similarities in descending order, and stored in a max heap H. (4) We compute the label similarity of node pairs from the top of H one bye one, and the pair with highest label similarity score is kept in a tuple R. If the upper bound of the label similarity of the pair in the top of H is smaller than the label similarity in R, the remaining pairs in H can be pruned safely and R is the *most similar node pair* from P to n_Q. (5) After all *most similar node pairs* are obtained, the nodes similarity between Q and P can be calculated by Definition 4.

Theorem 2. *Given a PPDG query Q, a PPDG P, and a threshold η, to ensure $Sim(Q, P) \geq \eta$, the nodes similarity $nSim(Q, P)$ is:*

$$nSim(Q, P) \geq \frac{\eta}{2 - \eta} \tag{6}$$

which is the threshold of the nodes similarity, denoted as τ.

Proof. According to Definition 6, we assume the structure similarity between Q and P is equal to 1, i.e. $sSim(Q, P) = 1$, then we have $nSim(Q, P) \geq \frac{\eta}{2-\eta}$. □

Pruning rule 2. When the nodes similarity between Q and P is calculated, we can compare it with threshold τ, obtained by Theorem 2. It is clear that if $nSim(Q, P) < \tau$, the graph P can be pruned safely. Note that the computing of nodes similarity is based on the label similarities of all *most similar node pairs*. Therefore, after getting the label similarity of each *most similar node pair*, we can compute the upper bound $nSim(Q, P)^{up}$ of the nodes similarity between Q and P by assuming the label similarities of remaining unprocessed nodes are equal to 1. If $nSim(Q, P)^{up} < \tau$, P can be pruned without further processing. Finally, all un-pruned graphs are stored in a candidate set $\mathcal{C}$ for the refinement step.

Algorithm 1 illustrates the details of the filtering step. To enable computing the nodes similarity in an iterative fashion, we use a tuple T to process the query. T is employed to maintain a node pair with the upper bound of its label similarity. Particularly, $T.pair$ stores a node pair, one node from the query Q and the other from a PPDG P, and $T.lsimUp$ stores the upper bound of the label similarity of the pair. We first initiate τ by Theorem 2 in Line 1 and store all the nodes of Q in a set N_Q in Line 2. Each node n_Q has a word set attribute n_Q. W to store its words extracted from its label. Next, we iterate the graphs $P \in \mathcal{P}$ one by one to compute the nodes similarity $nSim(Q, P)$ from Line 3 to Line 31. To begin with, a pair set $pairSet$ is initiated to store all the *most similar node pairs* between Q and P. For each P, we first assume all action nodes and data nodes in Q exactly match the nodes in P, and let M_A, M_D equal to the number of action nodes and data nodes in Q, respectively, in Lines 4. Then we also get

---

Algorithm 1. *Filtering by Nodes Similarity* $(Q, \mathcal{P}, \eta, u, v)$

Input : Query Q, A set $\mathcal{P}$ of PPDGs, Threshold η, Weight u, v
Output: Candidate set $\mathcal{C}$

1 $\mathcal{C} := NULL$, $\tau := \eta/(2 - \eta)$;
2 $N_Q \leftarrow$ All nodes in Q, $n_Q.W$ stores words extracted from the label of n_Q;
3 **for** *each* $P \in \mathcal{P}$ **do**
4 $\quad$ $pairSet := NULL$, $M_A := |Q.A|$, $M_D := |Q.D|$, $pruned := false$;
5 $\quad$ $N_P \leftarrow$ All nodes in P, $n_P.W$ stores words extracted from the label of n_P;
6 $\quad$ **for** *each* $n_Q \in N_Q$ **do**
7 $\quad\quad$ $H := NULL$;
8 $\quad\quad$ **for** *each* $n_P \in N_P$ *and* n_Q, n_P *are the same type of nodes* **do**
9 $\quad\quad\quad$ $T.pair := \{n_Q, n_P\}$;
10 $\quad\quad\quad$ $T.lsimUp := lSim(n_Q, n_P)^{u_P}$;
11 $\quad\quad\quad$ $H.push(T)$;
12 $\quad\quad$ $R.lsim := 0$;
13 $\quad\quad$ **while** $H \neq NULL$ **do**
14 $\quad\quad\quad$ $T := H.pop()$;
15 $\quad\quad\quad$ **if** $T.lsimUp < R.lsim$ **then**   // Pruning rule 1
16 $\quad\quad\quad\quad$ break;
17 $\quad\quad\quad$ **else if** $lSim(T.pair) > R.lsim$ **then**
18 $\quad\quad\quad\quad$ $R.pair := T.pair$;
19 $\quad\quad\quad\quad$ $R.lsim := lSim(R.pair)$;
20 $\quad\quad$ **if** *the nodes in* R *are action nodes* **then**
21 $\quad\quad\quad$ $M_A := M_A - 1 + R.lsim$;
22 $\quad\quad$ **else**
23 $\quad\quad\quad$ $M_D := M_D - 1 + R.lsim$;
24 $\quad\quad$ $nSimUp := (u * M_A + v * M_D)/(u * |A| + v * |D|)$;
25 $\quad\quad$ **if** $nSimUp < \tau$ **then**   // Pruning rule 2
26 $\quad\quad\quad$ $pruned := true$;
27 $\quad\quad\quad$ break;
28 $\quad\quad$ $pairSet \leftarrow R$;
29 $\quad$ **if** $!pruned$ **then**
30 $\quad\quad$ $P.pairSet := pairSet$; $P.nsim := nSim(Q, P)$;
31 $\quad\quad$ $\mathcal{C} \leftarrow P$;

32 **return** $\mathcal{C}$;

---

all the nodes of P with their words and store them in a set N_P in Line 5. From Line 6 to Line 11, we join each node $n_Q \in N_Q$ with all nodes in N_P to get node pairs and compute the upper bound of their label similarities, which are stored in $T.pair$ and $T.lsimUp$, respectively. We use a max heap H to store T. The pairs are sorted in H by $T.lsimUp$ in descending order. Line 12 initializes a tuple R to store the node pair which has the highest label similarity score after all pairs in H are iterated. From Line 13 to Line 19, we compute the label similarity of the pairs from the top of H and store the pair with the highest label similarity score in R. According to the Theorem 1, if the upper bound of the label similarity $T.lsimUp$ of a pair $T.pair$ is smaller than $R.lsim$, the iteration is stopped in Line 13, and R contains the *most similar node pair* with its label similarity. After we get one *most similar node pair* $R.pair$, M_A or M_D is updated based on its label similarity $R.lsim$ and the upper bound $nSimUp$ of the nodes similarity is computed in Lines 20–24. If $nSimUp$ is smaller than τ, P can be pruned (Line 25). All *most similar node pairs* are stored in the set *pairSet* (Line 28). If P is not pruned, after getting all the *most similar node pairs* between Q and P, the nodes similarity $nSim$ is computed by Eq. 2 and P is put into the candidate set C in Lines 29–31.

4.2   Refinement by Graph Structure

After processing Q with all PPDGs by Algorithm 1, we obtain a candidate set C of PPDGs. In this subsection, we utilize the graph structure to refine the candidates.

In the filtering step, we obtain the *most similar node pairs* stored in $P.pairSet$ for each candidate $P \in C$. Considering two pairs $\{n_Q, n_P\}$ and $\{n'_Q, n'_P\}$ in $P.pairSet$, there is an edge e_Q between n_Q and n'_Q. If there is also an edge e_P between n_P and n'_P, and e_Q and e_P are the same type of edges with the same direction, we can determine e_Q and e_P are matched. The straightforward approach is to check whether any two pairs in $P.pairSet$ have matched edges. However, the processing cost is very high, because we need check every two pairs. Even if there is no edge between n_Q and n'_Q, we still need to check whether the edge exists on Q or not.

We can use the graph structure to reduce the processing time. When we pick a pair $\{n_Q, n_P\}$ from $P.pairSet$, we get all the connected nodes of n_Q from graph Q. Next, for each of connected node n'_Q, we get pair $\{n'_Q, n'_P\}$ from $P.pairSet$. If there is an edge e_P between n_P and n'_P, and e_Q and e_P are the same type of edge with the same direction, we say the two edges are matched. Particularly, after one pair is iterated, the pair is removed to avoid the reverse checking from $\{n'_Q, n'_P\}$.

Theorem 3. *Given a PPDG query Q, a PPDG P, and a threshold η, we get the nodes similarity $nSim(Q, P)$. To let graph similarity $Sim(Q, P) \geq \eta$, the structure similarity is:*

$$sSim(Q, P) \geq \frac{nSim(Q, P) \cdot \eta}{2 \cdot nSim(Q, P) - \eta} \tag{7}$$

which is the threshold of the structure similarity, denoted as ϵ.

Proof. According to Definition 6, it is clear that the inequality holds. $\square$

Algorithm 2. *Refinement by Structure Similarity* (Q, C, η, u, v)

Input : Query Q, Candidate set C, Threshold η, Weight u, v
Output: search result $\mathcal{R}$

1   $\mathcal{R} := NULL$;
2   **for** *each* $P \in C$ **do**
3     $pruned$:=false;
4     $\epsilon = (P.nsim * \eta)/(2 * P.nsim - \eta)$;
5   M_{EA}:=$|Q.E_A|$, M_{ED}:=$|Q.E_D|$;
6     $unvisited := P.pairSet$;
7   **for** *each* most similar node pair $\{n_Q, n_P\} \in P.pairSet$ **do**
8       $unvisited.remove(\{n_Q, n_P\})$;
9       $neighbors \leftarrow$ the connected nodes of n_Q with edge types;
10   **for** *each* $n'_Q \in neighbors$ **do**
11   Find $\{n'_Q, n'_P\}$ in $unvisited$;
12   **if** *there is no same type of edge between* n_P *and* n'_P **then**
13   **if** $n'_Q.E$ *is an action edge* **then**
14            $M_{EA} := M_{EA} - 1$;
15   **else**
16            $M_{ED} := M_{ED} - 1$;
17       $sSimUp := (u * M_{EA} + v * M_{ED})/(u * |Q.E_A| + v * |Q.E_D|)$;
18   **if** $sSimUp < \epsilon$ **then**   // Pruning rule 3
19         $pruned$:=true;
20   break;
21   **if** !$pruned$ **then**
22       $P.sim \leftarrow$ the graph similarity;
23       $\mathcal{R} \leftarrow P$;
24 **return** $\mathcal{R}$;

Pruning rule 3. From the filtering step, we gain the graph nodes similarity $nSim(Q, P)$ for each candidate $P \in C$. Then we compute the threshold ϵ of the graph structure similarity according to Theorem 3. Like pruning rule 2, we assume the edges of Q and P are all matched in the beginning, and obtain the upper bound of the structure similarity $sSim(Q, P)^{up}$. When we process the graph structure similarity search, if one edge of Q cannot match any edge of P, $sSim(Q, P)^{up}$ decreases due to the unmatched edges. If $sSim(Q, P)^{up}$ is less than ϵ, P can be pruned safely without further processing.

Algorithm 2 illustrates the details of the refinement step. We iterate all PPDGs in the candidate set C. For each PPDG P, we first compute the threshold ϵ of its structure similarity based on its nodes similarity $P.nsim$ and graph similarity threshold η in Line 4. Then, we assume all edges are matched and set M_{EA} and M_{ED} as the number of action edges and data edges of query Q, respectively, in Line 5. Line 6 initializes a pairs set $unvisited$ to be filled with $P.pairSet$. Once a pair is visited, it is removed from $unvisited$ to avoid reverse checking. From Line 7 to Line 20, we iterate all *most similar node pairs* to match the edges between Q and P. Line 9 finds all neighbors of the node $n_Q \in \{n_Q, n_P\}$ with their edges, and store them into *neighbors*. For each neighbor $n'_Q \in neighbors$, we search *unvisited* to get the corresponding pair $\{n'_Q, n'_P\}$ (Line 11), and then check the PPDG P to identify if there is a same type of edge between n_P and n'_P and update M_{EA} and M_{ED} based on the result (Line 12–16). Next, we compute the upper bound of the graph structure similarity from the updated M_{EA} and M_{ED} and prune P if the upper bound is smaller than ϵ (Lines 17–20). Finally, if P is un-pruned, we compute its similarity score and put it in result set $\mathcal{R}$ (Line 22 and 23).

When the query Q has *path* edges, we consider it as a special PPDG. To process it, we make a little adaption on Algorithm 2 to refine the candidates. If there is a *path* from n_Q to n'_Q in Q, we obtain the pairs $\{n_Q, n_P\}$ and $\{n'_Q, n'_P\}$. Then we traverse the PPDG P from n_P to check whether n'_P can be reached, and the result of *path* searching is passed to the judging condition in Line 12 of Algorithm 2 to check whether the two *paths* are matched.

5   Experiments

Now we present the results of a comprehensive performance study to evaluate the efficiency and scalability of our proposed techniques. Following algorithms are evaluated.

- **NAIVE:** Techniques in Sect. 4 but without any pruning rule.
- **P1:** Techniques in Sect. 4 but using Pruning Rule 1 only.
- **P12:** Techniques in Sect. 4 but using Pruning Rule 1 and 2 only.
- **SIM:** Techniques presented in Sect. 4 to process similarity search with all pruning techniques.

Datasets. We have evaluated our similarity search techniques on both synthetic and real datasets.

The synthetic datasets were generated by randomization techniques. We create a word set containing 100 words: 50 different words and 25 pairs of synonymous words. A dictionary is built to store the mapping of synonymous words. Then we randomly choose n action nodes and $[0, 2n]$ data nodes to assemble p process graph. For each node, w words are randomly selected to make the label. After a graph is built, we make several small changes, such as changing the labels of nodes and adding/deleting nodes, to obtain 99 similar process graph. The number p varies from 2 K to 50 K (default value = 10 K). The number n of action nodes in each process is randomly chosen in a range varying from $[5, 10]$ to $[35, 40]$ (default value = $[15, 20]$). The number of data nodes is randomly chosen in $[0, 2n]$. For each node, there are up to w words randomly to be selected. The w varies from 10 to 25 (default value = 15). By the default setting, the total number of nodes is up to 600 K in our experiment. The threshold η varies from 0.2 to 0.8 (default value = 0.6). We choose 100 process graphs and get their subgraphs to make 100 query graphs, which are used in the experiment. The size s of query, i.e. number of action nodes in query, varies from 3 to 9 (default value = 5). The average processing time of the 100 queries on each dataset represents the performance of our query processing mechanism. The weight u and v are set to 1 in all experiments.

The real dataset consists of 42 PPDGs about PhD programs collected from the Web, manually created by the authors. The dataset includes personal process descriptions on processes such as research degree admission, scholarship applications, and attending graduation ceremony. In this dataset, the queries are chosen manually.

All algorithms are implemented in C++ and compiled by Cygwin GCC 4.3.4. The experiments are conducted on a PC with Intel i7 2.80 GHz CPU and 8 G memory on Windows 7 Professional SP1. All algorithms are run in main memory.

Performance Evaluation

We evaluate the performance of the four algorithms (NAIVE, P1, P12, SIM) in the experiment.

Real vs Synthetic. In the first experiment, we evaluate the performance of NAIVE, P1, P2 and SIM over the real and synthetic data. Due to the limited quantity of real process graphs, we magnify the result on the real data by 200 times in Fig. 3. It is shown that our techniques give the similar pruning power on both datasets, and each pruning rule is very effective and reduces the processing time. Particularly, pruning rule 2 has the best performance among the three pruning rules.

Impact of Threshold η. We evaluate the processing time of our algorithms as a function of the threshold η which varies from 0.2 to 0.8. Figure 4 shows the cost of P12 and SIM are reduced significantly when η increases, because the two techniques utilize η to perform the pruning. Comparing with P12, the processing time of SIM does not improve a lot when η is 0.8. The reason is that pruning rule 2 prunes a mass of graphs and leaves a few candidate graphs for the refinement step.

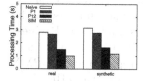

Fig. 3. Real vs Synthetic

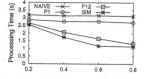

Fig. 4. Varying η

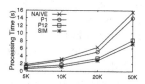

Fig. 5. Varying p

Evaluating Impacts by Different Setting. We study the scalability of our algorithms with regards to the different number of process graphs (p), number of words (w) in one graph node, number of action nodes (n), and the query size (s) in Figs. 5, 6, 7 and 8. The processing time increases with the increase of the

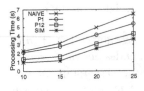
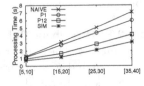
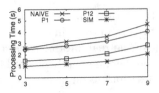

Fig. 6. Varying w **Fig. 7.** Varying n **Fig. 8.** Varying s

four parameters. However, the results also demonstrate that each pruning rule is effective and reduces the processing time in all settings. Clearly, the dataset size increases with the number of process graphs and action nodes thus the filtering and refinement processing becomes more expensive. Longer word size makes it difficult to process the label similarity match, which increases the processing time. When the query size grows, the processing cost increases because more nodes and edges are involved in the similarity search.

6   Related Work

One of the closely related work to ours is Cooking Graphs [11]. A cooking graph describes a cooking process with cooking actions and relevant ingredients information. However, cooking graphs are specialized to represent one domain and the action/data concepts are not as generic as PPDG. Also, the focus of the work is with implementing a graph mining technique to recognize cooking process patterns (by considering graph structures) and recommend a suitable cooking recipes for a user. Through PPDG and PPDG querying techniques, we aim to provide a platform to support various analysis tasks, not limited to recommendation. Besides, PPDG considers similarity matching in labels as well as the graph structures.

There are several work we can refer to in the area of BPM with regards to querying processes. In these work, queries are processed over BPMN (Business Process Modelling Notation) or equivalent notations. The main purpose of the languages is to extract actions (i.e., control flows). For example, The Business Process Query Language (BPQL) in [2] works on an abstract representation of BPEL[4] files. The BPMN-Q is a visual language to query repositories of BPMN models [1,8]. It processes the queries by converting both the query and BPMN to graphs. PPDG describes personal processes directly as graph (although mapping from/to BPMN is possible) and uses a query paradigm which takes both actions and data nodes into consideration with their labels and directed edges.

Recently, BPMN label matching techniques for querying similar process models is presented in [7]. However, the authors only consider action node labels. Also, they do not consider the directed edges (i.e., graph structure) present in the process models and simply refer to a process model as a set of activities.

---

[4] http://docs.oasis-open.org/wsbpel/2.0/OS/wsbpel-v2.0-OS.html.

In [3], the authors deal with the problem of retrieving process models in the repository that most closely resemble a given process model. This paper proposes three similarity metrics: (i) label matching similarity - it measures similarity based on words in the labels of business process model elements, (ii) structural similarity - it measures similarity based on graph edit distance of business process model, (iii) behavior similarity - it measures similarity based on the intended behavior (operation semantic) of process models. At present, authors have focused on developing the metrics rather than efficient implementation of algorithms.

Besides the BPM area, more general applications of graph similarity search have received considerable attention, such as Closure-Tree [6], K-AT [10], and SEGOS [12]. Specially, subgraph similarity search is to retrieve the data graphs that approximately contain the query. Grafil [14] proposes the problem, where similarity is defined as the number of missing edges regarding maximum common subgraph. GrafD-index [9] deals with similarity based on maximum connected common subgraph. [15] studies the problem of graph similarity search with edit distance constraints. However, they only consider the similarity of graph structure. Therefore, techniques proposed in [9,14,15] cannot be directly applied to PPDGs.

7   Conclusion

In this paper, we have investigated similarity search over Personal Process Description Graph (PPDG). We formally define the similarity between two PPDGs as a harmonic mean of two similarity scores: graph nodes similarity and graph structure similarity. By utilizing the features of PPDG nodes and structure, we develop effective and efficient algorithms with novel pruning techniques following the *filtering and refinement paradigm*. A comprehensive experimental study over both real and synthetic datasets demonstrates the efficiency and scalability of our techniques.

References

1. Awad, A., Sakr, S., Kunze, M., Weske, M.: Design by selection: a reuse-based approach for business process modeling. In: ER, pp. 332–345 (2011)
2. Beeri, C., Eyal, A., Kamenkovich, S., Milo, T.: Querying business processes. In: PVLDB, pp. 343–354 (2006)
3. Dijkman, R., Dumas, M., van Dongen, B.F., Käärik, R., Mendling, J.: Similarity of business process models: metrics and evaluation. Inf. Syst. **36**(2), 498–516 (2011)
4. Dumas, M., La Rosa, M., Mendling, J., Reijers, H.: Fundamentals of Business Process Management. Springer-Verlag, Berlin Heidelberg (2013)
5. Fellbaum, C.: WordNet: An Electronic Lexical Database. Language, Speech, and Communication. MIT Press, Cambridge (1998)
6. He, H., Singh, A.K.: Closure-tree: an index structure for graph queries. In: ICDE, p. 38 (2006)

7. Klinkmüller, C., Weber, I., Mendling, J., Leopold, H., Ludwig, A.: Increasing recall of process model matching by improved activity label matching. In: Daniel, F., Wang, J., Weber, B. (eds.) BPM 2013. LNCS, vol. 8094, pp. 211–218. Springer, Heidelberg (2013)

8. Sakr, S., Awad, A.: A framework for querying graph-based business process models. In: WWW, pp. 1297–1300 (2010)

9. Shang, H., Lin, X., Zhang, Y., Yu, J.X., Wang, W.: Connected substructure similarity search. In: SIGMOD, pp. 903–914 (2010)

10. Wang, G., Wang, B., Yang, X., Yu, G.: Efficiently indexing large sparse graphs for similarity search. IEEE Trans. Knowl. Data Eng. 24(3), 440–451 (2012)

11. Wang, L.: CookRecipe: towards a versatile and fully-fledged recipe analysis and learning system. Ph.D thesis, City University of Hong Kong (2008)

12. Wang, X., Ding, X., Tung, A.K.H., Ying, S., Jin, H.: An efficient graph indexing method. In: ICDE, pp. 210–221 (2012)

13. Xu, J., Paik, H., Ngu, A.H.H., Zhan, L.: Personal process description graph for describing and querying personal processes. In: ADC (2015)

14. Yan, X., Yu, P.S., Han, J.: Substructure similarity search in graph databases. In: SIGMOD, pp. 766–777 (2005)

15. Zhao, X., Xiao, C., Lin, X., Wang, W., Ishikawa, Y.: Efficient processing of graph similarity queries with edit distance constraints. VLDB J. 22(6), 727–752 (2013)

A Multilingual Approach to Discover Cross-Language Links in Wikipedia

Nacéra Bennacer, Mia Johnson Vioulès, Maximiliano Ariel López,
and Gianluca Quercini[✉]

Laboratoire de Recherche en Informatique (LRI), CentraleSupélec,
University of Paris-Saclay, 91405 Orsay Cedex, France
nacera.bennacer@lri.fr,mia.vioules@student.ecp.fr,
maximiliano.lopez@student.ecp.fr,gianluca.quercini@lri.fr

Abstract. Wikipedia is a well-known public and collaborative encyclopaedia consisting of millions of articles. Initially in English, the popular website has grown to include versions in over 288 languages. These versions and their articles are interconnected via cross-language links, which not only facilitate navigation and understanding of concepts in multiple languages, but have been used in natural language processing applications, developments in linked open data, and expansion of minor Wikipedia language versions. These applications are the motivation for an automatic, robust, and accurate technique to identify cross-language links. In this paper, we present a multilingual approach called *EurekaCL* to automatically identify missing cross-language links in Wikipedia. More precisely, given a Wikipedia article (the source) *EurekaCL* uses the multilingual and semantic features of BabelNet 2.0 in order to efficiently identify a set of candidate articles in a target language that are likely to cover the same topic as the source. The Wikipedia graph structure is then exploited both to prune and to rank the candidates. Our evaluation carried out on 42,000 pairs of articles in eight language versions of Wikipedia shows that our candidate selection and pruning procedures allow an effective selection of candidates which significantly helps the determination of the correct article in the target language version.

1 Introduction

Over the last 14 years, Wikipedia has grown to become the largest online encyclopaedia to date and one of the most popular websites worldwide. Its more than four million articles in English alone describe a wide range of topics, but the most interesting feature of this collaborative effort is that its vast amount of information is linked across more than 200 languages. For example, the article titled "Decision Theory" in the English Wikipedia has a cross-language link to the pages "Teoría de la decisión", "Teoria della decisione" and "决策论" in the Spanish, Italian, and Chinese versions of Wikipedia respectively. Over the past few years, attention has turned to this network and efforts have been made to build more robust and accurate links between Wikipedia versions. At its origin, cross-language links in Wikipedia were introduced to help users further

© Springer International Publishing Switzerland 2015
J. Wang et al. (Eds.): WISE 2015, Part I, LNCS 9418, pp. 539–553, 2015.
DOI: 10.1007/978-3-319-26190-4_36

their understanding by exploring concepts in multiple languages. However, as Wikipedia continues to grow, these links have been the means to achieve many other goals, including enriching linked open data platforms [2], introducing automatically new intra-language links in Wikipedia articles [10], and promoting cross-language information retrieval applications [1,3,9]. Typically, the cross-language links are manually added by authors of articles and are subject to being incomplete or erroneous due to the lack of an automatic verification tool. When the author of an article does not link to another Wikipedia language version for a specific reason, this is called a *missing cross-language link*. For instance, as of May 2015, the article titled "Caffettiera" in the Italian Wikipedia has a missing cross-link to the corresponding articles in French ("Cafetière") and Spanish ("Cafetera").

The goal of our paper is to find such articles with missing cross-language links and determine their appropriate respective articles in other languages, a challenging task which has already been addressed by other researchers [5,7,8]. Compared to existing approaches, we noticed a serious limitation in each algorithm at the stage of selecting the set of candidate articles in the desired target language (candidate set). Moreover, all these algorithms are language dependent. For instance, the algorithm defined in [8] needs to train a different SVM for each pair of languages under consideration. In [7] language features are used to determine the named entities in Wikipedia. In [5] language dependent textual features are also used to determine the number of common words belonging to two different languages.

In this paper, we present a multilingual approach called *Eureka Cross-Language Link (EurekaCL)* to automatically identify missing cross-language links in Wikipedia. *EurekaCL* works with a given article in a source language to identify the best possible corresponding articles in a target language. Moreover, *EurekaCL* is language independent in the sense that the algorithm needs no change or parameter tweak to be applied to any pair of language versions of Wikipedia. These languages can also be heterogeneous and based on different alphabets. For example we can search cross-links between Chinese and Greek articles.

Given a Wikipedia article (the source) *EurekaCL* uses the multilingual and semantic features of BabelNet 2.0 [6] in order to effectively identify a set of candidate articles in a target language that are likely to cover the same topic as the source. The resulting candidate set is further enhanced by exploiting the Wikipedia categories and the cross-language link paths between articles and is then reduced by an effective pruning procedure, which eliminates those candidates that are already connected via a path of cross-language links to other articles in the source Wikipedia. As a result, the size of the resulting candidate set is reasonably small while containing almost always the correct target article. Each candidate is assigned a score which reflects either the amount of paths of cross-links that connect it to the source or the size of their common neighbours. The score is then used to produce a ranked list of candidates.

Unlike existing approaches, which are usually tested on around 1,000 source articles, we evaluate *EurekaCL* on a dataset of 42,000 source articles. Moreover, the multilingual nature of *EurekaCL* allows us to evaluate it on a dataset including articles on multiple source languages and multiple target languages, while the other approaches only consider one source language and one target language at the same time. The results that we obtain show that *EurekaCL* performs very well.

The remainder of the paper is organized as follows: Sect. 2 reviews some previous approaches to discovering missing cross-language links. In Sect. 3 we outline the preliminary notations used to describe our algorithm, which is then detailed in Sect. 4. In Sect. 5, we evaluate *EurekaCL* and we conclude the discussion in Sect. 6.

2   Related Work

The challenge of finding missing cross-language links in Wikipedia has been addressed from many different angles since the first proposed solution by [8]. Among these different approaches, two main choices resonate: supervised versus unsupervised learning algorithms and graph-based versus text-based approaches. Overall, the comparison between the algorithms is difficult as each set of authors evaluated their algorithms with a different dataset, varying language versions of Wikipedia, and differing definitions of recall.

In [8], the authors opt for a classification-based approach with both graph-based and text-based features in order to find missing cross-language links between the German and English versions of Wikipedia. Their features rely heavily on the *chain link hypothesis*, which assumes that two equivalent (cross-language linked) Wikipedia articles are connected or should be connected via a chain of intra-language and inter-language links. The authors recognize the need to narrow down the candidate set for a source article and keep the top 1,000 candidate articles with the highest number of chain links to the source article. Then, they train their classifier using five graph-based features and two text-based features to predict whether or not a source article has a cross-language link to a candidate. They evaluate their classifier on a dataset of 1,000 source articles (RAND1000) and achieve a precision of 93.5 % and a recall of 69.9 %. As this approach relies on language features, it requires to train a classifier for each pair of languages. Furthermore, performing the morphological analysis on each article to generate the text-based features is very costly.

Another approach, *WikiCL*, is an unsupervised and graph-based algorithm [7]. Once again, *WikiCL* first reduces the number of possible candidates in the target language. To do so, they take a categorical approach to classify each article as representing: (i) a non-geographic named entity, (ii) a geographic named entity, or (iii) a non-named entity. The candidates for a source article in the target language should lie in the same category as the source article. The candidates are selected using graph-based features then ranked using the *semantic relatedness* which takes into account all links. The evaluation of *WikiCL* uses

English as the source language and find corresponding articles in Italian, French, and German (as target languages). The precision ranges from 89 % to 94 % and recall is between 89 % and 93 %.

CLLFinder is a supervised classification-based algorithm that uses a mix of four graph-based and text-based features [5]. The *CLLFinder* algorithm also uses the *chain link hypothesis* and adds further candidates in the target language that belong to the respective categories as the source article. Their experiments show that the size of the candidate set is too high. As in [8], the authors sort the articles by the number of times they appear in the set and reduce them to the 1,000 most frequent candidates. These candidates are then submitted to a classifier using graph-based and text-based features including cross-language links, title similarity (with translation) and text overlap (without translation). The cross-language link-based feature consists in looking for an intermediate language version article related via a cross-link both to the source and a candidate article. Their evaluation made on 1,000 articles in Portuguese for which their English counterparts are known (with French, Italian, and Spanish as intermediate languages) shows a precision between 85 % and 100 % and a recall between 97 % and 98 %. *CLLFinder* outperformed in precision and recall thanks to the cross-language link transitivity hypothesis.

As in the aforementioned approaches, our purpose is to determine, for a given article in a source language, the best possible corresponding article in a target language. However, *EurekaCL* does not require a training set and is both multilingual and language independent thanks to the use of the lexicographic and semantic knowledge of BabelNet. Moreover, unlike *CLLFinder*, *EurekaCL* completely exploits the transitivity nature of cross-link paths to enhance and prune the candidates and to determine the correct one.

3   Preliminaries

A language version α of Wikipedia is considered a directed graph W_α, where each node n_α represents a Wikipedia concept (article) in the language version α and has a set of categories $Cat(n_\alpha)$. An article n_α (e.g., the one titled "Paris" in the English Wikipedia) is usually connected to other related articles in the same language version (e.g., "Eiffel Tower", "Louvre") via *intra-language links*; an intra-language link between two articles n_α and m_α is denoted with $rl(n_\alpha, m_\alpha)$. Also, an article n_α (e.g., "Paris" in the English Wikipedia) can be connected to articles in other language versions covering the same topic (e.g., "Paris" and "Parigi" in the French and Italian Wikipedia respectively) via *cross-language links*; a cross-language link between the articles n_α and n_β is denoted with $cl(n_\alpha, n_\beta)$. There could be more than one path connecting two articles in different language versions. We denote with $path(n_\alpha, n_\beta)$ the set of paths connecting n_α to n_β considering cross-language links.

Theoretically, the *cl* links should be symmetric and transitive. Moreover, there should be at most one Wikipedia article in a given language version β that is directly linked to a given Wikipedia article n_α via a cross-language link. However, the reality is more complex than that, the cross-links between Wikipedia articles in different language versions could be inconsistent.

Fig. 1. Example of BabelNet synset and related synset senses

In addition, we can enrich the nodes of the graph W_α with additional information obtained from the multilingual semantic network BabelNet [6]. BabelNet results from the automatic integration of lexicographic and encyclopaedic knowledge from WordNet senses and Wikipedia pages. Each node n_α is described by a set of BabelNet *synsets*, each synset having one or more *senses*. The set $S(n_\alpha)$ represents all senses associated to n_α. Moreover, each synset has a set of related synsets. $R(n_\alpha)$ denotes all senses belonging to the related synsets for node n_α. We note that theses senses are multilingual.

Figure 1 shows an example of BabelNet synset senses we obtain for the French Wikipedia article *Lait*. In this case only one synset *milk* is returned by BabelNet with all its senses in multiple languages such as *leche de vaca* in Spanish (ES), and *leite de vaca* in Portuguese (PT). *evaporated milk, protein, chocolate milk, white, liquid, vitamin* are examples of related synsets and their multilingual senses. For the English Wikipedia article *Coffee percolator*, two synsets *percolator* and *coffee* are returned by BabelNet where *tea, maxwell, water* and *italy* are examples of related synsets.

4   *EurekaCL* Algorithm

EurekaCL is designed to optimize the candidate selection process for any language version pair. This selection exploits the extensive linguistic information found in BabelNet to retrieve the different senses associated to a Wikipedia article source and to reach the Wikipedia articles having similar senses in any language version. For greater efficiency, the selection is then followed by a pruning procedure to eliminate the wrong candidates by examining the available cross-link paths. Finally, a ranking procedure is applied to determine the best candidate for a Wikipedia source article. The ranking procedure also uses the cross-link paths to find the most probable target article in the candidate set.

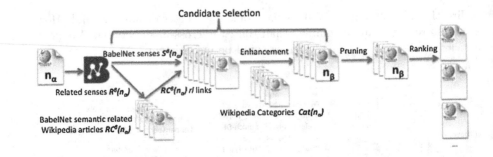

Fig. 2. *EurekaCL* algorithm

4.1   Candidate Selection

As shown in Fig. 2, *EurekaCL* builds the set of BabelNet synset senses $S(n_\alpha)$ for a given Wikipedia article source n_α. Then it selects the synset senses $S^\beta(n_\alpha)$ that are nouns and that represent Wikipedia articles in the target language version β to determine the set of candidate articles $C^\beta(n_\alpha)$.

If $C^\beta(n_\alpha) = \emptyset$, *EurekaCL* exploits the semantically related synset senses $R^\beta(n_\alpha)$ in the target language version β in order to retrieve the set $RC^\beta(n_\alpha)$ of Wikipedia articles semantically related to the source article n_α.

Based on the hypothesis that two Wikipedia articles covering the same topic share at least one semantically related Wikipedia article, *EurekaCL* adds to the candidate set $C^\beta(n_\alpha)$ all Wikipedia articles that have a symmetric intra-language link to the articles in $RC^\beta(n_\alpha)$. Formally:

$$\forall m_\beta \in RC^\beta(n_\alpha) \text{ if } \exists rl(m_\beta, n_\beta) \text{ and } \exists rl(n_\beta, m_\beta) \text{ then } n_\beta \in C^\beta(n_\alpha)$$

Enhancement. We define two main types of enhancements, one based on Wikipedia categories, and another based on the paths of cross-language links between the source article and the articles in the target language.

The first enhancement is meant to be an optimization of the candidate set, as it keeps only the candidates that share Wikipedia categories with the source article, based on the assumption that two Wikipedia articles covering the same topic share at least one Wikipedia category. Formally:

$$\forall c_\beta \in Cat(n_\beta), c_\alpha \in Cat(n_\alpha) \text{ if } \exists cl(c_\alpha, c_\beta), \text{ then } n_\beta \in C^\beta(n_\alpha)$$

We note that this enhancement is applied only if it does not reduce the set of candidates below a fixed size threshold t_{min} ($|C^\beta(n_\alpha)| > t_{min}$).

The second enhancement adds to the candidate set the articles of the target Wikipedia that are connected to the source article via a path of cross-links. Formally:

$$\text{if } \exists n_\beta \notin C^\beta(n_\alpha) \text{ and } path(n_\alpha, n_\beta) \neq \emptyset \text{ then } C^\beta(n_\alpha) \rightarrow C^\beta(n_\alpha) \cup \{n_\beta\}$$

Finally, *EurekaCL* adds to $C^\beta(n_\alpha)$ the Wikipedia article in language β whose title is the same as the title of source article n_α, if it exists. This will be useful for invariable named entities, whose name is the same across all latin-based alphabet language versions.

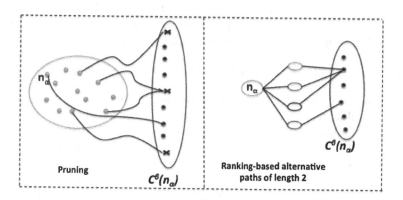

Fig. 3. Pruning and Ranking-based alternative paths illustration

4.2 Candidate Set Pruning

Once the candidate set is built, a pruning procedure is applied to remove wrong candidates. This pruning is based on the hypothesis that the cross-links are transitive. In other words, two Wikipedia articles connected by a cross-link path are likely to be about the same concept. Therefore, we define a wrong candidate as a Wikipedia article n_β not connected to the Wikipedia source article n_α but connected to another Wikipedia article m_α in language version α, as illustrated on the left of Fig. 3. More precisely:

$$\exists n_\beta \in C^\beta(n_\alpha), path(n_\alpha, n_\beta) = \emptyset \text{ and } \exists m_\alpha \neq n_\alpha, path(m_\alpha, n_\beta) \neq \emptyset \text{ then}$$
$$C^\beta(n_\alpha) \rightarrow C^\beta(n_\alpha) - \{n_\beta\}$$

To retrieve all the articles in the source language version α connected to a candidate n_β, *EurekaCL* goes through the *cl* paths using a depth first search. The complexity of this search for a candidate n_β is $\mathcal{O}(i+j)$ where i and j are respectively the number of nodes and links of the *cl*-connected component including the node n_β.

4.3 Candidate Set Ranking

EurekaCL uses successively two features to rank the pruned candidate set: the *alternative paths score* then the *neighborhood links score*. To begin, we exploit the transitivity assumption to rank the target candidates.

Alternative Paths Score. This feature is also based on the cross-link transitivity hypothesis. Unlike the pruning procedure where the transitivity is used to eliminate candidates, at this stage *EurekaCL* profits from the fact that a source Wikipedia article is connected to a candidate Wikipedia article by a cross-link path to determine the correct target article.

However, this could lead to a conflicting situation. Indeed, as explained in Sect. 3 an article in a language version α may be connected by cl paths to more than one Wikipedia article in the target language β such that:

$$\exists n_\beta \neq m_\beta \in C^\beta(n_\alpha), \text{ such that } path(n_\alpha, n_\beta) \neq \emptyset \text{ and } path(n_\alpha, m_\beta) \neq \emptyset$$

The question then arises as to how to resolve this conflict and to choose the best target. To deal with this issue, we consider all paths connecting the source article n_α to each candidate n_β and we assume that the candidate having the highest alternative path score is the most appropriate target. We define the *alternative path score aps* for a candidate n_β as:

$$aps = \frac{|path(n_\alpha, n_\beta)|}{\sum_{m_\beta \in C^\beta(n_\alpha)} |path(n_\alpha, m_\beta)|}$$

In our evaluation, we noted that considering all paths from n_α to the candidates leads to a high computational cost. For this reason, we decided to only use the paths consisting of three nodes: n_α, one intermediate node in a certain language version and the target candidate. The situation is better explained in the right side of Fig. 3. There are four paths leaving n_α, of which three lead to the upper node in the set $C^\beta(n_\alpha)$; that node is then ranked higher than the lower node.

Neighborhood Links Score. When the *alternative path score* does not allow any decision, *EurekaCL* exploits the neighboring articles connected to the source article and to a candidate article via rl intra-language links by considering both incoming and outgoing links. *EurekaCL* computes the ratio of common neighbors of each candidate $n_\beta \in C^\beta(n_\alpha)$. The common neighbor articles are retrieved thanks to the cross-links between n_α and n_β neighbors. The candidate having the largest score is ranked higher.

5   Evaluation

We downloaded from the Wikimedia Foundation website[1] the dump files of eight language versions of Wikipedia as of May 2014 and we transformed them into a Wikipedia graph, as described in Sect. 3, which we stored as a Neo4j 2.2.1 database. Neo4j is an increasingly popular open-source graph database that allows for easy modelling and fast traversals of large graphs, such as the Wikipedia graph on which we evaluated *EurekaCL*. To convert the Wikipedia dump files

---

[1] http://dumps.wikimedia.org/backup-index.html.

into a Neo4j database, we modified and used a tool named Graphipedia that is available on GitHub[2].

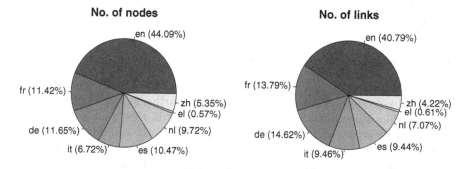

Fig. 4. The size of the Wikipedia graph

The resulting Wikipedia graph consists of 24,620,285 nodes and 329,485,368 links that are distributed among eight languages as shown in Fig. 4. Most of the nodes and links correspond to articles and links of the Wikipedia in English (en), which is by far the largest language version. The French (fr) and German (de) versions have a comparable size, as do the Spanish (es) and Dutch (nl) versions, at a smaller scale. The Italian (it) and Chinese (zh) versions have roughly the same amount of nodes, but considerable different links density; finally, the Wikipedia in Greek (el) includes the smallest percentage of nodes and links. We chose the languages to be evenly split between major (en, fr, es, de) and minor (it, nl, zh, el) versions, so as to assess the robustness of *EurekaCL* on graphs with a variable number of nodes and, most importantly, links.

For the evaluation of *EurekaCL*, we randomly selected a subset of 1,500 cross-links (n_α, n_β) for each language pair (α, β) in our Wikipedia graph, which resulted in a set G consisting of 42,000 cross-links (there are 28 possible language pairs). As the Wikipedia is contributed by millions of people across the world, we can assume with a certain confidence that each cross-link in G is correct, which means that it truly connects nodes corresponding to articles that cover the same concept in two different languages. Therefore, we can use G as the ground truth to evaluate *EurekaCL*.

More precisely, for each node pair $(n_\alpha, n_\beta) \in G$, *EurekaCL* is invoked to determine a *target node* m_β in the target language β for the *source node* n_α; if the output node m_β is the same as the *expected result* n_β, as per ground truth, the output of *EurekaCL* is considered to be correct for n_α. The purpose of the evaluation was to assess the number of correct answers of *EurekaCL* on all the input nodes. Obviously, we removed from the Wikipedia graph all the cross-links of G before running *EurekaCL*, and also we made sure that *EurekaCL* did not use indirectly those cross-links when querying BabelNet. To this extent, we

---

[2] https://github.com/gquercini/graphipedia.

configured *EurekaCL* so that it could not use Wikipedia as a data source when querying BabelNet for the synset senses of a source node.

All the experiments have been conducted on a computer running Linux Ubuntu 14.04 and equipped with a 8 core Intel Xeon Processor E5-2630 v3 running at 2.40 GHz, 32 GB of RAM and a 500 GB SCSI disk.

5.1   Evaluation of the Candidate Selection

The candidate selection is a crucial step for *EurekaCL* because it can affect its output in many ways, as we will show below. In particular, we identified three properties that a good candidate selection must fulfill. First and foremost, the expected result of any source node must be included in the candidate set; if not, *EurekaCL* will not be able to output the correct node, no matter how good its ranking strategy is. Second, the size of a candidate set should be reasonably limited, otherwise the ranking step will take too long to complete. Finally, the time required to select the candidates needs to be acceptably fast.

First of all, we define the *global recall* of the candidate selection as the ratio of source nodes of G for which the expected result is included in the candidate set. We observed the global recall for different values of the size threshold t_{min}, presented in Sect. 4.1. If the value of t_{min} is not set, which is equivalent to saying that we do not use the Wikipedia categories, the global recall is 98 %; for $t_{min} = 10$ and $t_{min} = 5$ the global recall is 95 % and 93 % respectively; if $t_{min} = +\infty$, which means that no threshold is set, the global recall is 88 %. In our evaluation we decided to set $t_{min} = 10$ to obtain a good compromise between the global recall and the size of the candidate set, which has an impact on the computational time.

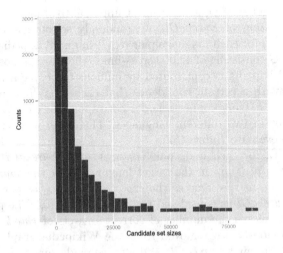

Fig. 5. The candidate set size histogram

As for the size of the candidate sets, the count histogram in Fig. 5 shows that for most of the source nodes the size of the candidate set is relatively small. Indeed, for 66 % of the source nodes the size is lower than 1,000, the average size is 1,372 and the maximum size is 85,154, which occurs in just one case. The first enhancement, which keeps only the candidates that share Wikipedia categories with the source article, greatly contributes to the reduction of the size of the candidate sets. Indeed without using this enhancement at all, the average size of a candidate set would be 2,042 nodes, which corresponds to passing 28,141,309 more candidates to the ranking step.

59 % of the expected results are included in the candidate set via BabelNet, while the others are included thanks to the enhancement. Among the candidates included via BabelNet only 11 % are found via the synset senses, while 89 % via the related synset senses. This is due to the fact that most of the source nodes in G correspond to Wikipedia articles about named entities (e.g., *Barack Obama*), for which the synset senses do not contain their translation.

Finally, the average time needed to select the candidates is 2.5 s; the maximum is 18 min and the minimum is 1 ms.

5.2   Evaluation of the Candidate Pruning

Pruning proved to be very effective in reducing the number of nodes in the candidate sets. It has also a highly positive impact on the final result of the ranking, as we will show later. In Fig. 6 we show the histogram of the candidate set sizes; compared to the one in Fig. 5 one can easily see that the average size is considerably lower than the size of the set before pruning (from 1,372 to 302); also, the maximum is now 47,545, which is almost a half of the maximum value before pruning.

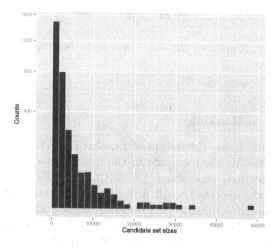

Fig. 6. The candidate set size histogram after applying pruning

As pruning involves deleting some candidates, there are chances that the expected result might be removed from the candidate set, although the pruning procedure has been conceived to remove only those candidates that are known to be connected by a path of cross-links to nodes other than the source node. We found that for 15 source nodes, a very small percentage of our entire dataset, the expected result was removed. It is interesting to note that in these cases the problem is that some cross-links in our Wikipedia graph are not correct. One such case is the source English article titled "Flight planning" whose expected result in the German Wikipedia is "Flugplanung"; when pruning, *EurekaCL* finds that "Flugplanung" is linked by transitivity through the Spanish article "Plan de Vuelo" to the English article "Flight Plan"; since this article is different than the source article, "Flugplanung" is then pruned from the candidate set.

Finally, we point out that the average pruning time is 710 ms, the maximum is 4 min and the minimum is 1 ms.

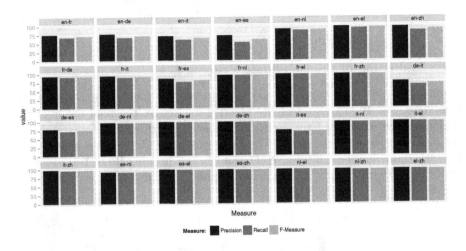

Fig. 7. Results for the Top-1 ranking by language pair

5.3 Evaluation of Ranking

For a given source node n_α, the output of *EurekaCL* is a list of nodes sorted by decreasing score; ideally, the top node in the list corresponds to the expected result n_β. In this section, we evaluate EurekaCL's ranking process. We look at its ability to rank the expected result first or among the top 10 candidates. To this extent, we use three measures - *precision*, *recall*, and *f-measure* - that we define as follows. Let G be the ground truth as previously defined and T_1, T_3, T_5, and T_{10} the sets of source nodes for which the expected result is ranked first, in the top 3, top 5, and top 10 respectively. Also, let I be the set of source nodes for which the expected result is included in the candidate set. We define a set of measures of precision P_i, recall R_i, f-measure F_i, for $i = 1, 3, 5, 10$ as follows:

$$P_i = \frac{|T_i|}{|I|} * 100 \quad R_i = \frac{|T_i|}{|G|} * 100 \quad F_i = \frac{2 * P_i * R_i}{P_i + R_i}$$

The results of the ranking for the Top-1 case for each language pair are shown in Fig. 7. First of all, we note that the average precision P_1, recall R_1 and f-measure F_1 across all language pairs are 94.24 %, 91.29 % and 92.68 % respectively. Based on the figure, it is evident that for some language pairs the results are sensibly lower than the average. This is especially true for the pairs en-fr, en-de, en-it and en-es. The problem here is that in most (1,248 out of 1,500) of the pairs $(n_\alpha, n_\beta) \in G$, such that α is the English language and β is the French language, the source node n_α is isolated, in that there is no cross-links incident with it (except the one that connects it to n_β, which has been removed for the evaluation). As a result, the ranking for these source nodes is based solely on the neighborhood score, while the alternative paths score never kicks in.

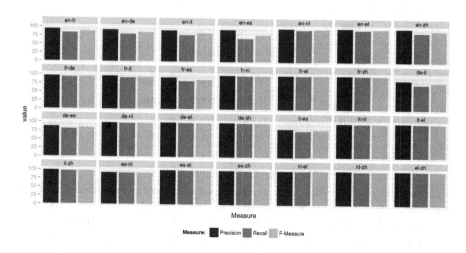

Fig. 8. Results for the Top-1 ranking by language pair after pruning

The benefits of the pruning on the ranking are shown in Fig. 8. We note that the precision/recall improve considerably, especially for those language pairs that we pointed out above. For the en-fr pair, for instance, the precision jumps from 77 % to 94 % and the recall from 68.93 % to 84.20 %. The average precision P_1, recall R_1 and f-measure F_1 across all language pairs are 97.04 %, 93.73 % and 95.28 % respectively.

Table 1 shows a summary of the results of the ranking. The value between parentheses in each cell refers to the result obtained after pruning.

Table 1. Results of the ranking

		Top-1	Top-3	Top-5	Top-10
Precision	**no pruning**	94.24	97.64	98.84	99.80
	pruning	97.04	98.83	99.34	99.62
Recall	**no pruning**	91.29	94.33	95.41	96.29
	pruning	93.73	95.36	95.83	96.10
F-measure	**no pruning**	92.68	95.88	97.02	97.93
	pruning	95.28	96.98	97.47	97.75

5.4 Comparison

We compare *EurekaCL* against two existing approaches, namely *WikiCL* [7] and Sorg&Cimiano [8], which have been previously evaluated on a publicly available dataset. This dataset, referred to as RAND1000, contains 1000 pairs of articles such that the source article is in the English Wikipedia and the expected result is in the German Wikipedia.

Table 2. Comparison

	Precision	Recall	F-measure
EurekaCL	**99.14**	**97.78**	**98.46**
WikiCL	89.00	88.00	89.00
Sorg&Cimiano	94.00	70.00	80.00

The results of the comparison, summarized in Table 2, show that *EurekaCL* clearly outperforms the other approaches, on both precision and recall. We note that the global recall for the candidate selection of *EurekaCL* is 98.62 %, while in Sorg&Cimiano is 86.5 %; the average candidate set size for *EurekaCL* is 753, while for Sorg&Cimiano is 1,000. These values are not available for *WikiCL*.

6 Conclusion

In this paper we presented and evaluated a multilingual approach called *EurekaCL* to automatically identify missing cross-language links in Wikipedia. The strongest points of *EurekaCL* are an effective candidate selection and pruning strategy, which reduces the number of candidates to rank and considerably improves the final precision and recall, as our experiments showed. Our evaluation is carried out on a large dataset, including 4 major and 4 minor Wikipedia language versions. We also ran *EurekaCL* on RAND1000, a publicly available dataset, to make a comparison with two existing approaches.

As future work, our goal is threefold. First of all, we intend to optimize the implementation of *EurekaCL* and the Wikipedia graph representation to speed up the ranking, which is the step that takes most of the time compared to the candidate selection step (9 s on average, with a maximum value of 45 min). Next, we plan to generalize *EurekaCL* to detect missing cross-links between Wikipedia pages belonging to other namespaces, in particular categories. Finally, we will also focus our attention on the problem of detecting erroneous cross-language links in Wikipedia, as done in [4].

References

1. Adafre, S.F., de Rijke, M.: Finding similar sentences across multiple languages in wikipedia. In: Proceedings of the 11th Conference of the European Chapter of the Association for Computational Linguistics, pp. 62–69 (2006)
2. Palmero Aprosio, A., Giuliano, C., Lavelli, A.: Automatic expansion of DBpedia exploiting wikipedia cross-language information. In: Cimiano, P., Corcho, O., Presutti, V., Hollink, L., Rudolph, S. (eds.) ESWC 2013. LNCS, vol. 7882, pp. 397–411. Springer, Heidelberg (2013)
3. de Melo, G., Weikum, G.: Menta: inducing multilingual taxonomies from wikipedia. In: Proceedings of the 19th ACM International Conference on Information and Knowledge Management, CIKM 2010, pp. 1099–1108. ACM (2010)
4. de Melo, G., Weikum, G.: Untangling the cross-lingual link structure of wikipedia. In: Proceedings of the 48th Annual Meeting of the Association for Computational Linguistics, ACL 2010, Uppsala, Sweden, 11–16 July 2010, pp. 844–853. Association for Computational Linguistics (2010)
5. Moreira, C.E.M., Moreira, V.P.: Finding missing cross-language links in wikipedia. JIDM J. Inform. Data Manage. 4(3), 251–265 (2013)
6. Navigli, R.: Babelnet and friends: a manifesto for multilingual semantic processing. Intelligenza Artificiale 7(2), 165–181 (2013)
7. Penta, A., Quercini, G., Reynaud, C., Shadbolt, N.: Discovering cross-language links in wikipedia through semantic relatedness. In: ECAI 2012–20th European Conference on Artificial Intelligence, pp. 642–647 (2012)
8. Sorg, P., Cimiano, P.: Enriching the crosslingual link structure of wikipedia -a classification-based approach. In: Proceedings of the AAAI 2008 Workshop on Wikipedia and Artificial Intelligence (WikiAI 2008) (2008, to appear)
9. Sorg, P., Cimiano, P.: Exploiting wikipedia for cross-lingual and multilingual information retrieval. Data Knowl. Eng. 74, 26–45 (2012)
10. Tsunakawa, T., Araya, M., Kaji, H.: Enriching wikipedia's intra-language links by their cross-language transfer. In: Proceedings of the 25th International Conference on Computational Linguistics, COLING 2014, pp. 1260–1268 (2014)

Adaptive Focused Crawling of Linked Data

Ran Yu$^{(\boxtimes)}$, Ujwal Gadiraju, Besnik Fetahu,
and Stefan Dietze

L3S Research Center, Leibniz Universität Hannover, Hannover, Germany
{yu,gadiraju,fetahu,dietze}@L3S.de

Abstract. Given the evolution of publicly available Linked Data, crawling and preservation have become increasingly important challenges. Due to the scale of available data on the Web, efficient focused crawling approaches which are able to capture the relevant semantic neighborhood of seed entities are required. Here, determining relevant entities for a given set of seed entities is a crucial problem. While the weight of seeds within a seed list vary significantly with respect to the crawl intent, we argue that an adaptive crawler is required, which considers such characteristics when configuring the crawling and relevance detection approach. To address this problem, we introduce a crawling configuration, which considers seed list-specific features as part of its crawling and ranking algorithm. We evaluate it through extensive experiments in comparison to a number of baseline methods and crawling parameters. We demonstrate that, configurations which consider seed list features outperform the baselines and present further insights gained from our experiments.

Keywords: Focused crawling · Linked data · Relevance assessment

1 Introduction

With the emergence and continuous evolution of Linked Data, crawling and preservation of entities has become an important challenge. Usually entities as part of Linked Datasets are interlinked with other entities into a semantic neighborhood, with links having specific semantics. However, entities and their semantic neighborhood evolve, what leads to a number of practical implications.

For instance, we consider the task of document annotation, namely *entity linking* or *named entity disambiguation*. In this task, specific *surface forms* are linked to entities appearing in the Web data. Often entity linking refers to restricted knowledge graphs such as DBpedia [1], YAGO [17]. A linked entity can be referred to from different surface forms (e.g. `Barack Obama` can be referred as 'Obama', 'US President' etc.). Considering the continuous evolution of Linked Data, capturing the temporal state and evolution of entities and their yet to be defined neighbourhood through dedicated crawling and preservation methods has become an increasingly important challenge.

Focused crawling, first introduced in [5] is a well studied field and has seen a wide application for web documents, specifically, especially in dataset creation

© Springer International Publishing Switzerland 2015
J. Wang et al. (Eds.): WISE 2015, Part I, LNCS 9418, pp. 554–569, 2015.
DOI: 10.1007/978-3-319-26190-4_37

and used by major search engines [3,4,6]. In the context of Linked Data, crawling is similar to that on the web and usually follow the links, i.e. object properties, of seed entities to find potential *candidate entities*.

Crawling for *candidate entities* in the Linked Data context is usually achieved through a breadth-first search (BFS) approach, as implemented by LDSpider [11], and can be tailored with respect to a prioritisation of crawling order and the considered distance, i.e. the maximum number of hops in the data graph.

Given the scale and dynamics of available Linked Data on the Web, more efficient *focused crawling* approaches are required which crawl and rank candidate entities for their relevance for a given *crawl intent*. The *crawl intent* usually is specified in the form of *seed entities*. Referring to the document annotation scenario, a typical seed list might be the set of entities extracted from a specific set of documents. The ranking of candidate entities can exploit two type of measures, *connectivity* or *similarity* between the candidate and seed entities. Such measures, exploit *lexical* as well as *structural*, respectively *graph-based* similarity metrics, which usually measure a notion of relatedness or similarity. As such, relevance assessment of entities is a critical challenge, of significant importance also with respect to other entity-centric tasks such as entity retrieval, semantic search or entity recommendation.

In our work, we show that the performance of entity ranking measures depends heavily on the nature of the seed list at hand, i.e., the coherence or crawl intent (see Sect. 6). We investigate these observations by comparing different crawl configurations which differ with respect to their maximum hop size and the chosen relevance detection metric. As one contribution, we analyse and compute the *crawl intent* of seed entities and reflect it in our premium crawling configuration. Here, we exploit an intuition documented in the work by Pound et al. [16], where the importance of individual entities of a composite query or seed list differs across seeds, with more specific entities usually reflecting the search intent to a higher degree. Exploiting this observation, we specifically boost entities closer to the crawl intent through a dedicated *attrition factor* as part of our candidate entity ranking.

To this end, our approach consists of the following steps: (i) seed list analysis, (ii) candidate crawling, (iii) seed list-specific candidate entity ranking, which are embedded into a crawling cycle. Our experiments show better performance and efficiency of our adaptive approach for a wide range of seed lists, compared with baseline ranking methods. In addition, insights gained from the above experiments are deemed applicable also in other scenarios, such as entity retrieval or entity recommendation, where candidate entity rankings need to be computed for a given query.

The paper is structured as follows. Section 2 introduces related works, followed by a problem definition and overview of the approach. Section 4 describes the seed list analysis, which provides a foundation for the adaptive crawling described later in the same Section. The experimental setup and results are described in Sects. 5 and 6 respectively. We discuss and conclude our work in Sect. 7.

2   Related Work

Focused Web Crawling. The purpose of web crawling is done for various reasons. Two of the widely spread use cases are the construction of datasets and that of vertical search engines like Google, Yahoo! etc. We focus our literature review mainly on the first case. Crawling data of a specific topic of information need, is also known as focused crawling. The term focused crawling was first introduced by de Bra et al. [5]. The focus has been widely shifted towards topic focused crawling.

Diligenti et al. [6] proposed a focused crawling approach that uses context graphs. A context graph in their case represents the link hierarchy between HTML pages and the co-occurring pages that are known to have as a target a relevant page. The main advantage of such an approach against traditional breadth-first-search crawls is that the model is optimized for the global relevance of pages rather than the immediate gains that are achieved from directly connected pages given a seed list. Chakrabarti et al. [4] proposed a topic focused crawling approach by assessing the relevance of a HTML page to a topic and further identifying sources (or web domains) that contain relevant documents closely connected to the seed pages. Later on, Chakrabarti et al. [3] proposed an improved approach with two main supervised modules. The modules are trained on the topic taxonomy Dmoz and the corresponding documents that are assigned to the topics. The first module serves as a source of generating new crawl targets that are passed as a priority list to the second module, which later on uses the DOM features from every proposed target crawl page to assess its final relevance for a given topic. McCallum et al. [13] presented a task oriented crawler for building domain-specific search engines. The approach crawls pages that are relevant to a specific topic and further extract information from the documents, i.e. title, date etc., which are later on used as query fields. Tang et al. [18] use external knowledge bases, specifically that of the medical domain, to improve both the relevance and quality of focused crawling.

Crawling of Structured Web Data and Linked Data. Meusel et al. [14] proposed a novel focused crawling approach for structured data. Their approach crawls microdata embedded in HTML pages. Further, they proposed a two-step approach. In the first step they use an online classifier that is trained on features extracted from the URLs of the pages, and in the second step assess the relevance of a group of pages (usually from the same source). In a closely related work, Isele et al. introduced an open-source crawling framework for Web data [11]. LDSpider traverses the Web of Linked Data by following RDF links between entities. LDSpider is less effective when a focused crawl is required, as depicted in later sections of this paper. In our previous work, we proposed a system that iteratively crawls and captures the evolution of Linked Data sets based on flexible crawl definitions [7].

Entity Relevance. Graph-based and lexical features of entities are used to assess the relevance of an entity in a range of tasks, such as *entity linking* or *entity* *retrieval.* Graph-based relevance has been used in entity linking [15]. Lexical features represent common features used in previous focused web crawling works [3,4,13]. In [8], the authors introduce an entity retrieval approach, which exploits specifically lexical features to cluster and retrieve entities from the Web of data.

The aforementioned works differ on several aspects w.r.t our work. The scope of the crawl techniques relies on HTML web pages, where the nature of the data and hence the features that can be extracted is different. For instance, DOM features are used in [3] to perform the focused crawling. Other works like [13,18] proposed specific task oriented crawls, that of domain-specific search engines and medical domain crawls based on respective knowledge bases. In contrast, our work focuses solely on focused entity crawling from the Web of Linked Data. While previous work on LD crawling, such as LDSpider, is focused in the sense that they use a set of seed entities as starting point, they do not compute the relevance of retrieved entities or provide a respective ranking. Another distinguishing feature is that, as opposed to our work, features of the seed list are not considered for configuring the crawl method. Instead, mostly static crawling approaches are used, which apply the same strategy and method on each seed list.

3   Problem Definition and Approach Overview

In this section, we first describe the motivation and use case of this work, then provide the formal problem definition as well as a brief overview of our approach.

3.1   Motivation and Problem Definition

Our work aims at providing a method for focused crawling of Linked Data, namely to crawl entities of relevance to a given seed list consisting of entities which describe the crawl intent. While this is a task of relevance for efficiently preserving a particular partition of the Linked Data cloud in general, we would like to emphasise its use in document annotation or data enrichment scenarios. Here, traditionally named entity disambiguation or entity linking techniques are deployed to annotate documents with entities from a specific knowledge graph. While such entity annotations enable structured queries to retrieve documents of relevance to specific entities, more sophisticated semantic search needs to exploit additional knowledge from the used reference graphs, such as DBpedia or related graphs. While the overall Linked Data graph might contain highly relevant information about the semantics of a specific entity, identifying the paths and entities of relevance, or the *semantic neighbourhood* of a given entity, is a non-trivial task.

Specifically, the challenge is to identify and crawl the most relevant entities for a given set of seed entities. We define *focused crawling* of Linked Data as follows. Given a specific seed list of entities $S = \{e_1, \ldots, e_n\}$, the aim is to crawl and rank relevant candidate entities $C = \{e'_1, \ldots, e'_n\}$, available from the Web. Though seeds could commonly be represented through terms which require a disambiguation step, for simplification purposes we assume that seed entities are represented by entity URIs, for instance, referring to instances within the DBpedia graph.

To illustrate the need for scalable and focused crawling approaches, note that our earlier experiments have shown that a 2-hop crawl of a given seed list results in 38,295 entities on average. Hence, we aim for a focused approach, where relevance of entities is computed as part of the crawling process and seeds for the following hop are determined based on their relevance to the seed list.

3.2 Approach Overview

In order to tackle the challenges of focused crawling, we adopt the following steps: (i) seed lists analysis, (ii) breadth-first search crawl (BFS) for candidates C, (iii) seed list-specific *candidate entity ranking*. The last two steps are embedded into a crawling cycle by using the relevant entities selected from step (iii) as next hop crawling queue for step (ii). Explained in more detail in the following section, we consider an *attrition factor* of each seed entity $e_i \in S$ to reflect the *crawl intent* during the focused crawling process. This is based on the assumption that entities within a seed list have varying importance for the *crawl intent*, and hence, their individual impact on the ideal result set differs.

The overall focused crawling framework is summarized in Algorithm 1.

Algorithm 1. Focused Crawling of Linked Data

Require: Seed list
Ensure: Related entities
 1: **function** FOCUSEDCRAWLING(*seedlist, depth*)
 2:     *coherence* ← COHERENCECOMPUTATION(*seedlist*)
 3:     *attritionFactors* ← ATTRITIONFACTORCOMPUTATION(*seedlist*)
 4:     *queue* ← *seedlist*
 5:
 6:     **for** *hop* = 0 → *depth* **do**
 7:         *candidates* ← BREADTHFIRSTCRAWL(*queue*)
 8:         *relatedEntities* ← RELEVANCEASSESSMENT(*candidates, attritionFactors*)
 9:         *queue* ← *relatedEntities*
10:     **end for**
11:     **return** *relatedEntities*
12: **end function**

A focused crawling *configuration* represents an implementation of Algorithm 1, dependent on specific attributes of the candidate crawling process and seed list analysis, such as *depth* for candidate crawling and *attrition factor* of seed entities in a seed list. The corresponding configurations are described in detail in Sect. 5.

4   Adaptive Focused Crawling

In this section we describe the seed list analysis and the adaptive crawling approach.

4.1   Seed List Analysis

Here we define and describe in detail the seed list analysis, specifically on how we compute the *seed list coherence* and *attrition factor* in Algorithm 1.

Seed List Coherence. We assume that crawl configurations require tailoring to the *seed list coherence*. The underlying assumption is that very specific and targeted seed lists will require different crawling and relevance computation methods than very broad and unspecific seed lists. An example of the former is, for instance, a seed list containing entities labeled with 'Barack Obama', 'President of the United States', 'George W. Bush', while 'John Lennon', 'Africa', 'Mahatma Gandhi' would constitute a very unspecific seed list.

We introduce a score to measure *seed list coherence*, Γ, computed as the reciprocal average shortest path between any seed entity pairs (see Eq. 1). The shortest path between any two seed entities is denoted by $\varphi(e_i, e_j)$ on a given knowledge graph.

$$\Gamma = \frac{n(n-1)}{2\sum\limits_{\substack{i,j \\ i<j}}^{n} \varphi(e_i,e_j)} \Leftarrow \begin{array}{c} \\ e_1 \\ e_2 \\ \vdots \\ e_n \end{array} \begin{pmatrix} \overset{e_1}{0} & \overset{e_2}{\varphi(e_1,e_2)} & \overset{\cdots}{\cdots} & \overset{e_n}{\varphi(e_1,e_n)} \\ & 0 & \cdots & \varphi(e_2,e_n) \\ & & & \vdots \\ & & & 0 \end{pmatrix} \tag{1}$$

The main intuition behind the computation of *seed list coherence* in Eq. 1 is that entities with shorter paths in a given knowledge base tend to be more closely related. Hence, the average shortest path between seed entities determines the *coherence* of a seed list. In our experiments, we identify two crawl scopes based on Γ: (i) *low coherence*, and (ii) *high coherence*. We consider seed lists that follow $0 \leq \Gamma \leq \gamma$ to exhibit low coherence, while $\Gamma > \gamma$ indicates high coherence. The value of threshold $\gamma = 0.5$ was identified during our experiments, the detailed setup is described in Sect. 5.1.

Seed Entity Attrition Factor. As shown in our experimental evaluation, specific entities within a seed list strongly reflect the crawl intent. For example, consider the seed list {Pulp Fiction, Film, Entertainment}, the most specific entity 'Pulp Fiction', reflects the most specific crawl intent, whereas the entities 'Film' and 'Entertainment' provide contextual information, namely that 'Pulp Fiction' is a movie. Motivated by this, we assume that the relevance of specific candidate entities is dependent on the seed entity they are related to. For example, candidate entities similar to entity 'Pulp Fiction' will be ranked higher than entities that are similar to other seed entities.

In order to improve the candidate entity ranking we define the *attrition factor* $\lambda(e_i)$ for each seed entity in S. The *attrition factor* $\lambda(e_i)$ is measured as the fraction of the Katz centrality score [12] of e_i over the sum of all scores from all other entities in S terms of the proportion of Katz centrality. The Katz score is computed w.r.t a reference dataset (in our case DBpedia), and entities with lower centrality score will have higher *attrition factor*, hence reflecting more strongly the *crawl intent*. The attrition factor $\lambda(e_i)$ is computed as in Eq. 2.

$$\lambda(e_i) = \log \frac{C_{Katz}(e_i)}{\sum_{j=1}^{n} C_{Katz}(e_j)} \tag{2}$$

where, e_i is the seed entity, n is the number of seed entities in S, and $C_{Katz}(e_i)$ is the Katz centrality of e_i, and $e_i \in S$.

$$C_{Katz}(e_i) = \sum_{k=1}^{5} \sum_{j=1}^{n} \alpha^k (A^k)_{ji} \tag{3}$$

where, A is the adjacency matrix of entity e_i, namely the connections of e_i in the reference dataset, whereas k reflects the number of hops in the reference graph that are used to compute the adjacency matrix. The second sum in Eq. 3 measures the connectivity degree between e_i and entities in the knowledge graph that are reachable within the hop k. Finally, α is an exponential penalization factor, which prefers higher connectivity degree on the earlier hops, and it takes values in the range of $\alpha \in [0, 1]$.

4.2   Candidate Crawling and Ranking

In this section, we describe the breadth-first search (BFS) crawl for candidates entities and the seed list-specific candidate entity ranking.

Crawl for Candidate Entities. The BFS crawl starts with a queue Q_i, $i = 0...depth$. This queue is populated by the seed list S in the first hop, and is in later hops replaced by the relevant candidate entities C (see Algorithm 1). With the increase of the number of hops, the number of candidate entities increases, hence we perceive a drop in crawling efficiency. The crawling *depth* defines the maximum number of hops, where we aim for a depth that provides high efficiency by means of sufficient candidates as well as short runtime.

Candidate Entity Ranking. Candidate entities are ranked according to their relevance to the seed entities in S. The relevance of a candidate entity to a seed list can be determined by distinguishing the following aspects.

– Pairwise relevance of a candidate entity with respect to each seed entity, i.e., $rel\{c_i, e_j\}$ where $c_i \in C$ and $e_j \in S$.

– Overall relevance of a candidate entity to the entire seed list, i.e., $rel\{c_i, S\}$ where $c_i \in C$.

Relevance of candidate entities c_i with respect to seed entities e_j is computed by using a distance measure, abridged with our weighting scheme consisting of the attrition factor λ, as introduced in Sect. 4.1.

This method involves using text-based similarity metrics such as *Jaccard Similarity*, to assess the similarity between the candidate entities and seed entities. Using labels and descriptions of entities, we construct term-reference vectors for each entity in the seed and candidate set, respectively S and C. This step is semi-automated as a user can configure the relevant datatype properties used to construct the term vector w_i for each seed entity e_i. The term-reference vector in our case represents a set of unigrams (single terms) extracted from the textual literal datatypes from the corresponding entities.

The following formulas show the adapted pairwise Jaccard Similarity ($PairwiseJaccardSim^+$), where the traditional Jaccard similarity has been adopted to consider the crawl intent, as well as the overall candidate relevance computed through the *Entity-based Jaccard Similarity* ($EJSim^+$).

$$PairwiseEJSim^+(e_i, c_j) = \frac{1}{\lambda(e_i)} \cdot \frac{|w_i \cap w_j'|}{|w_i \cup w_j'|} \tag{4}$$

$$EJSim^+(c_j) = \frac{\sum_{i=1}^{n} PairwiseEJSim^+(e_i, c_j)}{n} \tag{5}$$

where, $\lambda(e_i)$ is the attrition factor of the seed entity e_i, and w_i, w_j are the entity reference-vectors corresponding to e_i and c_j.

5 Experimental Setup

In this section, we introduce the datasets we consider for the crawling process and the approaches on generating the seed lists. Next we assess the performance of the different crawling configurations, constrained for the varying parameters: (i) *seed list coherence*, (ii) *hop-size* and (iii) *crawl-candidate ranking approaches*. Consequently, we draw conclusions about the optimal *crawl parameters*. Finally, we compare our approach against established baseline methods.

5.1 Seed Lists

Our experiments are following our original motivation, that is the document annotation scenario, where seed entities are directly linked to and extracted from documents in a corpus. As we aim to evaluate crawling performance for a variety of seed lists, we generate a range of seed lists from real-world documents. In order to eliminate noise, potentially introduced by extracting entities from an arbitrary corpus, we directly utilise Wikipedia, where DBpedia entities can be derived directly from the hyperlinks of a source page to other Wikipedia pages.

Since popular Wikipedia pages tend to be linked more consistently, we particularly consider Wikipedia pages with high page views during the first week of 2012 accumulating 1.6 billion records. We have specifically chosen a period in the past to ensure well-populated and popular pages, rather than new and insufficiently annotated pages. To generate seed lists of varying *coherence*, we follow two different strategies for creating seed lists.

- (1) top-k entities in single documents
- (2) k entities extracted from sets of related documents

The intuition is that (1) will produce more coherent and (2) more diverse seed lists. In (1) we start with the page of an original entity u, we extract the entities $u_1, u_2, ..., u_t$ appearing in the page and rank these according to their frequency. Assuming that the entities with the highest tf score are closer related to the original entity, we use the top k entities as seeds.

For (2) we start again with a set one most viewed wikipedia pages, i.e. entities. For each entity u, we randomly obtain a category from the set of categories associated with u and retrieve all entities $u_1, u_2, ..., u_k$ in the same category. We then extract the entities $u'_1, u'_2, ..., u'_k$ appearing in the pages of $u_1, u_2, ..., u_k$ as candidates. Then we randomly select k entities from the accumulated candidate set as low coherence seed list.

Applying this method on 3 wikipedia pages and 3 categories, we obtained 6 seed lists of size $k = 5$ with varying coherence Γ. The seed lists and the groundtruth are available at http://l3s.de/~fetahu/crawler_wise2015/.

5.2   Crowdsourced Ground Truth

We leverage crowdsourcing as a means to establish the ground truth in the form of a ranking of the most relevant entities within a specified n-hop neighborhood for a given seed list S. We adopt the following steps in order to establish the ground truth for evaluation.

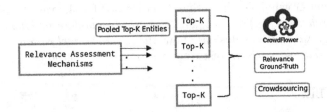

Fig. 1. Generation of ground truth for evaluation of relevance assessment measures.

First, we pool the top-k entities[1], as obtained by using the different relevance assessment measures. We model the activity of assessing the relevance

---

[1] In this case we pooled the $Top - 500$ entities resulting from all the different configurations for each seed list.

of an entity to the seed list, as an atomic microtask that can be deployed on a crowdsourcing platform such as CrowdFlower. In order to assist the crowd workers in assessing the relevance of an entity to a given seed list, we represent the seed list with a corresponding *topic* that describes the collective meaning projected by the seed list. In order to determine an apt *topic* that can represent a seed list and thereby the crawl intent, we follow a gamified approach such as the popular ESP game proposed by Von Ahn and Dabbish [19]. Each seed list is presented to 5 different experts, who are then asked to independently tag the seed lists with representative topics. When all 5 experts reach a concensus with respect to a particular description, that tag is chosen to be the representative *topic* of the seed list. For instance, for the seed list generated from `How I Met Your Mother (TV Series)` Wikipedia page, experts assigned the topic `How I Met Your Mother (TV Series) - Characters and Actors`.

Next, we present each entity gathered from the neighborhood of the seed list alongside the corresponding *topic* to crowd workers and request judgments of relevance according to a 5-point Likert-scale (ranging from *Not Relevant* to *Highly Relevant*). By aggregating the judgments from the workers into relevance scores, and ranking the pooled top-k entities based on these relevance scores, we thereby generate our ground truth of relevant entities corresponding to the seed list S. Finally, we use the ground truth thus established to evaluate each of the relevance assessment methods across the top-k entities.

In order to ensure that the ground truth is reliable, we take several precautions as recommended by our previous works [9,10] to curtail malicious activity and to avoid misinterpretations in the relevance scoring.

5.3 Crawl Configurations and Baselines

We distinguish *crawling configurations* by (a) the used candidate ranking approach, (b) the hop size (depth) for the crawling process, and (c) the consideration of the seed list coherence (or lack thereof). The parameters are listed below. Here the candidate ranking considers two cases: (i) candidate relevance scoring taking into account the attrition factor λ which is denote as $EJSim^+$, and (ii) candidate relevance scoring without the attrition factor, i.e. the baselines below.

The maximum hop size *depth* considered during the crawling process is one of the most significant impact factors for runtime and NDCG score. We run experiments with *depth* in $\{2, 3\}$, the corresponding methods are denoted with a subscript indicating the depth, where, for instance, $EJSim_2$ indicates a *depth* = 2.

We consider the following baseline and state-of-the-art approaches for the focused crawling task.

EJSim. Entity Jaccard Similarity, *EJSim* computed as *EJSim+* but without considering the *attrition factor*.

NJSim. We also consider the graph-based relatedness baseline, the *Neighborhood Jaccard Similarity* which is computed as follows.

$$PairwiseNJSim(e_i, c_j) = \frac{1}{\lambda(e_i)} \cdot \sum_{k=1}^{\tau} \alpha^k \frac{|N_k(e_i) \cap N_k(c_j)|}{|N_k(e_i) \cup N_k(c_j)|} \tag{6}$$

$$NJSim(c_j) = \frac{\sum_{i=1}^{n} PairwiseNJSim(e_i, c_j)}{n} \tag{7}$$

where, $N_k(e_i)$, $N_k(c_i)$ are the sets of neighboring vertices at the kth step of seed entity (e_i) and candidate entity (c_i) respectively, $\alpha \in [0,1]$ is a real number to ensure that closer neighbors have lower weight, and τ is the maximum length of the paths considered.

PageRank. Is a widely adopted approach for ranking crawling results (documents or entities). The computation of PageRank is formalized in Eq. 8.

$$PageRank(c_i) = \frac{1-d}{N} + d \sum_{c_j \in M(c_i)} \frac{PR(c_j)}{L(c_j)} \tag{8}$$

where $M(c_i)$ is the set of entities linked to candidate c_i in the entity graph, $L(c_j)$ is the number of outbound links to the entity c_j, N is the total number of entities, and d is the damping factor [2].

NB. We consider [3] as state-of-the-art approach for focused web crawling. However, while this approach originally implements focused crawling of Web documents as opposed to Linked Data, we introduce some adaptations. The DOM tree features cannot be considered in case of Linked Data, hence, we adopt lexical features for the NB classification. The state-of-the-art approach uses a predefined topic taxonomy with example URLs as training set. In our senario, the topic, reflected by the crawl intent in our work, is dependent on the seed list without predefined limitations. However, our groundtruth can be used as training set for a classifier. The classification of candidate entity as relevant is performed by the function formalized in Eq. 9.

$$NB(c_i) = P(y = 1|c_i) = \frac{P(y)P(c_i|y)}{P(c_i)} \tag{9}$$

where $P(y = 1|c_i)$ is the probability of candidate c_i belonging to class $y = 1$, which in our case translates as a *'relevant'* entity for a given seed list.

5.4   Evaluation Metrics

To evaluate the crawling performance of the different configurations, we consider the *Normalized Discounted Cumulative Gain (NDCG)* of the ranked set of candidate entities with respect to an established ground truth. The NDCG score is computed as follows:

$$nDCG@k = \frac{DCG@k}{iDCG@k} \quad DCG@k = rel_1 + \sum_{i=2}^{k} \frac{rel_i}{log_2 i}$$

where, $DCG@k$ represents the discounted cumulative gain at rank 'k', and $iDCG@k$ is the ideal $DCG@k$ computed with respect to the *ground truth* (described in the following section).

Another important factor to evaluate is the runtime performance of the different configurations. Due to the likelihood of large set of candidate entities existing for a particular seed list, runtime is a crucial factor. The runtime is simply measured in terms of the amount of time taken to complete a full-cycle of the crawling process.

6 Evaluation

6.1 Performance of Focused Crawling Configurations

In this section we report and discuss the performance of the different focused crawling configurations. On average, for the different seed lists we crawl approximately 491,425 triples. Hence, an important aspect is how well the different crawl configurations rank the candidate entities for their relevance w.r.t the seed lists. In Table 1 we report the performance of the different crawling configurations. The values are reported for the average NDCG scores at rank 100.

Table 1. Average NDCG@100 for different focused crawling configurations across seed lists with varying coherence.

Coherence	$EJSim_2^+$	$EJSim_2$	$NJSim_2$	NB_2	$PageRank_2$
$\Gamma < 0.5$	**0.7446**	0.7364	0.5752	0.6955	0.5346
$\Gamma \geq 0.5$	**0.6876**	0.6802	0.5193	0.5273	0.4408
	$EJSim_3^+$	$EJSim_3$	$NJSim_3$	NB_3	$PageRank_3$
$\Gamma < 0.5$	**0.6928**	0.6382	0.6608	0.6612	0.5722
$\Gamma \geq 0.5$	**0.5831**	0.5740	0.5197	0.4821	0.4166

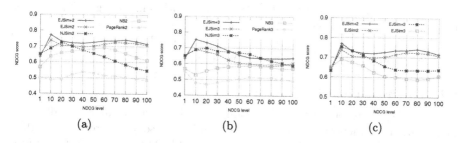

| (a) | (b) | (c) |

Fig. 2. (a) Performance of different configurations with depth 2 measured as average NDCG across all seed lists. (b) Performance of different configurations with depth 3 measured as average NDCG across all seed lists. (c) Performance of attrition factor (on depth 2 and 3) measured as average NDCG across all seed lists.

Performance. Table 1 presents the average NDCG@100 score of different configurations described in Sect. 5.3 with seed lists of varying coherence. Figure 2a and b present the NDCG score at different levels for the different configurations. Figure 2c presents a detail comparison between the performance of $EJSim$ and $EJSim^+$. It shows that the NDCG scores increased after introducing the attrition factor to EJSim. The average improvement across different NDCG levels is 1.6 % on $depth$ 2 and 4.3 % on $depth$ 3, suggesting a positive effect of the *attrition factor* for the cases of our seed lists. On the other hand, the *coherence* of the seed list appears to have no significant impact on the suitability of particular configuration.

From the results we can also observe that, the $EJSim^+$ outperforms baseline methods, specifically $NJSim$ and $PageRank$. The NDCG@100 of $EJSim^+$ is 16.9 % and 4.8 % higher than $NJSim$ for $depth$ 2 and 3 respectively. The NDCG scores at different levels (Fig. 2a and b) also support this insight. $PageRank$ shows the weakest performance since it does not take the crawl intent into consideration during candidate ranking. Furthermore, $EJSim^+$ shows 7.3 % and 9.8 % improvement in average for $depth$ 2 and 3 respectively compared to NB in our experiment. Based on the dataset, one of the reasons that NB and $EJsim^+$ have different performance while both use lexical features is that the classes within the training set are not independent from each other. The overlap between the feature set of different classes causes negative effects on the NB classification. The reason of the overlap is that the crawl intent is based on the seed lists instead of the topic from a predefined topic taxonomy. This also reveals that our method is more robust and flexible for a range of crawl intents.

Another useful and conclusive insight is the fact that $depth$ 3 has not led to performance gains with respect to *crawl depth* of 2. In fact we can observe the opposite from Fig. 2c, where for $EJSim^+$, increasing the crawl depth beyond 2 seems to lead to weaker NDCG scores on average.

Efficiency. Since the time-intensive step is the candidate crawling, there is no considerable difference between the crawling configurations with different ranking methods. However, the average runtime across seed lists and ranking methods increased from 548 s for depth 2 to 1936.6 s for depth 3 in average while there is no significant improvement of the NDCG score. Given the significantly increased runtime when crawling beyond hop 2, a crawl depth of 2 seems to provide optimal efficiency, and it is not advisable to crawl to a higher distance.

6.2   Discussion and Limitations

Next to the aforementioned observations, we also carried out an analysis on more varied characteristics of seed lists, leading to some interesting insights. After the seed list size reaches a certain threshold (20 in our case), the graph-based methods seem to outperform lexical ones in most cases. This might be explainable with the fact that the increasing heterogeneity of larger seed lists renders lexical methods less applicable.

Meanwhile, PageRank performance seems to improve with increasing seed list size and decreasing coherence. This can be explained since Page Rank is not query-biased in any way and the performance gap to the superior, seed list-based configurations is decreasing the more generic the seed lists become.

We evaluate the result of experiments based on a ground truth as described in Sect. 5.2. During the preliminary analysis of the dataset, we observe that the overall NDCG score for results of low coherence seed lists seems significantly higher than those of high coherence seed lists. This is due to the fact that high coherence seed lists have a more specific crawl intent, leading to narrow and often small result sets, and hence also a limited ground truth, while the low coherence lists have a much broader crawl intent as well as relevant entity set. This is reflected in our ground truth: the average number of entities labeled as related (score ≥ 3 and beyond) is 208 for low coherence seed list, and 145 for high coherence seed lists. Meanwhile, the narrow search intent also causes more disagreement among crowdsourcing workers for generating the ground truth, which makes the results for high coherence seed lists less consensual.

Another difficulty faced when evaluating the crawling task is the highly heterogeneous and varied nature of the possible result sets, originating from a highly heterogeneous Linked Data graph. While certain seed lists and crawl intents are well reflected in the Web of Linked Data, others are only sparsely represented, leading to highly varying quality of the retrieved result sets. Here, to some extent the availability of matching data, or the lack thereof, can have a significant impact on the measured NDCG scores. This problem is elevated due to the fact that crawl results rely on links, where the uneven distribution of links across entity types and partitions of the Linked Data cloud leads to result sets of varying size and quality.

7 Conclusions and Future Work

In our work, we have presented an adaptive focused crawling method which takes into account characteristic features of seed lists to improve the crawling and relevance ranking method. Our crawl configurations consider experimentally derived heuristics and thresholds for the crawl *depth* and relevance assessment method. Our experimental results support the underlying assumption that the crawl intent can be reflected dynamically through a dedicated *attrition factor* to improve the ranking of retrieved entities. Our results demonstrate that the attrition factor has a positive impact on the performance, showing significant performance improvement across a range of seed lists, compared to all considered baselines.

Additional insights from our experiments might be of practical value for focused Linked Data crawling approaches in general. For instance, our results suggest that crawls beyond depth 2, i.e. a distance of 2 hops in the Linked Data graph, seem to not improve the crawl quality but instead, introduce noise and significantly increase the crawl runtime, hence decreasing the efficiency.

One central obstacle when evaluating methods for the focused crawling task is the lack of a sufficient ground truth and the heterogeneity of the Linked Data graph, which naturally leads to highly diverse ground truth result sets for different seed lists. This might have a negative influence on the resulting ground truth and might dilute the performance evaluation to a certain extent.

However, given the scale of our experiments, the shown results provide conclusive indicators about the performance of the investigated configurations.

While our work aimed at achieving a general performance gain, specifically measured through NDCG, our ongoing and future work foresees in particular the on-the-fly optimisation towards particular performance goals with regards to precision, recall, or runtime. While the actual crawl aims may vary strongly between different crawls - in some cases, high precision may be mandatory, in others a broad crawl, i.e. high recall will be of higher priority - through further experiments we aim at identifying more specific heuristics which can lead to a more tailored adaptation.

In this context, we are also investigating the specific characteristics of graph-based relevance metrics and lexical metrics. The high divergence of result sets produced by these two different approaches documents that, naturally, lexical methods seem better suited to retrieve *similar* entities, while graph-based ones seems better suited to retrieve otherwise *connected* entities. This suggests that each approach might be specifically suited to certain kind of seed lists, or crawl intent, what might require an even more adaptive approach, involving the selection or configuration of suitable relevance ranking methods at runtime.

References

1. Auer, S., Bizer, C., Kobilarov, G., Lehmann, J., Cyganiak, R., Ives, Z.G.: DBpedia: a nucleus for a web of open data. In: Aberer, K., et al. (eds.) ISWC/ASWC 2007. LNCS, vol. 4825, pp. 722–735. Springer, Heidelberg (2007)
2. Brin, S., Page, L.: The anatomy of a large-scale hypertextual web search engine. Comput. Netw. ISDN Syst. **30**(1), 107–117 (1998)
3. Chakrabarti, S., Punera, K., Subramanyam, M.: Accelerated focused crawling through online relevance feedback. In: Proceedings of the 11th International Conference on World Wide Web, WWW, pp. 148–159. ACM, New York (2002)
4. Chakrabarti, S., Van den Berg, M., Dom, B.: Focused crawling: a new approach to topic-specific web resource discovery. Comput. Netw. **31**(11), 1623–1640 (1999)
5. De Bra, P., Houben, G.-J., Kornatzky, Y., Post, R.: Information retrieval in distributed hypertexts. In: RIAO, pp. 481–493 (1994)
6. Diligenti, M., Coetzee, F., Lawrence, S., Giles, C.L., Gori, M., et al.: Focused crawling using context graphs. In: VLDB, pp. 527–534 (2000)
7. Fetahu, B., Gadiraju, U., Dietze, S.: Crawl me maybe: iterative linked dataset preservation. In: Proceedings of the 13th International Semantic Web Conference (ISWC) Posters & Demonstrations Track, pp. 433–436 (2014)
8. Fetahu, B., Gadiraju, U., Dietze, S.: Improving entity retrieval on structured data. In: Proceedings of the 14th International Semantic Web Conference. Springer (2015)
9. Gadiraju, U., Demartini, G., Kawase, R., Dietze, S.: Human beyond the machine: challenges and opportunities of microtask crowdsourcing. IEEE Intell. Syst. **30**(4), 81–85 (2015)
10. Gadiraju, U., Kawase, R., Dietze, S., Demartini, G.: Understanding malicious behaviour in crowdsourcing platforms: the case of online surveys. In: Proceedings of CHI 2015 (2015)

11. Isele, R., Umbrich, J., Bizer, C., Harth, A.: Ldspider: an open-source crawling framework for the web of linked data. In 9th International Semantic Web Conference, ISWC. Citeseer (2010)
12. Katz, L.: A new status index derived from sociometric analysis. Psychometrika **18**(1), 39–43 (1953)
13. McCallumzy, A., Nigamy, K., Renniey, J., Seymorey, K.: Building domain-specific search engines with machine learning techniques (1999)
14. Meusel, R., Mika, P., Blanco, R.: Focused crawling for structured data. In: Proceedings of the 23rd ACM International Conference on Information and Knowledge Management, CIKM, pp. 1039–1048 (2014)
15. Pereira Nunes, B., Dietze, S., Casanova, M.A., Kawase, R., Fetahu, B., Nejdl, W.: Combining a co-occurrence-based and a semantic measure for entity linking. In: Cimiano, P., Corcho, O., Presutti, V., Hollink, L., Rudolph, S. (eds.) ESWC 2013. LNCS, vol. 7882, pp. 548–562. Springer, Heidelberg (2013)
16. Pound, J., Mika, P., Zaragoza, H.: Ad-hoc object retrieval in the web of data. In: Rappa, M., Jones, P., Freire, J., Chakrabarti, S. (eds.) WWW, pp. 771–780. ACM (2010)
17. Suchanek, F.M., Kasneci, G., Weikum, G.: Yago: a core of semantic knowledge. In: Proceedings of the 16th International Conference on World Wide Web, pp. 697–706. ACM (2007)
18. Tang, T.T., Hawking, D., Craswell, N., Griffiths, K.: Focused crawling for both topical relevance and quality of medical information. In: Proceedings of the 14th ACM International Conference on Information and Knowledge Management, pp. 147–154. ACM (2005)
19. Von Ahn, L., Dabbish, L.: Labeling images with a computer game. In: Proceedings of the SIGCHI Conference on Human Factors in Computing Systems, pp. 319–326. ACM (2004)

Trust-Based Coalition Formation for Dynamic Service Composition in Social Networks

Amine Louati[✉], Joyce El Haddad, and Suzanne Pinson

Université Paris-Dauphine, LAMSADE CNRS UMR 7243, Paris, France
{amine.louati,elhaddad,pinson}@lamsade.dauphine.fr

Abstract. With the growing number of published Web services, including those being generated in social networks, a lot of service composition approaches have been proposed in the literature. However, they often fail to take into consideration social dimensions between requesters and providers as well as the providers autonomy in deciding with whom to collaborate. To address these challenges, we propose the use of Multi-Agent Systems, as they have the ability to form coalitions of trusted partners. We present a trust-based dynamic coalition formation process for service composition in social networks. In particular, our coalition formation process engaging self-interested agents is incremental, dynamic and overlapping. Agents are equipped with a set of services with their advertised QoS values and cooperate to fulfill the requester query based on a decentralized decision-making process guided by trust. We also presents evaluation results of first experiments to demonstrate the validity of our approach.

Keywords: Web services selection · Trust · Coalition formation

1 Introduction

Recent advances in the field of service-oriented computing have made it possible to publish, locate and invoke Web services across social networks. More and more users are willing to use their social networks to propose as well as to search services. In the case where no single service can satisfy user query, it should be possible to combine existing atomic Web services to fulfill it. Web service composition is the process that creates a value-added application by combining atomic Web services into a composite service which has the potential to reduce effort and time of development [1]. Service composition problem has triggered a considerable number of research efforts. However, the results obtained often fail to comply with the requesters' expectations. This is because they do not take into consideration social dimensions such as trust and information resulting from previous experiences. As agents have demonstrated the capability to promote knowledge representation and interaction as well as social metaphors like trust, we propose the use of Multi-Agent Systems (MAS) in the context of Web service composition. We assign to each user in the social network an

© Springer International Publishing Switzerland 2015
J. Wang et al. (Eds.): WISE 2015, Part I, LNCS 9418, pp. 570–585, 2015.
DOI: 10.1007/978-3-319-26190-4_38

autonomous and self-interested agent equipped with a set of services along with their advertised QoS values. As several agents can provide different atomic services with different QoS values, cooperation among them is essential to achieve all possible service compositions. The coalition formation is well suited to model cooperation between agents.

The aim of this work is to perform service composition using coalition formation process (CFP). In this context, three challenges arise:

(1) Integrate *social dimension* in the composition process: usually, users prefer providers that do not only propose required services but also that can be trusted. Most of the previous work on coalition formation for service composition [2–4] do not take into account the contribution of social dimension during the coalition formation process. Agents may decide whether or not to participate in a coalition and whether to accept or to refuse a partner's membership without any trust consideration.

(2) Enable *providers autonomy* to decide with whom they cooperate: both requesters and providers are self-interested agents that is, they must have the ability to decide locally on the selection of their partners in the composite service. Griffiths and Luck have presented in [5] a coalition formation model that combines trust and motivation to generate successful cooperations between self-interested agents. In this work, only agents which are candidates to join a coalition decide whether or not to participate based on their trust in the initiator of the coalition. Members of the coalition do not decide whether to accept or to refuse a candidate membership.

(3) Incorporate *dynamism* in the CFP: dynamic coalition means that the current content of the composite service may evolve over time depending on the state of the involved members. Bourdon et al. have introduced in [6] a multi-agent model able to find provider-centered trusted coalitions achieving composite web services. Although their model is based on multi-agent trust management involving the broker as well as members in the decision-making process on partner selection, it does not allow agents to join and leave coalitions dynamically. A problem arises when a member is unsatisfied with the presence of a specific agent and no means to leave its current coalition to join another one.

In this paper, we propose a new broker-based multi-agent model for dynamic service composition. The composition process is performed by self-interested agents equipped with a set of services along with their QoS criteria values. They cooperate in a coalition formation to collectively provide multiple composite services that meet the complex user query. To address the above challenges, we firstly integrate social dimension using a trust mechanism in the sense that prior to cooperating, each agent considers its trustworthiness in the other involved agents. Secondly, we leave autonomy for providers to decide on the selection of their partners in the composite service based on trust. The satisfaction of all members results in the formation of a stable and efficient coalition. If some members are not satisfied, we propose a mechanism that allows these members to leave the coalition. Thirdly, we incorporate dynamism in our CFP by allowing not satisfied members to leave the coalition.

The rest of paper is organized as follows. Section 2 points out the main concepts that we use in this work and describes our broker-based multi-agent model. Section 3 describes our coalition formation process along with its three phases. Section 4 discusses the experimental setup and the performance evaluation of our approach. Finally, Sect. 5 concludes the paper and identifies future work.

2   Background

In this section, we first point out the concepts that we will use for the coalition formation process then, we give an overview of our broker-based multi-agent model.

2.1   Concepts Definition

A social network is modeled by a graph, where each individual user is assigned a self-interested agent and, an edge between two users represent a symmetric social relationship between them.

Definition 1 (Social network). *Given a set $A = \{a_1, a_2, ..., a_s\}$ of agents and a set $E \subseteq A \times A$ of edges, a social network is a connected graph $G = <A, E>$ where an edge $(a_k, a_j) \in E$ represents a symmetric social relationship between a_k and a_j.*

A social network can be viewed as a multi-agent systems in which agents are endowed with a set of services. They act autonomously on behalf of their users in a flexible manner in order to fulfill their queries.

Definition 2 (Agent). *An agent $a_k \in A$ is an autonomous entity such that $a_k = <S_k, Trust, CT, ET, \lambda Inf_k, \lambda Sup_k, \beta_k, Blist_k>$ where:*

- *$S_k = \{s_1, s_2, ..., s_{m_k}\}$ is the set of m_k offered services,*
- *$Trust(a_k, a_j)$ is the trust that a_k has in an agent a_j in the social network which is defined as the aggregation of trust in sociability and trust in recommendation (for more details see [7]),*
- *$CT(a_k, a_j)$ is the trust in cooperation that a_k has in an agent a_j,*
- *$ET(a_k, a_j, s)$ is the trust in expertise that an agent a_k has in a service s offered by an agent a_j which is defined as the aggregation of three quality of service criteria namely: specialization, reliability and quality rating (for more details see [7]);*
- *λInf_k and $\lambda Sup_k \in [0, 1]$ are respectively, the trust in cooperation lower and upper thresholds,*
- *$\beta_k \in [0, 1]$ is the trust in coalition threshold.*
- *$Blist_k$ is a blacklist containing a set of not cooperative agents that do not comply with its terms of use.*

A user query is a composite service defined by an abstract plan. It does not refer to specific services, but rather to a finite set of semantic interfaces describing service functionalities. The aim of the composition problem is to find services that instantiate the query.

Definition 3 (Service). *A service s is a tuple such as $s = (in, out, f, q^1, q^2, q^3)$ where in is a set of inputs required to use the service, out is a set of outputs delivered at the completion of the service, f is a semantic interface describing the provided functionality, and q^1, q^2, q^3 are the advertised values of criteria: specialization, reliability and quality rating.*

Definition 4 (User Query). *Let F be the description domain of available functionalities. A user query $Q = \{f_1, f_2, \ldots, f_n | \forall 1 \le i \le n, f_i \in F\}$ is a finite set of functionalities.*

We denote by $A_p \subseteq A$ the set of trustworthy provider agents, by $A_i = \{a_k | a_k \in A_p$ and $\exists s \in S_k$ such as $s.f \equiv f_i\}$ the set of providers offering a service with the functionality f_i and by $A_Q = \bigcup_{i=1}^{n} A_i$ the set of providers offering services for all required functionalities in Q. As several agents could offer services for required functionalities, coalition formation between them can address the composition problem.

Definition 5 (Coalition). *Let Q be a user query. A coalition $c = \{x_1, x_2, \ldots, x_n | \forall i \in [\![1, n]\!], \exists k \in [\![1, s]\!]$ such as $x_i = a_k$ and $a_k \in A_i\}$ is a set of agents that instantiate Q.*

During the coalition formation process providers are organized in coalitions where each of them is able to provide one or many required services. A coalition that does not contain a set of agents required to instantiate all functionalities in the query is called *intermediate coalition*. An intermediate coalition, denoted c_z, is a partial instantiation of a query such as $c_z = \{x_1, x_2, \ldots, x_n | x_i \in \{a_k, f_i\}$ and $a_k \in A_i\}$. The content of an intermediate coalition evolves. The transition from one intermediate coalition c_z to another c_{z+1} is done through *proposal*.

Definition 6 (Proposal). *A proposal $\phi = \{y_1, y_2, \ldots, y_n | y_i \in \{a_k, \emptyset\}$ and $a_k \in A_i\}$ represents either a membership request or a membership offer. In case of a membership request, $\phi = \{y_1, y_2, \ldots, y_n | \exists! y_i$ such as $y_i = a_k \in A_i$ and $\forall j \ne i, y_j = \emptyset\}$. In case of a membership offer, $\phi = \{y_1, y_2, \ldots, y_n | y_i = x_i$ if $x_i \equiv a_k, y_i = \emptyset$ otherwise$\}$.*

2.2   Broker-Based Multi-agent Model

Based on Klusch and Sycara work [8], our multi-agent model is a broker-based model as shown in Fig. 1. It encompasses three different roles.

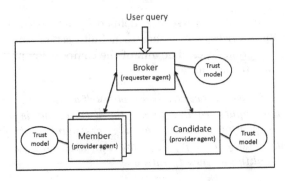

Fig. 1. Our broker-based multi-agent model

- *Candidate role:* a candidate is a service provider likely to join a coalition.
- *Member role:* a member is a service provider that is assigned to one coalition or several coalitions.
- *Broker role:* a broker is a service requester which is responsible for introducing candidates to members during the CFP and for orchestrating interactions between them. In case of several coalitions, it chooses also the coalition which has the best trust in expertise value.

Behaviors associated with each role will be described in detail in the following section using an interaction protocol.

3   Coalition Formation Process Description

Usually, service composition is done after a preliminary phase of service discovery. The service discovery phase described in our previous work [7] allows the requester agent a_r to identify a set of trustworthy providers A_p on which it will perform the coalition formation process (CFP). Considering a requester agent a_r in the social network (see Fig. 2(a)), after the service discovery phase each provider agent $a_k \in A_p$ is located with respect to a_r by a chain of trustworthy agents called *provider-recommender chain*. A provider-recommender chain is a father-child sequence of agents starting from the requester agent a_r and leading to a provider agent a_k in which all intermediate agents are either providers or recommenders. The output of the service discovery phase is a tree, called *Trust-Relation Social Network (TRSN)* (see Fig. 2(b)), built over the original social network (see Fig. 2(a)) where edges weights represent trust values between agents. Figure 3 illustrates the tree structure using a multilayered representation where the requester agent a_r is the root an each layer represents a level in the tree.

The goal of our CFP is not to find best providers that propose individually the required services, but rather to come up with a group of trustworthy agents that cooperate in a coalition formation to collectively provide composite services. Our CFP is original because it is:

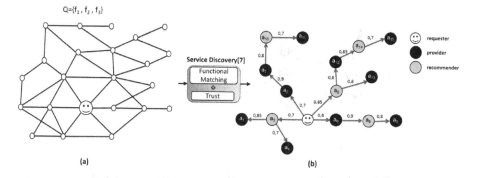

Fig. 2. (a) Social network and (b) Trust-relation social network: flat representation.

- *Incremental:* The CFP is achieved progressively by promoting first providers located in the first layer of the TRSN. It continues by seeking providers in the second layer and so one.
- *Dynamic:* Agents may join and leave autonomously their coalitions depending on the received messages and on their trust-based local decisions.
- *Overlapping:* Since a provider may offer more than one service, it can be member of several coalitions.

The CFP consists of three sequential phases namely: initial coalition generation phase, member selection phase and best coalition choice phase. In each of these phases, several actions are performed. Each phase produces an outcome that is used as input to the following phase.

3.1 Initial Coalition Generation Phase

The aim of this phase is to generate a set of initial coalition C in the TRSN. The inputs of the Algorithm 1 are Q the user query, A_p the set of providers and TRSN the trust-relation social network. Taking the broker role and setting C to $\emptyset$ (see Algorithm 1 line 1), the requester agent a_r starts with identifying providers in the first layer TRSN(1) which are at distance 1 from it. Then, it assigns each provider a_k to a new initial coalition c_z (for example, in Fig. 3, $C = \{c_1 = \{a_2\}, c_2 = \{a_6\}\}$) and sends to a_k an INFORM message containing the index of the coalition in which it is a member (see Algorithm 1 lines 4–5). We make the assumption that agents have the good will to join an empty coalition. However, afterwards they will be autonomous in their decision-making to join or to leave intermediate coalitions and to accept or refuse a membership request. On the reception of an INFORM message containing an index (see Algorithm 3 line 15), an agent a_k sets its role to member (see Algorithm 3 line 16). Each new initial coalition is added to C (see Algorithm 1 line 6).

At the end of this phase, the broker determines a set C of initial coalitions on which it will run the second phase of the CFP which is the member selection phase.

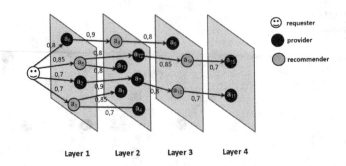

Fig. 3. Trust-relation social network: multilayered representation.

Algorithm 1. Initial Coalitions Generation Algorithm

Input: Q is a required query, A_p is the set of providers, TRSN is a multilayered
trust-relation social network where TRSN(1) is the first layer.
Variables: $C = \{c_z\}$ is the set of coalitions, $role_r$ is an array of $|C|$ values
$\in \{brok, mem, cand\}$ initially set to $brok$.

```
1  C ← ∅;
2  z ← 1;
3  for all (a_k ∈ TRSN(1) ∩ A_p) do
4  │   c_z = new initial_coalition(a_k, Q);
5  │   Inform(a_r, a_k, (z, mem));              /* see Algo. 3 line 15 */
6  │   C.add(c_z);
7  └   z ← z + 1;
8  return C;
```

3.2   Member Selection Phase

The aim of this phase is to complete sequentially each of the initial coalition
$c_z \in C$ by members providing required services to fulfill the user query. For sake
of simplicity, we limit the description of this phase to a single coalition formation
c_z. The member selection phase is an iterative phase handled by the broker a_r
where at the end of each iteration one member may join c_z. It is based on a
selection search strategy where a_r gives membership priority to providers close
to it in the tree because they are more trustworthy. Thus, a_r starts the member
selection in the first layer of the TRSN (see Algorithm 2 line 3). Recall that
our CFP is dynamic that is, agents can join and leave intermediate coalitions
at runtime. Therefore, it is essential to apply a timeout mechanism $timer_r[z]$
associated with c_z that allows the broker to decide when to stop the selection
process if it lasts too long (see Algorithm 2 line 4).

For each layer TRSN(l), a_r identifies a set of providers likely to join c_z. This
is done by first determining the set of required functionalities $func_r[z]$ in c_z
and then identifying providers $candP_r[z]$ that offer services with these required
functionalities (see Algorithm 2 lines 6–7). If there are only recommenders in
this layer or all identified providers do not offer services for required functionalities in c_z ($candP_r[z] = \emptyset$) then, a_r extends the providers identification to the
next layer TRSN($l + 1$). The transition from one layer l to another layer $l + 1$

Algorithm 2. Member Selection Algorithm

Input: A_p is the set of providers, $\mathcal{C} = \{c_z\}$ is the set of coalitions output of Algorithm 1, TRSN is a multilayered trust-relation social network, l_{max} is the maximum layer in TRSN.

Variables: $timer_r$ is an array of $|\mathcal{C}|$ integer, $candP_r$ is an array of $|\mathcal{C}|$ sets of candidate providers initially set to $\emptyset$, $func_r$ is an array of $|\mathcal{C}|$ sets of functionalities initially set to $\emptyset$, $count_r$ is an array of $|\mathcal{C}|$ integer initially set to 0, $reply_r$ is an array of $|\mathcal{C}|$ boolean initially set to False.

```
1   z ← 1;
2   for all (c_z ∈ C) do
3       l ← 1;
4       Activate(timer_r[z]);
5       while ((timer_r[z]) and (∃x ∈ c_z | x ≡ f) and (l ≤ l_max)) do
6           func_r[z] ← functionalities_identification(c_z);
7           candP_r[z] ← candidates_identification(func_r[z], TRSN(l) ∩ A_p);
8           while ((candP_r[z] ≠ ∅) and (∃f_i ∈ func_r[z])) do
9               a_j ← Argmax_{a_t ∈ candP_r[z] ∩ A_i} Trust(a_r, a_t);
10              for all (a_k ∈ c_z) do
11                  Propose(a_r, a_k, MembershipRequest(z, φ(a_j)));  /* see Algo. 3 line 1 */
12              wait (count_r[z] == |c_z|);
13              count_r[z] ← 0;
14              if (|OK| > |c_z|/2) then
15                  Propose(a_r, a_j, MembershipOffer(z, φ(c_z)));    /* see Algo. 4 line 1 */
16              else
17                  candP_r[z] ← candP_r[z] \ {a_j};
18                  reply_r[z] ← True;
19                  count_r[z] ← |c_z|;
20              wait (reply_r[z] == True);
21              reply_r[z] ← False;
22              wait (count_r[z] == |c_z|);
23              count_r[z] ← 0;
24          l ← l + 1;
25      z ← z + 1;
26  return C;
```

---

takes place if and only if all required functionalities in $func_r[z]$ are satisfied or the set of identified providers $candP_r[z]$ is empty (see Algorithm 2 lines 8 and 24). If a_r has a set of agents likely to join c_z ($candP_r[z] \neq \emptyset$) then, it will select based on trust one candidate among them to transmit its membership request to members of c_z. Trust is recognized as an effective mechanism for modeling and reasoning about others' honesty and reliability before cooperating with them [5,9–12]. More precisely, a_r ranks identified providers based on their trust values (computed in the discovery phase of our previous work [7]) and selects the most trustworthy one to be a candidate (see Algorithm 2 line 9). Then, it triggers a member selection conversation engaging members of c_z and the selected candidate a_j. The description of this conversation requires the definition of an interaction protocol that structure exchanged messages between agents. Figure 4 represents the member selection protocol using AUML formalism.

Having chosen a candidate a_j, the broker a_r sends to each member of the intermediate coalition c_z a PROPOSE message containing a membership request of a_j (see Algorithm 2 lines 10–11). The behavior of an agent is event-driven: on the reception of a PROPOSE message (see Algorithm 3 line 1), a member

---

Algorithm 3. Membership Request Algorithm

Variables: $role_k$ is an array of $|C|$ values $\in \{brok, mem, cand\}$ initially set to $cand$, $Blist_k$ is an array of blacklisted agents, $NbSoll_k[|A|]$ is an array of integer.

```
1   Procedure(RECEIVE_PROPOSE(a_r, a_k, MembershipRequest(z, φ(a_j))))
2       if (CT(a_k, a_j) > λSup_k and a_j ∉ Blist_k) then
3           Accept_Proposal(a_k, a_r, a_j);
4           NbSoll_k[j] ← NbSoll_k[j] + 1;
5       else
6           Reject_Proposal(a_k, a_r, a_j);
7
8   Procedure(RECEIVE_ACCEPT_PROPOSAL(a_k, a_r, a))
9       count_r[z] ← count_r[z] + 1;
10      |OK| ← |OK| + 1;
11
12  Procedure(RECEIVE_REJECT_PROPOSAL(a_k, a_r, a))
13      count_r[z] ← count_r[z] + 1;
14
15  Procedure(RECEIVE_INFORM(a_r, a_k, (z, mem))
16      role_k[z] ← mem;
17
```

---

a_k decides locally whether to accept or to reject a_j's membership request. As mentioned above, the decision-making process of an agent is based on trust evaluation. It could also be based on the result of the verification of its blacklist $Blist_k$ (see Algorithm 3 line 2): a member a_k may refuse to cooperate with a candidate agent a_j if it belongs to its blacklist. Thus, a membership request of a candidate included in this blacklist is directly rejected (see Algorithm 3 lines 5–6). Otherwise, a_k evaluates the trust in cooperation that it has in a_j based on the historical of the previous coalitions in which they were both involved. The trust in cooperation $CT(a_k, a_j)$ of an agent a_j from the point of view of another agent a_k is defined in formula 1.

Definition 7 (Trust in cooperation (CT)). *Let $NbSoll_k[j]$ be the total number of solicitations made by a_k to a_j to join its intermediate coalition and let $NbMem_k[j]$ be the number of a_j's memberships to the intermediate coalition in which a_k is member.*

$$CT(a_k, a_j) = \begin{cases} 1 & \text{if} \quad NbSoll_k[j] = 0 \\ \frac{NbMem_k[j]}{NbSoll_k[j]} & \text{otherwise} \end{cases} \tag{1}$$

The decision of a member a_k to accept or to reject the a_j membership request depends on its trust in cooperation value: if it exceeds λSup_k then a_k accepts a_j membership request and sends an ACCEPT_PROPOSAL message to the broker a_r. It also increments the $NbSoll_k[j]$ value (see Algorithm 3 lines 2–4). Otherwise, it rejects the membership and sends a REJECT_PROPOSAL message to the broker (see Algorithm 3 lines 5–6). On the reception of an ACCEPT_PROPOSAL message indicating a positive answer to the membership request, a_r increments the message counter $counter_r[z]$ and the number of positive answers (see Algorithm 3 lines 8–10). On the reception of a REJECT_PROPOSAL message indicating a negative answer to the membership request, a_r increments only the message

---

Algorithm 4. Membership Offer Algorithm

Variables: $evalC_k$ is an array of $|C|$ coalition trust evaluations, $NbMem_k[|A|]$ is an array of integer.

```
 1 Procedure(RECEIVE_PROPOSE(a_r, a_k, MembershipOffer(z, φ(c))))
 2 │   if (∃a_t ∈ c and a_t ∈ Blist_k) then
 3 │   └   Reject(a_k, a_r, (c, z));                                    /* see line 31 */
 4 │   else
 5 │   │   evalC_k[z] ← evalC(a_k, c);
 6 │   │   if (evalC_k[z] > β_k) then
 7 │   │   │   Accept(a_k, a_r, (c, z));                                /* see line 21 */
 8 │   │   └   role_k[z] ← mem;
 9 │   │   else
10 │   └   └   Reject(a_k, a_r, (c, z));                                /* see line 31 */
11
12 Procedure(RECEIVE_FAILURE(a_k, a_r, (a_k ∉ c)))
13 │   c_z.remove(a_k);
14 │   for all (t | a_k ∈ A_t) do
15 │   │   func_r[z] ← func_r[z] ∪ {f_t};
16 │   │   v ← 1;
17 │   │   for all (v ≤ l) do
18 │   └   └   candP_r[z] ← candP_r[z] ∪ {TRSN(v) ∩ A_t \ {a_k}};
19 │   count_r[z] ← count_r[z] + 1;
20
21 Procedure(RECEIVE_ACCEPT(a_k, a_r, (z, c)))
22 │   for all (a_t ∈ c_z) do
23 │   └   Inform(a_r, a_t, (z, a_k ∈ c_z));                            /* see line 36 */
24 │   c_z.add(a_k);
25 │   func_r[z] ← func_r[z] \ {f_i};
26 │   candP_r[z] ← candP_r[z] \ {a_k};
27 │   for all (a_t ∈ candP_r[z] ∩ A_i) do
28 │   └   candP_r[z] ← candP_r[z] \ {a_t};
29 │   reply_r[z] ← True;
30
31 Procedure(RECEIVE_REJECT(a_k, a_r, (z, c)))
32 │   candP_r[z] ← candP_r[z] \ {a_k};
33 │   count_r[z] ← |c_z|;
34 │   reply_r[z] ← True;
35
36 Procedure(RECEIVE_INFORM(a_r, a_k, (z, a_j ∈ c)))
37 │   if ((answer_k == KO) and (CT(a_k, a_j) < λInf_k)) then
38 │   │   role_k[z] ← cand;
39 │   │   Blist_k.add(a_j);
40 │   └   Failure(a_k, a_r, a_k ∉ c_z);                                /* see line 12 */
41 │   else
42 │   │   NbMem_k[l] ← NbMem_k[l] + 1;
43 │   └   Confirm(a_k, a_r, Yes);                                      /* see line 45 */
44
45 Procedure(RECEIVE_CONFIRM(a_k, a_r, Yes))
46 │   count_r[z] ← count_r[z] + 1;
47
```

---

counter $counter_r[z]$ (see Algorithm 3 lines 12–13). The broker waits for message from each member of c_z (see Algorithm 2 line 12). Once it received all messages (one message per member), it sets $counter_r[z]$ to 0 and makes a final decision based on the majority rule (see Algorithm 2 lines 13–14). In case the majority accepts the membership request of the candidate a_j, a_r initiates a conversation with the candidate agent a_j by sending a PROPOSE message containing a membership offer (see Algorithm 2 line 15). Otherwise, a_r removes the chosen candidate from $candP_r[z]$ and sets the synchronization parameters $counter_r[z]$ to 0 and $reply_r[z]$ to $True$ to unlock its waiting state (see Algorithm 2 lines 16–19).

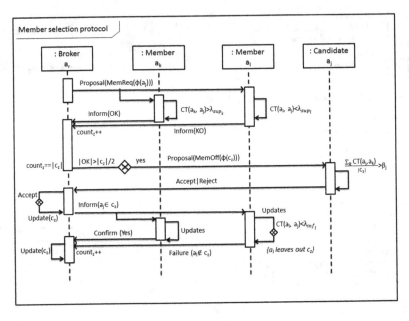

Fig. 4. AUML member selection protocol

On the reception of a PROPOSE message containing a membership offer (see Algorithm 4 line 1), an agent a_k checks whether the intermediate coalition c_z contains an agent in its blacklist. If so, it rejects the membership offer (see Algorithm 4 lines 2–3). Alternatively, it evaluates the trustworthiness of c_z using the measure defined in formula 2 (see Algorithm 4 line 5).

Definition 8 (Trust in a coalition (evalC)). *Let c_z be an intermediate coalition and a_k be a candidate. The trust in an intermediate coalition is defined as the arithmetic average of the trustworthiness of its members.*

$$evalC(a_k, c_z) = \frac{\sum_{a_t \in c_z} CT(a_k, a_t)}{|c_z|} \tag{2}$$

Once the evaluation is complete, a_k accepts the membership offer if its trust in the coalition c_z exceeds β_k. Consequently, it takes the member role ($role_k[z] = mem$) and replies to broker with ACCEPT message indicating the result of its decision (see Algorithm 4 lines 6–8). Otherwise, it replies to broker with REJECT message (see Algorithm 4 lines 9–10). In case of rejection (see Algorithm 4 line 31), the broker a_r removes a_k from $candP_r[z]$ and sets $counter_r[z]$ to $|c_z|$ (see Algorithm 4 lines 32–33) to unlock its waiting state (see Algorithm 2 line 22). In case of acceptance (see Algorithm 4 line 21), the broker a_r sends to all members of c_z an INFORM message announcing the successful selection of a_k (see Algorithm 4 lines 22–23). Then, it adds a_k to c_z, removes fulfilled functionality f_i from $func_r[z]$ and updates $candP_r[z]$ by removing all providers that offer a service with the functionality f_i (see Algorithm 4 lines 24–28). In both cases (see Algorithm 4 lines 29 and 35), a_r sets $reply_r[z]$ to $True$ to unlock its waiting state (see Algorithm 2 line 20).

On the reception of an INFORM message containing a successful selection (see Algorithm 4 line 36), a member $a_k \in c_z$ either has already accepted the membership request then, it increments $NbMem_k[j]$ value and sends an CONFIRM message to a_r to express its commitment (see Algorithm 4 lines 43–45) or has not accepted the membership request then, it may decide to leave c_z. This happens in case of unsatisfactory with the selection of the new member i.e., trust in cooperation value is lower than λInf_k. Dissatisfaction can result in an unhealthy and unstable group. Consequently, a_k leaves c_z by taking the candidate role ($role_k[z] = cand$), adds the new member to its blacklist and sends to the broker a FAILURE message indicating the result of its decision (see Algorithm 4 lines 39–42). Updating dynamically the agent blacklist, prevent our CFP from ping-pong effect where the coalition remains unstable i.e. agents leave and join coalition indefinitely. On the reception of an CONFIRM message containing a commitment (see Algorithm 4 line 45), the broker a_r increments $counter_r[z]$ (see Algorithm 4 line 46). On the reception of a FAILURE message from an agent a_k to leave the intermediate coalition c_z (see Algorithm 4 line 12), it first removes a_k from c_z then, it adds eliminated functionalities to the set of required functionalities in c_z (see Algorithm 4 lines 12–15). Finally, it enhances the set of candidates with providers offering the added functionalities from all previous layers and increments $counter_r[z]$ (see Algorithm 4 lines 16–19). Once a_r received all messages (see Algorithm 2 lines 21–22), it unlocks its waiting state and sets $counter_r[z]$ to 0. The member selection process of a coalition c_z repeats and continues until one of three termination conditions is met:

- The coalition is complete i.e. all required functionalities are instantiated,
- The timeout $timer_r[z]$ is due,
- The maximum layer of the TRSN tree is reached.

The result of this phase is a set $\mathcal{C}$ of coalitions which will be used as input for the next phase.

3.3 Best Coalition Choice Phase

Our approach addresses the aspect of finding multiple coalitions and then choosing the most appropriate for the requester. To do that, the broker first eliminates incomplete coalitions (see Algorithm 5 lines 1–3) then, performs a ranking of remained coalitions based on their trust in expertise values. Trust in expertise of a given coalition is defined as the average sum of the expertise of its members (see Algorithm 5 lines 4–5). Recall that trust in expertise of a member is a score established from the QoS values of its offered service(s) in the underlying coalition (see Definition 2). The coalition that has the best trust in expertise value is the composition service that will be transmitted to the user (see Algorithm 5 line 6).

Algorithm 5. Best Coalition Choice Algorithm

Input: $\mathcal{C} = \{c_z\}$ is the set of coalitions.
Variables: Exp_r is an expertise evaluation array, c is the best coalition.
1 **for all** $(c_z \in \mathcal{C})$ **do**
2 $\quad$ **if** $(\exists x \in c_z \mid x \equiv f)$ **then**
3 $\quad\quad$ $\mathcal{C}.remove(c_z)$;

4 **for all** $(c_z \in \mathcal{C})$ **do**
5 $\quad$ $Exp_r[z] \leftarrow \frac{\sum_{a_t \in c_z} ET(a_t)}{|c_z|}$;
6 $c \leftarrow Argmax_{1 \le z \le |\mathcal{C}|} Exp[z]$;
7 **return** c;

---

4   Experiments and Evaluation Results

In this section, we present evaluation results of the first experiments to demonstrate the validity of our approach. Experiments were conducted on a real dataset which is Facebook dataset[1] containing 4039 agents and 88234 edges. This dataset has been proved to possess the small-world and scale-free characteristics of social networks [13]. A prototype was developed using Java 1.7 and the Jade[2] multiagent platform. The TRSN graph data was stored in a GML (Graph Modeling Language) format[3]. Several experiments have been conducted, and all of them were run on a 3.1 GHz Core(TM) i5-2400 running windows 7, with a 8 G octet of RAM.

Table 1. Definition of test scenarios.

Scenario / Configuration	A	B	C	D	E
user query	2	3	4	5	6
service per functionality	5	5	5	5	5
timeout (ms)	5000	5000	5000	5000	5000

---

[1] http://snap.stanford.edu/data/egonets-Facebook.html.
[2] Telecom Italia Lab. JADE 4.3 http://jade.tilab.com/.
[3] http://www.fim.uni-passau.de/en/fim/faculty/chairs/theoretische-informatik/projects.html.

We created a pool of web services containing 10 categories of functionalities in each of them we defined 5 different services resulting in 50 services. We equipped each agent with three different services, randomly chosen among the pool. Initially, we set the blacklist $Blist_k$ size of each agent a_k to 100 randomly populated. Furthermore, we set trust thresholds as follows: $\lambda Sup_k = 0.75$, $\lambda Inf_k = 0.4$ and $\beta_k = 0.7$. Finally, we initialize the parameters $NbSoll_k[A]$ and $NbMem_k[A]$ with random values with $NbSoll_k[A] > NbMem_k[A]$.

Five test scenarios were created, as depicted in Table 1 where the user query line denotes the number of required functionalities in the query, the second line defines the number of service per functionality category representing a particular service and the third line specifies the applied timeout in millisecond. 100 coalition formation test runs were performed per scenario. In the first series of experiments, we study the communication cost of our approach for the different scenarios. In the second series of experiments, we discuss the completeness of our coalition formation process for the different scenarios.

Figure 5 shows the average number of exchanged messages per test scenario. As we expected, the higher the number of required functionalities is, the higher the number of exchanged messages is. This is particularly noticeable for scenarios D and E where more interactions are necessary to find providers offering services with the required functionalities. Moreover, for larger coalitions, there is more likely that an unstable state occurs. This observation will be confirmed in the next experiment (see Fig. 6 scenarios D and E). Instability requires more interactions to replace the missing members and complete the coalition.

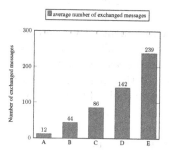

Fig. 5. Average number of exchanged messages per test scenario.

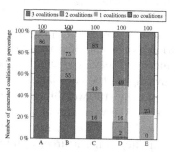

Fig. 6. Percentage of generated coalitions per test Scenario.

One of the key criteria reflecting the performance of web service composition is its capacity to deliver a composite service that fulfills user query. Figure 6 shows results averaged over 100 simulation runs, where the y axis represents the number of generated coalitions in percentage and the x axis the test scenario. We note that 86 % of scenario A test runs returned with the maximum number of generated coalitions. However, with the increasing number of required functionalities (as in scenarios B, C, D and E), the proportion of test runs with less than maximal number of generated coalitions also increases. For example, test runs on scenario E give 1 coalition in 23 % of time and no coalition in 77 % of time. This can be explained by the fact that members of large a coalition are more likely to be unsatisfied with the presence of a new member leading to an unstable coalition. Since our execution is limited in duration by a timeout, a single coalition is formed before the expiration of the timeout and the others remain incomplete.

Current findings prove that our proposed approach is effective for its ability to produce multiple composite services at a low cost of communications. However, we need to learn more about several issues such as how to determine appropriate values for trust thresholds used in decision-making in order to improve effectiveness and stability.

5   Conclusion and Future Work

In this work, we have presented an original trust-based dynamic coalition formation process for service composition in social networks. The coalition formation process is incremental, dynamic and overlapping where self-interested agents are equipped with a set of services along with their QoS criteria values. They cooperate based on a decentralized decision-making process guided by trust to collectively provide composite services that meet the complex user query. Our approach results in the formation of a stable coalition as it allows members to leave any coalition if they are no longer satisfied with the members of this coalition. Current findings prove that our proposed approach is effective for its ability to produce multiple composite services at a low cost of communications.

As future work, we plan to perform further experiments. By means of simulation, we would like to study agents behavior by monitoring their average frequency of leaving during the coalition formation process. We would also like to investigate the correlation between the quality of the chosen composite service and the trustworthiness of its members. We would also like to analyze the impact of the variation of the maximum layer value on the coalition formation process. Finally, we intend to extend the current model of trust in cooperation by including other sources of information such as reputation.

References

1. Ponnekanti, S.R., Fox, A.: Sword: a developer toolkit for web service composition. In: Proceedings of the 11th International WWW Conference, pp. 83–107 (2002)
2. Ermolayev, V., Keberle, N., Plaksin, S.: Towards agent-based rational service composition – RACING approach. In: Jeckle, M., Zhang, L.-J. (eds.) ICWS-Europe 2003. LNCS, vol. 2853, pp. 167–182. Springer, Heidelberg (2003)
3. Muller, I., Kowalczyk, R., Braun, P.: Towards agent-based coalition formation for service composition. In: Proceedings of the IEEE/WIC/ACM International Conference on Intelligent Agent Technology, pp. 73–80. IEEE Computer Society (2006)

4. Hongxia, T., Jian, C., Shensheng, Z., Minglu, L.: A distributed agent coalition algorithm for web service composition. In: 2014 IEEE World Congress on Services, pp. 62–69 (2009)
5. Griffiths, N., Luck, M.: Coalition formation through motivation and trust. In: AAMAS, pp. 17–24. ACM (2003)
6. Bourdon, J., Vercouter, L., Ishida, T.: A multiagent model for provider-centered trust in composite web services. In: Yang, J.-J., Yokoo, M., Ito, T., Jin, Z., Scerri, P. (eds.) PRIMA 2009. LNCS, vol. 5925, pp. 216–228. Springer, Heidelberg (2009)
7. Louati, A., El Haddad, J., Pinson, S.: Towards agent-based and trust-oriented service discovery approach in social networks. In: TRUST 2014 at AAMAS, pp. 1–12 (2014)
8. Klusch, M., Sycara, K.P.: Brokering and matchmaking for coordination of agent societies: a survey. In: Omicini, A., Zambonelli, F., Klusch, M., Tolksdorf, R. (eds.) Coordination of Internet Agents: Models, Technologies, and Applications, pp. 197–224. Springer, Heidelberg (2001)
9. Adler, P.: Market, hierarchy, and trust: the knowledge economy and the future of capitalism. Organ. Sci. **12**, 215–234 (2001)
10. Al-Sharawneh, J.: Social Networks: Service Selection and Recommendation. Ph. D. thesis, University of Technology, Sydney (2012)
11. Castelfranchi, C., Falcone, R.: Principles of trust for mas: cognitive anatomy, social importance, and quantification. In: ICMAS, pp. 72–79. IEEE Computer Society (1998)
12. Marsh, S.P.: Formalising trust as a computational Concept. Ph. D. thesis (1994)
13. Kleinberg, J.: The small-world phenomenon: an algorithmic perspective. In: Proceedings of the Thirty-Second Annual ACM Symposium on Theory of Computing, pp. 163–170 (2000)

Towards a Burglary Risk Profiler
Using Demographic and Spatial Factors

Cristina Kadar[(✉)], Grammatiki Zanni,
Thijs Vogels, and Irena Pletikosa Cvijikj

Information Management Chair, D-MTEC, ETH Zurich, Zurich, Switzerland
{ckadar,ipletikosa}@ethz.ch, {gzanni,tvogels}@student.ethz.ch
http://www.im.ethz.ch/

Abstract. According to modern crime victimization theories, the offender, the victim, and the spatial environment equally affect the likelihood of a crime getting committed, especially in the case of burglaries. With this in mind, we compile an extensive list of potential drivers of burglary by aggregating data from different open data sources, such as census statistics (social, demographic, and economic data), points of interest, and the national road network. Based on the underlying data distribution, we build statistical models that automatically select the risk factors affecting the burglary numbers in the Swiss municipalities and predict the level of future crimes. The gained information is integrated in a crime prevention information system providing its users a view of the current crime exposure in their neighborhood.

Keywords: Social issues · Open data · Risk factors · Data mining · Crime prevention information systems

1 Introduction

Consistently over time and nations, crime scores as one of the top public concerns [13]. Crime can have different forms and impacts but overall it disturbs the public welfare and reduces the citizens' quality of life [10]. Zooming in on Switzerland for example – once seen as the safest country in Europe, the country has succumbed in the last years to the continents average levels of street violence, burglaries and assault [8]. Within the pie of crime, residential burglary has one of the biggest shares: 70 % of all reported crimes represent property crimes, and 14 % of these are cases of household burglary [28]. According to the Swiss annual reports on criminal statistics [28], the year 2012 has seen an historical peak of 932 burglaries per year for every 100.000 inhabitants, making Switzerland the top target for burglars in Europe [3, 28].

Past research reveals that the number one fact influencing people's decision for buying or renting a real estate is the location of the property [26]. Properties located in regions where the crime rate is low or exhibits a downwards trend are considered as safer by the potential buyers. While existing information systems

© Springer International Publishing Switzerland 2015
J. Wang et al. (Eds.): WISE 2015, Part I, LNCS 9418, pp. 586–600, 2015.
DOI: 10.1007/978-3-319-26190-4_39

(IS) in the public and private sector offer basic visualizations and statistics over the types of different crimes in an area and their evolution over time [21,34], we are currently not aware of any solution providing a deep dive into the local crime factors, and offering its users an accurate and contextualized information of their current crime exposure.

Traditionally, criminology researchers looked at a standard set of demographic factors and built regression models to identify those factors that significantly correlate, either positively or negatively, with the crime rates [20]. But modern victimization theories hypothesize that the environment where the crime happens plays an at least as important role [38]. This is particularly true for burglaries, where the variation in risk of a location being victimized can be explained by factors like the surrounding area, the household and premise characteristics and other aspects of lifestyle affecting the location [19]. This creates an opportunity for non-conventional factors to be integrated in crime prediction models by tapping into open data sources describing our cities, like for example crowd-sourced maps and points of interests (POIs).

In this paper, we investigate a substantial set of heterogeneous factors, we determine if and how they contribute to the amount of residential burglary in the Swiss municipalities, and we make the results available to the public by means of a mobile application. The contributions of the paper are threefold:

1. informed by research in the field of criminology, we leverage various open data sources as relevant burglary predictors: (i) comprehensive census data of the Swiss population, (ii) different categories of POIs in the Swiss communes, and (iii) the national road network;
2. we build statistical models using specific probability distribution assumptions in order to select the model fitting best the data and returning the optimal list of factors affecting the burglary levels;
3. we develop, to the best of our knowledge, the first IS that provides a deep dive into the local crime driving factors, offerings its users a detailed view of the safety in their neighborhood in form of a burglary risk profiler.

The remainder of the paper is structured as follows: we first list existing crime prevention applications, provide a brief overview of background theories from criminological studies, and survey common burglary attractors and detractors in Sect. 2. We then present our method in Sect. 3 in terms of the datasets we have leveraged, the pre-processing and data manipulation we have conducted, and the statistical methods we have employed. We present and discuss the results obtained when applying our method to the data in Sect. 4. In Sect. 5, we describe the proposed implementation for visualizing the identified risk factors and the predicted crime levels. Finally, we conclude by discussing implications, limitations and future steps in Sect. 6.

2   Related Work

2.1   Crime Prevention Information Systems

Law Enforcement Solutions. To address the social issues of crime, police forces across the world are starting to involve citizens in crime prevention through community policing initiatives [25]. One approach towards citizen empowerment, leading to greater transparency and facilitating group problem-solving strategies, is sharing crime related information with the public in form of crime-tracking information systems, hosted mostly by the police departments themselves. Among the pioneers was the UK Home Office with the publication of monthly crime maps starting from December 2008 [4]. The interest in the Police. co.uk [21] website was so immense that the service crashed on its first day online due to the large number of visitors [12]. Internally, law enforcement agencies have also started integrating crime analytics and prediction components. In this space, one professional solution stands out: PredPol [24], deployed by different police departments in US and used to identify high risk areas in the cities so that patrols can be dispatched accordingly.

Commercial Solutions. Apart from the platforms created by the officials, numerous commercial web and mobile applications for crime prevention emerged in the last few years. These solutions fall into two broad categories: (i) crime mapping and (ii) crime sharing IS. Crime mapping IS offer visualization of past crime incidents from trustworthy sources like police departments, sheriff agencies, or news media, in form of individual points or heat maps, with the prime examples of CrimeReports [6], SpotCrime [27], and Trulia [34]. Crime sharing IS facilitate real-time reporting of personal experiences with other users or with the local authorities. The shared information can take the form of (a) personal incidents, like on WikiCrimes [36] and on CrimePush [5], or of (b) tips compiled by users, like on TipSubmit Mobile [33], or by some organization, like on Beat the Burglar Home Survey [1].

2.2   Crime Theories and Burglary Risk Factors

Socio-Demographic Risk Factors of Burglary. Early theories trying to explain crime activity in terms of cause or prevention have focused exclusively either on offenders or on victims characteristics [18].

Initially, criminology studies have focused solely on socio-demographic attributes as factors correlating with victimization and have noticed that specific groups of people were facing higher risk of victimization compared with other groups. For example: men, young adults, and African Americans experienced higher risk – while women, old residents and white people had relatively low risk of victimization [14]. Such socio-demographic factors can be included in the *lifestyle-exposure theory* as indicators of lifestyle activities of the victim [14]. In time, various studies falling within the lifestyle theory have found a plethora of factors driving the attractiveness of a property and its risk of being burgled.

Properties with high apparent value [40], properties with owners that have higher education and income levels [18], and properties with few people living inside and which are empty during the day or the night [35] – all have been shown to be more attractive for potential burglars.

On the offender's side, theories like the *social learning theory* and the *theory of anomie and disorganization* deem crime as a product of social disharmony [39]. For instance, population density, unemployment rate and poverty are highlighted as contributing variables to total crime, but also exclusively to burglary [7]. Furthermore, research has found that offenders do not travel far in order to strike and do so based on a crime template derived from their daily routines, traveling paths and overall awareness space [2]. Looking at the demographics of a given spatial unit of analysis, would therefore give insights not only into the victim's attributes, but also into the ones of potential criminals.

Socio-demographic factors cannot explain the complete context of crime, as we show below. Yet, they should be included in the models as attributes of both the victims and of the offenders, while not being relied on exclusively.

Spatial Risk Factors of Burglary. In an *integrative theory* on the forces affecting crime generation, Miethe and Meier highlight the importance of all three variables in the equation of crime: the offender, the victim, and the situation [18]. The social context works together with the offenders motivation and the victims vulnerability influencing the probability of a criminal event occurring. Even further, the specific environment (seen as the time and place supporting the victimization) is part of the social context where the victim and offender come together and plays a crucial role [38].

Based on a review of the empirical literature, a research brief of the Rutgers Center on Public Security [19] identifies four main aggravating or mitigating spatial correlates specifically of urban residential burglary:

1. proximity to public housing units as operationalization of social disorganization: disadvantaged and disorganized areas have been shown to have high levels of crime due to low levels of informal surveillance, security measures and socioeconomic status, and to high levels of ethnic heterogeneity and residential mobility.
2. proximity to pawn shops: burglars not only want to subtract goods in a quick fashion in order to avoid detection, but they also want to dispose of the stolen items as quickly as possible and convert them to cash.
3. proximity to bus stops: public transportation offers the means for accessing and exiting neighborhoods readily and anonymously.
4. proximity to police stations, fire stations and hospitals: the increased likelihood of authorities being present can be considered a mitigating factor resulting in a decrease of risk in the area.

In the case of suburban residential burglary, burglars may need to secure and use their own means of transportation to move larger products (e.g. televisions). Dwellings that are more permeable, easily accessible and located in big streets with well-established road network are indicated as highly targeted for burglary [9].

Furthermore, prior research has acknowledged that burglars first search for an appropriate offending area and only then scan for a specific target within that area [41]. It is therefore sensible to identify a high risk area on a city or neighborhood level based on the crime correlates creating an environmental setting conducive for residential burglary. Based on this information, it may be then possible at later stages to identify residences within the area that may be at even more risk based on individual characteristics (like employed security measures, lighting, proximity to alleyways etc.).

3   Methodology

3.1   Datasets and Pre-Processing

Burglaries Dataset. We extracted all burglary incidents with physical damage or after intrusion from an anonymized, aggregated digest of the official police records. This dataset is not public, and was made available to us via a data contract credited in the Acknowledgements section. As the extract was generated early 2014, some cases from 2013 were still not finalized by the police, so complete datasets of all burglaries in Switzerland were available only for years 2012 and older. Below, we will analyze both 2012 and 2011. In 2012, there were 72.996 burglary incidents reported in 2142 of the 2495 official municipalities registered that year. In 2011, there were 67.304 burglaries in 2103 out of 2551 municipalities. Figures 1 and 2 are exemplifying a choropleth map of the burglary density (in incidents/km^2) and a histogram of the burglary counts for 2011 – note that the long tail of the distribution continues until 5256 (city of Zurich) and was cut from the graph. For later analysis, we label every entry as high-crime or low-crime municipality by ordering the datasets on burglary counts and splitting them into two roughly equally-sized subsets.

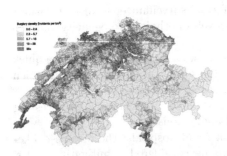

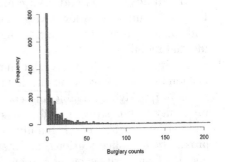

Fig. 1. Switzerland: Burglary density per municipality in 2011.

Fig. 2. Burglary counts per municipality in 2011.

Demographics Dataset. We acquired a total of 38 socio-demographic and economic factors describing the Swiss population from different governmental sources. Many of these factors were deemed in past studies as significantly influential of residential burglary. Most census data was derived from annual portraits provided by the Swiss Federal Statistical Office [29]: population density, population change (total, through migration, and through natural increase ratios), foreign nationals, age pyramid (young, adult, and old population ratios), spoken languages (German, French, Italian, and Romansh ratios), religion (Protestant, Catholic, and no religion ratios), area usage (settled and used for agriculture/forests/unused ratios), unemployment rate, residents employed in the different economy sectors (primary, secondary, and tertiary sector ratios), residential density (persons per apartment room), main means of transport (public and private ratios), and education level (less than obligatory, secondary, and upper education ratios). Housing (ownership housing, new housing, and vacant housing ratios), gender prevalence (man or woman), and marital status (single, married, and divorced ratios) data was compiled from the Population and Households Statistics [30], while wealth (net income) and social aid information were downloaded from the Interactive Statistical Atlas of Switzerland [31]. All data is already aggregated on the level of municipalities - the lowest administrative unit on which Swiss census data is publicly available - and was collected for years 2011 and 2012. Due to missing values, we remain in the end with 2364 and 2362, respectively, entries per year.

POIs Dataset. Widely used in cartography, especially in electronic variants including geographical information systems and GPS navigation software, POIs are specific point locations on a map that might be deemed as useful or interesting for specific activities. They are minimally described by the latitude and longitude or address of the location, type, name, and potentially, description.

We downloaded a POIs dump from the OSM database of Switzerland [22] on 19.02.2015. In total, 136.512 POIs were extracted spanning 167 basic categories, like bars, restaurants, schools, bus stations, or police stations [23]. Because the polygon buffers the Swiss border by 10 Km and covers Liechtenstein and most of the larger non-Swiss cities along the border, we first filter out all POIs that lay outside the country borders.

We then identify POI duplicates based on the following criteria: (i) being in the same municipality, (ii) having the same type, (iii) having similar names, and (iv) being closely located. For computing the name similarity, we transform the names to lowercase, and use the Smith-Waterman similarity score [32] to find a good alignment of the strings and count the number of identical characters. The computed score is a number between 0 (completely dissimilar names) and 1 (exact matching). In order to obtain a perfect similarity score of 1 for substrings, we normalize this score by dividing it by the string length of the shortest name. To find a good cut-off radius for duplicate search, we look at the proportion of exact name matches between every 2 POIs of the same type within a municipality as a function of their distance: 70 % of all POIs pairs within 10 m of each other are duplicates, 30 % at 50m, and 10 % at 100m. Based on this, we make the decision to mark as duplicate any pairs of POIs that are in the same municipality, have the same type, have an score of 1 by the metric defined above, and lay within 100 m of each other. Only the first point in the alphabetically sorted pair will be used in the further analysis steps.

Figure 3 contains a plot of the 98.585 remaining POIs after the preprocessing steps above. Based on the spatial burglary correlates identified in Sect. 2.2, we derive a set of four overarching categories: government buildings (including police stations, fire stations, embassies, etc.) restaurants/bars, shops, and public transportation (bus/tram/train stations, airports and so on). For every Swiss municipality, we then count the POIs falling into each of the previous categories; in case no POIs are found, we assume there are none (i.e. zero fill).

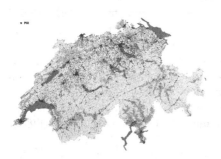

Fig. 3. Switzerland: POIs collected from OSM.

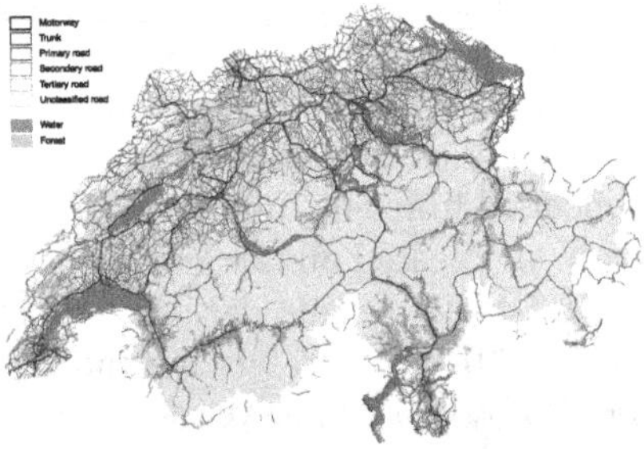

Fig. 4. Switzerland: national road network collected from OSM.

Roads Dataset. The meta-data of the Swiss road network was downloaded from OSM [22] on 09.05.2015. The extract contains 6 types of roads: motorway, trunk roads, primary road, secondary road, tertiary road, and unclassified roads, and is depicted in Fig. 4. As the literature review in Sect. 2.2 yielded main streets as escape roots and their proximity as an attractor for burglary, for each municipality we compute the motorway and total road density in road km/ha.

3.2 Statistical Approach

In this section, we propose two models for analyzing the effect of the various socio-demographic, economic, and spatial factors described above on the burglary rates of the Swiss municipalities. Our objective is twofold: (i) firstly, we want to build models that accurately predict future crime levels (in terms of expected count numbers and of expected high or low crime exposure), and (ii) equally importantly, understand the relevant risk factors and how they contribute to that.

For the first case, we deal with a response variable $y \in \mathbb{N}$ representing the number of occurrences of a specific event: the burglary counts within a municipality in the course of one year. We face a special case where the distribution of the response variable is discrete and is limited to non-negative integer values. Moreover, the distribution of the data is positively skewed with many observations having low count values – as seen in Fig. 2, the data follows a power law distribution. Applying an ordinary linear regression model to these data is problematic. To counteract these problems, Generalized Linear Models (GLMs) of the exponential family need to be employed. The Poisson regression assumes y has a Poisson distribution[1], and that the logarithm of its expected value $E(y) = \mu$ can be modeled as a linear combination of the input variables $\mathbf{x} \in \mathbb{R}^p$: $log(\mu) = \beta_0 + \beta_1 \mathbf{x}_1 + ... + \beta_p \mathbf{x}_p$.

On the side of the explanatory variables $\mathbf{x}$, we deal with many, potentially collinear factors. Including such a high number of variables in the model increases the risk of overfitting. Regularization methods help overcome this risk. Furthermore, we are interested in building a model with good accuracy (i.e. one that has high prediction power), but that still remains easily interpretable (i.e. has high inference power). Towards this goal, we exploit the power of the Least Absolute Shrinkage and Selection Operator (LASSO), a shrinkage method that imposes as regularizer a zero-mean Laplacian prior distribution on model parameters. Thus, LASSO performs automatically variable selection and yields sparse models, i.e. models that involve only a subset of the variables and are much easier to interpret [15].

Below, we deduct the steps necessary for estimating the regression coefficients. Let $\{\mathbf{x}_i, y_i\}_{i=1..n}$ be the input vector of n observation pairs in the training set, and $\{\beta_j\}_{j=0..p}$ the $p+1$ parameters of the model, where β_0 is the intercept and $\boldsymbol{\beta} = \{\beta_j\}_{j=1..p}$ for ease of notation. The log-likelihood of the training data in the Poisson case is given by:

$$l(\beta_0, \boldsymbol{\beta}|\mathbf{x}, y) = \sum_{i=1}^{n}(y_i(\beta_0 + \boldsymbol{\beta}^T \mathbf{x}_i) - e^{\beta_0 + \boldsymbol{\beta}^T \mathbf{x}_i})$$

The final objective function to be minimized is the LASSO penalized log-likelihood:

$$\min_{\beta_0, \boldsymbol{\beta}}(-\frac{1}{N}l(\beta_0, \boldsymbol{\beta}|\mathbf{x}, y) + \lambda\|\boldsymbol{\beta}\|_1)$$

where $\|\boldsymbol{\beta}\|_1 = \sum_{j=1}^{p}|\beta_j|$ is the l_1 norm of the parameters vector $\boldsymbol{\beta}$. This l_1 penalty has the effect of forcing some of the parameter estimates to be equal to zero when the tuning regularization parameter λ is sufficiently large [15].

For the classification of the crime in risk categories, we choose to implement the LASSO variant of the logistic regression classifier (also known in the literature as the maximum entropy classifier) and deduct below the classification function for the binomial case (i.e. two-class classification). Compared to above, the response variable $g \in G = \{low, high\}$ is a group membership, and not

---

[1] A discrete random variable Y is said to have a Poisson distribution with parameter $E(Y) = Var(Y) = \mu > 0$, if, for $k = 0, 1, 2,,$ the probability mass function of Y is given by: $P(Y = k) = \frac{\mu^k e^{-\mu}}{k!}$.

a natural value. The model is specified by the following logit transformation:
$\log \frac{P(G=high|X=x)}{P(G=low|X=x)} = \beta_0 + \beta^T \mathbf{x}$.

For each observation, it is convenient to code the class membership via a 0/1 response y_i with $y_i = 0$ for $g_i = low$, and $y_i = 1$ for $g_i = high$. Similarly to above, in order to learn the particular values of β (in other words, the weights of all predictors for each category), we compute the log-likelihood of the labels on the training data. The final objective function for the LASSO penalized logistic regression will prefer now more sparse models, reducing overfitting and improving performance [15]:

$$\min_{\beta_0, \beta}(-\frac{1}{N}\sum_{i=1}^{n}(y_i(\beta_0 + \beta^T \mathbf{x}_i) - \log(1 + e^{\beta_0 + \beta^T \mathbf{x}_i})) + \lambda\|\beta\|_1)$$

4   Experimental Results and Discussion

As the LASSO procedure is not invariant to linear transformations of the predictors because of the penalty based on absolute value of β, we normalize all data before beginning the analysis. In all experiments, we use cross-validation (CV) to compute the parameter estimates for many values of the penalty factor λ, and select the model with the minimal mean error. That means, for a given sequence of potential λs, the program does a k-fold cross-validation, with k = 10, to estimate the prediction error. Each k-times, a model is built based on a fraction $(k-1)/k$ of the data, and the prediction error is computed on the remaining fold. The average error across the k runs is reported and the model with the λ that yields the lowest average error is selected. Furthermore, we apply the obtained model to a dataset of new observations, to test the model's generalization ability on unseen data. Specifically, we will use below the 2011 burglary dataset to build the cross-validated models (i.e. as training and validation sets), and deploy these models to predict the 2012 crime levels (i.e. use the 2012 dataset as test set). For implementation we use the glmnet package, a very efficient R implementation of Lasso and Elastic-Net Regularized GLMs [11].

LASSO Poisson Regression. When running the CV experiments, we measure the goodness of fit of the Poisson model by its deviance: $D = 2\sum_{i=1}^{n}(y_i \log(\frac{y_i}{\mu_i}) - (y_i - \mu_i))$ where $\mu_i = e^{\beta_0 + \beta^T \mathbf{x}_i}$ denotes the predicted mean for observation i based on the estimated model parameters. If the model fits well, the observed values y_i will be close to their predicted means μ_i, causing both terms in D to be small, and so the deviance to be small. We show in Fig. 5 how the deviance varies with the different values of the regularization parameter λ when estimating a model based on the 2011 values. The optimal $\lambda = 0.7758532$ yields a minimum cross-validated Poisson deviance of 44.58 and selects 15 relevant input factors.

Listed in Table 1 are the corresponding coefficients for all initial $p = 44$ factors. These are to be interpreted as follows: one unit increase in the jth independent variable would lead to a multiplication of the dependent variable y by a factor of e^{β_j}, given the other variables are held constant in the model. For

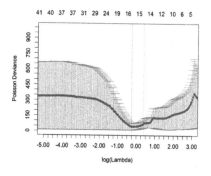

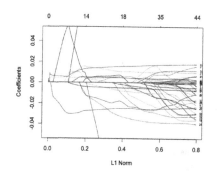

Fig. 5. Poisson regression CV: Poisson deviation as a function of the regularization parameter λ.

Fig. 6. Poisson regression CV: coefficient values as a function of the L1 norm of λ.

example: a 1 % increase in the unemployment rate of a municipality leads to an increase of the burglary counts (by a multiplication factor of 1.1031), while a 1 % increase in the ratio of houses privately owned decreases the burglary counts (the multiplication factor is 0.9750). Furthermore, Fig. 6 presents the exact regularization paths of the independent variables, and makes it visible how most get shrunken to 0 eventually.

After the automatic feature selection performed by the LASSO regressor, higher values of the following factors are found to increase the number of burglaries: population density, population change, unemployment rate, population ratio working in the services sector, population ratio using the public transport, population ratio with higher education, number of governmental buildings, and number of stations belonging to the public transport. On the other side, Swiss neighborhoods with the following characteristics exhibit lower burglary counts: higher number of Italian-speaking, Protestant, working in the raw materials industries, with secondary education, and male citizens; higher ratio of privately owned houses; and higher number of restaurants and bars. While most results are consistent with the criminology research presented in Sect. 2.2, the positive correlation with the number of governmental institutions comes as a surprise. We believe this is due to the bias of the OSM database towards more densely populated areas, where the users were more active and have tagged more POIs.

Moving on to prediction, we apply the learned regressor on new data and estimate the incident numbers for the year 2012. Figure 7 plots the distribution of the actual counts (in blue) vs. the distribution of the predicted counts (in transparent pink). It becomes clear that the empirical data shows more zeroes than would be expected under a regular Poisson distribution, and that we have a good match, but not a perfect one. To formally quantify the accuracy of the estimator we compute the root mean squared error $RMSE = \sqrt{\frac{1}{n}\sum_{i=1}^{n}(\hat{y}_i - y_i)^2}$: 171.42 in comparison to 34.88 on the training data.

Table 1. Poisson regression CV: coefficient values after shrinkage.

Predictor x_j	Coefficient β_j	Predictor x_j	Coefficient β_j	Predictor x_j	Coefficient β_j
population density	.00009597	settlement surface	.	high education	.00651758
population change	.00517831	empty surface	.	single	.
migration change	.	economic activity	.	married	.
natural change	.	unemployment rate	.09816673	widowed	.
foreigner	.	employed primary	-.02927544	divorced	.
young	.	employed secondary	.	man	-.1120552
adult	.	employed tertiary	.00508251	social aid	.
old	.	residential density	.	net income	.
german	.	ownership housing	-.02531557	government buildings	.00443092
french	.	new housing	.	restaurants and bars	-.004854803
italian	-.00634665	vacant housing	.	shops	.
romansh	.	public transportation	.01165278	transport. stations	.01526131
protestant	-.00013776	private transportation	.	motorway density	.
catholic	.	less oblig. education	.	total roads density	.
no religion	.	secondary education	-.01946257		

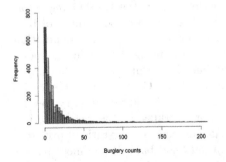

Fig. 7. Poisson regression prediction: histogram of actual vs. predicted burglary counts in the test set.

Table 2. Logistic regression prediction: classification confusion matrix in the test set.

		Predicted class	
		Low	High
Actual Class	Low	1020	110
	High	304	928

LASSO Logistic Regression. In a similar fashion, we run the binomial logistic regression CV experiments for year 2011 and obtain an optimal shrinkage parameter $\lambda = 0.0001961$. 31 input variables are preserved in the model, yielding a minimum cross-validated mean classification error of 14.89 %, i.e. a mean accuracy of 85.11 %.

We report in Table 2 the prediction results on the 2012 data in terms of a classification matrix containing the true positives tp, false positives fp, true negatives tn, and false negatives fn counts. The accuracy $Acc = \frac{tp+tn}{tp+tn+fp+fn}$, which represents the percentage of correct predictions made by the model when compared with the actual classifications, is 82.47 % – as expected, slightly worse than on the validation set above.

This prediction model performs well on a new target domain, making it a robust choice for deployment in a productive application, as we will show below. The logistic regression classifier is always fitted on the previous year's input factors and burglary data, and, once new input factors are received (e.g. the census releases the new statistics in the beginning of the year) the burglary risk for the current year is computed and shown in the burglary profiler of each municipality.

5 IS Implementation

In order to provide access to the results presented in the previous section to the Swiss citizens, we add a new component to a crime prevention IS we have previously designed and developed. CityWatch [16] intends to increase the safety of its users by raising the awareness about crime levels, as well as by providing concrete crime prevention tips. Figure 8 illustrates the system components, the main view of the crime risk profiler and the visualization of risk factors.

Fig. 8. System components (left), main application view showing aggregated crime map (center) and risk factors (right).

For the crime level information, CityWatch leverages data from two independent sources: (1) property claims data from a large Swiss insurance company, and (2) crowdsourced data. As such, the system provides trustworthy initial information, thus also overcoming the problem of unsustainable content generation and consumption typical for new crowdsourcing applications [17]. At the same time, it provides the possibility to its users to take on an active role in crime prevention, thus contributing to public good [25,37].

For each data source a separate crime map is generated which shows one year of historic data as well as crime prediction for the upcoming year. In order to address the privacy concerns and prevent potential identity disclosure of the victims, the trustworthy map is visualized in an aggregated form thus revealing only the number of incidents on a municipality level but not the exact locations where crime incidents occurred. At the same time, the crowdsourced map reveals the exact incident locations, including details such as date, time, a short description and a photo of the incident, but does not reveal the identity of the reporting user.

In addition to providing the number of incidents, the initial version of the system described in [16] was further extended to serve as a risk profiler by including the specific factors that influence the crime level for a location of interest and their significance based on the analysis presented in the previous section. Moreover, each area is further categorized as safe or unsafe, visualized using a semaphore scheme.

Apart from providing insights into the crime levels and predictors, CityWach counterbalances the potential negative effect that this information might have over individuals by informing its users about possible preventive actions. The preventive tips appear as (a) static tips, which are derived from the recommendations provided by government officials, and (b) dynamic tips, which are generated from the reported incidents and are personalized based on the user profile. Finally, depending on the user's preferences, notifications are triggered whenever an incident is reported in user's proximity or within any area of interest.

6 Conclusions, Limitations and Future Work

In this paper, we have exploited confidential police criminal records, as well as open data sources, such as census statistics, points of interest, and the national road network. Based on this data, we have built appropriate statistical models in order to identify the relevant risk factors and to predict future burglary rates, and we have shown how the derived information can be integrated in a crime prevention information system. The experimental results mirror some found by other studies in the field, but also reveal other correlates contributing to the empirical literature. Unemployment rates, population employed in the primary sector, houses in private ownership, and number of bus, tram, and train stations are some of the factors we found to influence burglary most strongly. Furthermore, we were able to predict high/low levels of crime exposure with an accuracy of roughly 82 % on unseen data.

This work has a broad impact. First, it gives individuals the means to understand the current level of criminality in their neighborhood and the type of risks they are facing. People living in places with higher criminal propensity can increase their awareness and undertake safety precautions. Second, the information could be leveraged by different government institutions. The Swiss police, for instance, could utilize the information by taking effective measures and increasing the patrols in the places predicted with high rates of burglary. Last, enterprises like insurance companies, could run awareness campaigns to inform their clients about the risk they face in order to safeguard their property. In general, such a service brings value to the society at large in form of the presented public platform, but would also find applications in the governmental and private sectors.

Our work is not without limitations and provides several opportunities for further research. We have presented here the results for only 2 consecutive years; to be able to claim that the burglary correlates hold over time, data from a longer time span needs to be analyzed. For more accurate predictions of incident counts, we plan to employ zero-inflated and over-dispersed poisson models in the future. What the IS implementation is concerned: integrating other sources of more dynamic data (like social media or human dynamics) will allow more granular predictions, while looking at other crime types (like pick-pocketing or vehicle theft) will open the way for new use cases.

Acknowledgments. The authors would like to acknowledge the data contract Nr. 140221 (Typ B) with Ref. 650.1-1 from September 2014 for the delivery of the confidential police criminal statistics.

References

1. Beat the Burglar Home Security Survey (2013). https://itunes.apple.com/us/app/beat-burglar-home-security/id562269446?mt=8. Accessed 24 April 2015
2. Bernasco, W.: A sentimental journey to crime: effects of residential history of crime location choice. Criminology **48**(2), 389–416 (2010)
3. Bradley, S.: Who's to blame for the Swiss crime wave? (2013). http://www.swissinfo.ch/eng/cops-and-robbers_who-s-to-blame-for-the-swiss-crime-wave-/35594462. Accessed 22 April 2015
4. Chainey, S., Tompson, L.: Engagement, empowerment and transparency: publishing crime statistics using online crime mapping. Policing **6**(3), 228–239 (2012)
5. CrimePush (2012). http://crimepush.com/. Accessed 22 April 2015
6. CrimeReports (2015). https://www.crimereports.com/. Accessed 22 April 2015
7. Cung, B.: Crime Demographics: An Analysis of LAPD Crime Data, Los Angeles, California, United States of America (2013)
8. Dacey, J.: Switzerland loses "safest country" accolade. http://www.swissinfo.ch/eng/switzerland-loses-safest-country-accolade/31023508. Accessed 22 April 2015
9. Davies, T.P., Bishop, S.R.: Modelling patterns of burglary on street networks. Crime Sci. **2**(1), 1–14 (2013). http://dx.doi.org/10.1186/2193-7680-2-10
10. Doran, B.J., Burgess, M.B.: Why Is Fear of Crime a Serious Social Problem? In: Putting Fear of Crime on the Map, pp. 9–23. Springer, New York (2012)
11. Friedman, J., Hastie, T., Tibshirani, R.: Regularization paths for generalized linear models via coordinate descent. J. Stat. Softw. **33**(1), 1–22 (2010)
12. Garbett, A., Wardman, J.K., Kirman, B., Linehan, C., Lawson, S.: Anti-social media: communicating risk through open data, crime maps and locative media. In: Proceedings of HCI Korea, pp. 145–152. Hanbit Media, Inc. (2014)
13. Global surveys show environment ranks low on public concerns (2013). http://news.uchicago.edu/article/2013/02/25/global-surveys-show-environment-ranks-low-public-concerns. Accessed 26 April 2015
14. Hindelang, M., Gottfredson, M., Garofalo, J.: Victims of Personal Crime: An Empirical Foundation for a Theory of Personal Victimization. Ballinger Publishing Company, Cambridge (1978)
15. Hastie, T., Tibshirani, R., Friedman, J.: The Elements of Statistical Learning: Data Mining, Inference and Prediction. Springer, New York (2009)
16. Kadar, C., Cvijikj, I.P.: CityWatch: the personalized crime prevention assistant. In: Proceedings of the 13th International Conference on Mobile and Ubiquitous Multimedia, pp. 260–261. ACM (2014)
17. Kumar, V.: Why do Consumers Contribute to Connected Goods? A Dynamic Game of Competition and Cooperation in Social Networks. Social Networks, pp. 1–54 (2009)

18. Miethe, T.D., Meier, R.F.: Crime and its social context: toward an integrated theory of offenders, victims and situations. In: Cullen, F.T., Wilcox, P. (eds.) Encyclopedia of Criminological Theory, pp. 631–635. Sage Publications, Inc. (1994)

19. Moreto, W.D.: Risk factors of urban residential burglary. RTM Insights **4**, 1–3 (2010)

20. Osgood, D.W.: Poisson-based regression analysis of aggregate crime rates. J. Quant. Criminol. **16**(1), 21–43 (2000)

21. London Crime Map (2015). http://www.police.uk/metropolitan/00BK17N/crime. Accessed 22 April 2015

22. OSM Data for Switzerland (2015). http://planet.osm.ch/

23. OSM Map Features (2015). http://wiki.openstreetmap.org/wiki/Map_Features. Accessed 22 April 2015

24. PredPol (2015). http://www.predpol.com/. Accessed 22 April 2015

25. Ratcliffe, J.H.: Damned if you don't, damned if you do: crime mapping and its implications in the real world. Policing Soc. **12**(3), 211–225 (2002)

26. Schwartz, A.E., Susin, S., Voicu, I.: Has falling crime driven New York city's real estate boom? J. Hous. Res. **14**, 101–135 (2003)

27. SpotCrime (2015). http://www.spotcrime.com/. Accessed 22 April 2015

28. Swiss Federal Statistical Office: Police Crime Statistics, Annual Reports 2011–2014 (2013). http://www.bfs.admin.ch/bfs/portal/de/index/news/publikationen. html. Accessed 22 April 2015

29. Swiss Federal Statistical Office: Portraits of Communes 2011–2012 (2015). http://www.bfs.admin.ch/bfs/portal/en/index/regionen/02/key.html. Accessed 10 January 2015

30. Swiss Federal Statistical Office: STAT-TAB - the FSO's interactive database (2015). http://www.bfs.admin.ch/bfs/portal/en/index/infothek/onlinedb/stattab.html. Accessed 3 March 2015

31. Swiss Federal Statistical Office: Statatlas Switzerland (2015). http://www.bfs.admin.ch/bfs/portal/en/index/regionen/thematische_karten/02.html. Accessed 16 February 2015

32. Smith, Temple F., Waterman, Michael S.: Identification of common molecular subsequences. J. Mol. Biol. **147**, 195–197 (1981)

33. TipSubmit Mobile (2014). https://itunes.apple.com/us/app/tipsubmit-mobile/id533472612?mt=8. Accessed 24 April 2015

34. Trulia Crime Map (2015). http://www.trulia.com/local. Accessed 24 April 2015

35. Tseloni, A., Wittebrood, K., Farrell, G., Pease, K.: Burglary victimization in England and Wales, the united stages and the Netherlands. Br. J. Criminol. **44**, 66–91 (2004)

36. WikiCrimes (2015). http://www.wikicrimes.org/. Accessed 22 April 2015

37. Wallace, A.: Mapping city crime and the new aesthetic of danger. J. Vis. Cult. **8**(1), 5–24 (2009)

38. Wilcox, P.: Theories of Victimization. In: Fisher, B., Lab, S. (eds.) Encyclopedia of Victimology and Crime Prevention, pp. 978–986. Sage Publications, Inc. (2010)

39. Williams, F.P., McShane, M.D.: Criminology Theory. Elsevier, New York (1993)

40. Wittebrood, K., Nieuwbeerta, P.: Criminal victimization during one's life course: the effects of previous victimization and patterns of routine activities. J. Res. Crime Delinq. **37**, 91–122 (2000)

41. Wright, R.T., Decker, S.H.: Burglars on the Job: Streetlife and Residential Breakins. Northeastern University Press, Boston (1994)

Extract Knowledge About Geo-Location Using Context and Content Information of Geo-Tagged Social Media

Sheetal Takale[1]([⊠]) and Prakash Kulkarni[2]

[1] Vidya Pratishthan's COE, Baramati, Savitribai Phule Pune University, Pune, India
sheetaltakale@gmail.com
[2] Walchand College of Engineering, Sangli, India
pjk_walchand@rediffmail.com

Abstract. Recent years have witnessed a phenomenal growth of social sharing websites. These social sharing web sites host a large volume of geo-referenced and community contributed media resources. However, due to the size and inability of computers to understand the content of the images, it is inherently difficult to organize, browse or search these collections. The work proposed here addresses an interesting task of organizing these geo-referenced media on Flickr to generate visual and thematic summarization of specified geo-location. The major challenge is: how to use the unstructured and unrestricted community contributed media and annotations to generate knowledge? We propose the modified PLSA algorithm to generate representative sets of images and tags for given geo-location. We have used location metadata, tags associated with images, and the visual features of images to understand context and content of images. To deal with social tagging accuracy problem, we consider ternary interrelations and multiple intra-relations among user, image and tag and model the relations using HOSVD. The novel visualization scheme summarizes the geo-location with a rich display landscape and provides location description using location representative tags. Experiments are performed on geo-tagged Flickr images for various geo-locations. The experimental results have validated the proposed method.

Keywords: Geo-referenced photographs · Social media · Tensor factorization · Summarization · PLSA · Clustering · Location visualization

1 Introduction

Due to the advent of geo-referenced digital photography, photos are now associated with the metadata describing the geographic location in which they were taken. The rapid growth of content sharing websites has resulted in large volume of location data stored on social media websites. Geo-location specific multimedia content is accompanied by user generated content (UGC) such as title, tags, comments and description. These annotations are in the form of free keywords, also

© Springer International Publishing Switzerland 2015
J. Wang et al. (Eds.): WISE 2015, Part I, LNCS 9418, pp. 601–615, 2015.
DOI: 10.1007/978-3-319-26190-4_40

known as social tags. Using these annotations, users can express their personal opinion to describe items. So, social annotations can be utilized to understand users interest and context of the image.

Various photo sharing websites allow users to be owner, tagger or commentator or to mark favorite for their community contributed images. They also allow users to interact and collaborate (share, chat) with each other in a social media dialog. The typical structure of photo sharing social media involves three types of interrelated entities: user, image and tag. The largest repositories of user location histories are in fact photo sharing web sites like Flickr. Viewing and interacting with such collections has a broad social and practical importance. However, these collections are inherently difficult to organize, browse and search due to their size and the inability of computers to understand the content of the images. This is leading to interesting tasks like geographically organizing photos and location visualization [3,5,8–10]. These geo referenced photos can be organized geographically. But, we need to filter, sort and summarize the collection of photos.

The work proposed here addresses an interesting task of organizing these geo-referenced media on Flickr to generate visual and thematic summarization of specified geo-location. Content and context information of geo-location images and modified PLSA algorithm is used for location visualization and theme generation. Content information for images is obtained using visual analysis. Context information for geo-location images is generated using users potential annotations. To improve the underlying associations between the images and tags, user affinity, image affinity and tag affinity are modeled in the form of graph to represent intra-relations among them. Multiple intra-relations and interrelations among User, Image and Tag are incorporated into 3-order tensor. We utilize tensor factorization framework [4,7,11] for tag refinement. We introduce the problem of Location visualization using content, context and geo-reference information of social images. The unstructured and unrestricted community contributed images and annotations are utilized to generate the knowledge. Annotation refinement or recommendation of tags for un-tagged geo-referenced images is achieved. We propose a modified PLSA model to discover the location theme by combining textual and visual content of images. To summarize, major contributions include:

- An outline of the new approach for generating visual and textual summarization of geo-location using content, context and geo-reference information of social images (Sect. 3).
- An implementation of algorithm for refinement of image tag association by tag recommendation using Multiple Inter and Intra relations of U, I, T and HOSVD (Sect. 4).
- An Implementation of modified or extended PLSA for generation of image clusters. Additional Conditions to be satisfied by Extended PLSA include visual cluster connectivity and users coverage (Sect. 5).
- Tag cleaning, Tag scoring and Tag selection steps are implemented to select representative tags and demote personal and general tags (Sect. 5).
- Evaluation of geo-location summarization algorithm. We compare this method with baseline methods (Sect. 6).

2   Related Work

The rich User Generated Content (UGC) on social sharing web sites has opened up great possibilities for novel multimedia research and applications. Geo-referenced multimedia data mining is the most attractive and important filed of research. Recent research and applications on online geo-referenced media [6] can be grouped into different classes such as: (a) Organizing media resources geographically, (b) Social knowledge extraction from geo-referenced media, (c) Learning landmarks in the world, (d) Estimating Geographic location of photo.

Visual summary generated by any algorithm should be having representative and diverse set images. The most intuitive approach to organizing social media resources is to perform image clustering and then select a representative image of each cluster. State-of-the-art clustering approaches can be divided into several categories including visual content, text associated with images, automatically generated metadata such as geo-tags, users activity statistics or a combination of these resources.

Rudinac et al. [10] developed an approach for automatic visual summarization of a geographic area that exploits user contributed images and related explicit and implicit metadata. It is based on the random walk with restarts over a graph that models relations between images, visual features extracted from them, associated text, as well as the information on the uploader and commentators.

Kennedy et al. [5] proposed a multimodal approach for selecting diverse and representative image search results for landmarks. They also rely on both the visual information in the images and the user-contributed tags for these images. The two step method extracts the representative tags first, which are further utilized to automatically find place and event semantics.

Fang et al. [9] investigated the problem of location visualization from multiple themes. They identify the highly photographed place (POI) and discover their distributed themes.

Hao et al. [8] proposed an approach for location overview generation approach, which first mines location representative tags from travelogue and then uses these tags to retrieve images from web.

Despite the importance and benefits of social tagging, it suffers from reduced recommendation accuracy problem. Reasons behind are: Free Nature: Tags assigned are free keywords. They are subject to multiple interpretations and results in polysemy and synonymy problems. Cold Start or Tag Sparsity: Many a times user rarely participates in tagging process. This results in untagged photos. Visual-Textual Relevance: How to interpret the relevance of user contributed tag with respect to the visual content? Ternary Relation: Social tagging data forms a ternary relation between users, resources, and tags.

In order to capture the ternary relation among users, tags and items in social tagging systems, previous methods like [7] focused on generating recommendations based on tensor factorization (TF) techniques. Such methods are able to (1) solve problems like polysemy and synonymity (2) preserve the ternary relation (3) reveal the latent associations among users, tags, and items (4) provide more accurate recommendation. These methods do not solve the problem of sparsity, and the cold start problem. To deal with these problems multiple inter and intra relations among U, I, T are utilized by the method proposed in [4].

3   System Architecture

Figure 1 outlines the framework of the system. The input into our first module is a location, which is specified by its geo-tag. We utilize set of images taken at given geo-location and the associated metadata as input to our system. The interrelations, multiple intra relations among U, I, and T are incorporated into tensor factorization framework for tag refinement and potential annotations for images are generated in second module. In the third module of the proposed work, context and content information of image is incorporated into probabilistic generative model to generate visual summary of the location. Unlike to the standard probabilistic model, the extended PLSA proposed here combines, image-tag association, location context information represented by annotations and image content information for images to generate themes. Finally, for each visual theme, generated in the third module, representative tags are selected.

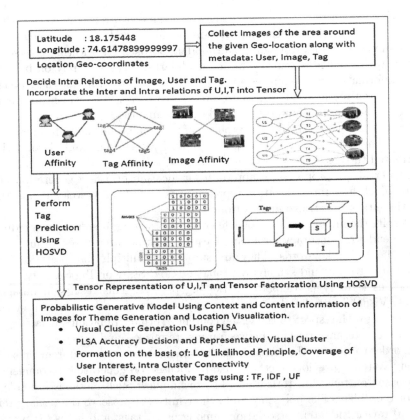

Fig. 1. System architecture

4 Image Tag Refinement Using HOSVD

4.1 HOSVD Algorithm Steps

Let U, I, T denote the sets of users, images, tags. The set of observed tagging data is denoted as, $O \subset U \times I \times T$. The ternary interrelations can then constitute a three dimensional tensor, $Y \epsilon R^{|U| \times |I| \times |T|}$, which is defined as:

$$Y_{u,i,t} = \begin{cases} 1 \; if(u,i,t) \\ 0 \; otherwise \end{cases} \tag{1}$$

To jointly model the three factors of user, image, and tag, we employ the general tensor factorization model, Tucker decomposition. In Tucker Decomposition [4, 7,11], the tagging data Y are estimated by three low rank matrices and one core tensor.

$$\hat{Y} = C \times_u U \times_i I \times_t T \tag{2}$$

where, $\times_n$ is the tensor product of multiplying a matrix on mode n.
The core tensor C governs the interactions among user, item and tag entities.

$$C = Y \times_1 U_{c1}^{(1)^T} \times_2 U_{c2}^{(2)^T} \times_3 U_{c3}^{(3)^T} \tag{3}$$

The Tensor decomposition problem is reduced to minimizing a point-wise loss on $\hat{Y}$, defined as :

$$min_{U,I,T,C} \sum_{(u,i,t)\epsilon|U|\times|I|\times|T|} (\hat{Y}_{u,i,t} - Y_{u,i,t})^2 \tag{4}$$

where, Y is the original observed tagging data and $\hat{Y}$ is the result of tensor factorization.

4.2 Multiple Intra Relations of User, Image and Tag

To handle the problem of sparsity and to increase the quality of recommendations, multiple Intra relations [4] between U, I, and T are incorporated into the Tensor Factorization. The multiple intra relations between U, I, and T are modeled and represented in the form of graphs. Graph based clustering algorithm is further utilized. Tag refinement problem addressed here is divided into two steps: incorporation of intra relations of U, I, and T into tensor and, tensor Factorization using Tucker Decomposition.

User Affinity. Each image is owned, shared, annotated or commented by user. For example if the image owned by user U_1 is commented by the user U_2, then the user U_1 and user U_2 are associated. Therefore, we measure the affinity relationship between two users using the number of images shared, annotated, commented and liked or marked favorite by them.

$$Sim^U_{i,j} = \frac{n(u_i, u_j)}{n(u_i) + n(u_j)} \tag{5}$$

Image Affinity. For each image we extract three types of features to capture 81-dimensional global color and 120-dimensional texture content and 512-dimensional GIST features of image [1]. Visual similarity between images is defined by using RBF kernel.

$$Sim^I_{i,j} = \exp(-\frac{\| x_i - x_j \|^2}{2\sigma^2}) \tag{6}$$

Tag Affinity. Tag affinity graph is constructed based on the tag context and semantic relevance [4]. The context relevance of two tags is simply encoded by their weighted co-occurrence in the image collection. Tag semantic relevance information is obtained using WordNet. Contextual Similarity Between two tags t_i and t_j is computed as:

$$CSim^T_{i,j} = \frac{n(t_i, t_j)}{n(t_i) + n(t_j)} \tag{7}$$

Semantic similarity between two tags t_i and t_j represented as is computed using WordNet. If two tags share the semantic relations such as synonym, hypernym, hyponym, meronym, and holonym, the words are considered as related. We assume λ_c and λ_s are the weights of context relevance and semantic relevance $Sim^T_{m,n}$ represents similarity between two tags t_m and t_n.

$$Sim^T_{i,j} = \lambda_c CSim^T_{i,j} + \lambda_s SSim^T_{i,j} \tag{8}$$

Regeneration of Tensor by Incorporating Multiple Intra Relations. The multiple intra relations of user, image and tag are incorporated into Tensor Factorization framework [2]. The rank of tensor will be $R^{5\times4\times4}$, if there are 5 Users, 4 images and 4 tags. Based on Tag affinity definition, if clustering result is: T_1, T_2, T_3 and T_4. Then, we regenerate initial triplets in the form of tag clusters: $CT_1 - \{T_1, T_2, T_3\}, CT_2 - \{T_4\}$. Based on regenerated triplets, the initial tensor is reconstructed as, $Y \epsilon R^{|U|\times|CT|\times|I|}$. Similarly, based on image affinity and user affinity, the initial tensor is reconstructed as, $Y \epsilon R^{|U|\times|T|\times|CI|}$ and $Y \epsilon R^{|CU|\times|T|\times|I|}$.

5 Geo-Location Summarization

5.1 Visual Cluster Generation

For document clustering and theme generation, we propose the use of probabilistic generative model which extends standard PLSA [12]. Assume, for a given geo-location, a set of N images $D = \{D_1, D_2, ... D_N\}$ is retrieved. Each image d is represented as a vector of word occurrences, $W = \{w_1, w_2, ... w_N\}$, which are collected from associated tags. By considering image as virtual document and tags as terms, we obtain Document-Term matrix. We apply PLSA to model the generation of location images and tag occurrences. The Document-Term Matrix is constructed as follows: The rank of matrix is $DT \epsilon R^{|I| \times |T|}$ where,

$$Dt_{i,j} = \begin{cases} 1 \ if \ tag \ j \ belongs \ to \ image \ i \\ 0 \ otherwise \end{cases}$$

PLSA associates an unobserved class variable $z \epsilon Z = \{z_1, z_2, ... z_k\}$ with each occurrence of $w \epsilon W = \{w_1, w_2, ... w_M\}$ in a document $D = \{d_1, d_2, ... d_N\}$. Standard Expectation-Maximization Approach:

Expectation (E) Step: We compute the posterior probabilities of latent variables from the previous estimate of the model parameters (randomly initialized).

$$P(Z_k|d_i, w_j) = \frac{P(z_k)P(d_i|z_k)P(w_j|z_k)}{\sum_{t=1}^{K} P(z_1)P(d_i|z_t)P(w_j|z_t)} \tag{9}$$

Maximization (M) Step: Here model parameters are updated for given posterior probabilities (computed in previous E-step).

$$P(z_k|d_i) = \frac{\sum_{i=1}^{N} n(d_i, w_j)P(z_k|d_i, w_j)}{\sum_{i=1}^{N} \sum_{j=1}^{M} n(d_i, w_j)P(z_k|d_iw_j)} \tag{10}$$

$$P(w_j|z_k) = \frac{\sum_{i=1}^{N} n(d_i, w_j)P(z_k|d_i, w_j)}{\sum_{i=1}^{N} \sum_{j=1}^{M} n(d_i, w_j)P(z_k|d_i, w_j)} \tag{11}$$

$$P(z_k) = \frac{\sum_{i=1}^{N} \sum_{j=1}^{M} n(d_i, w_j)P(z_k|d_i, w_j)}{\sum_{i=1}^{N} \sum_{j=1}^{M} n(d_i, w_j)} \tag{12}$$

By following the likelihood principle, $P(z_k|d_i)$ and $P(w_j|z_k)$ are determined by maximization of the log-likelihood function. $n(d_i, w_j)$ denotes the number of times the word w_j occurs in document d_i.

$$L = \sum_{i=1}^{N} \sum_{j=1}^{N} n(d_i, w_j) \log \sum_{k=1}^{K} P(w_j|z_k)P(z_k|d_i) \tag{13}$$

5.2   Extended PLSA for Representative Cluster Generation

Standard PLSA results in k number of image clusters. The above mentioned likelihood function considers only the textual associations among the images and tags and does not include the content level information for images. Fact is images with similar content should share common topics. Therefore, we can use the image similarity as a constraint over images to learn the latent topics of interest more accurately. Similarly, the image cluster generated by the standard PLSA may have all the images from same user. Ideally, the image cluster for specified geo-location should be covering a broad interest. The extended PLSA proposed here aims at maximization of following two conditions in addition to the log-likelihood principle.

(1) *Coverage of Users Interest:* Aim is maximization of number of users $|U_k|$ that are represented in photos from cluster z_k (2) *Intra Cluster Connectivity:* If cluster's photos are linked to many other photos in the same cluster, then the cluster is more likely to be representative. The links between photos represent that the photos are visually similar and share same context (set of tags). Aim is maximization of average number of links per photo in the cluster. Visual similarity between photos is decided by following the method described for image affinity computation. Image context information will be decided as follows: Assume T_u and T_v are the set of tags assigned to the two images D_u and D_v, respectively. Using Jaccard similarity measure:

$$ConxtSim = \frac{n(T_u \cap T_v)}{n(T_u) \cup n(T_v)} \tag{14}$$

where, $n(T_u)$ and $n(T_v)$ represent the number of tags assigned to image u and v respectively and $n(T_u \cap T_v)$ represents the number of tags common for image u and v.

Ranking of Cluster. The score of the cluster is decided for ranking of results. The score is computed as [9]:

$$score(z_k) = \sum_{i=1}^{|U|} \log(N_{images}(u_i) + 1) \tag{15}$$

where, $N_{images}(u_i)$ is the number of images in the cluster z_k which are: (a) owned, tagged, commented, shared, annotated or marked favorite by the *ith* user. (b) number of links per image are more than average number of links per image in the cluster Once the generated clusters are ranked according to cluster score, from each cluster we select representative images.

5.3   Location Representative Tag Generation

Tag Cleaning. Once the clusters have been determined, a lot of tags are collected from all the images in the cluster. Many of these tags are noisy and need cleaning: (1) Some general frequent tags and very rarely occurring tags are removed by following Luhn's idea. (2) Tags are identified as irrelevant if they are: stop words, meaningless words, time and number related words, camera related words, abbreviations or acronyms, words with hyphens or dashes, or misspelling. (3) Tags which are not utilized by more than one user are removed. (4) Suffix removal is done using Porters Algorithm.

Tag Score. The system computes scores for each clusters tags to extract representative tags. In other words, we consider each cluster z_k, and the set of tags T_k that appear with photos from the cluster. We assign a score to each tag $t \epsilon T_k$ according to the three factors: (a) Term Frequency (TF) (b) Inverse Document

Frequency (IDF) and (c) User Frequency (UF). One of the factors we use is TF-IDF (term frequency, Inverse Document Frequency). This metric assigns a higher score to tags that have a larger frequency within a cluster compared to the rest of the area under consideration. The assumption is that the more unique a tag is for a specific cluster, the more representative the tag is for that cluster. To avoid tags that appear only a few times in the cluster, the term frequency element prefers popular tags. The term frequency $tf(z_k, t)$ for a given tag t within a cluster z_k is the count of the number of times t was used within the cluster.

$$tf(z_k, t) = n(z_k, t) \qquad (16)$$

The inverse document frequency for a tag t, computes the overall ratio of the tag t amongst all photos D in the geo-location region under consideration:

$$idf(t) = \frac{|D|}{n(D, t)} \qquad (17)$$

While the tag weight is a valuable measure of the popularity of the tag, it can often be affected by a single user who accesses the image a large number of times. To guard against this scenario, we include a user element in our scoring, that also reflects the heuristic that a tag is more valuable if a number of different users use it. In particular, we factor in the percentage of users in the cluster z_k that use the tag t.

$$uf(t, z_k) = \frac{n(U, t)}{|U|} \qquad (18)$$

The final score for tag t in cluster z_k is computed by

$$score(z_k, t) = tf(z_k, t) \cdot idf(t) \cdot uf(t) \qquad (19)$$

The higher the TF-IDF score and the UF score, the more distinctive the tag is within a cluster. For each cluster, we retain only the tags that score above a certain threshold. The threshold is needed to ensure that the selected tags are meaningful and valuable for the aggregate representation. We use an absolute threshold for all computed clusters to ensure that tags that are picked are representative of the cluster.

Tag Selection. The goal of tag selection algorithm is that (a) important textual concepts that are related to specific location are selected and (b) unimportant or highly personal tags are demoted. For user specified geo-location, given a set of images, $D = \{d_1, d_2, ... d_N\}$ is a set of images for topic k. We aim to extract representative tags for this topic from the complete set of tags W_k associated with the topic k. We follow the following two conditions for extraction of representative tags: *Condition 1:* If a tag t is representative tag for theme/topic k, then the probability of observing the tag t among images D_k of theme k is larger than the probability of observing it among all images in D. *Condition 2:* A tag t is a visually representative tag if its annotated images are visually similar to each other. *Condition 3:* A tag t is representative as per user's interest if same tag is utilized by maximum number of users. *Conditions 2* and *3* are already covered according ranking of clusters. Occurrence probability of tag t_i in the set of images D_k of theme k is computed as:

$$p(t_i|D_k) = \frac{n(t_i \cap D_k)}{n(D_k)} \tag{20}$$

In the same way, occurrence probability of tag t_i in the complete set of images D is computed as:

$$p(t_i|D) = \frac{n(t_i \cap D)}{n(D)} \tag{21}$$

Rank of tag t_i for topic k is decided by the condition:

$$p(t_i|D_k) - p(t_i|D) > 0 \tag{22}$$

6 Experimental Work

The goal of our system is to generate a set of location representative images and a set of location descriptive tags for user specified geo-location. Our implementation utilizes three features: location (geo-tags), user, and tags. The system clusters the set of images using extended PLSA. These clusters are formed on the basis of visual-textual relation, visual connectivity, and users interest. Representative tags are selected by first ranking them on the basis of TF, IDF and UF. Selection of representative tags is done using three conditions: separation, cohesion and users interest.

The goals of evaluation are to: (a) Verify the performance of newly proposed Extended PLSA algorithm for image clustering. (b) Determine the representativeness of selected location tags. (c) Test whether the visual and textual location summary generated by our algorithm is satisfactory or not. The goals are directly dependent on subjective means. Therefore, we performed our evaluation by user tests. We executed three experiments to accomplish these goals.

6.1 Dataset

Our dataset was collected by crawling images and associated metadata using Flickr API. To test our approach we selected only those locations for which we can retrieve at least 100 CC-licensed Flickr images. For the practical implementation of our algorithm we constrained this selection to the range of the radius of 1 Km around the input location. Together with image, the accompanying metadata, i.e. tags and user information are also collected. The system is highly dependent on the geo-tagged photos uploaded by different users on Flickr. (a) More photographs are taken at locations that provide views of some interesting object or landmark. (b) Photos are taken or photos on social sharing web site are accessed by a large number of users. (c) Textual tags assigned must reflect the presence of interesting landmarks in a location.

6.2   Clustering Performance Test

We evaluate the effectiveness of our Extended PLSA based clustering algorithm with respect to four baseline methods. *Visual Clustering:* K-means clustering is applied to visual features to cluster images into K clusters. We adopted visual clustering mechanism based on clustering approach carried out in past such as [5]. *Geo-tag Based Clustering:* GPS coordinates or Geo-tag based image clustering approach is utilized by Fang et al. [9], Jaffe et al. [3], and Liu [13]. Mean shift clustering is employed to cluster the photos based on GPS coordinates. *Tag Based Clustering:* Semantic and contextual meaning of tags is utilized for clustering of associated images. Graph based clustering algorithm is utilized. *Users Interest Based Clustering:* If any user uploads, comments, annotates, shares, likes, or marks favorite to the image, the user is considered to be interested in the image. Based on users interest images are clustered.

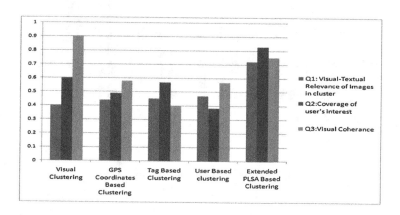

Fig. 2. Clustering performance test

In this test, we showed to our subjects clusters of images formed by each of these four baseline methods and our method. We performed within subject evaluation with a set of 20 subjects. Subjects were asked to rate each cluster on criterias such as: visual-textual relevance of images in cluster, coverage of users interest, and visual coherence. Figure 2 shows the results. From these experiments we can conclude as follows: (a) Only visual features are poor at understanding the content and context of the image, which results in making visual relevance insufficient for generating summarization. (b) Clustering on the basis of only GPS coordinates or Geo-tags can never be covering content or contextual information or users interest. One major possibility is GPS coordinates may be inaccurate (e.g. users may take photos from a long distance of the sights). (c) User provided tags for community contributed images are far from perfection for image clustering. Tags suffer from ambiguity, knowledge and terminology limits of users, tags assigned may not be the actual descriptive words for the image. (d) Clustering based on users interest may be misleading and may skew the selections towards generally insignificant subjects.

6.3   Summary Relevance Test

The goals of summary relevance test are (a) to confirm that the visual and textual summary generated by the system outperforms the baseline methods, (b) to verify that the generated visual summary for given geo location is representative, but still diverse and precise, (c) to confirm the user satisfaction by evaluating user feedback for a set of questions, (d) to confirm that the tags selected to generate textual summary of geo location are descriptive and representative for the location and personal or unimportant tags are ignored while summarizing the location.

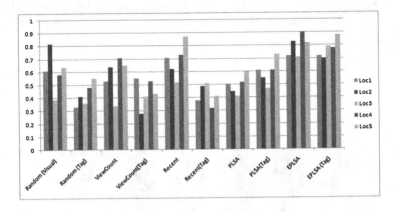

Fig. 3. Precision for images and tags

The baseline methods utilized are: *Random(B1):* Random selection of images and tags. *View Count(B2):* Images are sorted according to highest view count. Top ranking images with no more than one photo per user are selected to generate visual summary of geo-location. Tags assigned to these images are selected using TF-IDF method to generate textual summary. *Recent(B3):* The most recent photos with no more than one photo per user are selected to generate visual summary. Tags assigned to these images are selected using TF-IDF method to generate textual summary. *PLSA Based Summarization(B4):* Images and tags are selected by our system without applying the extended PLSA conditions. *Extended PLSA Based Summarization(B5):* Images and tags are selected by our system.

Using these baseline methods, we select ten representative images and ten representative tags for five different geo-locations. The ground truth judgments of image and tag representativeness are defined by human evaluator. Using ground truth judgments, we evaluate precision for each of visual and textual representation. The precision metric measures the percentage of images and tags that are indeed representative of location. Figure 3 shows precision values for images and tags for five locations.

It is observed that performance of Random, Recent and ViewCount methods is not consistent over time and location. Due to probabilistic nature, PLSA based summarization performance is also not consistent over time. In case of extended PLSA, the probabilistic nature is governed by the two conditions of cluster connectivity and user coverage. So, the Extended PLSA approach performance exceeded the baseline methods that have proven to be less effective, less consistent and less robust in the face of changing data and time.

A fact with precision based evaluation method is that precision does not capture all the aspects that could impact the perceived quality of a set of representative images. In the precision based evaluation, each image is identified as representative (1) or not representative (0). But the fact is representativeness is not binary. Repetition of similar or nearly identical images in the summary could affect the quality of summary. These issues of relative representativeness can be evaluated by human judges. Next we describe a wider evaluation that was designed to measure the relative representativeness of images.

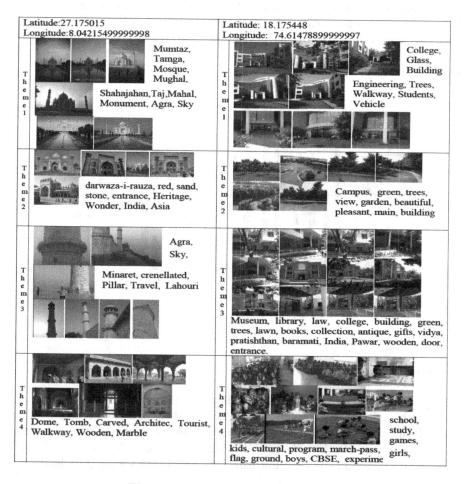

Fig. 4. Location summarization results

6.4   User Survey

We conduct a small-scale user study to evaluate the effectiveness of the proposed method and the user experience of the novel visualization form. The experiment was conducted with two different locations which were known to the user. For each location, images are grouped into four different themes as shown in Fig. 4. Four criteria are considered: (a) Representativeness: the level of representativeness of the visual cluster (0: Worst Representative, 10: Best Representative). The visual cluster representativeness is decided by the total number of images in the cluster and number of representative images in the cluster. (b) Coverage: the extent that the mined visual themes and representative tags provide sufficient information about the location (0: Insufficient, 10: Best). (c) Uniqueness: The extent of uniqueness of images in the visual cluster. Uniqueness of the images in the cluster is decided as the number of representative images minus redundant photos in the cluster (0: Not Unique, 10: Unique). (d) Satisfaction: how satisfactory are the aggregated multiple themes for location visualization (0: Not Satisfied, 10: Very Satisfied). We invited 20 participants, who are well known to the given geo-location for the user study experiment. The eight themes depicted in Fig. 4 are selected for evaluation. The results are averaged over all participants for each theme and shown in Fig. 5. The participants gave positive feedback to the novel location visualization scheme.

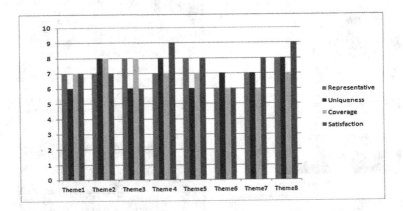

Fig. 5. User survey results

7   Conclusion

The phenomenal growth of personal and shared digital photo collections presents considerable challenges in building navigation and summarization applications. By utilizing our method for location visualization, we enable users to view the most relevant samples from large-scale photo collections. We have presented a novel location visualization scheme using extended PLSA for geographically and thematically organizing photos into multiple themes. The proposed tensor factorization HOSVD using intra-relations between use, image and tag helps deal with the tagging problems of, visual textual relevance, cold start, ternary relationship among user, image and tag. Experiments on Flickr datasets for various known locations show that the proposed framework greatly outperforms the baseline and also shown its advantage in deriving compact location visualization and themes for improving user experiences.

References

1. Oliva, A., Torralba, A.: Modeling the shape of the scene: a holistic representation of the spatial envelope. Int. J. Comput. Vis. (IJCV) **42**(3), 145–175 (2001)
2. Rafailidis, D., Daras, P.: The TFC model: tensor factorization and tag clustering for item recommendation in social tagging systems. IEEE Trans. Syst. Man Cybern. Part A Syst. Hum. **43**(3), 673–688 (2013)
3. Jaffe, A., Naaman, M., Tassa, T., Davis, M.: Generating summaries and visualization for large collections of geo-referenced photographs. In: Proceedings of the 8th ACM International Workshop on Multimedia Information Retrieval, pp. 89–98. ACM, New York (2006)
4. Sang, J., Xu, C., Lu, D.: User-aware image tag refinement via ternary semantic analysis. IEEE Trans. Multimedia **14**(3), 883–895 (2012)
5. Kennedy, L., Naaman, M.: Generating diverse and representative image search results for landmarks. In: Proceeding of the 17th International Conference on World Wide Web, pp. 297–306. ACM, New York (2008)
6. Zheng, Y.-T., Zha, Z.-J., Chua, T.-S.: Research and applications on georeferenced multimedia: a survey. Multimed Tools Appl. **51**, 77–98 (2011). doi:10.1007/s11042-010-0630-z
7. Symeonidis, P., Nanopoulos, A., Manolopoulos, Y.: A unified framework for providing recommendations in social tagging systems based on ternary semantic analysis. IEEE Trans. Knowl. Data Eng. **22**(2), 179–192 (2010)
8. Hao, Q., Cai, R., Wang, X.-J., Yang, J.-M., Pang, Y., Zhang, L.: Generating location overviews with images and tags by mining user-generated travelogues. In: MM 2009, 19–24 October 2009, Beijing, China (2009)
9. Fang, Q., Sang, J., Xu, C., Lu, K.: Paint the city colorfully: location visualization from multiple themes. In: Li, S., El Saddik, A., Wang, M., Mei, T., Sebe, N., Yan, S., Hong, R., Gurrin, C. (eds.) MMM 2013, Part I. LNCS, vol. 7732, pp. 92–105. Springer, Heidelberg (2013)
10. Rudinac, S., Hanjalic, A., Larson, M.: Generating visual summaries of geographic areas using community-contributed images. IEEE Trans. Multimedia **15**(4), 921–932 (2013)
11. Kolda, T.G., Bader, B.W.: Tensor decompositions and applications. SIAM Rev. **51**(3), 455–500 (2009)
12. Hofmann, T.: Probabilistic latent semantic indexing. In: Proceedings of the 22nd Annual International ACM SIGIR Conference on Research and Development in Information Retrieval, SIGIR 1999, pp. 50–57, New York, NY, USA (1999)
13. Liu, X., Liu, R., Cao, Q., Li, F.: Geo-referenced tourist attraction photo tagging by mining community photo collections. In: Huet, B., Ngo, C.-W., Tang, J., Zhou, Z.-H., Hauptmann, A.G., Yan, S. (eds.) PCM 2013. LNCS, vol. 8294, pp. 257–268. Springer, Heidelberg (2013)

Building Secure Web Systems Architectures Using Security Patterns

Eduardo B. Fernandez

Department of Computer Science and Engineering, Florida Atlantic University,
Boca Raton FL, 33431, USA
ed@cse.fau.edu
http://www.cse.fau.edu/~ed

Abstract. Software patterns are encapsulated solutions to recurrent problems in a context. Patterns combine experience and good practices to develop basic models that can be used to build new systems, to evaluate existing systems, and as a communication medium for designers. *Security patterns* provide guidelines for secure system requirements, design, and evaluation. We consider their structure, show a variety of them, and illustrate their use in the construction of secure web-based systems. These patterns include among others Authentication, Authorization/Access Control, Firewalls, Secure Broker, Web Services Security, and Cloud Security patterns. We have built a catalog of over 100 security patterns. We complement these patterns with *misuse patterns*, which describe how an attack is performed from the point of view of the attacker, we show XSS as an example. We integrate patterns in the form of *security reference architectures* that represent complete systems. Reference architectures have not been used much in security and we explore their possibilities. We show how to apply these patterns through a secure system development methodology. We introduce patterns in a conceptual way, relating them to their purposes and to the functional parts of the architecture. Example architectures include a browser and a cloud computing system. The use of patterns can provide a holistic view of security, which is a fundamental principle to build secure systems. Patterns can be applied throughout the software lifecycle and provide an ideal communication tool for the builders of the system. They are also useful to record design decisions. The patterns and reference architectures are shown using UML models and examples are taken from my book: "*Security patterns in practice: Building secure architectures using software patterns*", Wiley Series on Software Design Patterns, 2013. The talk also includes some recent patterns, e.g. Network Function Virtualization. Security is a fundamental quality for any web system. Most proposed solutions are ad hoc or partial; regretfully security must be holistic and systematic. Patterns provide the basis for systematic and holistic approaches and are becoming more important every day. Attendees will be able to understand the idea behind security patterns and security reference architectures, get acquainted with some of them, and use them to build architectures for secure systems.

© Springer International Publishing Switzerland 2015
J. Wang et al. (Eds.): WISE 2015, Part I, LNCS 9418, p. 617, 2015.
DOI: 10.1007/978-3-319-26190-4

The WordNet Database:
Form, Function, and Use

Mark A. Finlayson

Assistant Professor of Computer Science, School of Information and Computing
Sciences, Florida International University, 11200 S.W. 8th Street,
ECS Room 362, Miami FL, 33199, USA
markaf@fiu.edu

Wordnet[1] is a large lexical database of the English language. Like a regular dictionary, it indexes base form words (such as the word *run*) to meanings (e.g., "move fast by using one's feet" as well as "a score in baseball"). Unlike a regular dictionary, it encodes significant amounts of additional information about the interrelationships of word meanings and lexical forms. Perhaps most helpfully, it marks what words are almost exactly synonymous, and so can be used as a thesaurus in addition to a dictionary. Beyond this, however, Wordnet encodes a number of other relationships, such as the fact that an *animal* (synonymous with *animate being*, *creature*, or *fauna*) is a type of *organism*, which is in turn a type *living thing*. This is called the semantic relationship of type-subtype, and Wordnet encodes semantic and lexical relationships between its entries such as type-subtype, part-whole, substance-whole, member-set, domain-topic, antonymy, derivationally related forms, among others. In addition to this rich repository of language meaning, Wordnet is further notable for its size, containing over 155,000 base wordforms, 117,000 meanings, and 188,000 relationships beyond synonymy, including over 46,000 lexical relationships and 142,000 semantic relationships.

Wordnet can be of great use to any application that has to interact with natural language text. In this tutorial, we will first learn about the form of the Wordnet database: the core concepts, what kinds of relationships are encoded in the database, and some caveats about the database contents. We will also examine a small selection of tasks enabled by each type of information encoded in the database. These tasks are provided only as a sample of potential applications, as the range of uses is limited only by one's imagination. Tasks we will learn about include low-level NLP tasks such as lemmatization or root finding (given the inflected form "running" return the root "run", or given the irregular form "is" return the root "be"), all the way up to conceptual processing tasks such as determining that *cats* and *dogs* are more similar to one another than to *turtles*, *plants*, or *cars*.

In addition to the form and utility of the database, we will learn how to interact with the database programmatically. We will first review ways of loading Wordnet into common databases such as MySQL, Sqlite, PostgresSQL, and the like, such that it can be. After this we will examine how to interface with the database directly within a Java programming language environment, focusing on the library the MIT Java Wordnet Interface (JWI)[2]. JWI is small, extremely fast, easy to use, and provides API access to all available Wordnet database information.

---

[1] http://wordnet.princeton.edu/

[2] http://projects.csail.mit.edu/jwi

© Springer International Publishing Switzerland 2015
J. Wang et al. (Eds.): WISE 2015, Part I, LNCS 9418, p. 618, 2015.
DOI: 10.1007/978-3-319-26190-4

Author Index

Printed in the United States
By Bookmasters

Printed in the United States
By Bookmasters